LANDSCAPE ARCHAEOLOGY BETWEEN ART AND SCIENCE

Landscape Archaeology between Art and Science

From a Multi- to an Interdisciplinary Approach

S.J. Kluiving and E.B. Guttmann-Bond (Eds.)

LANDSCAPE AND HERITAGE SERIES

AMSTERDAM UNIVERSITY PRESS

The publication of this book is made possible by grants from:

Cultural Heritage Agency
Ministry of Education, Culture and Science

Cover illustration: The site of Corral Corral, a village with characteristic circular structures in the Peruvian Andes. Photo is taken by N. Goepfert. Picture from Deodat and LeCocq (this volume, their fig. 3)
Cover design and lay-out: Magenta Ontwerpers, Bussum, the Netherlands

ISBN	978 90 8964 418 3
e-ISBN	978 90 4851 607 0 (pdf)
e-ISBN	978 90 4851 608 7 (ePub)
NUR	682

Printed and bound by CPI Group (UK) Ltd, Croydon, CR0 4YY

Contents

THEME VI   HOW WILL LANDSCAPE ARCHAEOLOGY DEVELOP IN THE FUTURE?

Preface

Since the 1960s, 'landscape' has been a key topic of archaeological research all over the world. Initially drawing on environmental archaeology, and using models from the earth sciences as well as cultural ecology, landscapes have been conceptualised predominantly as the natural environments determining human behaviour or as a backdrop to human action. In the New Archaeology of the 1960s, ecology and settlement patterns were studied together with anthropology, with the aim of piecing together information on past economic and social systems (Trigger 1989, 295). Lewis Binford argued that the goal of archaeology should be to understand the range of human behaviours and the differences in culture, based on a belief that cultures were adaptive responses to our environment (Binford 1962), and archaeologists at the time were optimistic that culture and culture change were rational and could be predicted based on archaeological assemblages and settlement patterns.

In the 1980s, a new theoretical perspective, post-processualism, rejected most of the tenets of processualism. Ian Hodder, a key proponent of the new thinking, argued that cultures are not predictable and that artefacts and symbols have different meanings depending on context and culture (Hodder 1986). Within this new school of research, it is not so much the mechanisms of human adaptation to changing natural circumstances that deserve attention, as the different ways in which people in the past perceived and ordered their environments according to space, time and culture. New diachronic approaches were developed that highlight the continuous reuse of monuments and the constant reordering of landscapes within subsequent societies with different social, ritual and mnemonic systems. A similar development took place in the field of Historical Geography from the 1980s, mainly based on the ideas of the New Cultural Geography, with Denis Cosgrove and Stephen Daniels as its main exponents (Cosgrove & Daniels, 1984). Landscape Archaeology in the 21st century is divided between, on the one hand, various interdisciplinary approaches based on intensive fieldwork, aimed at mapping and documenting landscapes and using quantitative methods for predictive modelling (Verhagen, this volume), and on the other hand, post-processualist approaches which aim to understand landscapes as reflections of past societies (e.g. David & Thomas, 2008).

In January 2010, the first international Landscape Archaeology Conference (LAC2010) was organised at the VU University, Amsterdam. The mission of the congress was to have multiple sessions within which scholars from the different disciplines could exchange and discuss research experiences, theories and ideas. The conference attracted many more visitors than originally expected, with over 200 abstracts of which 154, compiled by more than 180 authors, were accepted. In addition to seven keynote lectures, 41 abstracts were selected for oral presentation and 106 abstracts were accepted for poster presentations. In this volume the proceedings of the LAC2010 are presented in 35 papers, covering the wide field of landscape archaeology. In addition, a number of papers derived from abstracts of LAC2010 are published elsewhere in a special issue in Quaternary International (Kluiving, Lehmkuhl & Schuett 2012, see also).

This volume will begin with a discussion of Landscape Archaeology and its history, followed by a discussion of themes and summaries of the papers in this volume. We end with some concluding remarks and suggestions for future research. Although the editors have put a lot of energy into editing the text of this volume, please note that many contributors are writing in English as a second language. We have endeavoured to smooth out the English, but we apologise for the awkwardness of some phrasing in the text. All 35 papers of the LAC2010 proceedings have been peer-reviewed by at least two reviewers. The editors have approached a large number of peer reviewers within the participating institution within the VU University, the Cultural Heritage Agency and the international Advisory Board of LAC2010. We wish to thank the following colleagues who have made the realisation of the proceedings of LAC2010 possible.

Henk Baas (Cultural Heritage Agency, The Netherlands), Jos Bazelmans (Cultural Heritage Agency, The Netherlands), Sjoerd Bohncke (Earth Sciences, VU University, The Netherlands), Judith Bunbury (University of Cambridge, United Kingdom), Richard Chiverell (University of Liverpool, United Kingdom), Adrie de Kraker (IGBA, VU University, The Netherlands), Andrew Fleming (University of Wales Lampeter, United Kingdom), Eric Fouache (Université de Paris, France), Simon Holdaway (University of Auckland, New Zealand), Matthew Johnson (University of Southampton, United Kingdom), Henk Kars (IGBA, VU University, The Netherlands), Kees Kasse (Earth Sciences, VU University, The Netherlands), Jan Kolen (Faculty of Arts, VU University, The Netherlands), Menne Kosian (Cultural Heritage Agency, The Netherlands), Michel Lascaris (Cultural Heritage Agency, The Netherlands), Frank van der Meulen (Earth Sciences, VU University, The Netherlands), Antoine Mientjes (Faculty of Arts, VU University, The Netherlands), Hans Renes (Utrecht University, The Netherlands), Nico Roymans (Faculty of Arts, VU University, The Netherlands), Steven Soetens (IGBA, VU University, The Netherlands), Theo Spek (University of Groningen, The Netherlands), Simon Troelstra (Earth Sciences, VU University, The Netherlands), Ronald van Balen (Earth Sciences, VU University, The Netherlands), Jef Vandenberghe (Earth Sciences, VU University, The Netherlands), Philip Verhagen (Faculty of Arts, VU University, The Netherlands), Mats Widgren (Stockholm University, Sweden).

Sjoerd Kluiving & Erika Guttmann-Bond
Institute for Geo- and Bioarchaeology
VU University Amsterdam, The Netherlands, April 2012

LAC2010: First International Landscape Archaeology Conference

Authors
Sjoerd Kluiving[1,2] and Erika Guttmann-Bond[1,2]

1. Institute for Geo- and Bioarchaeology, Faculty of Earth and Life Sciences, VU University Amsterdam, Amsterdam, The Netherlands
2. Research Institute for the Cultural Landscape and Urban Environment, VU University Amsterdam, Amsterdam, The Netherlands

Contact: s.j.kluiving@vu.nl

INTRODUCTION

The study of landscape archaeology has historically drawn on two different groups of definitions of the term 'landscape' (Olwig 1993, 1996). On the one hand, the original, medieval meaning of landscape is 'territory', including the institutions that govern and manage it. Landscapes according to this definition can be observed subjectively, but also objectively by research based on fieldwork and studies in archives and laboratories (cf. Renes 2011). The second definition developed when artists painted rural scenes and called them 'landscapes'. In the latter, not only the paintings, but also their subjects became known as landscapes. Dutch painters re-introduced the word 'landscape' into the English language, and the word therefore gained a more visual meaning than it had on the Continent. The visual definition turns landscape into a composition that is made within the mind of the individual, so using this definition it could be argued that there is no landscape without an observer (Renes 2011).

While in the latter definition the term 'landscape' originates from the Dutch 'landschap' (Schama 1995; David & Thomas 2008), it is probably more accurate to state that the study of 'territorial' landscapes originated as the study of historical geography and physical geography. This can be traced back to the classical authors, with Strabo noting that 'geography (...) regards knowledge both of the heavens and of things on land and sea, animals, plants, fruits, and everything else to be seen in various regions' (Strabo 1.1.1.). Physical geography is by nature an interdisciplinary field (geology, botany, soil science etc.) and in the late 18th and early 19th centuries it continued to focus on the study of the physical environment, for example in the work of the German researcher Alexander Von Humboldt. During the 19th century, most geographers saw human activities in the landscape as strongly defined by the physical landscape (such as in 'Anthropogeographie' in Germany: Ratzel 1882).

This approach changed in the early 20th century, when the human element was introduced. During the early 20th century in France, a new generation of geographers defined a growing role for human societies (Vidal de la Blache 1922). Carl Sauer, an American geographer, argued in 1925 that geography cannot be simply the study of the natural environment, because people have effected major changes on the landscape. He pointed out that there is a historical element: 'we cannot form an idea of a landscape except in terms of its time relations as well as of its space relations. It is in a continuous process of development or of dissolution and replacement' (Sauer 1925). Sauer is credited with introducing the concept of the cultural landscape, i.e. a landscape that owes much of its character to human intervention. The cultural landscape became the focus for the Berkeley school of American Cultural Geography, founded in the 1920s. The geography of this school included studies of the effects of people on the landscape and considered issues such as population density and mobility, the structure and extent of housing and settlement, the nature of production (e.g. farms, forests and mines) and the communication networks (Sauer 1925). Sauer made the point that landscapes are not static, but are continually changing, and therefore a study of the cultural landscape is a study of landscape *history*. Sauer said that one of the first steps that ought to be taken in historical geography is to try to understand 'the former cultural landscape concealed behind the present one' (Sauer 1941), a statement which more or less defines the aims of Landscape History and Landscape Archaeology. In fact, Landscape History and Landscape Archaeology are so inextricably linked that they are often regarded as being more or less the same thing (Barker & Darvill 1997; Johnson 2007a; Fleming, this volume).

The 20th century has been characterised by the creation of new interdisciplinary studies, the blending of different disciplines and the emergence of new fields of research. Sauer noted that geography was becoming increasingly interdisciplinary; economics were becoming an important element of landscape studies, and 'sociologists have been swarming all over the precincts of human ecology' (Sauer 1941). An important step in interdisciplinary thinking was Eugene Odum's development of the field of Ecology in the 1940s. He suggested that, rather than teaching Natural History as a range of separate subjects (botany, biology, geology etc.), the subjects should be taught together. When he proposed that these subjects were part of interacting systems and should be taught under the all-encompassing field of 'Ecology', he was laughed out of the room by his colleagues at the University of Georgia. Odum (being a strong-minded individual) went on to become one of the most influential leaders in interdisciplinary teaching and research, publishing the first textbook on Ecology in 1953 (Odum 1953). The interdependence of different species is now taken for granted, and has radically changed the way we look at the world today. This has had enormous repercussions for the way we consider the preservation of ecosystems, but also of physical and cultural landscapes and the application of landscape archaeology today.

PROGRESS IN 'INTERDISCIPLINARITY' AND LANDSCAPE ARCHAEOLOGY

The drive to carry out interdisciplinary research in physical geography and historical geography continued to grow and develop in the latter half of the 20th century, and the field of landscape archaeology grew out of the continuing exchange of ideas. The French historian Fernand Braudel published an 'histoire totale' of the Mediterranean world in the time of Philips II, in which he combined archive research, landscape topography and cultural, economic and political developments (Braudel 1949). Braudel was preceded by the historian Marc Bloch, who gave much attention to landscapes in his history of French rural society (Bloch 1932).

In 1955, William George Hoskins published the groundbreaking book, *The Making of the English Landscape*, which drew together physical geography, economic and social history and aerial photographs, which were becoming more widely available from around the early 1950s (Hoskins 1955). Hoskins noted that, although there were many books on English landscape and scenery, this was the first to look systematically into the historical evolution of the landscape.

The discovery of deserted medieval settlements also made an important contribution to landscape archaeology. Hoskins was involved in this discovery, but the theme was largely developed by the economic historian Maurice Beresford and the archaeologist John Hurst (Beresford 1954; Beresford & Hurst 1971). The research combined fieldwork, airial photographs (Beresford & St Joseph 1958) and archival study, and was probably the main basis for the development of the study of the history and archaeology of medieval landscapes in the United Kingdom.

As we have seen, the interdisciplinary approach had already been implemented by historical geographers in America, but aerial photography was accelerating the thinking and the understanding of palimpsest landscapes (Crawford 1953), which Sauer described as 'the former cultural landscape concealed behind the present one' (Sauer 1941). In 1957, Bradford argued in *Ancient Landscapes* that it is important to explore 'complete social units, advancing from single sites to regions' (Bradford 1957). This point was also well demonstrated in *A Matter of Time* (1960), which published the extensive cropmarks seen in aerial photographs of some of England's major river valleys (RCHME 1960). The survey, undertaken by the Royal Commission on the Historical Monuments of England (RCHME), identified cropmarks covering immense areas of land. The survey made it clear that settlement sites were set within field systems and were linked by trackways that were visible for many kilometres. It was also clear that the landscape was a palimpsest of prehistoric and historic landscapes, one on top of the other, with traces going back to the Neolithic and with many features from different periods still surviving today. This book was an important factor in the emergence of Rescue Archaeology, as these massive river valley landscapes were rapidly being quarried away for gravel used in road-building.

Archaeological theory made another great breakthrough in the 1960s, with the 'New Archaeology', or 'Processual Archaeology' (Caldwell 1959; Binford 1962; Trigger 1989). This was essentially the introduction of scientific techniques into archaeology in a more rigorous way. Archaeology had always been multidisciplinary, linked inextricably with geology but also with the classics and art history – but Caldwell argues that before World War II, archaeologists in America concentrated on describing sites and defining cultures, and that classification did not lead further, to interpretation (Caldwell 1959). The New Archaeology, or Processualism, aimed to take archaeology into more rigorous territory. Processualism emphasised a scientific approach that was built into the research project designs and brought together archaeology, anthropology and natural history, including (eventually) the studies of botany (both charred seeds and pollen), entomology, molluscs, zoology, sedimentology and soil science.

Two important publications from the Council for British Archaeology illustrate the new thinking: *The Effect of Man on the Landscape: the Highland Zone* was published in 1975 (Evans et al. 1975) and the companion volume on the Lowland Zone was published in 1978 (Limbrey et al. 1978). Both included papers on a range of scientific techniques that were aimed at understanding the environment and people's place within it, together with traditional subjects such as place name evidence. Environmental archaeology considered the landscape in which people lived; how they interacted with the landscape, how resources were used and how the environmental parameters of the landscape affected the people who lived in it. Such studies included analyses of broad changes in the landscape, such as woodland clearance, down to detailed studies, such as how particular rooms were used based on the botanical and entomological evidence.

Over time, the questions that environmental scientists were asking became ever more varied and more imaginative. Archaeological theory was also moving on, and the emergence of Post-Processual Archaeology in the 1980s, aimed to 'get into the minds' of people in the past. Hodder (1986) makes a convincing argument that symbols mean different things in different cultures, but many proponents of post-processualism simply reject science altogether and settle instead for speculation. Barbara Bender argues that 'one cannot be objective' (Bender 1998, 5) and seems to suggest that we should therefore give up trying.

Scientists, on the other hand, argue that scientific methodology enables us to set up the parameters to test hypotheses under controlled conditions which can be reproduced by others. If subjective observations are to be part of archaeological landscape research, it is important that these observations are properly identified and weighed for their importance. When new ideas, thoughts or opinions are introduced into archaeological research, it provides the opportunity for others to weigh up the potential and perhaps set up experiments to test some of the more grounded speculations. Within the new school of post-processual landscape research, it is not so much the mechanisms of human adaptation to changing natural circumstances that are the focus of attention, but rather the different ways in which people in the past perceived and ordered their environments according to space, time and culture.

This apparent split within landscape archaeological research, with the division between processualists, earth scientists, geographers and environmental archaeologists on the one hand and post-processualists, new cultural geographers and social scientists on the other hand, has been one of the driving forces for the organisation of the LAC2010 conference. International scientific conferences as well as publications have shown an old-fashioned disciplinary division in themes and approaches that contradict the progress in interdisciplinarity that has been made so far.

LANDSCAPE ARCHAEOLOGY TODAY

Landscape Archaeology today is the result of the interdisciplinary developments in archaeology, historical and physical geography discussed in the previous sections. The basic division appears to be grounded in the two types of landscape definitions. The 'territorial' definition for landscape encompasses both natural and cultural types and is adopted by processual archaeologists, physical geographers and most historical geographers. The 'perceived' definition of landscape is claimed by the post-processual archaeologists, the new cultural geographers and the social scientists. The most important challenge of today is to develop fruitful conceptual and practical combinations of classical studies of natural and cultural landscapes in archaeology, physical geography and historical geography that can methodologically and conceptually be integrated into the cultural approaches of the New Cultural Geography and the Post-Processual Archaeology (Fleming 2007, 2008; Johnson 2007a, b). This question was raised during LAC2010 and is addressed in a number of the contributions within these proceedings.

What makes Landscape Archaeology such an interesting field of study today is the wide array of scientific disciplines involved and the further leaps forward in interdisciplinary work. The lectures and poster presentations at LAC2010 drew on more than twenty different disciplines, including social, cultural, historical and natural sciences (fig. 1). It was particularly interesting to note the leaps forward in technology, which are helping us to see and analyse the landscape in new ways. Geophysical techniques and aerial photographs have helped us to 'see' beneath the soils since the 1940s, but LiDAR images now help us to see through forested land and allow us to identify features on the ground with remarkable clarity. Geographical Information Systems enable us to pull together the great swathes of information that we are accumulating from the range of interdisciplinary techniques, helping to integrate the information but also to process it in new ways, such as viewshed analyses. Google Earth has also provided a new way to survey remote areas, with simple, visual satellite data providing vast quantities of new data. Looking at different electromagnetic spectra enhances the detail further, and recent studies in desert areas in the Middle East and Egypt have identified tens of thousands of new sites using this new method (Parcak 2009; Kennedy & Bishop 2011).

The LAC2010 conference in Amsterdam highlighted many of the new techniques used in landscape archaeology, but also demonstrated the value of traditional methods, especially in developing countries that have not yet experienced the extensive archaeological research that has been undertaken in the Western World. Above all, LAC2010 showed that the integration of disciplines advocated by the conference themes has led to various stages of collaboration and combinations in multi- and interdisciplinary research in landscape archaeology today.

Landscape archaeology is the science of material traces of past peoples within the context of their interactions with the wider natural and social environment they inhabited. The mission of LAC2010 was to have multiple sessions with papers that were complimentary to each other in terms of natural and cultural themes, so that cultural and natural landscape archaeologists, as well as archaeologists, historical geographers and earth scientists, could mingle and discuss relevant themes with each other. The 154 abstracts of LAC2010 are the basis of this first international initiative that is published digitally (VU IGBA Publications 2010). The conference was divided into six thematic sessions that are fully represented in the 35 papers of these proceedings (Table 1). Ten papers derived from abstracts of LAC2010 are published elsewhere in a special issue on LAC2010 in Quaternary International (Kluiving, Lehmkuhl & Schuett 2012).

Figure 1. Fragment of notebook of one of the participants during the LAC2010 conference 26-28 January 2010.

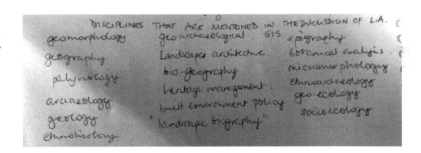

LAC2010: THEMES AND SUMMARIES

Below is a summary of the themes and subthemes of LAC2010, which are dealt with in the papers that belong to the specific sections (Table 1). In the following, all six themes will be discussed, along with short summaries of the corresponding papers. In this introduction a few papers have been regrouped within the additional theme of Rescue Archaeology. Finally, in this section a discussion on a post-processual theme will be presented: Landscape and Memory.

Table 1. LAC2010 themes, abstracts and papers in this volume.

LAC2010 theme number	Themes and subthemes	Number of abstracts	Papers in this volume
1	**How did landscape change?** Natural and perceived landscapes Landscape evolution Human interaction with environment Social theory Global awareness and mutability Peopling of landscape zones Cultural forces in constructed landscapes	50	10
2	**Improving temporal, chronological and transformational frameworks** Chronologies in landscape evolution with respect to stratigraphy, geochronology, paleoclimatology and paleogeography Relation between landscape transformations by humans and landscape evolution by natural processes	23	5
3	**Linking landscapes of lowlands to mountainous areas** Impact of relative sea level changes to landscapes and their inhabitants Landscape change by mobile human groups and sedentary people Behavioural continuity on a changing lowland landscape Dynamic interactions between people and their mountainous landscapes Relation between mountainous landforms and human presence Seasonal connections between landscapes (transhumance etc.)	18	3
4	**Applying concepts of scale** Use of multiscale datasets, test pit observations, geo(morpho)logical mapping Scale of historical landscapes and its spatial contribution Changes in theorising space in archaeology	16	2
5	**New directions in digital prospection and modelling techniques** GIS and archaeology: Integrating GIS into landscape archaeology Satellite imagery as a resource in the prospection for archaeological sites Predictive modelling of archaeological sites using geomorphology and geoarchaeology, and using models of human behaviour and land use Landscape change, human occupation and archaeological site preservation	35	10

Table 1. LAC2010 themes, abstracts and papers in this volume (continued).

LAC2010 theme number	Themes and subthemes	Number of abstracts	Papers in this volume
6	**How will landscape archaeology develop in the future?**	12	5
	Contribution of landscape archaeology to addressing transition periods in archaeology		
	Effect of climate change on expected developments in landscape archaeology		
	How might landscape archaeology be taught in the future?		
	What role can landscape archaeology have in heritage planning?		
	Sum	154	35

HOW DID LANDSCAPES CHANGE?

LAC2010 has shown that the theme of landscape change is dealt with in many diverse ways and disciplines. This general theme attracted the highest number of abstracts, as well as the highest number (10) of papers in this volume, together with theme 5: New directions in digital prospection and modelling techniques (table 1).

The majority of Theme 1 papers can be classified in terms of change in natural and perceived landscapes due to human interaction with the environment. Sometimes such changes in the landscape are discussed together with preservation issues, as well as with settlement location choices. Reher et al. (this volume) investigated the alterations made to an environment following the introduction of Roman gold mines. The geomorphological alterations are described, complemented by pollen analysis. Together, these methods show how the local population was drawn into the mining work, relocated to sites near the mines and given specialised work to do to support the industry. Agriculture already had an impact on the landscape in the Iron Age, but the removal of villages into the mountains meant that agriculture expanded in this unlikely area, rather than in the more promising agricultural land lower down in the valleys. The paper demonstrates clearly, in an interdisciplinary landscape archaeology project, how the Romans radically reordered the landscape, settlement patterns and economy of this region.

In a river basin in Extremadura, Spain, a combination of field walking and geomorphological survey was employed, using GIS to draw together the dynamic natural river sedimentation and erosion processes in relation to settlement patterns (Mayoral et al., this volume). The results produced a predictive model showing where archaeology is likely to be found in situ, and where natural and man-made changes will have removed archaeological sediments and finds from their original location. The research demonstrates that the First Iron Age landscape was in fact densely settled, so much so that it was already spreading off of the best agricultural land when the Romans colonised the area.

Demetradze & Kipiani (this volume) have surveyed an area of connecting river valleys in eastern

Georgia, gathering together information from historical records and published excavations. The aim was to integrate information from individual sites with the wider landscape, to consider the features in relation to each other and to the natural environment, and to place the landscape in a historical context, from the 4th century BC to the 4th century AD. Classical and medieval sources are also integrated, adding to the long-term perspective of this landscape study by providing evidence for the area's decline.

Kampa & Ipsikoudis (this volume) demonstrate the importance of relic Byzantine monastic remains in Rhodope, Greece. Remnants of Byzantine structures, mills, cisterns, water towers, churches, settlements and a bathhouse have been discovered in this wooded landscape, but also the ecological traces left by the Byzantine methods of land management. The monasteries were destroyed in the 16th century, but the new Muslim settlers settled down near the monastic remains and they seem to have carried on some of the early agro-forestry systems. Much of the complex early landscape has now disappeared, largely due to modern afforestation.

In a study of English town commons, the archaeological content is investigated and reported on by Nicky Smith (this volume). The past, present and future of English town commons are discussed. Smith argues that town commons should be recognised as a valid historical entity and a valued part of the modern urban environment. In this study, landscape archaeology is landscape evolution: it is about the fabric of the land as created and modified over a long period, in which people's current activities are part of that continuum.

In another cultural landscape change project, Nurme et al. (this volume) carried out an investigation of Estonian manor parks, researching historical archives together with field survey and cartographic analysis. Estonian parks are unusual in that the Baroque style arrived late to Estonia and lasted long after it had gone out of fashion in the rest of Europe. The manors built between 1730-1770 were in the Baroque style, characterised by symmetrical planning, straight lines, avenues of trees and water features. After 1770, the new, more naturalistic English style gardens were created, but often mixed with the more formal (and out-moded) earlier style. The combination has created a type of cultural landscape that the authors argue is unique to Estonia.

The last two papers in this section focus not so much on landscape change, but rather on settlement choices based on human interaction with the environment. Woltinge (this volume) combines archaeological data with geological features of pingo scars in the northern Netherlands in a statistical test. Pingo scars have historically been regarded as good habitation locations for hunter-gatherers, because of their relatively high position in the landscape and their easy access to water. Pingo scars also provide an excellent base for environmental reconstruction based on palynological research. The study demonstrates that human occupation and its interaction with the landscape seem to be based on much larger environmental elements than just the specific features of pingo scars.

In an interdisciplinary paper on irrigation and landscape, Maurits Ertsen (this volume) discusses the ancient irrigation systems in the Zerqa triangle in the Jordan Valley. How might they have functioned? Physical aspects of the irrigated landscape were explored by basic hydrological and hydraulic modelling. Here, water tapped from the Zerqa River was transported to the fields through open canals by gravity. The settlement patterns found in the valley suggest close connections to the canal system from the Iron Age onwards. Climatic conditions typically would give irrigation a supplemental character. The paper concludes that the application of physically based models yield realistic results when margins of uncertainty for the physical parameters are being constrained by nature.

The papers on this theme by Hajek et al. (this volume) and Kart Aktaş (this volume) are discussed in the additional theme of 'rescue archaeology and threats to the cultural landscape'.

IMPROVING TEMPORAL, CHRONOLOGICAL AND TRANSFORMATIONAL FRAMEWORKS

In this theme we have encouraged the participants of LAC2010 to send in abstracts that deal with a) chronologies in landscape evolution, with respect to geological subdisciplines, and b) the relation between landscape transformations by humans and landscape evolution by natural processes (table 1). Traditionally, there have been two contrasting views on landscape change. On the one hand, there are many archaeologists who believe that cultural processes have been the dominant factor in landscape transformation. By contrast, those with a natural science background have tended to view natural processes such as climate change as the key factors influencing landscape change. All of the papers submitted under this theme illustrate the progress that has been made in interdisciplinary research, in which natural scientists and archaeologists now work together to determine the effects of both cultural and natural transformations on the landscape.

Interdisciplinary research in the coastal region of ancient Etruria (Tuscany, Italy) interlinks historical, environmental and archaeological data in a detailed long-term overview of landscape change. The research is drawn from a vast range of techniques including geomorphology, palaeogeography, remote sensing, geophysical surveys, archaeological research (including intensive surveys and monitoring, excavations and underwater archaeology), archaeometric and archaeological studies of finds (metals and pottery), bioarchaeology, the study of ancient and medieval epigraphy, literary sources, toponyms and historical cartography. This 'total archaeology' approach has ensured that a range of natural and cultural landscape features have been identified (Pasquinucci & Menchelli, this volume).

On a smaller scale, changes of vegetation connected to the erection and use of megalithic graves resulted in a high-resolution reconstruction of land use and forest history at Krähenberg in Schleswig-Holstein, Northern Germany. Archaeological data corroborated by the pollen record suggests that human impact occurred periodically in a period from the end of the Atlantic period (Neolithic), with an increasing intensity during the Bronze Age (Sadovnik et al., this volume). The landscape of south-western Lazio, Italy, was studied in a combined geo- and settlement archaeological approach, focusing on the changes caused by humans in the Roman Republican and Imperial Age. The first results of a brief assessment of the nature and extent of man-made landscape alterations in a part of the suburban hinterland of Rome are presented, and the methodological framework is outlined (Teichman & Bjork, this volume).

An interdisciplinary study of the medieval city of Brussels showed that in its early stages the city was not densely settled, contrary to an image in an illuminated manuscript. Evidence from early maps, excavations and analysis of the soil and phytoliths indicate that agriculture took place in open spaces within the medieval city itself, and as the population rose, the city became more densely populated and the open spaces within it were slowly filled in (Vannieuwenhuyze et al., this volume).

The paper in this theme by Rössler et al. (this volume) is discussed in the additional theme of 'Rescue archaeology and threats to the cultural landscape'.

LINKING LOWLANDS TO MOUNTAINOUS AREAS

The third theme addresses the links between lowlands and mountainous areas. The underlying assumption while formulating the theme was that the contrasts between the two types of natural landscapes are present in almost every part of the globe in variable scales and proportions. Moreover, lowlands and mountainous areas form the background for mobile human groups as well as sedentary people. In a qualitative model of human-environment interactions in a western Andean valley, Peru, the adaptation to the constraints of one limited resource (irrigable land) lead to the suboptimal exploitation of another limited resource (water), which lead to an overall decline in agricultural productivity per unit of arable land and to increasing vulnerability (Hesse & Baader, this volume). The paper shows that the inherent dynamics of cultural landscape evolution in the context of irrigation-based societies may explain the observed changes in the archaeological record. The paper is a contribution to the ongoing debate on potential causes for cultural change in pre-Hispanic coastal Peru.

Mientjes (this volume) investigates the complexities of pastoral transhumance, using an ethnographical approach coupled with historical research in order to understand the nature of pastoralism in Sardinia, Italy, over the past 200 years. Understanding pastoralism ethnographically provides clues to investigating it archaeologically, and shows that we can identify early transhumance routeways through the mountainous landscape, as well as animal shelters and pens, wells and shepherds' huts. This paper is also discussed in the additional theme of 'rescue archaeology and threats to the cultural landscape'.

An interdisciplinary landscape analysis, focusing on the Grosetto area of southern Tuscany, Italy enabled Pizziolo (this volume) to define ancient shorelines, identifying where dry land would have occurred in the different phases based on geological, historical and topographical criteria. The relation between lowlands and mountainous areas is nicely illustrated by the fact that prehistoric funerary activity took place in the surrounding hills, while the settlements which are related to them are difficult to trace because of the complex geological history of the low-lying alluvial plain.

APPLYING CONCEPTS OF SCALES

Theme 4 incorporates the use of multiscale datasets, from test pit observations to geological and geomorphological mapping, as well as how to apply concepts of scale in landscape archaeology. This theme also addresses a classic difference between concepts of scales in geological and archaeological research. Archaeological research often departs from find spots or excavation sites by synthesising material finds in economic, cultural and social networks that can extend over large areas. By contrast, geological research combines information from drill holes and sedimentary sections (including laboratory analyses) into e.g. regional-scale maps showing paleogeography or specific soil information. Often national scale maps in palaeogeography cannot be used in regional or local archaeological prospection (Bazelmans et al. 2011). Palaeogeographic maps usually integrate archaeological data only when it is useful for providing dating evidence, e.g. when archaeological finds in certain stratigraphic layers can be used to date specific landforms (e.g. Vos & van Heeringen 1997). On the other hand, is it still a challenge to explore an integration of reconstructed former landscapes including palaeovegetation with e.g. economic traffic on variable transport routes of former civilisations? Two papers that reflect the 'scale' theme have been incorporated within these proceedings.

In arid parts of Australia hunter-gatherers moved over substantial areas, complicating the task of archaeologists interested in understanding behavioural changes related to human-environment interactions. Holdaway et al. (this volume) analysed archaeological remains in a fifteen-kilometre long drainage system and provided information on the nature of land use and movement over areas that are orders of magnitude larger than those actually studied. The movement of flakes, interpreted as examples of 'gearing up behaviour', occurred across distances that moved people well beyond the limits of the lake in the study area. In the longer term, the accumulation of hearths indicates how the accumulations of behavioural events relate to long-term shifts in the climate and therefore resource availability.

Surface assemblages of prehistoric sites usually reveal only part of the actual extent of the archaeological activity below ground. The difference between surface scatters and buried sites was assessed in two archaeological case studies in Thesprotia, Western-Greece, which demonstrated discrepancies between surface and subsurface assemblages. Several multiscale data sets were acquired, including phosphorous sampling, trial excavations and geophysical techniques. Surface assemblages are biased mainly through erosion and modern landscaping, but also due to so-called 'walker effects' (Forsen & Forsen, this volume).

NEW TECHNOLOGY IN DIGITAL PROSPECTION AND MODELLING TECHNIQUES

The tradition of interdisciplinary methods is continuing and is now broadening to include new technologies, such as LiDAR and satellite imaging. GIS is enabling researchers to integrate different layers of data, which further enhances the potential for interdisciplinary research and for quantification and mapping. This was very well demonstrated in the fifth session of LAC2010: 'New Directions in Digital Prospection and Modelling Techniques'. A useful introduction to this theme is given in a broad overview of the recent development of digital techniques within a 'revolution' in current landscape archaeology (Verhagen, this volume). As a warning, Verhagen concludes that we are caught somewhere between the shining perspective of using quantitative models as exploratory, heuristic devices that can be used for a number of different research questions and the technical and theoretical limitations of what we are trying to do with them.

Deodat & Lecoq (this volume) have used GIS and Google Earth satellite data to survey an area of the Peruvian Andes for archaeological villages and structures, together with an analysis of the ecological zones, rivers and roads through the mountains. Escobar has carried out a GIS-based investigation into the Romanisation of the Antequera Depression in Malaga, southern Spain, in a statistical analysis of settlement patterns. Viewshed analysis showed for the first time that pre-Roman, Iberian settlements were predominantly on high ground, with a high degree of intervisibility, while in the Roman period this pattern changed and inter-visibility became much less important (Moreno Escobar, this volume).

Opitz et al. (this volume) point out that LiDAR is a technique which is complementary to the traditional method of fieldwalking: although LiDAR works poorly on ploughsoils, it is very effective on many areas that cannot be fieldwalked, such as uplands, marshes and particularly forests. The case study makes the point very clearly: LiDAR is an extremely rapid and effective technique which is allowing archaeologists to see through forest cover and to map the archaeology below.

Van der Zee & Zuidhoff (this volume) argue for the use of LiDAR as a standard method to be used in archaeological planning within a lowland area within the Netherlands, a theme which runs through many of the papers of Theme 5. Hesse (this volume) has used high-resolution LiDAR to carry out the complete archaeological mapping of Baden-Württemberg, and argues convincingly that extracting Local Relief Models (LRM) from high-resolution LiDAR data has the potential to greatly improve the potential of such data for the prospection, mapping and monitoring of archaeological sites.

Van Roode et al. (this volume) have combined a suite of methods for the survey of complex urban sites, including intensive fieldwalking, aerial photography, LiDAR, geomorphological studies and a range of geophysical methods, followed by limited excavation. The varied data is then integrated using GIS and computer modelling. The team aims to develop a standard set of methods which others can adopt, and which they hope to see become standard practice. The focus is on management and presentation and they argue that computer reconstructions are not only a very effective way of providing a visual impression of the site, but they are also a rapid way of communicating with the public, in that they can be produced early on in the excavation and survey process. This project demonstrates a very high level of co-operation and focus between three academic and research institutes and three private companies.

New directions in archaeological prospection are shown in two papers. The settlement probability mapping and model uncertainty assessment provided by Fernandes et al. (this volume) provides an opportunity to test the statistical model and a more profound insight into the settlement location choices for the Malia-Lasithi region in Crete during the Protopalatial period (1900-1650 BC). The methods presented here are totally novel within the field of archaeology. This quantitative assessment of model uncertainty based on confidence intervals is usually lacking in settlement probability mapping.

In southern Germany, Posluschny et al. (this volume) present a GIS and database system with which they model the agricultural potential of settlements within their natural surroundings based on topography and soil quality. The model is used to calculate the maximum number of people that can be fed from within the hinterland of both princely sites and 'regular' settlements by cattle and crops. The area around the settlements can be defined by cost-based calculations, which suggests a territorial border based on walking time. The model is then used to compare the agricultural potential of different settlement sites as well as of sites from different periods. The research shows the potential for landscape archaeology in general, but also bridges the gap between environment, culture and social behaviour.

The paper in this theme by Klagyivik (this volume) is discussed in the additional theme of 'Rescue archaeology and threats to the cultural landscape'.

What all of the papers in this session demonstrate is that there is a wide range of new tools that can be applied to archaeological research, and integration of these methods has brought forth new ways of investigating and analysing the landscape. While van Roode et al. (this volume) would like to see a standardised practice, it is also worth considering on a case-by-case basis the range of possible methods, old and new, that would be most appropriate for any survey and excavation. There are broad ranges of scale, from mapping extensive areas using satellite data, to very detailed observation of the land surface, and finally, there is detailed excavation. We would be reluctant to be too prescriptive about the methods that any individual project or researcher ought to use; rather, we would argue that we should all be aware of the range of possibilities of these new developments and make informed decisions when designing new research projects.

THE FUTURE OF LANDSCAPE ARCHAEOLOGY

The last session of LAC2010 (theme 6) discusses the future of landscape archaeology and offers a wide range of different papers. Most authors discuss the interdisciplinarity of landscape archaeology and its ambiguous meaning in society and academia. The role, history and future of landscape archaeology in studies mainly situated in the United Kingdom are widely discussed. The future prospect of landscape archaeology is presented by Fleming (this volume) in the form of a SWOT Analysis (Strengths, Weaknesses, Opportunities and Threats), in which the flexibility of landscape archaeology and its capacity to operate at different scales are classified as strengths, next to contributions to human historical ecology. Weaknesses are formed by the dangers of localism, considered on geographical as well as historical scales and in political and social context. Opportunities within the field of landscape archaeology include discoveries that challenge our perceptions of the world, while threats are formed by limitations of digital resources as well as 'the postmodern challenge' that has introduced an unhelpful and unnecessarily polarised debate. Meier (this volume) discusses the shapeless use of the term landscape archaeology in actual academia. With regard to the history of the term 'landscape', in Meier's opinion 'landscape archaeology' is the proper term to refer to any academic approach which concentrates on the social construction of space. Interdisciplinary research necessarily has to communicate not only its results, but also the presuppositions and rationalities of the participating disciplines. The author concludes that interdisciplinarity is a specific form of academic social behaviour.

Most studies about landscape archaeology are studies of archaeology or geology, and are not about landscape in the way it is meant to be (Fairclough, this volume). Landscape Archaeology, as Fairclough argues, could become an important part of broader landscape research in addition to being a subdiscipline of archaeology. Landscape archaeology can bring special and unique expertise to landscape studies, and can in its turn benefit from exposure to the different horizons, theories and aims of other landscape disciplines. Herring (this volume) argues for the landscape archaeologist's role to be widely inclusive in academia and society. If all society matters, then all landscape matters and all stories have relevance. Herring stated that all disciplines that are shown in inquisitive, theoretical, empirical and phenomenological approaches are drawn into this, as well as all actors. In addition, historic landscape characterisation can be used strategically as a spatial framework in which to assess the historic environment's sensitivity to types of change, or its capacity to accommodate it. A future prospect is that landscape archaeology's key role may be not only to demonstrate how landscape diversity, the outcome of long, medium and short-term change, came into being, but also how that understanding can most effectively inform a sustainable future.

Finally, Johnson (this volume) states that interpretations based on the direct field experience of an archaeologist should be revised. He argues that if landscape archaeology is to develop into a responsible and rigorous field science, it must abandon theories based simply on direct experience and should instead reflect more seriously on the relationship of evidence to interpretation.

RESCUE ARCHAEOLOGY AND THREATS TO THE CULTURAL LANDSCAPE

One thing all of the papers in the section on New Technology have demonstrated is that the new techniques for survey, mapping and quantifying archaeological landscapes have a great deal to offer in the area of planning and conservation. In order to schedule or protect sites or landscapes, we have to first map and understand their character. Our techniques for doing so have increased exponentially. This is important, because the papers also demonstrate that we are not all doing research for pure empirical or scientific reasons: the new techniques can also be applied to rescue archaeology. The use of rescue archaeology as a vehicle for research has been demonstrated in a number of papers in this volume and raises the subject of the threat of destruction of archaeological sites and landscapes.

There are a number of threats to our cultural landscapes, some ongoing and some new. The way that we react to these threats is an insight into what we value most highly. Archaeological sites and cultural landscapes are threatened by urban sprawl, as exemplified in Kart Aktaş's paper on Istanbul (theme 1, Kart Aktaş, this volume); poor planning legislature can allow sites to be built upon or quarried away. Gravel extraction has destroyed large areas of intact prehistoric landscape in England, including some remarkably well preserved landscapes that were buried under alluvium (RCHME 1960). The Valetta Convention has protected the archaeology in many European countries by regulating planning legislation so that archaeology is taken into consideration, but agriculture is not specifically covered by the agreement (Valletta Convention 1992). Agriculture is a major force for destruction, and plough damage has destroyed countless sites across Europe. In Wiltshire (United Kingdom) a vast area of Salisbury Plain is held as military land used for bombing, tank exercises and other destructive practices, and yet the aerial photographs and LiDAR show that the archaeology has survived much better in this area than in the surrounding, heavily cultivated farmland (McOmish 1998; Barnes 2003).

At the same time, agriculture is an essential part of our cultural landscape and agricultural practices have made our landscapes what they are today. The form of fields and field boundaries, drainage and irrigation are all part of our national character, for instance hedgerows in Britain, drainage dykes in the Netherlands and high pastures in the Alps and Mediterranean. Both arable and pastoral agriculture play an important role in maintaining the landscape. The foot-and-mouth crisis in European agriculture caused huge numbers of livestock to be destroyed, and one fear was that the landscape would change dramatically without grazing animals maintaining the ecosystem. Chalk grassland was once regarded as a natural ecosystem, but after experiments in grazing reduction on England's South Downs, it was discovered that when sheep are removed from chalk grassland, it will revert to woodland (Hope-Simpson 1940) – as will many of our moors and heathland (Svenning 2002). The degree of openness in the woodland due to wild herbivores and other natural occurrences is debated (Vera 2000), but there can be no doubt that pastoralism has had a major impact on our cultural landscapes.

Mientjes (theme 3, this volume) has investigated the complexities of pastoral transhumance in Sardinia over the past 200 years, showing that ethnographic studies can shed light on the cultural complexities of past and present land use, ownership and grazing rights. He argues that understanding pastoralism ethnographically provides clues to investigating it archaeologically, and shows that we can identify early transhumance routeways through the landscape, animal shelters and pens, wells and shepherds' huts. Ethnographic evidence helps us to put such features into context, adding to our understanding of the culture, history, economy and physical geography of the cultural landscape.

In some regions, there have been active efforts to eradicate the landscapes of the past. During the Cultural Revolution in China, attempts were made to sweep aside the art, culture and landscapes of earlier regimes. In the USSR and other former Eastern Bloc countries, the government attempted to eradicate religious sites and symbols, or they were disregarded in the planning system. This issue was discussed in Hájek et al. (theme 1, this volume). The authors describe the problem of restoring a landscape of great historical importance, which has been largely destroyed by open cast mining. In the Middle Ages the region had a religious element which dominated the way the landscape would have been seen and understood. Whole villages were destroyed by coal mining, and what is possibly even more poignant, pilgrimage routes – dotted with chapels, crosses and other markers – have been obliterated. The authors pose the question: how do you restore such a landscape? They argue that minor landmarks such as crosses, chapels, statues and milestones should be preserved, even if they are preserved out of their original context, which no longer exists. It is not ideal, but it is preferable to losing these monuments altogether, along with all that they represent.

Klagyivik (theme 5, this volume) tells a similar story of destruction. She has undertaken interdisciplinary research on monastic gardens in Hungary and the vicinity, using historical maps and GIS to identify the surviving remains of garden features. She went on to apply geophysical methods (particularly resistivity) to survey any areas which have not been built upon or destroyed. A particular problem with such research in Eastern Europe is the disregard of the former Soviet regimes for religious structures and the wilful destruction of so many monastic features in the landscape. An example is the hermitage at Majk, where there are surviving remains of Baroque, English, 19th-century and early 20th-century garden styles – a whole history of monastic garden design on one site. The tragedy of this example is that after hundreds of years of both preservation and development, it suffered the most devastating destruction in the period *after* World War II. On a more hopeful note, research into garden design is providing the information needed to reconstruct and preserve such monuments, and Klagyivik has provided a fine example of how this should be done.

Rösler et al. (theme 2, this volume) have also addressed problems of widespread landscape destruction. Their study, set in Eastern Germany, uses rescue excavation in advance of open cast mining as a vehicle for archaeological landscape research. Charcoal was created in this area on a small scale in the Middle Ages and on a large scale from 1567 until the 19th century, when it was used for fuel in the nearby iron manufactory. Over 400 charcoal-burning hearths have been excavated, and the impact on the environment has been analysed in terms of geomorphology and vegetation change. A key impact of deforestation for charcoal burning has been the erosion of the underlying Pleistocene sands, which have blown away and buried areas of the medieval and later landscape. Pre-medieval sites have also been discovered, including Mesolithic flint scatters, Neolithic and Bronze Age burials, Bronze Age post structures and a 3rd/4th-century Germanic village. This study is a good example of rescue archaeology, turning round a destructive process and using it to gain information.

LANDSCAPE AND MEMORY IN POST-PROCESSUAL ARCHAEOLOGY

The notion that our landscapes are made up of elements of past and present is hardly new. The past has always been a source of fascination, and we can see evidence for this in religious structures and burial traditions from earliest prehistory to the present day (Bradley 1987, 2002). Bronze Age barrows cluster around earlier, Neolithic barrows, seeking an association with what came before (Parker Pearson 1993). Early Bronze Age barrows become incorporated into field boundaries in the Late Bronze Age (Bradley 1978), or later shrines were built on top of them. In 601 Pope Gregory advised the English Abbot, Mellitus, to build Christian altars on pagan religious sites, and there are many examples of this in England (Muir 2000, 153). In the city of Bath, for instance, the Roman bath sits upon the Celtic Shrine to Sulis, and beside the two, providing further religious continuity, is the Abbey church of Saint Peter and Saint Paul. This phenomenon is widespread (Bradley 1987), a key example being the Dome of the Rock in Jerusalem, which is built on top of the Temple of Solomon.

The degree to which people are aware of the age of their landscape differs; some farmers can be good sources of information, while others scoff at the notion that there is anything of interest below their ploughsoils, unaware of the extensive remains that exist just below the surface. Knowledge of local history and folk memory are variable; there are still villages in England with rivalries going back to the Civil War. The way that people perceive the landscape around them, and the age of the landscape, is an area that spans both archaeology (particularly post-processual archaeology) and Environmental Psychology. Environmental Psychologists look at the responses of people to particular landscapes, their perceptions and how they experience the landscape; influences include culture, personality, role (e.g. farmer, painter or hillwalker) (Craik 1986). An interesting new discovery is that not only do views of natural landscapes or vegetation actually reduce stress, but landscape *paintings* reduce the anger and stress of workers in office settings (Kweon et al. 2008). This is another interesting overlap between the arts and sciences within the field of Landscape Archaeology.

CONCLUDING REMARKS AND FUTURE RESEARCH

The LAC2010 has shown that an interest in landscape archaeology is worldwide and is emerging in new places which have not previously had much support for such studies. LAC2010 has made a first thorough inventory of international landscape archaeological research that has been carried out in the first decade of the 21st century. Overseeing the results of the papers presented in this volume, while comparing them with the initial LAC2010 conference themes (table 1), shows that quite a number of subthemes have been covered, while some remain still to be explored. In natural and cultural processes of landscape change, true integration of social change and processes in natural landscape evolution is still a challenge (e.g. Reher Diez et al., this volume). We need more integrated examples of landscape transformations by humans and landscape evolution by natural processes (e.g. Pasquinucci & Menchelli, this volume; or Vanniewenhuyze et al., this volume). Variable scale concepts can be further applied on lowland and mountainous landscape systems, while also the integration of information on the scale of historical geographical landscape research with archaeology and physical (palaeo)geographical data (e.g. Holdaway et al., this volume) will be a next step in future research in landscape archaeology. The papers presenting new technology illustrate the new possibilities for integrating interdisciplinary data (e.g. Deodat & Lecocq, this volume; or Escobar, this volume), and are therefore an important driving force into the future of landscape archaeology. In the 'Future' session the interdisciplinary nature of landscape archaeology and its ambiguous meaning in society and academia is well covered by all papers in this theme (Fleming, this volume; Fairclough, this volume; Herring, this volume; Johnson, this volume; or Meier, this volume), and should set the stage for the future research agenda.

Based on the results of LAC2010, we call for more integration and collaboration with respect to the various (sub)disciplines involved in landscape archaeology. In the United Kingdom the distance in academia between the 'field' and material landscape studies with post-processualism is extremely large (Fleming 2007, 2008; Johnson, 2007a, 2007b). In this volume the discussion goes on (Johnson, this volume; Meier, this volume), and is also described as an unhelpful and unnecessarily polarised debate (Fleming, this volume). Within continental Europe (Germany, The Netherlands) the division appears to be moderate, although also in these regions a formal division in academic educational systems hinders integration between the archaeological sciences, such as environmental archaeology and geoarchaeology, with humanities-based archaeology departments where archaeological theory plays a greater role (see also Meier, this volume). The objectives of LAC2010 were set out to form a medium of discussion between disciplines. Within the contributions of these proceedings it has been shown that the methodology in many papers covers multiple disciplines, and also that various examples of collaboration exist between archaeologists, historical geographers and earth scientists, which gives rise to the hope that future developments will show more interdisciplinary studies and therefore better integrated landscape archaeological research.

The course of the discussion at LAC2010 in all six thematic sessions also gave rise to a discussion on what other themes in Landscape Archaeology might be pursued in the future. Some suggestions were:

- Can the understanding of the relationship between other cultures (past and present) and their landscapes inform our community as we seek to manage changing modern landscapes in a sustainable way?

- Can the behaviour of past people in changing landscapes inform our decisions and help us to adapt to climate change in various changing landscapes in the world today?

- What is the impact of relative sea level change on former landscapes, and how did people in the past react to such processes of natural landscape change?

- What is the relationship between landscape transformations by human causes and landscape evolution by natural processes? How dependent are people on their natural landscape?

- How are landscapes described by a) natives with traditional values, and b) geographers and archaeologists? Is there a connection between landscape perceptions in cultural traditions and the natural conditions of landscape processes and/or geological resources?

- How are landscapes perceived by people today, and how might this have differed in the past?

- How are landscapes influenced by anthropogenic, historical, social, and/or economical processes through time and space?

Some of these ideas are already being pursued, which demonstrates that we are sharing a worldwide 'climate of opinion' in which many of us are thinking along the same lines. The final question has been a key area of landscape research since the 1920s, but the climate change issues are more recent. When we look into the future of the 21st century we see that the impact of climate change on the global landscape can only be considered in a wide interdisciplinary approach, such as that taken in the latest reports of the IPCC (Pachauri & Reisinger 2007). Today, for example, there is much focus on the Inuit and the ways that they are coping with climate change (e.g. Ford et al. 2008). The United Nations and many NGOs are supporting the use and development of indigenous technology and engineers are promoting 'intermediate technology' which often relies on traditional methods of agriculture, water management and engineering (Scialabba & Hattam 2002; IAASTD 2009). Attempts are being made to correlate indigenous soil classification techniques with modern soil analysis and mapping, with some success (Gray & Morant 2003), although it is difficult to reconcile the two systems which rely on different variables and observations (Payton et al. 2003). This is certainly a forward-looking approach, however, and using GIS to map the different soil types has had some success.

LAC2010 has attracted a wide variety of relevant disciplines in landscape archaeology today, has made an inventory of the current situation and presented its potential for future research in these proceedings. It is useful to have ideas for further research, and to identify lacunae in our understanding, but we are also very aware that our colleagues at the LAC are a group of dynamic researchers who have ideas of their own. With the Proceedings of LAC2010 established, we can now look forward to hearing about the new research directions and results that our colleagues come up with at the next international Landscape Archaeology Conference (LAC2012), to be held at the Freie Universität in Berlin in June 2012. Landscape archaeology today has been positioned clearly within this new developing field of research between art and science.

ACKNOWLEDGMENTS

We thank Hans Renes of CLUE VU University Amsterdam (NL) and Theo Spek of University of Groningen (NL) for reviewing our paper, which has greatly improved its content. Dr Judith Bunbury of Cambridge University (UK) is acknowledged for sharing her notebook with us as well as discussing new research themes in landscape archaeology.

REFERENCES

Barker, K. & Darvill, T. 1997. Landscape Old and New. In K. Barker & T. Darvill (eds.), *Making English Landscapes: Changing Perspectives*. Oxbow Books, Oxford.

Barnes, I. 2003. Aerial remote-sensing techniques used in the management of archaeological monuments on the British Army's Salisbury Plain Training Area, Wiltshire, UK. *Archaeological Prospection* 10 (2), 83-90.

Bazelmans, J. Meulen, M. van der, & Weerts, H. 2011. *Atlas van Nederland in het Holoceen. Landschap en bewoning vanaf de laatste ijstijd tot nu*. Bert Bakker, Amsterdam.

Bender, B. 1998. *Stonehenge: Making Space*. Berg, Oxford.

Beresford, M.W. 1954. *The lost villages of England*. Lutterworth Press, London.

Beresford, M. & J.G. Hurst (ed.), 1971. *Deserted medieval villages*. Lutterworth, London.

Beresford, M.W. & J.K.S. St Joseph, 1958. *Medieval England; an aerial survey*. Cambridge University Press, Cambridge.

Binford, L.R. 1962. Archaeology as anthropology. *American Antiquity* 28 (2), 217-225.

Bloch, M. 1932. Les caractères originaux de l'histoire rurale française. In Arbos, P. (ed.) *Revue de géographie alpine* 20 (3), 609-614.

Bradford, J. 1957. *Ancient Landscapes: Studies in field archaeology*. Bell, London.

Bradley, R. 1978. Prehistoric Field Systems in Britain and North-West Europe – A Review of Some Recent Work. *World Archaeology* 9 (3), 265-280.

Bradley, R. 1987. Time regained: the creation of continuity. *Journal of the British Archaeological Association* 140, 1-17.

Bradley, R. 2002. *The Past in Prehistoric Societies*. Routledge, London.

Braudel, F. 1949. *La Méditerranée et le monde méditerranéen à l'époque de Philippe II*. Colin, Paris.

Caldwell, J.R. 1959. The New American Archeology. *Science* 129 (3345), 303-307.

Craik, K.H. 1986. Psychological reflection on landscape. In E.C. Penning-Rowsell & D. Lowenthal (eds.), *Landscape Meanings and Values*, 48-64. Allen and Unwin, London.

Crawford, O.G.S. 1953. *Archaeology in the field*. Phoenix House, London.

David, B. & Thomas, J. 2008. Landscape Archaeology: Introduction. In B. David & J. Thomas (eds.), *Handbook of Landscape Archaeology*, 27-43. Left Coast Press, Walnut Creek, California.

Evans, J.G., Limbrey, S. & Cleere, H. (eds.) 1975. *The Effect of Man on the Landscape: the Highland Zone*. Council for British Archaeology Research Report 11, London.

Ford, J.D., Smit, B., Wandel, J., Allurut, M., Shappa, K, Ittusarjuat, H. & Qrunnut, K. 2008. Climate change in the Arctic: current and future vulnerability in two Inuit communities in Canada. *The Geographical Journal* 174, (1) 2008, 45-62.

Fleming, A., 2007. Don't bin your boots! *Landscapes* 8, 85-99.

Fleming, A., 2008. Debating landscape archaeology. *Landscapes* 9, 74-76.

Gray, L.C. & Morant, P. 2003. Reconciling indigenous knowledge with scientific assessment of soil fertility changes in southwestern Burkina Faso. *Geoderma* 111 (3-4), 425-437.

Hodder, I. 1986. *Reading the Past: current approaches to interpretation in archaeology*. Cambridge University Press, Cambridge.

Hope-Simpson, J.F. 1940. Studies of the vegetation of the English chalk: VI. Late stages in succession leading to chalk grassland. *Journal of Ecology* 28 (2), 386-402.

Hoskins, W.G. 1955. *The Making of the English Landscape*. Hodder and Stoughton, London.

IAASTD 2009. International Assessment of Agricultural Knowledge, Science and Technology for Development. Summary for decision makers of the Sub-Saharan Africa (SSA) report. Viewed from http:// www.agassessment. org/index.cfm?Page =Press _ Materials&ItemID =11

Johnson, M., 2007a. *Ideas of landscape*. Blackwell, Malden.

Johnson, M., 2007b. Don't bin your brain! *Landscapes* 8, 126-128.

Kennedy, D. & M.C. Bishop. 2011. Google earth and the archaeology of Saudi Arabia. A case study from the Jeddah area. *Journal of Archaeological Science* 38, 1284-1293.

Kluiving, S.J. Lehmkuhl, F. & B. Schütt 2012. Landscape archaeology at the LAC2010 conference. *Quaternary International*, Volume 251, 1-6.

Kweon, B-S, Ulrich, R.S., Walker, V.D. & Tassinary, L.G. 2008. Anger and Stress: The Role of landscape posters in an office setting. *Environment and Behavior* 40 (3), 355-381.

Limbrey, S. & Evans, J.G. (eds.) 1978. *The Effect of Man on the Landscape: the Lowland Zone*. Council for British Archaeology Research Report 21, London.

McOmish, D. 1998. Landscapes preserved by men of war. *British Archaeology* 34, 12-13.

Muir, R. 2000. *The New Reading the Landscape: Fieldwork in landscape history*. University of Exeter Press, Exeter.

Odum, E.P. 1953. *Fundamentals of Ecology*. Saunders, Philadelphia.

Olwig, K.R. 1993. Sexual cosmology: Nation and landscape at the conceptual interstices of nature and culture; or, what does landscape really mean? In B. Bender (ed.) *Landscape; politics and perspectives*, 307-343. Providence/ Oxford: Berg.

Olwig, K.R. 1996. Recovering the Substantive Nature of Landscape. *Annals of the Association of American Geographers* 86, 630-653

Parker Pearson, M. 1993. *Bronze Age Britain*. Batsford, London.

Pachauri, R.K. & Reisinger, A. (eds.). 2007. *Contribution of Working Groups I, II and III to the Fourth Assessment Report of the Intergovernmental Panel on Climate Change*. IPCC, Geneva, Switzerland, 104.

Parcak, S.H. 2009. *Satellite Remote Sensing for Archaeology*. Routledge, Oxford/New York.

Payton, R.W., Barr, J.J.F., Martin, A., Sillitoe, P., Deckers, J.F., Gowing, J.W., Hatibu, N., Naseem, S.B., Tenywa, M. & Zuberi, M.I. 2003. Contrasting approaches to integrating indigenous knowledge about soils and scientific soil survey in East Africa and Bangladesh. *Geoderma* 111 (3-4), 355-386.

Pretty, J. 1998. *The Living Land: Agriculture, food and community regeneration in rural Europe*. Earthscan Publications Ltd., London.

RCHME 1960. *A Matter of Time*. HMSO, London.

Ratzel, F. 1882. *Anthropo-geographie oder Grundzüge der Anwendung der Erdkunde auf die Geschichte*. Stuttgart: Engelhorn

Renes, J. 2011. European landscapes; continuity and change. In Z. Roca, P. Claval & J. Agnew (eds). *Landscapes, identities and development*. Ashgate, London, 117-136.

Sauer, C.O. 1925. The morphology of landscape. In J. Leighly (ed.), *Land and Life: a selection from the writings of Carl Ortwin Sauer*. University of California Press, Berkeley.

Sauer, C.O. 1941. Forward to historical geography. In J. Leighly (ed.), *Land and Life: a selection from the writings of Carl Ortwin Sauer*. University of California Press, Berkeley.

Schama, S. 1995, *Landscape and Memory*. Fontana Press, London.

Scialabba, N.E. & Hattam, C. (eds.) 2002. *Organic Agriculture, Environment and Food Security*. Rome: FAO.

Svenning, J.-C. 2002. A review of natural vegetation openness in north-western Europe. *Biological Conservation* 104, 133-148.

Tilman, D. 1998. The greening of the green revolution. *Nature* 396, 211-212.

Trigger, B.G. 1989. *A History of Archaeological Thought*. Cambridge University Press, Cambridge.

Valletta Convention 1992. http://conventions.coe.int/Treaty/en/Treaties/Html/143.htm

Vidal de la Blache. 1922. *Principes de géographie humaine*. Armand Colin, Paris (also translated as *Principles of Human geography*, E. de Martonne (ed.), Holt, New York, 1922).

VU IGBA Publications Abstracts LAC2010. http://www.falw.vu.nl/en/research/geo-and-bioarchaeology/publications/index.asp

Vera, F.W.M. 2000. *Grazing Ecology and Forest History*. CAB International, Oxford/New York.

Vos, P.C. & R.M. van Heeringen 1997: *Holocene geology and occupation history of the Province of Zeeland (SW Netherlands)*. In M.M. Fischer (ed.): Holocene evolution of Zeeland (SW Netherlands), Haarlem (Mededelingen NITG-TNO 59), 5-109.

How did landscape change?

1.1 Cultural landscapes of Seusamora in Eastern Georgia

Authors

Irina Demetradze and Guram Kipiani

School of Graduate Studies, Ilia State University, Tbilisi, Georgia
Contact: idemetra@iliauni.edu.ge

ABSTRACT

The interpretation of a number of previously excavated Classical period archaeological sites has advanced our understanding of the formation processes of cultural landscapes in Iberia (east Georgia). In the 1950s, remnants of fortifications discovered in the close vicinity of the modern village of Tsitsamuri (east Georgia) were identified as part of the defensive system of Seusamora, one of the two fortified cities of Iberia mentioned by Strabo. Archaeological sites with different functions subsequently uncovered in the vicinity were also linked with Seusamora. According to the archaeological record, this fortified city with its agricultural lands played an important role in the socio-cultural life of Hellenistic Iberia. However, ancient Georgian written sources do not provide any information about Seusamora itself. The written records mention another contemporary fortified city, Armaztsikhe, referred to as Harmozice by Strabo, and highlight its significance while neglecting Seusamora. It is probable that this fortified settlement lost its importance in the Roman and early medieval periods, but its agricultural lands remained vital. As a result, Seusamora appears to have been subordinate to the newly developed polis of Mtskheta, which became the capital of Iberia in Roman times.

This paper adopts a landscape archaeology approach that is innovative for Georgian archaeology, and aims to define a provisional model of the sequence of the development and decline of the Seusamora settlement, and to bring to light the natural and cultural prerequisites that stimulated the changing configurations of settlement networks through time. The reconstruction of natural and cultural landscape history deals with the integration of human perceptions, written records and cultural factors rebuilt from the artefacts and their natural counterparts. This approach revealed the Seusamora settlement system and its spatial and temporal changes in diverse chronological, environmental and social frameworks.

KEYWORDS

cultural landscape, settlement system, field system, tile, Seusamora

INTRODUCTION

Studies of the origin, development and decline of past human settlement are of great importance in archaeology. From prehistoric times, humans tried to adjust their dwellings to their environment. Presumably, at an early stage of human development, adaptation was more instinctive than cognitive. However, experience obtained over time allowed an ancient human community to settle down in an area with a favourable environmental setting and shape, and to construct a surrounding landscape. Settlement development is deeply connected to a proper human understanding of the environment. In the past, humans gradually acknowledged micro and macro environments and, as a result, a dynamic process of their constant interaction emerged. Relationships between past human communities and the environment affected natural site formation processes and, subsequently, human cultural stability and change over intervals of time. These processes have stimulated consideration of archaeology as a human ecology and one of the components of the environment as a whole (Butzer 1984).

This concept prompted Georgian archaeology to take a new and different approach: to place archaeological sites in a wider environmental context, to consider them as creations that emerged from the setting of an area, and to study their relationship to the changing environment in which they existed. In this context several research questions arose:

- How may we explain settlement development and decline at Seusamora?
- How have natural and cultural factors stimulated changes in the settlement network configuration?
- How are the new patterns of settlement related to changes in the landscape itself?
- When did exploitation of the Aragvi river banks begin?
- What was the nature of the exploitation of the countryside in Late Antiquity?

This paper aims to answer these research questions through the methodologies employed, some of which were innovative for Georgian archaeology. The main innovative approach was that attention was drawn to the landscape in the broadest meaning of this term. The concept of landscape was employed as widely as possible, and it was considered as a continuum of constructed and conceptualised landscapes. These non-exclusive, even overlapping, aspects specified a range of cultural landscapes (Ashmore & Knapp 1999). The interpretation of aerial photographs and the application of GIS (Geographic Information System) and of GPS (Global Positioning System) tools were also innovative. These were used to complement the investigation of the physical landscape by means of a surface survey, field mapping and a consideration of the geology of the area and a study of historical maps. Ancient written sources were analysed and compared to the archaeological record and to landscape evidence. Multidisciplinary data integration provided indicators of human activities on the one hand and landscape change on the other. Landscape, variously measured, is simply that very context in which objects are found, and from which meaning can be sought

(Stoddart 2000, 3). The human-land interactions that took place on this particular landscape of eastern Georgia were considered over the period from the fourth century BC to the fourth century AD.

THE SURVEY ZONE AND THE SEUSAMORA SETTLEMENT SYSTEM

The area chosen for our study is located two to three kilometres from the modern village of Tsitsamuri in the east Georgian historical province of Shida Kartli (Inner Kartli). This is an area of approximately 5 km² ascending from the left bank of the Aragvi River at 460 metres above sea level and rising to 680 metres on a fortified hill. The Aragvi River flows from the southern slope of the Caucasus Mountains to join the Kura River 1.5 km south of Tsitsamuri village. It is an important feature in the area, and presently occupies a relatively shallow cutting. A modern highway crosses the lowlands on the left bank of the Aragvi. The whole area is bordered on the north-east by Saguramo Mountain. The landscape is characterised by steep and gentle slopes, hills and flat areas, and a flood plain. A number of relief units can be defined: mountain slopes covered by deciduous trees, hills partially covered by bushes, low and open valleys that are tran-

Figure 1. Map showing location of study area, Seusamora, Mtskheta,
Armaztsikhe. Inset shows location of Georgia within the Caucasus.

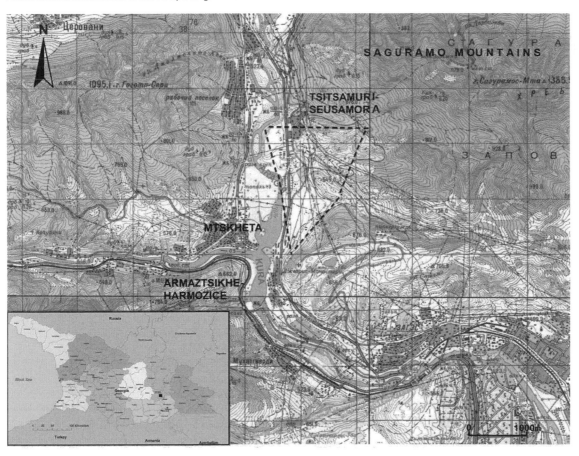

sected by several gullies and small rivers. The small rivers join the Aragvi River (fig. 1). The positions of watercourses shaped the landscape. Alluvial land, which is situated on the left bank of the Aragvi River, is used for agriculture and horticulture. The landscape as a whole is a homogenous sector due to cultural and historical factors that will be considered below.

Archaeological studies of the area over the last 50 years involved a series of excavations which were, however, neither systematic nor intensive. Some excavations were limited to small-scale or salvage operations that exposed archaeological sites or features separately. The archaeological sites discovered in the area in question during different time spans will be discussed in chronological order.

The earliest archaeological site discovered within the area was funerary in character. Two parts of a cemetery were excavated in the lowlands on the left bank of the Aragvi River (figs. 2. 4a, 4b). The cemetery yielded pit graves, some of which were roofed by stone slabs and covered with piles of stone, and cists. The first part was dated to the fifth to third centuries BC on the basis of the grave goods (Ramishvili 1959). The second part was attributed to the fourth to third centuries BC (Djgarkava 1982). Since then, however, the archaeological data have been revised and a more precise date of the end of the fourth to the beginning of the third centuries BC has been suggested for the lower chronological range.

Figure 2.
1. a Fortification wall
   b Round Structure
2. Settlement
3. Post Structure
4. a, b, c Cemetery
5. a, b, c Field System

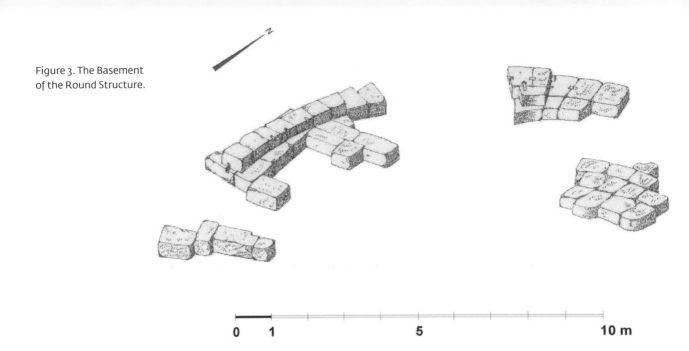

Figure 3. The Basement of the Round Structure.

0 1   5   10 m

The discovery of a round structure and fragments of a defensive wall on the hilltop led to the identification of the fortification system of Seusamora city, as mentioned by Strabo. Archaeological investigations were stimulated by a passage from Strabo's 'Geography': 'The passes from Armenia into Iberia are the defiles on the Cyrus and those on the Aragus. For, before the two rivers meet, they have on their banks fortified cities that are situated upon rocks, these being about sixteen stadia distant from each other – I mean Harmozice on the Cyrus and Seusamora on the other river. These passes were used first by Pompey when he set out from the country of the Armenians, and afterwards by Canidius' (Strabo XI, III, 5). Strabo mentioned two fortified cities at the confluence of the Kura (Cyrus) and Aragvi (Aragus) Rivers. Harmozice was discovered first (in Georgian written sources it is known as Armaztsikhe). On the basis of its architecture, wall paintings, baths and the luxury grave goods unearthed there, Harmozice was thought to be the residence of the kings of Iberia (Apakidze 1968). The discovery of Harmozice facilitated the identification of another fortified city, Seusamora. Strabo described precisely the location of the two cities and the distance between them. Both cities were situated on naturally fortified hills at a distance of three kilometres from each other. In addition, the etymology of the toponym Seusamora has been studied and it has been suggested that it is a modified Greek form of the Georgian Tsitsamuri (Melikset-Bekov 1917, 252; Apakidze 1958, 130).

Archaeological evidence assigned to Seusamora city comprises the remains of a dry-stone fortification wall and a round structure built with grey sandstone blocks and adobe. The foundation of this structure was cut into the rock and was overlaid by block courses connected to each other vertically and horizontally, probably by means of iron and wooden fasteners (fig. 3). The structure was thought to be a tower. A substantial number of tile fragments were scattered around, some of which were painted red and had arrow-shaped markings (Apakidze 1958). This type of tile was widespread in Iberia and is attributed to the first century AD (Dzneladze 1996). Excavations had revealed parts of the defensive system of the city but, owing to a lack of excavated data, the location of the settlement itself remained unknown at that stage.

A structure with an octagonal shape and a rectangular annex was excavated in the north-eastern area of '5b' valley (figs. 2, 3). The basement of the structure was constructed with cobblestones and adobe (fig. 4). The remains of plastered timbers were found in the centre. The structure had two doors: one on the north side and the other on the west, connecting the octagonal part with the annex. Piles of tiles were found above and around the basement of this structure. Some of them were painted red and had arrow-shaped markings. The structure was probably used for guarding the valley and also for storing crops or plants (Kipiani 2003).

The third part of the cemetery was located about 300 metres east of the first part, on the opposite side of the modern highway (fig. 2, 4c).This section consisted exclusively of cist-type burials. The burials yielded grave goods, such as pots and bronze and iron pins, which found close parallels with items from the Samtavro cemetery located 1.5 km to the west, on the right bank of the Aragvi River (Abutidze & Glonti 1999). Among these burials should be mentioned one cist made of worked stone slabs. A circled cross was engraved on the inner part of the eastern slab (fig. 5). This cross appeared to provide a date for the cist. This shape of cross, which is usually assigned to Constantine the Great, was a popular image in early Christian Georgia. However, by the beginning of the fifth century AD it had lost its emblematic character and was not reproduced after this time. Thus this cist and, correspondingly, this section of the cemetery could be attributed to the late fourth century. The chronology of the cemetery suggested that it functioned over at least eight centuries – from the fourth century BC to the fourth century AD. The presence of graves suggests the existence of a synchronous settlement.

Figure 4. The Basement of the Octagonal Structure.

Figure 5. The Cist with the Engraved Circled Cross. *See also the full colour section in this book*

The research design called for in the identification of the settlement included a careful archaeological study of the landscape. This involved a walkover survey, the collection of surface scatters (potsherds, tiles), the observation of cultural layers, site sampling and site mapping (GIS maps) in conjunction with available aerial photography. However, there is no 'right' sampling strategy for such a survey, just as there is no single strategy appropriate to all excavations (Barker 1991, 4). The principles of site selection and location within the landscape were also addressed. The landscape was used as a tool and the sites in question were considered as creations that emerged from the environmental setting of the area. The observation of specific cultural and landscape features, the exploration of how the sites related to their local landscape and the interpretation of site locations suggested that the settlement associated with this cemetery was concentrated on the higher terrain, which has a predominance of open vegetation, light and well-drained soils. The settlement was defined by its natural setting, which correlated with specific landscape forms. A high terrace was located directly below the fortified hill and appeared a perfectly suitable area for permanent settlement and land use (figs. 2, 6).

The site survey also identified agricultural lands, an integral part of the built environment. The field system consisted of three valleys that were suitable for a subsistence strategy based on both pastoral and arable agriculture (figs. 2, 5a, 5b, 5c).The irrigation canals exposed in the first valley suggested that the inhabitants of the area had practised arable agriculture. There is no clear evidence for the dates of the arable lands or irrigation canals themselves, but we can assume that the fields were exploited from the time that the settlement was established. The canals could also have been introduced in those times. The exploitation of these agricultural fields could have continued into the medieval period and even until modern

Figure 6. The Permanent Settlement located below the Fortified Hill.

times, before they were used as a storage space for a Soviet military base. In spite of these findings, a process of agricultural intensification could not be inferred.

Previous Georgian archaeological studies conducted in this area were mainly limited to excavations and descriptions of sites and artefacts. Findings were never placed in any kind of context nor even interpreted. They were just compared to analogous materials from other parts of Georgia. The present research aimed to link and complement existing data by new site surveys, analyses and interpretation supported by innovative approaches. As a result, the dwelling area of Seusamora city has been defined. Previously excavated archaeological sites and features with differing functions have been placed in a wider environmental context and linked to each other, and an archaeological map of the site has been created. A holistic view illustrated the whole concept of the landscape and the Seusamora settlement system was revealed. The Seusamora settlement system consisted of a fortified city, attendant agricultural lands, a pasture area and a cemetery (fig. 2). The fortified walls and a structure were located on the highest point, on the hilltop (680 metres AMSL). A settled area was found more than a hundred metres below (540 metres AMSL). A cemetery, post structure and field system were located at the lowest points (480-460 metres AMSL). The location of the archaeological sites and similar construction materials stressed their linkage. This was reinforced by the fact that all components were guarded by a single defensive system that was in keeping with the natural and cultural situation of the area.

Spatial manifestations were stressed in a study of components of the state administration of Urartu, a kingdom of which parts were located in southern Caucasia (in the modern Republic of Armenia). Physical features and their relations to each other were described as fundamental to the formation and operation of the state apparatus. Three axes of architectonics – settlement location, size dynamics and topography – were discussed. The location of the settlements were described in terms of two primary metric dimensions, elevation and position, each of which interact to define architectonic relations. Location was understood as a regional measure of built environments that defined patterns of movement between them and highlighted the spatial definition of subject-state interactions (Smith 1999).

While considering the landscapes of Seusamora, the geology of the area should also be taken into account. The area has been in its present condition from the late Pleistocene. However, in the classical period, natural changes still affected the environment of the area as evidenced by modifications of the Aragvi River banks. The Aragvi and Kura flood plains were densely covered by deciduous forests, parts of which are still visible in areas adjacent to the cities of Tbilisi and Rustavi (Maruashvili 1971, 286-290). These cities are located 20-30 km to the east of the Tsitsamuri and Mtskheta areas. The soils in these areas also indicate that there was a process of deforestation (Sabashvili 1965, 286-292). Archaeological records recovered from the areas under question suggest that intensive human occupation of this neighbourhood took place in the beginning of the first millennium AD. The process of deforestation at this time probably caused increased sedimentation and a rise in the Aragvi river bed, and consequently a rise in its water level. We can argue this based on the fact that the parts of the fourth-century-BC cemetery located on the edge of the Aragvi River are being washed by its waters. Nineteenth-century topographic maps and modern ones also show a difference between the lines of the banks of the Aragvi River today and more than a century ago. Cultural modifications of this locale, together with natural factors, caused transformations in landscape in the classical period. These developments probably resulted in the creation of an additional, third area of agricultural land (figs. 2, 5a).

The inhabitants of Seusamora settlement were mainly engaged in agriculture and cattle breeding. Objects from burials of differing dates suggest that pottery was developed within the settlement as well. Bronze and iron items were present in the graves. However, due to an absence of evidence, nothing can be said about activities connected with metallurgy. To understand the composition of the settlement and the activities of its inhabitants, three interdependent factors should be studied. Three proposed levels comprise: (a) individual buildings or structures, (b) the settlement layout, and (c) the settlement distribution or the spatial relationships between different communities on a zonal scale (Tringham 1972: xxiii). In the case of Seusamora we possess neither structure nor settlement layouts. However, the spatial relationships can be presumed on a settlement level, but not on a community level. Consequently, the inhabitants' lifestyle could not be detailed.

DISCUSSION

At this stage of our research we are not in a position to discuss the dates of the establishment and decline of the fortified city of Seusamora. Archaeological and written records and landscape do not provide adequate information. In particular, archaeological evidence and the Georgian records are not fully aligned. According to Strabo, the fortified city of Seusamora existed in the first century BC. It had a favourable natural and strategic position. The Seusamora settlement system and its fortifications suggest that the status of the city was self-regulating and it probably possessed a special significance in late Hellenistic – early Roman times. However, Georgian medieval written records do not provide any information on this. Presumably, the city lost its advantageous position with the rise of Mtskheta. The city of Mtskheta emerged on the opposite side of Seusamora, at the confluence of the Aragvi and Kura rivers. A significant change occurred in the organisation of the landscape in the area during the Roman period and Mtskheta became the capital of Iberia. The city grew into a major urban centre as a result of its central position and trading privileges. It was located at the intersection of roads from Colchis, the North Caucasus, Armenia and Albania and gained control of the area.

The nature of the relationship between these two cities is rather obscure. It depended on a variety of political and economic factors that are disregarded by the Georgian written sources. Landscape evidence is more reliable. Population growth and the establishment of the new polis increased the demand on the food supply. Most probably, the Seusamora urban settlement declined gradually in a competitive environment, but continued its existence as a rural settlement due to its fertile lands which supplied the metropolis of Mtskheta. The creation of the third agricultural field in the lowlands on the left bank of the Aragvi River might be attributed to this time period. The fact that Seusamora is not mentioned in Georgian written records might be explained as follows. By the seventh to eleventh centuries AD, when the Georgian sources were written, Seusamora was already rural land, of insufficient importance to be mentioned in the historical records and, besides, Georgian authors were obviously not familiar with Strabo's work. Thus, we presume that Seusamora took on the meaning of the countryside that surrounded the central polis, and in Roman and medieval times its lowlands were defined as agricultural with permanent fields. We are now in a position to understand the meaning of the term *dabnebi* (a place or an area) repeatedly mentioned in the Georgian historical sources in relation to cities. Apparently what is meant is a chora or rural settlement. The similarities between Iberian settlements and the settlement networks that existed in Hellenistic and Roman Greece are noteworthy (Alcock 1993, 2002).

We have evaluated and integrated multidisciplinary data: historical, archaeological/cultural, and natural. This approach has allowed the changing configuration of the Seusamora settlement system and the corresponding modifications in the landscape under human impact to be tracked. The creation of the Seusamora settlement system shaped the cultural landscape of the area, and we have attempted to reconstruct the development of this cultural landscape. From the beginning of the fourth century BC, the high terrain was occupied by the inhabitant community, while the lowlands were used as a cemetery. According to Strabo, the hill was already fortified in the first century BC. The displacement of the population towards the valley areas probably began at the beginning of the first millennium AD, an indication that the principle around which the landscape was organised had changed. Before this time the lowlands were uninhabited due to the dense tree cover. The shift of the settlers to the lowlands and intensive production strategies resulted in woodland clearance. More fertile soils were exploited. As a result, a more stable landscape developed, which was oriented towards the central polis and which possessed specific, well-defined territories. The central polis appeared to be Mtskheta, which emerged after the final modification of the right bank of the Aragvi River. The form and function of the landscape had changed.

This state of affairs relates to the appearance of permanent settlements in general (Patterson 2006). These processes, reinforced by the nature of human interactions with the landscape, supported the creation of the cultural landscapes of Seusamora. It is acknowledged that the three interpretive descriptors of meaning-laden landscapes – constructed, conceptualised and ideational – are often difficult to differentiate. Furthermore, landscape is essentially all of these things at all times: it is the arena in which and through which memory, identity, social order and transformation are constructed, played out, reinvented and changed (Ashmore & Knapp 1999, 9-10).

Similar approaches have been employed in studies of the political landscapes of Urartu. Regime, economy and the built environment of the kingdom have been used to identify the landscape of the polity – the spaces real, perceived and imagined that were created by political communities as sovereign territories. Politics lives in landscapes where the experiential order of spatial practices are part of a multidimensional reorientation of special practices that also attempt to transform the perceived links between people and place and the imagined sources of the landscape. Pictorial images of built environments are more ambiguous in origin, leaving room for an account of how distinct imagined landscapes mediated institutional relationships (Smith 2003).

CONCLUSIONS

Our approach to landscape archaeology studies has revealed the importance of environmental and cultural interrelations. It has emphasised the development of better interdisciplinary procedures in order to achieve more objective interpretations in Georgian archaeology. In the Seusamora example discussed here an attempt was made to reconstruct landscape history, to model interactions between human communities and the landscape, and to establish parameters for settlement patterns. A settlement system with agricultural lands has been identified for the first time in Georgian archaeology. This may serve as a provisional model for settlement patterns that will facilitate a better understanding of the interactions between site networks and the utilisation and modification of the landscape.

REFERENCES

Abutidze, A. & N. Glonti. 1999. *Excavations at Tsitsamuri, The Results of Investigations of Mtskheta Archaeological Institute* 3 (in Georgian), 27-28.

Alcock, S.E. 1993. *Graecia Capta. The Landscapes of Roman Greece*. Cambridge University Press, Cambridge.

Alcock, S.E. 2002. *Archaeologies of the Greek Past. Landscape, Monuments and Memories*. Cambridge University Press, Cambridge.

Apakidze, A. 1958. *Seusamora – Tsitsamuri of Antique Sources, Works of Djavakhishvili Institute of History* 4 (in Georgian), 119-140.

Apakidze, A. 1968. *Cities of Ancient Georgia* (in Russian), Metsniereba, Tbilisi.

Ashmore, W. & A.B. Knapp. 1999. Archaeological Landscapes: Constructed, Conceptualized, Idealized. In Ashmore W. & Knapp A.B. (eds.) *Archaeologies of Landscapes: Contemporary Perspectives (Social Archaeology)*, 1-30. Wiley-Blackwell, Malden, MA.

Barker, G. 1991. Approaches to Archaeological Survey. In Barker G. & Lloyd J. (eds.), *Roman Landscapes: Archaeological Survey in the Mediterranean Region*, 1-9. British School at Rome, London.

Birkeland, P.W. 1999. *Soils and Geomorphology*. Oxford University Press, New York.

Butzer, K. 1984. *Archaeology as Human Ecology*. Cambridge University Press, Cambridge.

Djgarkava, T. 1982. *Kamarakhevi Cemetery*. Catalogue, Mtskheta 6 (in Georgian), 189-91.

Dzneladze, M. 1996. *Clay Construction Material of Classical Georgia* (Tile) (in Georgian), Bermukha, Tbilisi.

Johnson, M. 2007. *Ideas of Landscape*. Blackwell Publishing, Malden, MA.

Kipiani, G. 2003. *Agricultural Lands of Mtskheta: Tsitsamuri, Iberia-Colchis* (in Georgian), 110-117.

Layton, R. & Ucko, J.P. 1999. Introduction: Gazing on the landscape and encountering the environment. In Ucko, J.P. & Layton R. (eds.), *The Archaeology and Anthropology of Landscape, Shaping Your Landscape*, 1-20. Routledge, London.

Maruashvili, L. 1971. *Geomorphology of Georgia* (in Russian). Metsniereba, Tbilisi.

Melikset-Bekov, L. 1917. *Discovering Strabo's Seusamora*, IKORGO I-3 (Reports of Caucasian Department of Russian Geographic Society) (in Russian), 249-255.

Patterson, J.R. 2006. *Landscapes and Cities. Rural Settlement and Civic Transformation in Early Imperial Italy*. Oxford University Press, New York.

Ramishvili, R. 1959. *Kamarakhevi Cemetery, Materials for Georgian and Caucasian Archaeology 2* (in Georgian), 5-54.

Sabashvili, M. 1965. *Soils of Georgia* (in Georgia). Metsniereba, Tbilisi.

Smith, A.T. 1999. The Making of an Urban Landscape in Southern Transcaucasia: A Study of Political Architectonics. *American Journal of Archaeology* 3 (1), 45-71.

Smith, A.T. 2003. *The Political Landscape*. University of California Press, Berkeley/Los Angeles, CA.

Stoddart, S. 2000. Introduction. In Stoddart, S. (ed.), *Landscapes from Antiquity, Antiquity Papers 1*, 1-10. Antiquity Puplications LTD, Cambridge.

Strabo. 1988. The Geography of Strabo Vol. V, translated by Jones, H.L.. Harvard University Press, Cambridge, MA.

Tringham, R. 1972. Introduction: Settlement patterns and urbanization. In Ucko, J.P., Tringham, R. & Dimbely, G.W. (eds.), *Man, Settlement and Urbanism: xix-xxvii*. The Garden City Press Limited, Hertfordshire.

1.2 Irrigation and landscape: An interdisciplinary approach

Author
Maurits Ertsen

Department of Water Resources Management, Delft University of Technology, the Netherlands
Contact: m.w.ertsen@tudelft.nl

ABSTRACT

Studying irrigation history is studying the history of civilisation in dry areas where the natural environment and water infrastructure are closely connected. Yet, surprisingly little is known about the ways irrigation provided the material base for civilisations to prosper. Our knowledge how irrigation developed as interplay between hydraulic and humans is limited, presumably because that kind of knowledge is highly interdisciplinary in nature. Several aspects of water use in irrigation systems need to be explored, including timing and distribution, in order to study how water fluxes on different time scales could be incorporated in archaeological research. This paper discusses irrigation in the Zerqa triangle in the Jordan Valley where water tapped from the Zerqa River was transported to the fields through open canals under gravity. The settlement patterns found in the valley suggest close connections to the canal system from the Iron Age onwards. Physical aspects of the irrigated landscape will be explored by basic hydrological and hydraulic modelling.

KEYWORDS

irrigation, Jordan, explorative modelling, water management

INTRODUCTION

A most fascinating aspect of irrigation is the close connection between civilisation and the natural environment in which water infrastructure acts as interplay. Many civilisations of the past have used irrigation to feed their population. Intensified production provided a relatively secure food source for a larger population as it enabled the peasant population to produce a surplus to support the non-peasant population. Food security enabled development of city kingdoms in many regions: Mesopotamia, Egypt, the Indus-valley, China, Mexico and (coastal) Peru (Scarborough 2003). Because of the importance of irrigation, it has been well studied. The well-known Hydraulic Hypothesis is based on manipulation of water from larger streams for irrigating areas without reliable rainfall. Despite this valuable research, what has been stated by one author more than 25 years ago is still a valid comment. 'Much of the archaeological literature on irrigation canals merely notes their presence at a certain period(s) and records their association with sites. It may describe or even classify the main features of an irrigation system [...] but does not fully elucidate the technology, hydraulics or hydrology of the canals. Even where excavations into canals have been carried out, the researchers have not used their data to its full potential [...].' (Farrington 1980).

In other words, how the irrigation systems underlying the ancient civilisations may have functioned, is not clear at all. Irrigation systems, spatial conglomerates of artefacts, are supposed to supply crops with water. This requires both physical distribution facilities to transport water and socio-political arrangements to coordinate between actors in dealing with water flows. User strategies have an impact on the system and the system constrains user actions. Hydraulic behaviour of irrigation systems resulting from human action is partly constraining and partly enabling human action. Irrigation systems have structuring properties (see Sewell 2005; Ertsen 2010; Ertsen & Van Nooijen 2009). Analysing irrigation development through hydraulic and hydrological modelling will build understanding on how these systems may have been operated at the various stages of their development, how water was distributed and how agriculture was organised, in relation to the settlement patterns of the regions.

As much as elsewhere, human survival in the Zerqa triangle in the Jordan Valley depended on the human ability to adapt to the natural environment, not just in the reactive sense, but also in the proactive sense of shaping the environment, with a major instrument being irrigation. Temperatures are very high because of the low altitude of 300 metres below mean sea level. Average winter temperature is 15 degrees Centigrade, while the summer average is 32 and maximum daily temperatures of 40 to 45 degrees are quite common. Rain falls only between November and April, with an average of about 290 millimetres a year, although annual variability is quite large. Theoretically the average falls just within the limit that is considered the minimum for dry-farming, but frequent droughts would make it impossible to sustain a stable society of some size without a more secured water supply from irrigation. The Zerqa triangle has been extensively studied within the Settling the Steppe-project, and I use the extensive data available through these studies in this paper (Kaptijn 2009; see also Kaptijn et al. 2005; Van der Kooij 2007). My interest is to develop ideas on how irrigation in the Zerqa triangle may have functioned. In this paper I focus on the relation between crop water needs, rainfall and irrigation strategies. I start with a little background on irrigation in the Zerqa triangle, followed by a discussion on crop water requirements and water availability. Then I develop some first ideas about irrigation system management and discuss some first modelling results. The paper ends with an outlook to future work.

IRRIGATION IN THE ZERQA TRIANGLE

Modern irrigation in the Jordan Valley is generally drip irrigation, but before the 1960s, a canal system existed that may well have a long history. Available 19th- and early 20th-century itineraries suggest that the few people who lived in the Valley at this time used canals that tapped water from the river Zerqa (Kaptijn 2009). Aerial photographs taken in the 1940s clearly show a number of small canals that brought water to fields located at considerable distance from the river. On unpublished blueprints from the East Ghor canal, a construction project after World War II, the old canal irrigation system is depicted in great detail (fig. 1; Kaptijn 2009). Three main channels tapped water from the Zerqa, with a series of secondary and tertiary canals irrigating a considerable area. Maintenance and construction of these canals will have been labour intensive and required a communal effort as a large area was dependent on a single primary channel. The Jordan Valley was only sparsely populated and the aerial photographs show that only a small part of the entire system was in use. It is unlikely that 19th-century farmers developed a large irrigation system which they used only partially. In 1920, inhabitants of the Zerqa Triangle stated that 'neither they nor their fathers made these channels; they only cleaned existing ones' (Kaptijn 2009).

As several sources suggest that the Jordan Valley was devoid of sedentary population during the 17th and 18th century, the only likely large-scale farming society from which 20th-century people could

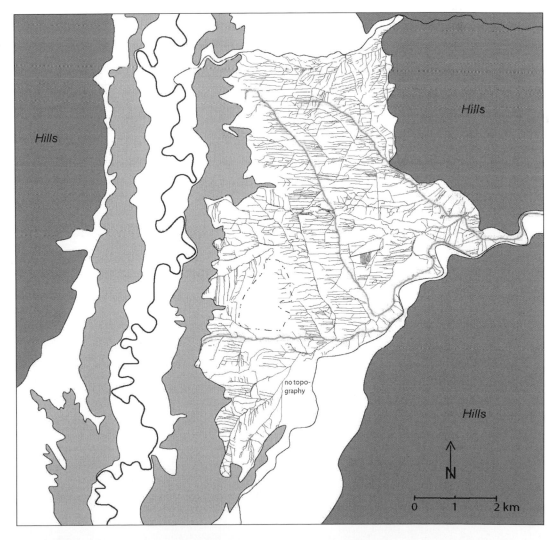

Figure 1. Schematic overview of the irrigation system of the 19th and 20th century (Kaptijn 2009).

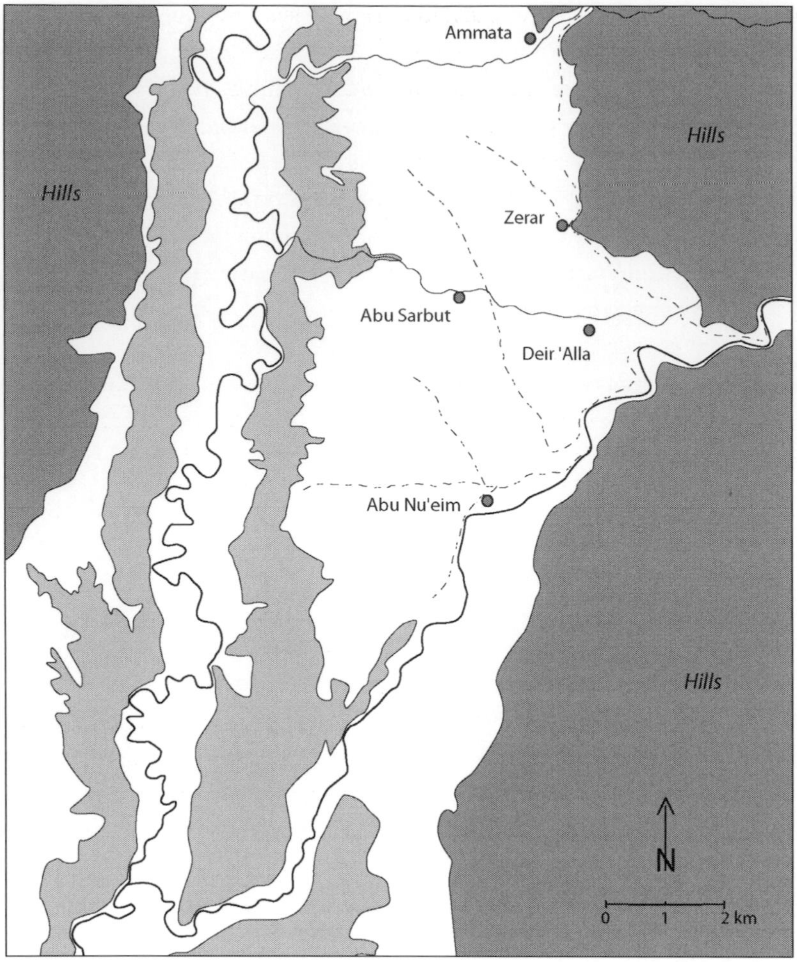

Figure 2. Irrigation canals and associated sugar mills in the Mamluk period (Kaptijn 2009).

have inherited the irrigation system was that the Mamluk period (1260-1500 AD), when the Valley was widely used for sugar cane cultivation. Sugar cane is a tropical crop that grows in summer and needs large amounts of water. In the Jordan Valley climate, irrigation would have been vital. Sugar was produced by crushing the cane and boiling the juice down until raw sugar remained. This was done locally in water-mills. Several of these watermills or sugar-related sites have been excavated. All known mills are located along the three main irrigation channels known in the 20th century or along known wadis (fig. 2). It is thus likely that the 20th-century irrigation system dates back at least to the Mamluk period.

Although there is no direct evidence in the form of excavated canals it is likely that a similar system of canal irrigation existed in the Iron Age II period (1000-540 BC). From the number of tells it is clear that the Zerqa Triangle was rather densely populated during this period (Kaptijn 2009). Although the large number of tells which were found are usually small, it is unlikely that many small contemporary communities could have existed in this dry area over a prolonged time. Furthermore, archaeobotanical data show crops were grown that cannot have been cultivated without irrigation in this area, for example large quantities of flax. There is no direct evidence on how irrigation was practised but the presence of tells in the middle of the plain in combination with the layout of the Zerqa triangle suggests that canals would have provided the water. The potential routes of canals are limited by several features in the landscape, and the main channels may have been located at more or less the same place as during pre-modern times.

A different system of cultivation was probably practised in the Late Chalcolithic and Early Bronze Age (3600-2300 BC). It is clear that during the Early Bronze I period many small sites were located in the plain and along permanent water courses like the river Zerqa and the Wadi al-Ghor. In the Early Bronze II and III periods this system changed and habitation shifted from the valley plain to the foothills. Instead of several small villages people grouped together in larger walled settlements. It is clear that during the Early Bronze Age the Zerqa Triangle was quite densely occupied. The climate was probably slightly more humid than today, but not so much that higher evaporation rates, and thus crop growth, could be sustained by rainfall only. As the rivers and wadis were at that time not as deeply incised as today, they would have submerged the area regularly. The hypothesis that Early Bronze Age communities practised farming in the floodplains could very well apply to the Zerqa Triangle (Kaptijn 2009).

CROPS AND WATER

Let us turn first to the orders of magnitude of water needed for crop production. How much of this water need was met by rain and how much by river flow? What are the differences per crop and what can be said about the temporal pattern of water required? As far as can be established, the agricultural techniques open to the different communities in the different periods discussed above would have been comparable. This means that the agricultural base for which crop water requirement can be determined are comparable. Cropping patterns for the different periods were determined using archaeobotanical data from excavations, ethnographical analogies and the few ancient texts available (Kaptijn 2009). For the Iron Age, the data show high amounts of cereals, both in seeds as in number of samples. Cereals need significantly more water when they are almost mature than when they are germinating. The Mamluk period is in some respects considerably different to any other period. Within other periods all evidence points to simple subsistence farming for the local market, whereas the Mamluk period saw large-scale sugar cane farming in the valley. Sugar cane may have taken as much land as possible, even if this came at the expense of the local population's subsistence (Kaptijn 2009). The villagers probably worked as employees on the plantations, with food crops being grown on fields that lay fallow from sugar cane.

For all cropping patterns, crop water calculations were made using the simple standard approach as given by the Food and Agricultural Organisation (FAO). In this approach, crop development is divided in four different stages, from emerging to maturing. The duration of each developmental stage differs per crop and the total amount of time needed to reach maturity. In the Jordan Valley the high temperatures cause plants to mature more rapidly than average. Agricultural reports from the early 20th century have documented the growing seasons of the most common crops. These timing aspects have been used to make the FAO data applicable to the specifics of the Jordan Valley. In each phase, a so-called crop coefficient expresses how the crop transpiration relates to potential evapotranspiration. For example, when emerging, crops have shorter roots and cannot evaporate as much as when mature. The climatic data on potential evapotranspiration and rainfall were available at the agricultural station in the Zerqa triangle. River flow data were taken from Nedeco (1969). Although the climate in the region definitely changed over time and discharge and rainfall will have differed, these differences are supposed to be not extremely large. Climatic proxy data suggest that the Iron Age II period may have been comparable to today, but that the Mamluk period might have been more humid (Kaptijn 2009). There is much more to be discussed about changes in climatic conditions in the region and the impact on human activities (see for example, Issar & Zohar 2004), but for the present study, modern data on rainfall, temperature and river discharge were used for analysis.

Taking the modern climate data and combining them with the crop data, it can be determined how much water a crop needs (Kaptijn 2009). These water demands can be connected to a real stretch of land actually cultivated when the total cropping pattern is known. For the cropping system documented in the 1950s in this part of the valley, frequencies and amounts of cultivated crops were determined. First, it must be noted that as much as 30% of the land lay fallow. A large proportion was taken up by cereals, while fruit and vegetable took up only small plots of land. Using the relative importance of crops, the water demands can be translated into irrigable hectares given the water availability in the river. On average, the base flow of the Zerqa River is usually sufficient to irrigate the area potentially served by irrigation canals tapping from the river Zerqa, which is about 45 km². However, this simple estimation assumes that people were able to use all of the water in the river, which is unlikely. Furthermore, when taking the driest year of the data series, much lower areas of about 18 km² appear to be available for irrigation. The Mamluk period is again different, as a dry year would typically sustain only twelve irrigated km². Sugar cane is a highly water-demanding crop.

WATER REQUIREMENT AND IRRIGATION STRATEGIES

Existing studies on physical properties of ancient irrigation tend to apply straightforward calculation routines, similar like the one presented above. A simple example shows the potential of adding a little more real-world complexity. For an ancient system in Peru, water requirements for cotton were calculated with a similar approach (in the FAO Cropwat model; Ertsen & Van der Spek 2009) which proved to be comparable to what Nordt et al. (2004) found. The resulting water use scenario, which was supposed to be optimal, as it sustained maximum crop transpiration throughout the growing period, was used as input

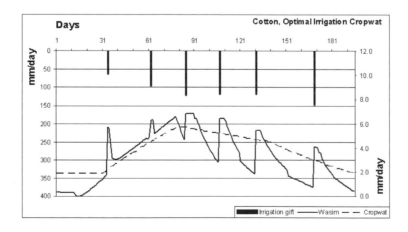

Figure 3. Comparing two different models (Ertsen & Van der Spek 2009).

for WaSim, a one-dimensional soil-water balance model, which includes soil water movements, even on a very basic level (Hess et al. 2000). Figure 3 shows that the physically more realistic results suggest that the crop cannot sustain evapotranspiration for full growth, as water moves down from the root zone to deeper soil layers. The downside is that WaSim applies a daily time step and thus needs daily values for forcing inputs like rainfall, evaporation or other data. Obviously, daily data for 3,000 years ago are hardly available. In case of the Peruvian case modelled above, this can be overcome, as there is no rainfall in summer and evaporation does not change that quickly.

In the Jordan case, however, rainfall is an important aspect of the water balance. Rainfall data are available on a monthly basis. The question then is how to translate these monthly values to daily values. Usually some kind of rainfall generating algorithm would be applied for that (as in Whitehead et al. 2008). For this paper I applied two simple, different methods, as I only want to make the point that it matters how rainfall data are included in the analysis. Figure 4 shows these two methods for a two-year time frame of the total sample. The lower graph shows rainfall and resulting evapotranspiration (in WaSim) for rain-fed wheat when rainfall is proportionally distributed over a month; the upper graph shows what happens in case rainfall is randomly distributed over a month. It is worth noting that the total amount of water available for crops is the same, although the shape of the graphs may suggest otherwise. The

Figure 4. Two different daily rainfall patterns.

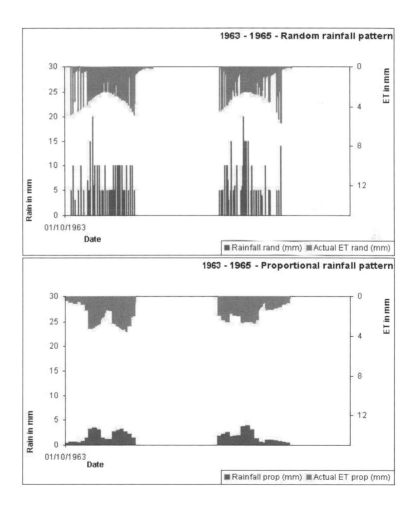

sketchy rainfall pattern of the upper graph is reflected in sketchy evapotranspiration, as crop and soil only evaporate what is actually available. However, sketchy evapotranspiration may mean that the crop develops less smoothly than average rainfall may suggest. Especially young crops, with their underdeveloped root system, are vulnerable for water stress. When water stress occurs early in the growing season, the crop may actually not develop and as such not profit from better water availability later in the season.

An extreme case of such a situation is presented in figure 5. The figure shows crop evapotranspiration, and thus growth, of a wheat crop planted in November with rainfall averaged over the period. Calculations were made in Aquacrop, a crop water model developed by FAO. One of the crops was modelled starting in dry soil at wilting point (wp), when crops cannot extract the water in the soil anymore. The other crop was modelled with the soil being at field capacity (fc), meaning that the soil contains the maximum amount of water for crops to be extracted. As the continuous and growing evapotranspiration suggests, this crop managed to develop. The model indicated that the crop produced a biomass of 8.3 ton per hectare in Aquacrop, with an actual yield of wheat of 2.5 tons per hectare. The dry crop did produce biomass, but only 0.135 tons per hectare, which never managed to produce any yield. The exact numbers are not that important, but the difference between considerable biomass and no biomass is. The figure not only shows the importance of starting conditions when studying irrigation and yields, but also suggests that irrigation may not have been needed, as rain fed agriculture is an option on the Zerqa plain when soils can be brought to field capacity.

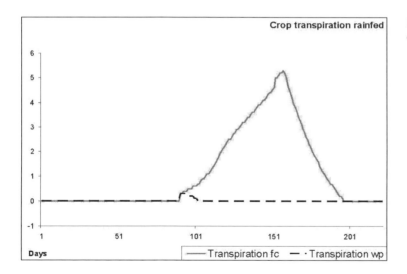

Figure 5. Dry and wet starting conditions.

In many canal irrigation situations comparable with the Zerqa triangle, the start of the growing season is largely determined by rainfall. Typically, irrigation is not the single source of water in these situations, but is used to supplement the rainfall which will be available most of the years in the growing season. The crop is sown when the rain has started, with irrigation starting once the river flow has increased because of the rains. This means that in the early season crops would need to survive on rainfall. For farmers, determining the optimal starting moments can be difficult. If one is too early, because of some good rains which are not sustained, the young crop may not prosper because of drought conditions in the early stage. In case of late sowing, the water availability is probably fine, but the crop may grow too long in the dry season after the wet months. As shown above, the difference between a good choice and a bad one may actually be defined as having a harvest or not. Calculations confirmed that a single 'rescue' irrigation gift relatively late in the season, 100 days after sowing and with dry starting conditions, is too late, as the crop already died. Full irrigation during the entire season will not have been needed, at least not to sustain crops like wheat, which can grow on deeper soil moisture as soon as its roots can reach deeper layers. For crops like vegetables, which have roots that go less deep and need more frequent watering, irrigation may have been more vital during the growing cycle, with less water required per gift. Issues like starting date, rainfall patterns, crops and irrigation needs are highly dependent on each other, and thus need to be carefully studied before conclusions on feasibility of crop strategies can be drawn. Further and more detailed analysis would need to include statistically developed scenarios of rainfall and other variables. Based on the analysis so far, I will take one step further in another direction and develop some ideas about irrigation in the Zerqa triangle, especially about system management.

IRRIGATION MANAGEMENT

Climatic conditions as in the Zerqa triangle typically would give irrigation a supplemental character. Typically, in supplemental irrigation the issue is not to match the crop water requirements exactly, but to give the crop a boost every now and then during the season to avoid too high and continuous water stress. Typically, a pre-sowing irrigation gift to ensure starting conditions could be applied. It may be possible that such a gift is arranged through flooding, either arranged or natural. Although I will not pursue this now, such agricultural water management might have been typical for Bronze Age agriculture in the Zerqa triangle. Furthermore, the first gift after sowing should not be too soon, as this could damage the young plants. Periods between gifts would be in terms of weeks, not days. A scenario developed in Aquacrop, in which a wheat crop receives three irrigation gifts during the season, after 30, 60 and 90 days, each 75 mm, showed that unfavourable, dry starting conditions could be overcome. Wet starting conditions would still produce more crop transpiration, but dry starting conditions would not lead to problems with crop survival. These three gifts of 75 mm would be delivered within 30 days to all farmers. When we assume that water should be available to the entire area covered by the irrigation canals of figure 1, required flows can be calculated (table 1; fig. 6). With upstream canal number 1 as the largest potential irrigable area by far, and thus the limiting area, a continuous flow in the canal system of 0.22 m³ /s would be needed during 30 days. Assuming irrigation during day and night time, as is not uncommon, this flow would allow the whole area to be irrigated in 30 days with the required 75 mm. Canals 2 (middle) and 3 (downstream), each considerably smaller than canal 1, together cover about the same area as canal 1.

As such, canal 1 would need to be operated continuously during the total irrigation period of 90 days. Canals 2 and 3, however, could be operated in at least two different ways. In one scenario, the canals could be operated at the same time. In another scenario, the canals could be operated one after the other, as their total operation time is about 30 days. This second scenario requires less total flow for the three canals together. When comparing base flow in the Zerqa river with the water demands of the two scenarios, it becomes clear that operations after each other, thus some kind of cooperation between canals 2 and 3, pays off, as the system as a whole seems to be under much less stress. Obviously, the relative positions of

Table 1. Irrigation gift of 75 mm every 30 days

Canal	Irrigation gift m	Estimated surface m²	Irrigation time Days	Discharge m³/s
1	0,075	7500000	30	0,22
2	0,075	2500000	10	0,22
3	0,075	5250000	21	0,22

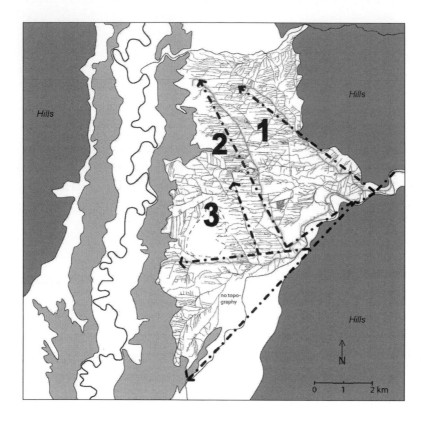

Figure 6. Simplified canal system. *See also the full colour section in this book*

canals 1, 2 and 3 need to be accounted for. It is clear that canal 1 as the upstream user has potentially less problems, but physical upstream positions do not necessarily coincide with social upstream positions (Ertsen & Van Nooijen 2009).

To analyse this issue of spatial dependency in irrigation further, and to include more realistic processes like river fluctuations, other types of analysis including canal modelling is needed. For example, in a study on Mesopotamia, Altaweel (2007) concluded that, amongst others, 'gravity flow irrigation' promoted stable yields. When taking a closer look, however, these calculations conceptualised this gravity flow as a 150 mm water volume in one time step. One wonders whether such water gifts could actually be delivered, as inflows would for example fluctuate. A first attempt to analyse actual water fluxes in the same Peruvian system mentioned earlier suggested uneven distributions of flows and thus volumes over the irrigation system (Ertsen 2009). The pattern was consistent with different material remains in different parts of the irrigated the area, and could be associated with differences in water availability and management practices (Hayashida 2006).

To include this complexity from the irrigation-management perspective in the Zerqa triangle, a canal model was developed using the software package SOBEK. First, each main canal detected on the base map (fig.1) was simplified (fig. 6). Each of the three canal areas was further simplified by equipping the main canal with offtakes each irrigating sub-areas of 50 hectares. This yielded fifteen offtakes along canal 1, five along canal 2 and ten along canal 3. Given the length of the canals, the distances between these model offtakes were different for each canal area. Canal cross-sections and slopes were based on data available on the 20th-century situation. With this model set-up, different scenarios were tested, including starting irrigation upstream or downstream, including infiltration, gate settings, etc. For each scenario it was calculated how much water would enter each 50 hectare sub-area during the 30-day interval. The total volume delivered to each sub-area was compared to the reference volume shown in Table 1. As it is unknown what management was applied in the Zerqa triangle, the relative certainty of water delivery to each sub-area, defined as the average amount of water arriving in a sub-area for all scenarios compared to the reference amount expressed in a percentage, was plotted on the canal map.

The canal system included in the model was based upon the canals from the 20th century. As such, results cannot be used directly for other periods. Especially the Mamluk period, with its irrigation of sugar cane and the need to divert water to the mills, will need a different treatment in terms of hydraulics. In figure 7, the relative certainty mentioned above is shown as an overlay to the Iron Age settlement map. As was to be expected, the upstream areas of the main canals have a more reliable water delivery than the downstream areas. Canal 3 is an interesting case, as the entire area appears to have a relatively high water

Figure 7. Relative certainty of water delivery. *See also the full colour section in this book*

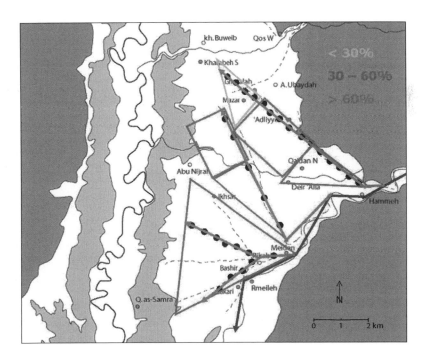

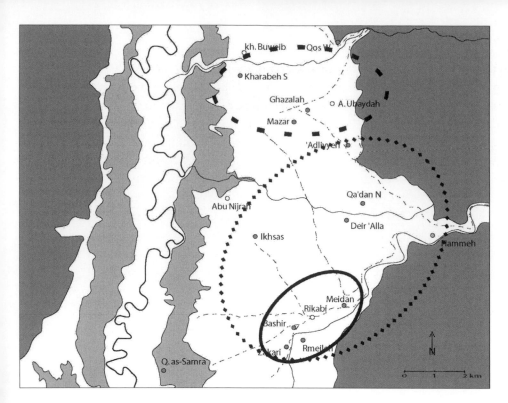

Figure 8. Linking water security to settlement.

security, even though the area is downstream of the other two. When one simplifies figure 7 into likely areas of influence of the three main canals together in relation to Iron Age settlements (fig. 8), it becomes apparent that the area in the grey circle must have been able to draw water from the canals to the north of the system in the Zerqa triangle. Furthermore, where the red area would represent a relatively certain area of influence of Zerqa water, the blue area may be the most interesting. In this area, which roughly co-incides with canal 3, many settlements were found. Although these are close to the river, and as such not necessarily linked to canal irrigation, the settlements come awfully close to the canals projected on the Iron Age map. Is this a construct of my approach or an indication of the relative water security brought to these settlements through canal-based irrigation?

FIRST CONCLUSIONS AND OUTLOOK

The calculations and modelling discussed above are still relatively simple. However, I think I have shown that it is worthwhile to develop such a modelling-based approach to generate the fluxes and balances in ancient irrigated environments, as it can yield new insights in these environments. The modelling results for the Zerqa area indicated that rain-fed irrigation would have been possible, in the case that starting conditions for the crops would be favourable. Furthermore, it was shown that irrigation in the growing season could overcome dry starting conditions. In both cases, irrigation would have been needed, but could result in quite different management arrangements. However, in all irrigation management arrangements, sharing the water from the river would have been an issue. The model results suggest that cooperation between users of the different canals would have been more beneficial than conflictual.

These conclusions need to be taken with care. A fundamental aspect of the modelling methodology is that the reference situation is unknown, which in my view demands modelling approaches explicitly based on the physical processes creating surface and subsurface water fluxes. A recent initiative of the International Association of Hydrological Sciences, Predicting in Ungauged Basins (PUB), shows the strength of new, largely understanding-based methods instead of calibration-based methods. PUB aims to improve existing hydrological models in terms of their ability to predict in ungauged basins, with a focus on predictive uncertainty, alternative data sources and the links between these two issues (Sivapalan et al. 2003). It is true that simple, conceptual models do allow for quick analysis of water systems. Simple models, however, cannot provide the detailed analysis of the material context for irrigation needed in this research, precisely because there is no information to validate the models of ancient irrigation systems. The drawback of using conceptual models is their inherent uncertainty in the values for the model parameters, which are not based on realistic physical behaviour. These model parameters need to be calibrated against real data, which are not available. Applying conceptual models only will increase the uncertainty of model outcomes. Compare the irrigated landscape with a bath tub, for which we can establish physical properties and set realistic filling and emptying mechanisms, including a shower (rainfall), tap (river) or leak (groundwater) – compare with Harrower (2008) who used a digital elevation model (the bath tub) to define the channels when analysing irrigation in Arabia. Setting parameters in such a model requires good quality data on aspects like soils, climate, geology and groundwater. Applying physically based models can yield realistic results, even in catchments without measured data (Lange et al. 1999, as margins of uncertainty of the physical parameters are being constrained by nature.

ACKNOWLEDGEMENTS

I am extremely thankful to Eva Kaptijn for the discussions leading to this paper and allowing me to use her basic material (which I partially helped to develop). We presented part of the material in the session on 'Responses of Complex Societies to Climatic Variation' at the Stine Rossel Memorial Conference, Climate and Ancient Societies: Causes and Human Responses. Other elements were developed during the workshop on Water and Power, University of Durham, UK, organised by Tony Wilkinson.

REFERENCES

Altaweel, M. 2007. Investigating agricultural sustainability and strategies in northern Mesopotamia: results produced using a socio-ecological modeling approach. *Journal of Archeological Science* 34, 1-15.

Ertsen, M.W. (accepted 2010): Structuring properties of irrigation systems. Understanding relations between humans and hydraulics through modelling. *Water History* 2.

Ertsen, M.W. & J. van der Spek. 2009. Modeling an irrigation ditch opens up the world. Hydrology and hydraulics of an ancient irrigation system in Peru. *Physics and Chemistry of the Earth* 34, 176-191.

Ertsen, M.W. & R. van Nooijen. 2009. The man swimming against the stream knows the strength of it. Hydraulics and social relations in an Argentinean irrigation system. *Physics and Chemistry of the Earth* 34, 2000-2008.

Farrington, I.S. 1980. The archaeology of irrigation canals, with special reference to Peru. *World Archaeology* 11, 287-305.

Harrower, M.J. 2009. Is the hydraulic hypothesis dead yet? Irrigation and social change in ancient Yemen. *World Archaeology* 41, 58-72.

Hayashida, F.M. 2006. The Pampa de Chaparrí: water, land, and politics on the north coast of Peru. *Latin American Antiquity* 13, 243-263.

Hess, T.M., P. Leeds-Harrison & C. Counsell. 2000. *WaSim Manual*. Institute of Water and Environment, Cranfield University, Silsoe, UK.

Issar, A.S. & Zohar M. 2004. *Climate Change – Environment and Civilization in the Middle East*. Springer, Berlin.

Kaptijn, E. 2009. *Life on the watershed. Reconstructing subsistence in a steppe region using archaeological survey: a diachronic perspective on habitation in the Jordan Valley*. Sidestone Press, Leiden.

Kaptijn, E., L.P. Petit, E.B. Grootveld, F.M. Hourani, G. van der Kooij & O. al-Ghul. 2005. *Dayr 'Alla Regional Project: Settling the Steppe (first campaign 2004)*. Annual of the Department of Antiquities of Jordan 49, 89-99.

Kooij, G. van der. 2007. Irrigation systems at Dayr 'Alla. In F. al-Khraysheh (ed.), *Studies in the history and archaeology of Jordan IX*. Amman, 133-144.

Nedeco 1969: *Climate and Hydrology. Annex B of the Jordan Valley Project*. Final Report. Hashemite Kingdom of Jordan.

Nordt L., F. Hayashida, T. Hallmark & C. Crawford. 2004. Late prehistoric soil fertility, irrigation management, and agricultural production in northwest coastal Peru. *Geoarcheology* 19, 21-46.

Scarborough, V.L. 2003. *The flow of power. Ancient water systems and landscape*. SAR Press, Santa Fe, New Mexico.

Sewell, W.H. 2005. *Logics of History. Social Theory and Social Transformation*, Chicago Studies in Practices of Meaning. Chicago University Press, Chicago.

Whitehead, P.G., S.J. Smith, A.J. Wade, S.J. Mithen, B.L. Finlayson, B. Sellwood & P.J. Valdes. 2008. Modelling of hydrology and potential population levels at Bronze Age Jawa, northern Jordan: a Monte Carlo approach to cope with uncertainty. *Journal of Archeological Science* 35, 517-529.

1.3 Principles of preservation and recalling of memory traces in an industrial landscape: A case study of decayed monument recreation in the brown-coal mining area of Bílina, Czech Republic

Authors

Tomáš Hájek[1], Barbora Matáková[2], Kristina Langarová[3] and Ondřej Přerovský[4]

1. Project Solutions, Ltd., Praha, Czech Republic
2. Mendel University in Brno, Faculty of Horticulture in Lednice, Department of Landscape Planning, Lednice, Czech Republic
3. Institut für Landespflege, Albert-Ludwigs-Universitaet Freiburg, Freiburg, Germany
4. Atelier Gingo, Praha, Czech Republic

Contact: tomas.hajek@projectsolutions.cz

ABSTRACT

The landscape of north-western Bohemia covers the area between the towns of Duchcov, Most and Bílina and its immense historical value could hardly be overestimated. Unfortunately the landscape was heavily hit by the industrial era of coal mining (especially on the surface) and it was exfoliated on a large scale. Mining is in decline, and we are facing the task of restoring almost annihilated landscapes, including their memory traces. It is important to summarize the possibilities and methods of recalling, or 're-creating' landscape memory, and even to search for new ones.

The paper works with the symbols of a specific predicative value – 'minor landscape monument', e.g. crosses, statues, small wayside chapels, etc. These widely disappeared because of brown-coal mining in the mining area of Bílina, and thus are endowed with extended symbolic strength. The leading idea of this project of preservation and recalling of memory traces in an industrial landscape is that of regarding small landscape monuments as the main source of historical identity and an equally essential tool for re-creating it. This project was worked out as a case study investigating the decayed small monuments as well as their re-calling in the strip-mine area of Bílina. The retrieval of information regarding essential cultural, historical, and natural values as well as a general development plan were implemented on the basis of multi-disciplinary cooperation including philosophy, cultural heritage theory, landscape architecture, landscape planning and horticulture. As a result 'the principles of preservation and recalling of memory traces' themselves were proposed as the general attitude to the preservation and recalling of historically precious areas within industrial, deeply changed landscapes afflicted by aberrations of modernity.

A methodical procedure has been thus evolved to be used for a wide range of landscapes of high historical value that have been roughly modified by industrial activities.

KEYWORDS

Middle Ages, modernity, post-modernity, landscape memory, minor landmark (small landscape monuments), cultural heritage theory, industrial landscape, landscape planning

AREA DESCRIPTION

The north Bohemian brown-coal basin covers the area between the Krušné hory mountains (Erzgebirge) and České středohoří (Central Czech Highlands). The land was heavily hit by the industrial era of coal mining, which dates back at least 200 years in this area. Today, coal mining has declined. The Bílina Mine, one of the last persisting active mines, covering an area of 7,441.59 ha, produces boiler coal with annual production of 9 million tons and with 50 million m3 of overburden soils stripped (Bílina Mines, 2009). The core area is situated behind the western border of the active mine, between the villages of Mariánské Radčice and Lom (fig. 1) and it is supposed to be the 'Bílina Nord Mine Multifunctional Open-Air Museum' (Hájek et al. 2009).

Figure 1. Orthophotomap of the area (©GEODIS Brno, Ltd.): Krušné hory mountains in the north-western part. Bílina Mine, one of the last active mines, covering the area of 7,441.59 ha. The Bílina Nord Mine Multifunctional Open-Air Museum is proposed in the core area. ArcGIS map: Matáková 2009.

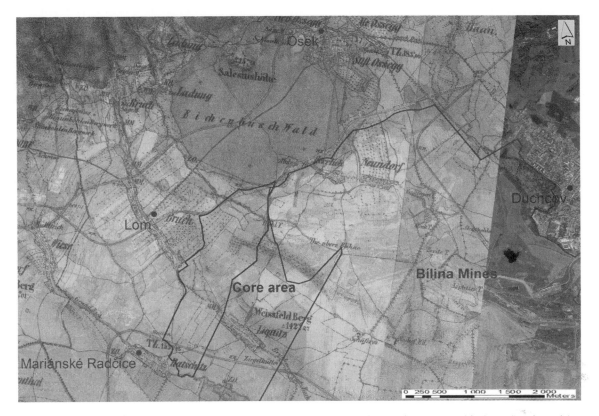

Figure 2. Historical map from the 2nd military survey (©Austrian State Archive) overlapping with the present state (©GEODIS Brno, Ltd.). Intentionally composed landscape with the structure of lanes, pilgrims' routes, composition axes, etc. The map depicts the villages destroyed because of brown-coal mining. ArcGIS map: Matáková 2009.

The immense historical value of the area between the towns of Duchcov (Dux), Most (Brücks) and Bílina (Bilin) could hardly be overestimated (fig. 2). The basic landscape structure is derived from the early Gothic period in association with the mission of the Cistercian Monastery in Osek (Ossegg). The Baroque layout of the landscape is at least of the same importance; moreover, pilgrimage routes create an obvious axis within the landscape structure. The pilgrimage route Mariánské Radčice (Maria Ratschitz) – Libkovice (Liquitz) – Hrdlovka (Herrlich) – Osek (Ossegg) was created just within the framework of foundation of the monastery in Osek (Ossegg). Another landscape-forming axis is the route between Osek (Ossegg) and Duchcov (Dux), which became the centre of the domain in the 15th century. Dux later became the property of the Waldsteins and this enabled successive remodelling of the landscape structures, namely axes and viewpoints. The pilgrim's site of Mariánské Radčice (Maria-Radschitz) with Hněvín Castle above the city of Most form a unique part of the view axis, spreading within the landscape (fig. 3). A unique archaeological site is situated nearby. Archaeological excavations from the partially destroyed village of Světec are considered to be among the most beautiful archaeological excavation collections from the 6th century in Central Europe. The excavations were presented on the occasion of the exhibition 'Schauefenster, der erste: Kelten und Germanen an der Elbe' in Dresden in 2009 (Poppová 2009). The 18th century was a turning point, since that was when the brown coal mining developed and

Figure 3. One of the last quite well-preserved examples of the historical concept of the composed landscape, village of Mariánské Radčice with church and cloister, Hněvín Castle on the hill above the town of Most (in the background). Photo: Matáková 2009.

is still going on now. The onset of mining changed the landscape quite substantially. The landscape was heavily hit by the industrial era of coal mining (especially on the surface) and exfoliated on a large scale.

EXCLUSIVENESS

Two forces oppose each other within the Bílina Mine landscape. On the one hand the landscape is one of the most precious in terms of its historical values. On the other hand its inner sense and the task are rather modern: you see remnants of villages and towns, abolished due to coal mining requirements (fig. 4), you see immense sites of mine dumps as well as the profiles of baggers and excavators in the background of the big mountains of the Krušné hory (Erzgebirge) and České středohoří (Central Czech Highlands) nearby. These contrasts between the Middle Ages and modernity, and modernity and post-modernity are considered to be the main source of exclusiveness of the area; moreover these opposing principles merge into each other, fighting each other, paying deference to each other. The contrast as a message is highly visible, and therefore symbolically strong. The proposed 'Bílina Nord Mine Multifunctional Open-Air Museum' in close contact with the active Bílina strip-mine itself is therefore a fortunate idea highlighting commemorative functions of the landscape and merging them together with other possible functions, especially recreational. There is no effort on the part of this submitted project to put the industrial development aside, to define it as principles going astray, as principles fallen into oblivion.

GOALS

The sense of the 'Bílina Nord Mine Multifunctional Open-Air Museum' landscape project is focused on two main goals: to enhance the quality of the environment especially for local people and, above all, from

Figure 4. Fragments of destroyed villages: mine-pit lake with electric pole. Krušné hory mountains in the background. Photo: Matáková 2009.

the point of view of this contribution, to recall the lost landscape memory. The common theory of recultivation is worked-out in detail, but the same cannot be said about the 'recultivation' of cultural traces. Outlined project principles are a small contribution to the general methodology of 'recultivation of cultural traces' and could be used on other landscapes too. In general it has to be said in advance: To restore almost annihilated landscapes absolutely, including their memory, cultural and historical traces, is not possible because the landscape itself represents 'Change', fully linked with time and entropy.

SUMMARY OF INTRODUCTORY ANALYSES AND EVALUATIONS

The methodical process of introductory analyses and evaluations is linked to the common principles of landscape character assessment that are in use in the Czech Republic (Salašová 2007, Bukáček & Matějka 1999, Löw & Míchal 2003). For the purpose of the presented case study, several steps have been proposed for a comprehensive evaluation:

- Analysis of contemporary landscape layout values
- Analysis of visual and perceptive characteristics (fig. 5)
- Analysis of vegetation layer
- Registering of historical and cultural values
- Registering of the road system
- Evaluation according to criteria (visual and perceptive characteristics, vegetation, historical and cultural values, etc.)

Figure 5. Analysis of visual and perceptive characteristics – panorama: South-eastern view from the reference point no. 17. (Hájek et al. 2009). Corel Draw X4: Matáková 2009. *See also the full colour section in this book*

METHODOLOGICAL PROBLEMS

The restoration of a landscape of such uniqueness raises a number of methodological issues. Theoretically, research and design principles rest on the question of how one area, though familiar, could be the heir of old living networks and stories of a different, though familiar area. In our approach we regard minor landscape monuments as those which could represent the entire fund of lost monuments. Why do we think that small landscape monuments fully deserve to bear such a broad task? The answer is given by the theory of landscape and cultural heritage theory. They gather, rally, concentrate the meanings, randomly scattered within the landscape, in one spot (place = topos). Their outstanding feature seems to be a tight bond with the venue as their highest value, and maybe more than big monuments they express the harmony of death, change and the eternity of habitual rather than extraordinary landscapes. Therefore they could be seen as a good tool for restoring the memory traces of landscapes and landscapes themselves if, of course, the landscape contains such minor monuments, mostly of catholic origin. But very rarely, if ever, is there landscape which deserves this name that is bare and inexpressive regarding symbols, whether visible, or archaeological.

Re-creation will not take place in the original venues, thus another scientific issue arises: whether to re-create the monuments through the exact space symmetry or symbolical symmetry in the designed area of the multifunctional open-air museum? Another methodological problem rests in the way of re-creating them: whether to utilise the full shapes and material identity, or to work only with a symbolic glimpse of the previous shapes. But these are questions for the next step of the research as well as analysis and survey of architectural values and vegetations that determine the venues worthy of installation of re-created monuments, filling the criteria regarding the symbolic value, visual value, etc. To sum up the results to date: The so-called re-calling of decayed or destroyed minor monuments from the areas exfoliated by mining, as a method of remembering the landscape as such, is an important methodological step put forward, explained and evidenced by the scientific team.

Forty-nine minor landmarks that were originally situated in the area and disappeared as a consequence of brown coal mining have been discovered and identified. The method of investigation was based on the analysis of old maps and archive materials (fig. 6) and supplemented with field surveys. From the total number 49 of identified disappeared minor monuments, old pictures of 20 of them were discovered (see table 1 and figs. 7, 8, 9). Pictures often do not provide enough information to gain full knowledge of each monument, though great effort was exerted to obtain them.

Table 1: Examples from the inventory of identified disappeared minor monuments within the Bilina Mine region.

Identified disappeared minor landmarks, basic description	The previous site	Picture available?
Statue of St. John of Nepomuk (date of origin unknown) (fig.7)	Libkovice (Liquitz)	Yes
Baptistry from the Church of St. Michael, 1666	Libkovice (Liquitz)	No
Iron cross on a stone socle 1739	Libkovice (Liquitz)	No
Four baroque pilgrimage chapels (date of origin unknown)	On the way from Libkovice (Liquitz) to Mariánské Radčice (Maria Radschitz)	No
Two columns (one with embossment, second with statue) (date of origin unknown)	Libkovice (Liquitz), next to cemetery	No
A stone for resting, 1674	Hrdlovka (Herrlich)	Yes
Remnant of the so-called Hofmann's Cross (date of origin unknown) (fig.8)	Hrdlovka (Herrlich)	Yes
Baroque Chapel of St. Mary (date of origin unknown) (fig.9)	Hrdlovka (Herrlich), next to Rieger's Inn	Yes
Altar stone (date of origin unknown)	Hrdlovka (Herrlich)	Yes
Kindermann's Chapel (date of origin unknown)	Hrdlovka (Herrlich)	Yes
Cross in Osek (Osseg) (date of origin unknown)	Hrdlovka (Herrlich) to Lom (Bruch)	Yes
Stone cross(date of origin unknown)	Hrdlovka (Herrlich), by sulphur spring	Yes
Two crosses (or chapels) (date of origin unknown)	Neighbourhood of Nový Dvůr (Neuhof)	No
Round column made from sand-stone (1550)	The road from Liptice (Liptitz)	Yes
Plague column (date of origin unknown)	The road from Duchcov (Dux) to Liptice (Liptitz)	Yes

Table 1: Examples from the inventory of identified disappeared minor monuments within the Bilina Mine region (continued).

Identified disappeared minor landmarks, basic description	The previous site	Picture available?
Rectangular column of St. Mary made from sand-stone (1550)	The road from Duchcov to Teplice (Toeplitz)	No
Round plague column (date of origin unknown)	The road from Duchcov (Dux)	Yes
Cross (date of origin unknown)	The road from Duchcov (Dux) to Bilina (Bilin)	Yes
Round column (date of origin unknown)	The road form Duchcov (Dux)to Bílina (Bilin)	Yes
Road-toll column (date of origin unknown)	The way from Duchcov (Dux) to Bílina (Bilin)	Yes
The Jan Hus Memorial (date of origin unknown)	Břežánky, (Briesen), the village square	Yes
The Kaiser Joseph II Memorial (date of origin unknown)	Hrdlovka (Herrlich)	Yes
The T. G. Masaryk Memorial (the era of the first Czechoslovak Republic)	Libkovice (Liquitz), village square	No
The Nelson Mine Disaster Memorial (1934)	Hrdlovka (Herrlich)	Yes
Stone fountain, 1737	Hrdlovka (Herrlich)	Yes
The 'Štengr' sulphur spring (date of origin unknown)	Hrdlovka (Herrlich)	Yes

A minor remark: Regarding the reconnaissance of frequent occurrence of minor monuments, the pilgrimage route from the Osek Monastery to the Mariánské Radčice holy shrine should be mentioned again. Through this medieval route, which now leads through thoroughly changed scenery, a legend about 'Change' and 'the Landscape' could be narrated, in some sense martyrological, but as a whole prone to the future.

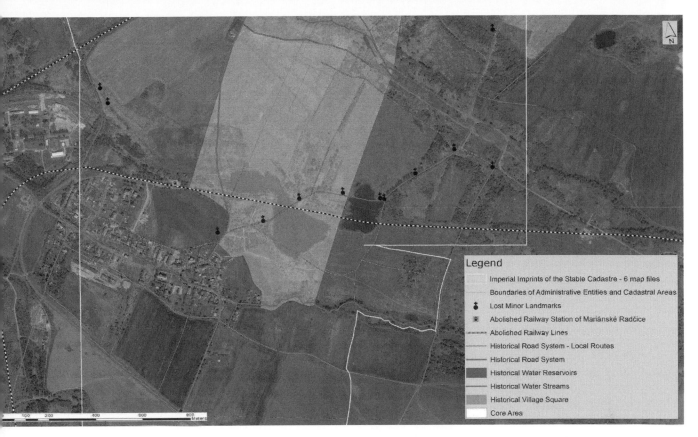

Figure 6. Register of historical and cultural values, register of road system based on analysis of old maps and archive materials. Part of the Map of the historical communication network and minor landmarks (Hájek et al. 2009), ArcGIS map: Matáková 2009. *See also the full colour section in this book*

Figure 7. Statue of St John of Nepomuk, archive material, Přerovský 2009.

Figure 8. Remnant of the so-called Hofmann's Cross, archive material Přerovský 2009.

Figure 9. The baroque Chapel of St Mary, archive material Přerovský 2009.

Figure 10. Preserved monuments in Mariánské Radčice, transferred from the destroyed village of Holešice: statues of Saints in front of the cloister: St Felix (1720) and St Lawrence (1722-1730). Photo: Matáková 2009.

KEY: MINOR LANDMARK

The landscape perceptions are closely interconnected with explications of the landscape. The symbols, inherited images as well as personal experiences enable this. The small landscape monuments determine the landscape as well explaining it. The landscape memory is founded in the presence of the firm points within the landscape, especially static ones (Sádlo, 2005, 226). The landscape is represented by symbols that are eloquent and absolute. The Czech landscape with its different meanings is concentrated into symbols, monuments defined by the 'place', topos (from the Greek τοπος). These topos monuments (crosses, chapels, statues, milestones, etc.) have topos in their designation (Hájek & Bukačová 2004). Minor landmarks are considered to be unique elements identifying European cultural landscapes (fig.10).

CONCLUSION

Research and design principles of vanished landscapes restoration require thorough debate at international level. The principles of this approach, as discussed in this paper, were thoroughly discussed at the Institut für Landespflege (Konold 2007, Langarová 2009), Albert-Ludwigs-Universität Freiburg (Germany) and at the Department of Landscape Planning, Mendel University in Brno (Czech Republic).

ACKNOWLEDGMENTS

With special thanks to Ing. Sylvie Majerová (horticulture specialist). This paper benefitted from the criticisms of two anonymous reviewers.

REFERENCES

Bílina Mines. 2009.http://www.sdas.cz/showdoc.do?docid=736. (Accessed on 17 March 2010)

Bukáček, R. & Matějka, P. 1999. *Landscape character assessment*. [Hodnocení krajinného rázu]. Metodika SCHKO ČR, Praha.

Hájek, T. & Bukačová, I. 2004. *Geschichte der kleiner Denkmäler: Von der Interesselosigkeit zur Faszination*.Eckhard Bodner, Pressath.

Hájek, T., Matáková, B., Langarová, K., Majerová, S. & Přerovský, O. 2009. *Bílina Nord Mine Multifunctional Open-Air Museum* [Polyfunkční muzeum v otevřené krajině – Dul Bílina Nord]. Project Solutions, Praha.

Konold, W. 2007. Dynamik und Wandel von Kulturlandschaften: Was können Biosphärenreservate leisten? *UNESCO heute: Zeitschrift der Deutschen UNESCO-Kommission* 2, 19-22.

Langarová, K. 2009. *Bewertung des Landschaftsbildes in Tschechien: Ein Methodenvergleich*. Dissertation, Institut für Landespflege, Albert-Ludwigs-Universität Freiburg.

Löw, J. & Míchal, I. 2003. *Landscape character*. [Krajinný ráz]. Lesnická práce, Kostelec nad Černými lesy.

Poppová, K. 2009. Schauefenster, der erste: Kelten und Germanen an der Elbe. *Eine Ausstellung des Landesamtes für Archäologie* 1 (7).

Press release. 22 December 2009. The National Museum in Prague. http-://www.nm.cz/press.php?m=1&F_START_LIMIT=3 (Accessed on 15 February 2010).

Sádlo, J. et al. 2005. *The landscape and revolution* [Krajina a revoluce]. Malá Skála, Praha.

Salašová, A. 2007. *Landscape character – theoretical background and methodical principles of preventive assessment*. [Krajinný ráz – teoretická východiska a metodické principy preventivního posuzování]. Mendel University, Brno.

1.4 Cultural forces in the creation of landscapes of south-eastern Rhodope: Evolution of the Byzantine monastic landscape

Authors
Maria Kampa and I. Ispikoudis

Laboratory of Rangeland Ecology, School of Forestry and Natural Environment, Aristotle University of Thessaloniki, Thessaloniki, Greece
Contact: mariakampa97@hotmail.com

ABSTRACT

Cultural landscapes constitute the cultural, social, ecological and economical heritage of the local population. The mountainous area of South-eastern Rhodope, from the Byzantine times till the 1970s, includes a complicated palimpsest characterised by multiplicity and density of natural and cultural elements. The purpose of this paper is the research, identification and evaluation of the cultural landscape of the area of Mt Papikion in south-eastern Rhodope. Information and data were collected and analysed. The consecutive producers of landscape, with different cultural identities, acted within a historical and ecological outline. The monks and later the Pomaks (part of the Muslim minority in Greece as recognised by the Lausanne Treaty in 1923) preserved the basic landscape's structures, while adapting them to their cultural reality. The ruins located in the area indicate the existence of a renowned centre of Byzantine monasticism which is mentioned in the ancient sources from as early as the 11th century.

Since the 1970s, new stakeholders and social changes have had their impact on the landscape, and modern methods of landscape management have replaced the traditional practices. Although it is recognised that traditional knowledge can contribute to sustainable resource use in general (Schmink et al. 1992), modern management rarely takes it into account. Due to extensive afforestations and to the loss of traditional practices of woodland management, the landscape has become more homogenised. Historical, demographic, social, natural and economic changes affect the evolution of the landscape, so it should be mapped, registered and evaluated before its complete disappearance. The cultural landscape of the area is a significant source of knowledge of the traditional environmental know-how.

KEYWORDS

cultural landscape, Byzantium, monasticism, Pomaks

INTRODUCTION

The region of the Papikion Mountain in Rhodope has a variety of traditional landscapes. It was first inhabited by hermits and the communal monastic system was introduced later. During the 11th century AD monks began developing a monastic centre, adopting techniques to shape the landscape according to their needs. After the 13th century, monasteries began to decline, mostly due to fires. During post-Byzantine times, Papikion was no longer considered to be a monastic centre. However, close to the ruins of the monasteries the Islamised residents of the uplands of Rhodope, the Pomaks, built small settlements in order to exploit the arable land around them. The Pomaks are one of the few remaining traditional societies in Greece (Dalègre 1997). They are part of the Muslim minority in Greece, and their landscape practices contributed to the formation of the post-monastic Byzantine landscape.

STUDY AREA

The study area is situated in Western Thrace (fig. 1). It comprises the mountainous area of the north-eastern Rhodope and covers around 13 km². The centre of the area is situated between the parallels, 25º 17' longitude and 41º 13' latitude. Its north-eastern edge is shaped by the border between Greece and Bulgaria, while its south-eastern border is formed by the edge of the Thracian plain. The climate is Mediterranean. The altitude varies from 200 to 1501m. The relief is shaped by numerous flows and slopes. The substratum is gneiss. The vegetation belongs to the subzone of Quercion confertae and the subzone of Fagion moesiacae.

The Pomaks have probably survived as a traditional society because, for historical and political reasons, they were almost completely isolated until the 1970s (Dalègre 1997). Even today, access to the Pomaks' villages is difficult. The Pomaks still identify themselves as 'mountain people' and carry out practices which date back to the Byzantine period, such as pollarding, shredding, multicultures, apiculture, semi-transhumance and conservation of their environment.

METHODOLOGY

In order to identify and register the evolution of the landscape, data was gathered from:

- Landscape archaeology: study of elements visible at the surface (traces of human impact on the soil or other activities, terraces, aqueducts, mills, 'wool washing machines' etc.).
- Vegetation, species of plants and appearance of trees: composition, annual rings, nature of growth, their way of handling.

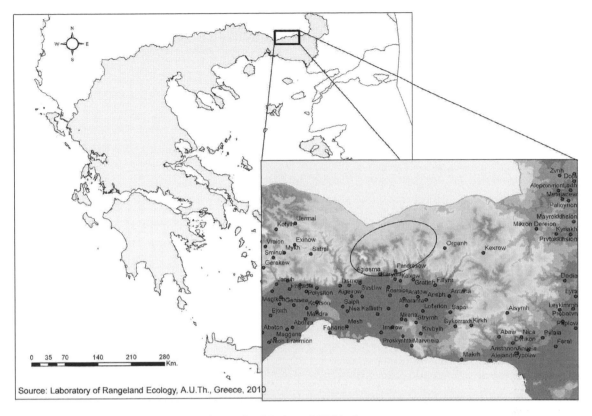

Figure 1. Study area. Source: Laboratory of Rangeland Ecology, A.U.TH., Greece 2010.

- Etymology of the names of villages, castles, rivers, valleys etc.
- Written files, charts, cadastres, delimitations, books local authorities, historical books (descriptions by sightseers).
- Interviews. Emphasis was given to the elder shepherds and farmers, since their testimonies were valuable for the registration of the traditional practices.
- Extensive fieldwork, literature reviews.
- Statistical data analysis, digitalisation of the 1970, 1994 and 2004 forest maps. The main sources of statistical data were the National Statistical Service of Greece (N.S.S.G.) and the Forestry Service (National Statistic Service of Greece 1961, 1971, 1981, 1991), (Venetis 1985).
- Use of spatial analysts ArcMap 9.2 to produce map showing distance zones from rivers.

RESULTS

Byzantine monastic landscapes developed out of the interaction between man and nature. They are the cultural, social, ecological and economic heritage of the people. Due to historical, demographic, social, natural and economic changes the evolution of these landscapes has been rapid and is usually irreversible, so they should be mapped, registered and evaluated before their complete disappearance.

The earliest historical testimony referring to Mt Papikion is found in the 'Typikon', (Foundation Charter) of Gregory Pakourianos, a Byzantine official, drawn up in 1083 for the monastery of Theotokos Petritzonitisa (Batskovo) (Zikos 1984). The Byzantine historians and chronographers, Nicetas Choniates and John Cinnamus, refer to the existence of several 'holy places and monasteries' in their account of the events of the 12th century (Kourilas & Halkin 1936.). Further important information about Mt Papikion is provided in the biography of two Athenian monks, Barnabas and Sophronios, written by Akakius Sabaites in the first quarter of the 13th century. Mt Papikion, according to Akakius, 'got its name from the existence there of many large monasteries'; 370 are mentioned, which may be an exaggeration, but it does reflect the importance of the monastic centre (Zikos 2001). The etymology of the name 'Papikion' probably derives from the term *pappos*, which means elder or monk (Domparakis 1989).

The excavations showed that the first establishment on the mountain possibly dates back to the period of Iconomachy (726-843 AD), when the icon worshippers had taken to mountainous areas far from the urban centres where icon worshipping had been forbidden (Bakirtzis 1989). In the region of West Thrace is located a line of Byzantine cities: Komotini (Koymoutzina), Xanthi, Didymoteicho, Maroneia, Mousunoupoli, Anastasioupoli, etc. that constituted the wider urban environment of Papikion and with which the monasteries had close contact (Asdracha 1976; Papazotos 1980). On Mt Papikion, as in other monastic centres, monastic life would initially have been one of solitude. The coenobitic (communal)

Figure 2. Map with the monuments located in areas where remnants of vine and wheat cultures and pastures were spotted (red= cultivated areas, purple= pastures). *See also the full colour section in this book*

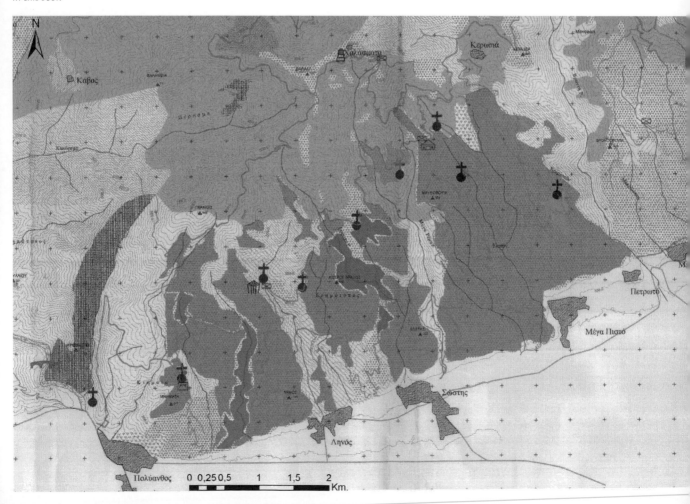

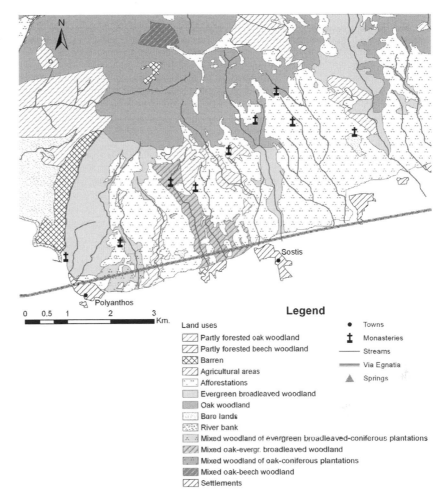

Figure 3. Vegetation map showing the location of the Byzantine monuments, the monastic centres located near water springs and streams and the afforestated areas. Source: The Research Program Pythagoras II – Environment. *See also the full colour section in this book*

Legend

Land uses

▨ Partly forested oak woodland
▨ Partly forested beech woodland
▧ Barren
▨ Agricultural areas
⋯ Afforestations
▩ Evergreen broadleaved woodland
▩ Oak woodland
Bare lands
River bank
Mixed woodland of evergreen broadleaved-coniferous plantations
▨ Mixed oak-evergr. broadleaved woodland
Mixed woodland of oak-coniferous plantations
▨ Mixed oak-beech woodland
▨ Settlements

● Towns
✝ Monasteries
— Streams
═ Via Egnatia
▲ Springs

0 0.5 1   2   3
Km.

Sostis
Polyanthos

system was introduced later, as it was on Mt Athos. Mt Athos is in the region of Halkidiki, a peninsula south-east of the city of Thessaloniki (in northern Greece). Athos is the last prong of Halkidiki's three characteristic peninsulas. Papikion was organised along the lines of Mount Athos which continues to be, until today, Greece's biggest and most important monastic centre. (Papachrysantou 2004). It is certain that the monastic centre of Papikio imitated the athonic system and kept communication with Mont Athos (Charizanis 2003). Excavated organised monastic complexes confirm the coenobitic system in Papikion which did not however, destroy the hermetic way of life. The coexistence of the two monastic ways can be deduced from a text by Akakius Savaites. The historical testimony indicates that monastic life at Mt Papikion reached its acme in the 11th and 12th centuries (Zikos 2001). The inhabiting of Mount Papikion primarily by anchorites (hermits), and later on the development of the monastic centre, can be explained on the one hand by the proximity to Byzantine cities, the existence of the Via Egnatia and the proximity to Papikion mountain, and on the other hand by the abundance of natural resources (Papazotos 1980; Cultural tourist guide 1999) (figs. 2 and 3). Monks took advantage of the natural environment to cultivate the land and create pastures. Pomaks exploit the same areas as shown in the first map (Charisiadis 1963) but afforestations during the 1970s destroyed evidence of both monastic and Pomak land uses.

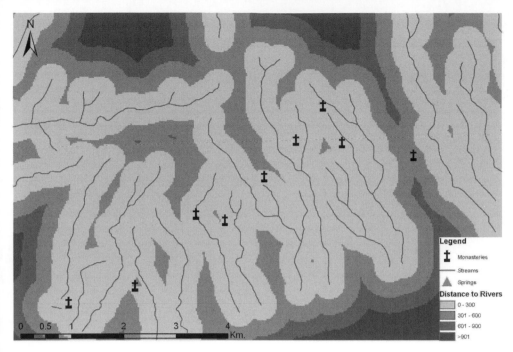

Figure 4. Map with classes of distance zones from streams. Source: The Research Program Pythagoras II – Environment.

Legend

† Monasteries
— Streams
▲ Springs

Distance to Rivers
0 - 300
301 - 600
601 - 900
>901

Page 75, left >
Figure 5. A vine on a tree, a traditional way of vine culture.

Page 75, right >
Figure 6. Ancient willows used for basketry by the monks.

Mt Papikion, along with the entire mountainous zone north of Komotini, is made up of metamorphic rock (gneiss, schist etc.) which forms the Rhodope Massif. Stones, pebbles and sand were available in the local rivers and were used as building materials during the Byzantine era. The excavated monasteries are located near springs, and close to either ravines or rivers (fig. 4). This figure demonstrates that eight out of nine monasteries were located extremely close to the rivers. The monks had benefited from the water, using it for the operation of watermills and of other buildings, whose operation relies on water (baths, water towers, cisterns). Furthermore, the rivers supplied the monastic communities with Lenten food, fish and oysters.

Everything we knew about Papikion until the 1980s was based on historical sources. In 1983, however, we acquired new information from the excavations that took place until the early 1990s. Excavations near the villages of Kerasia, Sostis, Linos and Mischos brought to light the remains of three single-aisled, cruciform, vaulted Byzantine churches, large parts of two monastic settlements and the ruins of a Byzantine bathhouse. All these remains date between the end of the 11th century and the middle of the 13th century.

Another element related to the physical planning/location of the excavated monasteries is their proximity to villages or settlements of modern times. It seems that after the destruction of the monasteries during the 16th century, the local Islamised people developed their villages or settlements near or on monastery complexes. An abandoned settlement was Chotzalar (hoca=clergymen, priests) its Greek name being Monachoi (monks). These names indicate the subsequent use of the place. More precisely, 6 km from the village Linos, the excavations brought to light a monastic complex and a little higher up, in the location Kilise Dere, a cistern and a watermill. The old Linos – known in Turkish as Eski Kiouplou – is situated at a higher altitude, directly next to the Byzantine monastic complex of Linos. Both names are reminiscent of vine culture: Linos in Greek is the wine press, while Eski Kiouplou (Küplü) can be translated into Greek as the old wine shop (Tunkay & Karatzas 2000). Most monuments were found in areas where remnants of agro forestry systems related to vine, and wheat cultivations have been attested. De-

spite the afforestations in the 1980s, landscape units covered by these cultivations still exist, due to traditional systems of cultivation by the Pomaks. There are descriptions of how to use trees to support vines, reported as *anadendrades ampeli* (= vines on trees), during classical times and in the Byzantine period, or as *ypoklima dendra* (= under-vine trees), in the Byzantine period and later (Delivoria 2002) (fig. 5). According to the inhabitants of the area, this used to be a common form of multiculture. However, it is important to note that, despite the fact that Muslims preserved the cultivations and traditional practices, the use of certain products was diversified due to their religious beliefs. Instead of producing wine, they produced *petmez* (treacle). Traditional techniques like pollarding and shredding continued to be carried out, until their prohibition by the Forestry Service (Venetis 1985).

A water mill and cistern are located in the Kilise Dere ravine; Kilise means temple, church, while Dere means brook. Certain tree species in the area are relics of the monastic plantings. The walnut trees, in particular, are a versatile kind of plant, and in addition to the timber supply, they were used for fruit, oil and pharmaceutical use. Moreover, willow trees (fig. 6) indicate basket weaving, but they are also related to pharmaceutics. Through the interviews, today's inhabitants explain that the willow leaves are also used for these purposes. The Kilise Dere cistern is located near the north-western corner of the exonarthex, and is one of the most important buildings of the complex. It is semi-subterranean and constructed of undressed masonry alternating with courses of brick. Four strong piers, which are linked to each other, and the outer walls with brick built arches, divide the building into nine sections. Each is covered with a small elliptical dome, of which the three northern most survive. A small stone staircase in the south-west corner provided access to the cistern.

In 1987, 3 km north of Mischos, on Mt Papikion, a complete Byzantine bath complex was uncovered, situated on a north-south axis. The monks made use of the local water supply and the slope of the ground in order to create a bath (figs. 7, 8). The oak woodland might well have been used for gathering firewood, essential for bathing, and might subsequently have supplied firewood and leafy hay for the Pomaks. The

Figure 7. Byzantine bath.

Figure 8. Oak forest and water point of the Pomaks close to the bath.

numerous iron horseshoes, also found in the area, demonstrate the use of animals for transportation and agricultural purposes.

CONCLUSIONS

The afforestations of the 1980s covered the majority of pastures used by the monks and later by the Pomaks. Modern forests have also replaced the areas where firewood was gathered by the monks (for bathing), and subsequently by the Pomaks. Today, areas of traditional cultural landscape still survive, but some are on the verge of disappearance. The remains include structures, such as monastic ruins, water mills, a bath and cistern, as well as ecological remnants such as walnut trees, pollards, the remains of viticulture and pastures. The landscapes of the study area are the results of a complex religious, rural and ecological history; they are cultural landscapes, derived from a complex interweaving of human and ecological factors.

ACKNOWLEDGEMENTS

This research was funded by the research programme Pythagoras II-Environment, partially funded by the EU.

REFERENCES

Asdracha C. 1976. Les Rhodopes au XIVe siècle: Histoire administrative et prosopographie. *Revue des études Sud- Est Européennes* 34, 175-209.

Bakirtzis Ch. 1989. *Western Thrace in the early Christian and the Byzantine periods: Results of archeological research and the prospects (1973-1987)*. International Symposium for Byntine Thrace, Komotini 28-31 May 1987, 41-58.

Charisiadis N. 1963. *Management plan Public Woodland Complex of Amaxades- Iasmos- Sostis- Asomati. Period 1961- 1970*. Komotini Department of Forestry of Rhodope (in Greek).

Charizanis Ch.G. 2007. *The establishment and operation of the beginnings of Mount Athos*. Stamoulis, Athens (in Greek).

Charizanis Ch.G. 2003. *Monasticism in Thrace during the Byzantine times*. Giahoudis, Thessaloniki (in Greek).

Cultural-Tourist Guide. 1999. *Egnatia Monasteries of Street (Streets of Orthodox monasticism) 2, Central and Western Macedonia*, Thrace – Former South C.D. of Macedonia, southern Bulgaria. Ministry of Culture.

Dalègre J. 1997. *La Thrace grecque: populations et territoire*. l'Harmattan, Paris.

Delivoria V. 2002. *Phytologion* (2nd edition). Bank of Attica, Militos Genus Press Ltd, 347 (in Greek).

Dormparakis P. 1989. *Abridged Glossary of Classic Greek Language. Etymologicon – Hermeneutic* (5th edition). I.D. Kollaros Ltd, Athens, 926 (in Greek).

Kourilas E. & Halkin F. 1936. *Deux Vies de S. Maxime le Kausokalybe ermite au Mont Athos (XIVe siècle)*. Analecta Bollandia 54, 63-109, (in French).

N.S.S.G. (National Statistical Service of Greece) 1961, 1971, 1981, 1991. *Katanomi tis ektaseos tis ellados kata basikes katigories gis* (classification of the Greek area into basic land uses). Hellenic Statistical Authority, Athens.

Papachrysantou G. 2004. *The athonic monasticism: Disciplines and Organisation*. Institute of Byzantine Studies, Athens (in Greek).

Papazotos Th. 1980. *Pre-excavational Researches in Mount Papikion*. Thracian Annals (in Greek).

Schmink, M., Redford, K. H., & Padoch, C. 1992. *Traditional peoples and the biosphere: framing the issues and defining the terms, Conservation of neotropical forests: working from traditional resource use*. Columbia University Press, New York, 3-13.

Tunkay F. & Karatzas L. 2000. *Turkish Greek dictionary*. Centre of Eastern Languages and Culture, Athens, 865.

Venetis, A. 1985. *Forest management plan for forest area of Amaksades-Iasmos-Sostis- Asomatos, 1985-1994*. Komotini, Forestry Service Komotini (in Greek).

Zaharis, S. & A. 1977. *The forests of Crete from antiquity until today*. Athens, Aspioti-ELKA (in Greek).

Zikos, N. 1984. Byzantine Walk in Thrace. *Archaeology* 13, 71-77 (in Greek).

Zikos, N. 2001. *Mount Papikion, Archaeological guide*. Region and Regional bureau of Eastern Macedonia, Thrace (in Greek).

Petit L. 1904. *Typicon de Grégoire Pacourianos pour le monastère le Pétritzos* (Baškovo), Vizantijskij Vremennik 11.

1.5 The change analysis of the green spaces of the Historical Peninsula in Istanbul, Turkey

Author

Nilüfer Kart Aktaş

Department of Landscape Planning and Design, Faculty of Forestry, Istanbul University, Istanbul, Turkey
Contact: niluferk@istanbul.edu.tr

ABSTRACT

The Historical Peninsula, having a nearly 8500 year historical background, serving the capital city of three empires, having a strategic position, offering unique natural beauties, architectural and archaeological values and a stunning skyline, was declared as a protected area in 1985. However, the area has been subject to planning legislation several times, although the planning did not address some crucial issues of the city, and caused the conversion of the historical downtown into a suburban area. This is one of the common problems of historical cities. As a consequence, the Historical Peninsula, through both social and physical changes, has experienced changes in entity and identity – and still it is changing.

The purpose of this study is to investigate the spatial formation of the Historical Peninsula from past to present through literature review and visual materials (such as maps and photographs), using both qualitative and quantitative research methods. I will aim to establish the location of green spaces, and I will consider the alteration of those spaces due to changing social structures, political decisions, planning legislation and physical conditions. Several maps of the peninsula, prepared at different dates and time periods, were selected to conduct the study. These maps vary in resolution and appearance, but the aspects and size of the green spaces are specified. As a result of the study, functional and spatial changes of green areas in Historical Peninsula were examined in its historical process.

KEYWORDS

urban landscape, urbanisation, change analysis, green spaces, Historical Peninsula

INTRODUCTION

The function of cities, in general, has changed and developed rapidly over time, depending upon various factors. Cities have also expanded, and as a result of these changes they have gained different identities, in terms of history and politics. The political, physical and technological changes have resulted in the destruction of the urban textures, largely because of changes in social values which have lead to new demands. Because of all these factors, the development areas in big cities have become, in a historical sense, collapsed areas.

The transformation process still continues in the city of Istanbul today, as it has occurred in other developed cities around the world. The city of Istanbul has attracted a huge number of people from the Anatolian part of the country, who arrived as immigrant workers. This has caused a poorly planned urban sprawl and a rise in population from 1,000,000 to 9,000,000 in a very short time. Ultimately, the city of Istanbul has passed through a stage of considerable change and development in terms of architecture in the last twenty years (Ergun 2003).

As a result, we see a different architectural structure in the city, which has impacted on the Historical Peninsula. The city is now full of small shops and manufacturing workshops, which have been integrated somehow into the historical environment around them. This feature of the city makes it uncommon and unique among all the other big cities. The so-called central inner region of Istanbul, which was separated from the outer parts of the city by a great wall, is the old, historical, culturally valuable and main part of the city. This is called the Historical Peninsula. Due to poorly controlled urbanisation, this central part of the historical city faces some serious problems both structurally and architecturally. After having survived major physical, political and cultural changes, it is easily possible, now, for this region to collapse with a serious earthquake.

The social and cultural structure of the inhabitants has caused dramatic and unwanted physical changes, too. The increasing demand for land in the city has resulted in soaring prices of land parcels, lots, houses and buildings. The high cost of lots suitable for building construction, particularly in the central part of the city, has caused a serious shortage of green spaces and recreational areas, which are now very rare in the city. And yet, in ancient times and in the recent past, Istanbul was a city full of vegetable gardens and agricultural fields. It was teeming with ethnic minorities from different origins, living peacefully together, while it was also a centre for trade, transport and the administration of the country. Unfortunately, there are no longer enough empty areas in the city with which to create new green spaces, and worst of all, the existing green and recreational areas in the city are simply of poor quality.

MATERIAL

The Historical Peninsula is a densely populated area containing the most important historical opuses in Istanbul. Haghia Sophia Mosque, Blue Mosque, Topkapı Palace, Süleymaniye Mosque, the Hippodrome, Sultanahmet Square, the world famous Covered Bazaar, the Beyazıt Complex, Cemberlitas, the Museum of Basilica Cistern, the Mosaics Museum, the Kariye (Chora) Museum, the Archaeological Museum and many others are located in this peninsula. The Historical Peninsula is also the centre of government, with the Governorship of Istanbul and many offices of tax administration. Many faculties of Istanbul University, important libraries and hospitals are located in this area. This region also has the most important archaeological sites in Istanbul. The total area of the Historical Peninsula covers 1,562 ha.

HISTORICAL PERSPECTIVE OF THE HISTORICAL PENINSULA

The history of the Historical Peninsula, having a nearly 8500 year historical background and serving the capital city of three empires, can be broken down into five periods.

1. Greek Period
2. East Roman Period
3. East Roman-Byzantine Period
4. Ottoman Period
5. Republic Period

INTRODUCTION OF THE STUDY AREA AND ITS ENVIRONMENT

In this section, the location, topography, archaeological situation, cultural situation and visual situation are investigated.

Location and topography

The Historical Peninsula has a geographical location delimited by the Haliç to the north, the Marmara Sea to the south and east and land walls to the west (fig. 1).

Figure 1. Location of Historical Peninsula.

The landscape of the city has played an important role in terms of the architectural development of the city so far. An old and historical city built on the seven hills; it has unique monuments, historical and modern buildings. The different features of the city serve to create an impressive skyline (Anonymous 2004).

Archaeological situation

The Historical Peninsula, spreading through and including the Fatih district, hides precious archaeological remains. It is therefore under protection by law and it has been declared a historic site by the regulation, taken by the First Council for the Protection of the Cultural and the Natural properties in Istanbul, dated 12 July 1995, numbered 6848 (Anonymous 1990). As the capital of both the Byzantine and Ottoman Empires, the city is one of the most important in the modern Republic of Turkey. However, the architectural structure of the city makes it impossible to carry out secure archaeological excavations (Tekin 1996).

The core of the city was essentially the centre of the Byzantine Empire, where the Topkapi Palace and the Hagia Sophia still stand today. The area on which the Topkapi Palace was built was formerly the acropolis of the city, where there were temples. It is known that there were two or three seaports in the Golden Horn (the peninsula to the north) at the time. In some ancient records, it was reported that there was an agora, which had a statue of Helios on it, encompassed by four galleries supported by four columns. The Greek historian Xenophon told of the presence of a large square called Thrarion. On the northern part of this square there was an area called Strategion where the leading citizens of the city lived. There were also said to be gymnasiums and cisterns. In the Roman Empire period, the city was not bounded by the Sarayburnu region and it spread out of its borders. It is known that a viaduct built for carrying water to the city was made during the reign of the emperor Hadrian between the years 117-138 AD. In the meantime, the building of the city's hippodrome was begun during the reign of the emperor Constantine between the years 306-337 AD. There was a theatre in the city, which was built during the reign of the emperor Septimus Severus. There were also baths, including one called Akhilleus near the Strategion. Unfortunately, all of these structures were almost totally destroyed and there are hardly any remnants surviving today (Tekin 1996).

The Historical Peninsula, the region called Sultanahmet-Cankurtaran and the region Fener-Balat are the rich parts of Istanbul today, with historical buildings, sites, cultural heritage and antiques both on the surface and below ground. (Anonymous 2004).

Cultural situation

The Historical Peninsula, which is nearly 8500 years old, has a rich and unique cultural heritage. The whole Historical Peninsula was registered on the World Cultural Heritage List by UNESCO in 1985. The protected area includes the Sultanahmet Archaeological Park, the Süleymaniye Mosque, the mosque in Zeyrek (which was converted from a church), the Walls of Istanbul and the surrounding area. Istanbul contains many historical buildings from the periods of the Roman Empire, Byzantium and the Ottoman Empire. The most impressive and valuable old buildings of the city are the wooden houses, which still exist in the wider region, including the provinces Zeyrek, Süleymaniye, Fener, Balat and their surroundings. Many fountains, baths, etc. may be considered as the good examples of the works of civil architecture of their periods.

The famous Wall of the city, which was built to protect Byzantium, is regarded as one of the strongest walls in the first historic age. Although the walls were partially dismantled in the 19th century, parts are still remaining and are now being conserved.

Visual Situation

The Historical Peninsula has features of both historical and modern architecture, with its old and new buildings, houses and green spaces, side by side. This gives the city an impressive and unique silhouette. The Historical Peninsula was built on the seven hills, and on each hill some monumental buildings were constructed. The monumental and historical buildings on each hill create a different ambience, which makes the hills more visible and striking.

Method

The purpose of this study is to investigate the spatial formation of the Historical Peninsula from past to present, through literature review and visual materials (e.g. maps, photographs, gravures, etc.). This includes establishing the location of green spaces and the alterations of those spaces formed by political decisions and planning policies, physical and social conditions.

To conduct the study, several maps of the referred peninsula, prepared at different dates and time periods, were selected by considering their time period, resolution and appearance. The aspects and size of the green spaces have been specified. The characteristics and size of the green spaces is also noted in various planning proposals through history. The maps which were consulted were prepared between 1887-1882 by Ekrem Hakkı Ayverdi and by Necip in 1918, 1965 and 1999. In the research process, Autocad 2008, Netcad 4.0 ve ArcGIS 9.1 (GIS – Geographic Information System) softwares were used. To determine changes of green spaces in different time periods, the maps were digitised in a vector format, and used as a base for GIS. In the GIS context, the qualitative and quantitative changes were determined using overlay analysis.

RESULTS

The Process of Change on the Historical Peninsula

The Historical Peninsula has passed through major architectural changes due to changing social, physical and economic needs. Social changes may be the most salient reason for the alteration, but fires, earthquakes and epidemics have also taken their toll. The main calamities on the Historical Peninsula have been due to big fires. Over 90 major fire incidents occurred in Istanbul in the 18th century, including fires in the years 1718, 1756 and 1782. These three conflagrations caused great damage to the area from Cibali, the Golden Horn, to the shores of Langa in the Marmara Sea, destroying the wooden houses around the region. It is estimated that the old and worn-out wooden houses on the Historical Peninsula that survive today are remnants of the period when those big fires occurred. After these fires, the people of Istanbul were encouraged to build their houses from stone instead of wood. However, there was a lack of time and funding, and many people were in a desperate situation and in need of shelter, which prevented them from building in stone. Furthermore, the people of Istanbul preferred to build wooden houses in a haphazard way. These structures are called 'Istanbul Type Houses' in today's architectural jargon. Although these wooden houses were easily and quickly built, unfortunately they were burned easily and quickly too (Yerasimos 1996).

People fled from Anatolia in the 16th-17th centuries due to the Celali rebellion, and came to Istanbul. Many others in the Ottoman Empire also immigrated to Istanbul, due to other rebellions and wars. The population of the city increased enormously and suddenly decreased again during recurrent episodes of the bubonic plagues in those times (Yerasimos 1996).

Earthquakes have also made a great impact on the landscape of Istanbul, which is located on a fault line. We learn from different records that nearly 54 major earthquakes have occurred in the city so far. As we have a large number of records such as chronicles and manuscripts, we know a lot about the earthquakes in the 6th century. The 54 earthquakes between the years 342-1454 AD damaged the city on a wide scale. However, many of the changes that have been described occurred gradually, as habits of life and economic conditions changed (Mango 1997).

The changes in the layout of green spaces in the Historical Peninsula have been investigated under two headings: social changes and physical changes.

Social changes

In 700 BC, there were people called the Megara who lived on the Historical Peninsula and were chiefly fishermen, fishmongers and traders. They lived in the small area now called Sur-u Sultani (The Wall of the Sultanate). The population of Byzantium was nearly 40,000 and its people mostly dealt in trade. On the hill, where later the Topkapi Palace would be built, stood the city's main acropolis. On the north-west of the Acropolis, near the coastal area, there was a gymnasium, a temple, a stadium and some other buildings. On the south-west of the Acropolis there were units for the administration, and on the south of the Acropolis there were private premises that belonged to the wealthy people of the time. Beneath them there were the premises belonging to the people of the middle class. In 196 AD, the first Wall of the city was built under the reign of the Roman Emperor Septimus Severus, and the city spread out to the south and south-west (Anonymous 2004).

In 330 AD, by the decree of the emperor Constantine, the city was rebuilt and renamed Constantinopolis. He built a new seaport and also extended the line of the Wall to the west, in order to create a new settlement area. Consequently, during his reign the population of the city rose from 150,000 to 200,000. The new seaport was built on the coast of Marmara near the Wall, parallel with the growth of the city. His aim was mainly to create an important and strong capital city (Kuban 1993).

Following the construction of many further city walls, a new double wall line was built in the 5th century, called the Theodosian walls. After this change, the borders of the Historical Peninsula took their present shape. However, the area between the walls built during the reigns of Theodosius and Constantine was not regarded as an attractive settlement area by the people of Byzantium. Therefore, the population in this area was sparse. In the area behind the Walls, the monasteries, the villas and the vegetable gardens were established. Yet, after moving the palace of Byzantium to the Blahernai, as an exception, the region between Blahernia and Tekfur Palace became a densely settled area. During the invasion of the Latins, the city was destroyed and looted. In the 15th century, the population decreased drastically as a result of diseases and plagues, decreasing to around 30,000-50,000 (Çelik 1998).

In 1453 BC, Constantinople was conquered by Sultan Mehmet II of the Ottoman Empire, which effectively put an end to the Byzantine Empire. Many people were forced to move from different parts of the

Empire and to settle in Istanbul. These people were Muslim, Christian and Jewish subjects of the empire. After Crimean War, the Armenian subjects were settled in the region between the Unkapanı Balat, and Gedikpaşa. In the reign of the Sultan Yavuz Selim, the Armenian subjects were settled to the region called Samatya and the Jewish people fled from the Spanish inquisition were settled to the region called Balat. All these movements caused the cosmopolitan formation of the city, demographically.

The Jewish subjects of the Ottoman Empire, who lived in the region called Balat in the 15th century, had lived in the Bahçekapı and Eminönü before that. After the big fire in the year 1660, they resettled in the region called Balat because of the extension work of the city's seaport and the building of a new mosque called Yeni Camii by the Mother of Sultan Turhan. The Greek subjects lived in the region called Fener, which was the spiritual centre of the Orthodox Church. The Fener region was the place of the most important subjects of the Ottoman Empire; as the translators for the empire in the Foreign Office Department, they were both the aristocrats and the favourites of the Palace (Ortaylı 2007).

The neighbourhood called Karagümrük of the Ottoman Empire was one of the most important neighbourhoods with its respected Turkish people, who were tradesmen, artisans and civil servants of the Empire, educated in the madrassas or religious schools. These Turkish-speaking people created the Istanbul dialect of today. There were a lot of madrassas in this area, and consequently, most of the religious clerics, 'ulemas', were of this neighbourhood. The presence of the madrassas and the location near to the ceremonial road gave an advantage to the neighbourhood, which grew rapidly. The area became one of the most important centres, with its dervish lodges and the private premises of the Muslim clerics (Ortaylı 2007).

The neighbourhoods called Kumkapı and Samatya were the settlement places for the Armenian people. These regions were well known for their storage sheds for wood and the bars for the fishermen. Kumkapı is famous for its churches and the office building belonged to the Armenian patriarchate. Kumkapı, as the centre for the fishmongers and vegetable sellers, became a neighbourhood, supplying food for the city (Ortaylı 2007).

Zeyrek was one of the most elegant neighbourhoods of the city; it was full of beautiful villas, with a beautiful panorama. The monastery and church called Pantokrator in the period of Byzantium was converted to mosque, and called Molla Zeyrek Mosque (Ortaylı 2007).

Over the 18th century, there were increasing problems of public disorder, overcrowding, big fires and plagues, and consequently the people of Istanbul began to prefer to live north of the Golden Horn. In the 19th century, there was great development in the regions called Galata and Pera, with the building of new foreign office buildings including embassies and banks. During this period, the empire was making an effort to westernise itself. Because of this new development, the Dolmabahçe Palace was built in order to be close to the new city centre and the Sultan moved to his new palace, and never returned to Topkapı Palace. In other words, the Ottoman dynasty left the old Historical Peninsula. During this period, the Muslim, Greek, Armenian and Jewish families preferred to move to the north of the Golden Horn, leaving their old neighbourhoods on the Historical Peninsula, while the empire was trying to open itself both culturally and economically towards the Occident (Yerasimos 1996).

In the 20th century, people from the wealthier classes moved to the Province, which is called Beyoglu today, leaving Balat, Fener and Samatya. The poor people from the lower classes were then able to occupy the settlement places they abandoned on the Historical Peninsula. As a result of this change, new poor neighbourhoods began to appear close to the Wall of the city on the Historical Peninsula. Consequently,

Balat and Fener regions became an area of slums, and this poor condition and outlook is still the same to-day (Ortaylı 2007).

Physical changes

In the historical process from past to present, the characteristics of users and the usage properties have changed. In accordance with to the user changes, green spaces have also altered. In the scope of this unit, the physical changes of Historical Peninsula are determined in the historical process. The maps were prepared between 1887-1882 by Ekrem Hakkı Ayverdi, by Necip in 1918, 1965 and 1999 were evaluated.

Map of Ekrem Hakkı Ayverdi: These maps were drawn by engineers between 1875-1882 and published by Ekrem Hakkı Ayverdi. The scale of these maps is 1/2000 and the language of the original maps is Ottoman Turkish. The green spaces have been grouped as vegetable garden, mosque garden, square, graveyard, garden, pasture, home garden, public garden, school garden, private garden, palace garden and empty space. Between 1875-1882, the total area of green spaces was 277.2 hectares and the largest portion of the green spaces (63,7 %) was composed of vegetable gardens (table 1 and fig. 2).

Table 1. Green spaces, their functions and quantity on the map of Ekrem Hakkı.

Functions	Quaility	Quantity (ha)	Percent (%)
Vegetable garden	passive	176,6	63,7
Mosque garden	passive	14,6	5,3
Square	passive	28,6	10,3
Graveyard	passive	8,5	3
Garden	passive	13,3	4,8
Pasture	passive	2,2	0,8
Home garden	passive	1,5	0,5
Public garden	passive	4	1,4
School garden	passive	2,2	0,8
Private garden	passive	0,008	0,003
Palace garden	passive	5,8	2,1
Empty space	passive	20,1	7,3
Total		277,2	100

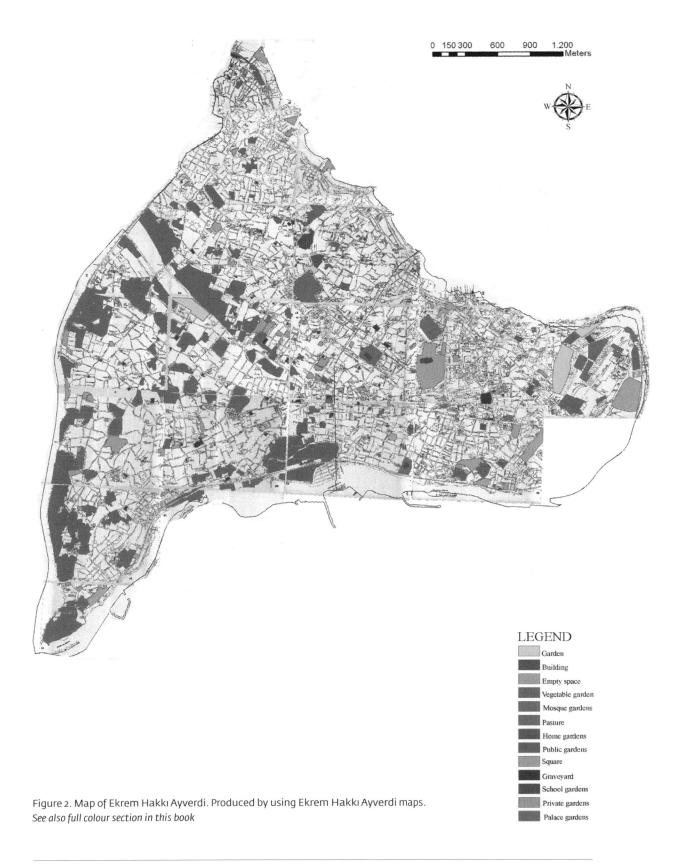

0 150 300   600   900   1.200
Meters

LEGEND
Garden
Building
Empty space
Vegetable garden
Mosque gardens
Pasture
Home gardens
Public gardens
Square
Graveyard
School gardens
Private gardens
Palace gardens

Figure 2. Map of Ekrem Hakkı Ayverdi. Produced by using Ekrem Hakkı Ayverdi maps.
See also full colour section in this book

Map of Necip: This map is the topographical urban map that was drawn by Mr. Necip with the scale of 1/5000 in 1918. The green spaces were grouped as vegetable garden, mosque garden, square, graveyard, garden, pasture, public garden, school garden, private garden, vacant lot, empty space, park refuge and military zone (table 2 and fig. 3).

Table 2. Green spaces, their functions and quantity on the map of Necip.

Functions	Quality	Quantity (ha)	Percent (%)
Vegetable garden	passive	187,8	40,2
Mosque garden	passive	27	6
Square	passive	10	2,1
Graveyard	passive	0,05	0,01
Garden	passive	13,6	3
Pasture	passive	10,62	2,3
Public garden	passive	1,9	0,4
School garden	passive	0,5	0,1
Private garden	passive	0,3	0,06
Vacant lot	passive	77,3	17
Empty spaces	passive	0,1	0,02
Park	passive	123,1	26,5
Refuge	passive	0,1	0,02
Military zone	passive	0,1	0,02
Other green spaces	passive	3,7	1
Total		467,7	100

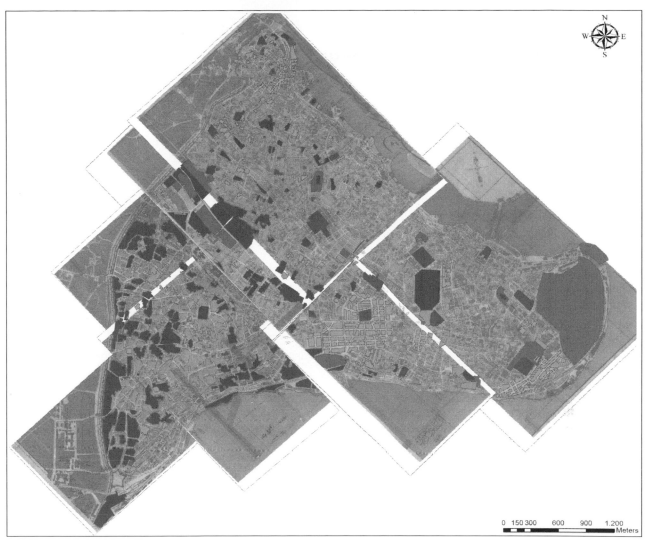

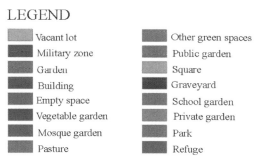

Figure 3. Map of Necip. Produced by using Necip maps.
See also full colour section of this book

LEGEND

	Vacant lot		Other green spaces
	Military zone		Public garden
	Garden		Square
	Building		Graveyard
	Empty space		School garden
	Vegetable garden		Private garden
	Mosque garden		Park
	Pasture		Refuge

Map-1965: These maps are made up of seven pieces; however, they do not contain the whole peninsula. The green spaces have been grouped as mosque garden, graveyard, sporting areas, refuge and empty spaces. One can see sports areas as an active green space for the first time (table 3 and fig. 4).

Table 3. Green spaces, their functions and quantity on the map of 1965.

Functions	Quality	Quantity (ha)	Percent (%)
Green spaces	passive	412,7	91
Graveyard	passive	4	0,9
Mosque garden	passive	16	3,5
Refuge	passive	11	2,4
Court	passive	6,2	1,4
Sporting area	active	2,5	0,6
Empty spaces	passive	0,1	0,2
Total		452,5	100

Figure 4. Produced by using present maps (1965). *See also full colour section of this book*

LEGEND
- Court
- Building
- Mosque garden
- Other green spaces
- Graveyard
- Refuge
- Sporting area

Map-1999: These maps were prepared from air photographs taken in 1996 and orthophotos certified in 1998 with the scale of 1/5000 and 1/1000. These maps are made up of 63 pieces in the scale of 1/1000 and ten pieces in the scale of 1/5000. There are various types of open and green spaces, reflecting the variety of existing functions in the Historical Peninsula. The total area of green spaces such as botanical garden, parks, sports areas, vegetable garden, square, was 153.8 hectares (table 4) (fig. 5). However, military zones, mosques, church and synagogue areas, hospital areas, university areas, parking areas, archaeological areas and empty spaces should be added to these green spaces.

Table 4. Distribution of existing green spaces in the Historical Peninsula (Planning and Development Department, 2003).

Green spaces	Quality	Eminönü Area (ha)	Percent (%)	Fatih Area (ha)	Percent (%)	Total Area (ha)	Percent (%)
Botanic garden	passive	1,6	0,3	–	–	1,6	0,1
Parks and green spaces	active	49,5	9,7	74,5	7,2	124,0	7,9
Vegetable garden	passive	–	–	9,8	0,9	9,8	0,7
Sporting area	active	2,3	0,5	10,5	1,0	12,8	0,8
Square	passive	5,6	1,1	–	–	5,6	0,4
Total		59	11,6	94,8	9,1	153,8	9,9

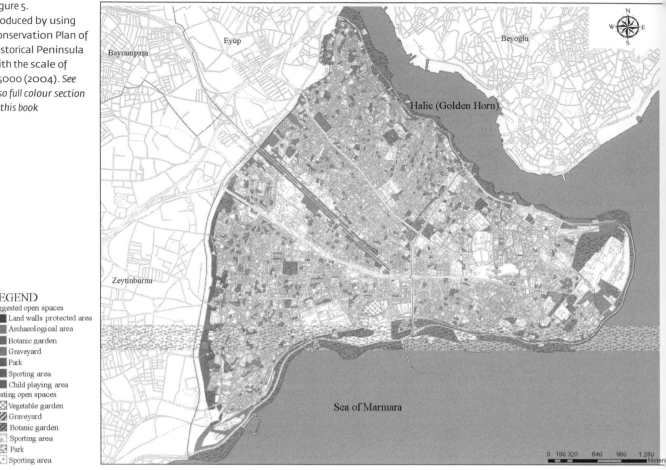

Figure 5.
Produced by using
Conservation Plan of
Historical Peninsula
with the scale of
1/5000 (2004). *See
also full colour section
of this book*

LEGEND
Suggested open spaces
▨ Land walls protected area
▨ Archaeological area
▨ Botanic garden
▨ Graveyard
▨ Park
▨ Sporting area
▨ Child playing area
Existing open spaces
▨ Vegetable garden
▨ Graveyard
▨ Botanic garden
▨ Sporting area
▨ Park
▨ Sporting area

CONCLUSION

Cities are in the process of changing to meet the needs of modern life. However, many of these changes have become problems, particularly in historical city centres. These problems include social problems such as intensive migration, increasing crime, insecurity and unemployment, and physical problems such as intensive urbanisation, transportation problems, poor infrastructure and lack of planning.

Throughout the 18th century, people emigrated from the city centre to the north of Haliç, because of the lack of public security, overcrowding, fire and epidemic diseases. People began to prefer the gardens on the Bosphorus for leisure and recreation, and the Sultans built summer houses along the Bosphorus coast.

In the 19th century, as a result of the major fires which dramatically affected the Historical Peninsula, the quantity and size of the existing green spaces decreased. Not only fires, but also intensive immigration to the peninsula caused the transformation of houses with gardens into apartments. As a result of this, the concept of house gardens began to disappear. This can be seen on the map of Ekrem Hakkı Ayverdi. The empty spaces (making up 7.3% of the area) are the third-biggest green spaces after vegetable gardens and squares, and may be the result of the fires.

The Historical Peninsula had a population of 14,805 in 1477, but in 1927 the population had reached 900,000, with 250,000 people living inside the city walls. There were, however, still houses with gardens and vegetable gardens inside these walls. People gathered in mosque squares and used their gardens for recreational activities. Tall trees planted along the garden walls provided security. There were all types of fruit trees, designed areas and water elements in the gardens.

From the middle of the 19th century onwards, the city began to expand as rich people moved away from the Historical Peninsula to areas such as Karagümrük, Haseki, Fener and Balat. The reason for this was the gradual transformation of the Historical Peninsula into a slum. In this period, vegetable gardens and houses with gardens were transformed into apartments, and as a result of this the quantity of open space decreased. Then, the concept of the 'municipality park' appeared.

The empty spaces that can be seen on the map of Ekrem Hakkı Ayverdi, which make up 7.3%, decreased to 0.02% in the 1918 map of Necip. The percentage of vacant lots became 17%. This can be considered as the result of the movement of improvement. On this map, the 'park' as an active green zone makes its first appearance. The Gülhane Park was prepared and opened to public in 1912. It was also, clearly, on the map of Necip. Because of the increase in the number of the people on the Historical Peninsula, a lot of mosques were built and this increased the number of yard space belonging to the mosques.

When examined closely in terms of the total distribution of green areas on the maps, it is obvious that the map of Ekrem Hakkı (dated 1875-1882) contains the least amount of green area. But, it can be explained that this was not a plan for the region at that time, and the borders along the roads were not included on the plan as empty spaces for greening. The year 1918 was the year with the most green areas in the city. We also see a large number of green areas on the map drawn in 1965, although this does not cover the whole Historical Peninsula. Until 1950, Istanbul had been protected on a large scale. From 1950 onwards, Istanbul began to change as a result of rapid urbanisation and movement of improvement. Insufficient attention was paid to the historical value of the city while redesigning it.

Green areas total 354.6 hectares on the map drawn in 1999. Only a few of the old fields and vegetable gardens have survived, and they are mostly in the area surrounding the Wall, which is declared as green space under protection. Today, there is no room for green spaces in the Historical Peninsula. In the complex of functions, the peninsula's existing green spaces are composed of open spaces of historical buildings (mosque gardens, palace gardens, etc.), squares, large coastal parks, land-wall protected areas, the coast of Halic and Marmara, and on a small scale, neighbourhood parks and sporting areas. Gülhane Park, Eminonu, Beyazıt and Sultanahmet Squares and the coastal parks are the main open areas for people. But, especially in Eminonu, small-scale parks cannot meet the people's recreational needs.

The changes in the function and density of the population have affected the Historical Peninsula deeply, resulting in a lack of empty areas to green the city. The high population density of the city has caused an extreme increase in the price of the lots. As a result, only some small areas can now be greened and used as recreational areas in the city. Consequently, with its high population density, the city has become a very difficult place to live comfortably, due to the lack of green and recreational areas.

In conclusion, the Historical Peninsula has suffered terrible environmental degradation which is not appropriate, especially given its extraordinary history. The buildings and monuments of this area should form a harmonious layout with each other and with the surviving green spaces. The main goal should be to protect the historic environment, architectural properties, archaeological opuses, environment and skyline. The real identity of the Historical Peninsula should be uncovered and maintained.

REFERENCES

Anonymous. 1990. *Historical Peninsula Master Plan Report*. Istanbul Metropolitan Municipality, Directorate of Planning.

Anonymous. 2004. *Historical Peninsula Master Plan Report*. Istanbul Metropolitan Municipality, Directorate of Planning.

Çelik, Z. 1998. 19. *Yüzyılda Osmanlı Başkenti Değişen İstanbul*. History Foundation, Yurt Press, Istanbul.

Ergun, N. 2003. İstanbul Şehir Merkezi Yakın Çevresinde Kullanım ve Kullanıcı Farklılaşması. *International 14*. Urban Design and Implementation Symposium, Urban Regeneration and Urban Design, 28-29 May 2003, Istanbul, Mimar Sinan University, 346-356.

Kuban, D. 1993. Koloni Şehrinden İmparatorluk Başkentine. *Journal of Istanbul 4*, 10-25.

Mango, C. 1997. The Urbanism of Byzantium-Constantinople, Rassegna (İstanbul, Constantinople, Byzantium). *Quarterly year XIX, 72-1997/IV*, Bologna, İtaly, Editrice Compository, 16-24.

Ortaylı, İ. 2007. *İstanbul'dan Sayfalar*. Alkım Press, Istanbul.

Tekin, O. 1996. *Eski Çağda İstanbul: Byzantion. Dünya Kenti İstanbul*. History Foundation, Istanbul, 108-130.

Yerasimos, S. 1996. *Batılılaşma Sürecinde İstanbul, Dünya Kenti İstanbul. Habitat II. Urban Meeting Istanbul 1996*. International Scientific Assembly, 3-12 June 1996.

1.6 The evolution of an agrarian landscape. Methodological proposals for the archaeological study of the alluvial plain of Medellin (Guadiana basin, Spain)

Authors

Victorino Mayoral[1], Francisco Borja Barrera[2], César Borja Barrera[3], José Ángel Martínez del Pozo[1] and Maite de Tena[4]

1. Merida Institute of Archaeology, Mérida, Spain
2. University of Huelva, Huelva, Spaim
3. University of Seville, Seville, Spain
4. University of Extremadura, Mérida, Spain

Contact: vmayoral@iam.csic.es

ABSTRACT

In this paper we introduce the background, objectives and initial results of a regional research project focused on the evolution of the agrarian landscape in the Medellin alluvial plain (Badajoz province, Spain). From several different fields we face the task of analysing the changing role of this settlement as a political and administrative centre placed at a key point in communications of the Peninsular southwest. At the same time, we assess the methodological challenges of defining distinct territories based on the archaeological record in an area which has been severely affected by geomorphological processes, and which has also been disturbed by intensive agricultural activity in recent years. Firstly, basic criteria for the surface survey are explained. The workflow is explained from the definition of survey areas to the analysis of spatial distribution of finds. Secondly, we will show the preliminary results of the geomorphological interpretation and mapping of the study area. This is the starting point for a suitability model for the analysis of the interaction between factors that determine our perception of surface archaeological evidence. The objective is the elaboration of a map that would help to evaluate the degree of terrain stability, as a means to assess the representativeness of the surface record and potential risks for conservation.

KEYWORDS

agrarian landscapes, Archaeological survey, alluvial geoarchaeology, protohistory, Romanisation, Medellin (Spain)

INTRODUCTION: ALLUVIAL LANDSCAPES: GEOMORPHOLOGICAL PROCESSES, HUMAN IMPACT AND THEIR EFFECT ON THE PRESERVATION OF THE ARCHAEOLOGICAL RECORD

It has been established that the visibility and preservation of the archaeological record is strongly influenced by the complex interaction of many different factors, both natural and induced by human intervention. As a result, it is unwise to undertake any survey project without providing a detailed study of the dynamics generated by these elements. In this regard, Mediterranean landscapes present a strong challenge because of the complexity of geomorphic processes, soil erosion and a long history of intensive land use (see for example Bintliff 2005; van Andel et al. 1990). In particular, alluvial landscapes offer a changing environment, sometimes at a rapid pace, making it difficult to interpret the meaning of surface finds. Simply assuming that these areas are too modified for systematic study is unsatisfactory, since extreme cases of hidden or vanished remains coexist with areas of better conservation.

Our experiences in dealing with the difficult constraints of river valleys lead us to test a new methodology that could be useful for similar environments in other regions. We understand that it is essential to develop procedures that in the end allow comparative studies. What we are proposing here is an approach from a surface survey informed by the knowledge of the geomorphological dynamics. Although this kind of work has ample precedent in other areas of the Mediterranean (see seminal references like the work by Vita-Finzi 1969), in the Iberian Peninsula the evolution of fluvial environments is an issue which is not often considered. As an exception we can quote the wide scientific literature on the geoarchaeological interpretation of some areas in the Peninsular hinterland like Aragon, where an arid environment has produced intense erosion processes dramatically affecting land use and preservation of archaeological sites (Wilkinson et al. 2005). There is also remarkable scientific work focused on the reconstruction of the shoreline and geomorphological changes on the coast of the south of Spain (see for example the results of interdisciplinary projects like Arteaga & Hoffman 1999) and many specific studies of coastal sectors like the Cadiz Bay (Alonso et al. 2003; Borja et al. 1999; Dabrio et al. 2000).

Another reason to bring our particular experience to a wider community of researchers is the procedure we developed for the incorporation and management of information. The contribution of diverse recording methods and data types is typical of this kind of study, and increases the need for a technological platform that can easily integrate and handle these sources. The usefulness of GIS for such tasks has been well established (see for example Howard et al. 2008 as a case study in a fluvial context). Nevertheless, there is still a path to explore in the design of new methods for addressing the understanding of complex processes with these tools. Perhaps the more challenging topic in this regard is to build quantitative models for the simulation of a temporal sequence. Weighing the contribution of many different aspects, it is possible to revisit the past by creating a regression simulation of past scenarios. But by using such a model, it is also possible to make predictions on the future evolution of factors affecting the preservation of archaeological remains. It should be a key contribution for the development of effective heritage management in territorial policies.

PREVIOUS ARCHAEOLOGICAL RESEARCH IN MEDELLIN

Although there are several references in the descriptions of travellers and scholars of the 16th to 19th centuries, archaeological research in Medellin had its real start in the early 1970s. Excavations by Almagro Gorbea in 1969 marked the beginning of a long period of research on prehistoric and protohistoric occupation of this place (see as the most recent reference Almagro Gorbea 2009). These interventions established with some certainty a continuous chronological sequence that extends from the 8th century BC until the Roman period. Excavations and casual finds have helped to establish the location of the protohistoric settlements (Almagro Gorbea & Martin Bravo 1994; Jimenez Ávila & Haba Quirós 1995). A complete review of the evidence can be found in the first cited work.

As a whole, this research has allowed the characterisation of Medellin as a key site for the understanding of the process of social change, economic and cultural development in the peninsular southwest between the Late Bronze and the beginning of the Iron Age. Regarding the Roman period, the year 1970 marks the beginning of the first systematic work, with the excavation carried out by Mariano del Amo in the theatre (Del Amo y De La Hera, 1973, 1982). This scholar is also responsible for the first survey carried out in the surrounding area of Medellin (Del Amo y De La Hera, 1973). But we had to wait till 1998 for the publication of the first systematic approach to the study of the Roman city of *Metellinum*, thanks to Salva-

Figure 1. Location of the study area in the Iberian Peninsula. Detail map of the Medellin environment. Black stars indicate previously known sites.

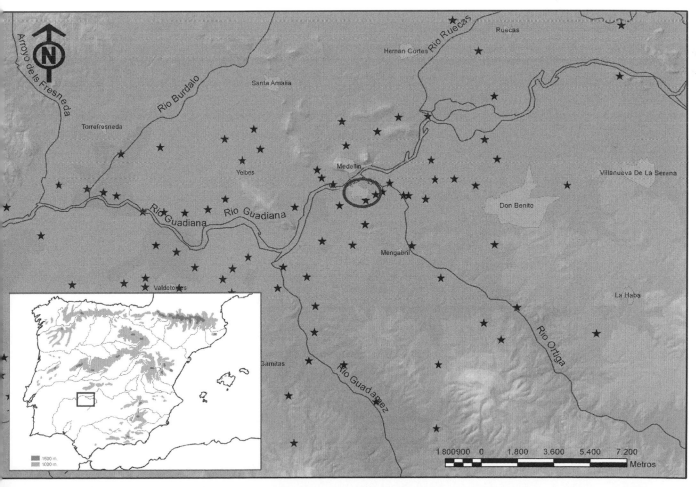

Figure 2. View from the north of the Medellin castle.

dora Haba (Haba 1998). This book is a synthesis of the large amount of dispersed information (epigraphy, numismatics, historical sources, news of finds) and presents a state of the art that changed little until the beginning of the 21st century. Now, research has been revitalised as part of a series of interventions for the revalorisation of the archaeological zone of Medellin. The most striking result of this activity has been the establishment of the excellent state of preservation of the Roman theatre. Recent research on the architectural remains in the area seems to show that this building was part of a public monumental complex that could be identified as the forum of the Roman colony.

In contrast to the great attention paid to the Medellin nucleus, efforts to investigate human occupation in the surrounding countryside have been scarce. Most of the available data come from the report of casual finds, often during agricultural work. The article by Mariano del Amo is a first attempt to systematise these references (Del Amo y De La Hera 1973). The job was continued in the 1980s by Jose Suarez de Venegas, who undertook a more thorough review of the known sites and conducted new survey work. Following the research trend marked by other specialists in Roman archaeology of the region (Fernandez Corrales & Cerrillo Martin de Caceres 1980; Fernandez Corrales 1986; Cerrillo Martin de Cáceres 1988; Fernandez et al. 1992), this author addressed the problem of rural settlements from the theoretical perspective of spatial archaeology.

All this information was gathered and reviewed in the aforementioned work of Haba, which presents an inventory of all the evidence on Roman settlements around Medellin. More recent surveys before the beginning of the current project correspond to the selective prospection in 2006-2007 around the Cerro Manzanillo site; a First Iron Age farm was located 14 km north-east of Medellin (Rodriguez Diaz et al. 2009).

The archaeological analysis of the territory of Medellin offers an excellent opportunity to advance the characterisation of ancient rural landscapes in the middle Guadiana basin, as an example of the processes of social complexity and increasing human pressure in the south-west peninsular between prehistory and the Roman period. For us, this project is also a chance to work on the definition of the objectives and priorities of an agrarian archaeology of the pre-industrial world.

THE KNOWLEDGE GAINED AND CHALLENGES FOR THE CURRENT RESEARCH

All this previous research raises the hypothesis that there was a succession of colonising events in the Guadiana basin from prehistoric to Roman times. These peaks in economic activity and demographic growth could have created an increasing pressure on natural resources. That is what apparently happened during the Copper Age, and again in the First Iron Age and the Roman Imperial Age. Regarding the protohistoric stage, discussion is currently very intense thanks to several recent publications of surveys and excavation reports, as quoted above around El Manzanillo, or la Mata de Campanario (Rodríguez Diaz 2004). These finds demonstrate the existence of an intense, albeit scattered occupation of the Guadiana valley landscape, which started in the mid-7th century BC, having its maximum development during the 5th century BC (Rodriguez Diaz et al. 2006; Jiménez Ávila et al. 2002). Almagro Gorbea (1977) was the first author to argue the importance of the phenomenon of small rural sites. According to his proposal they were the dwellings of an aristocracy who established patronage links and the formalisation of private ownership of land. The largest settlements in the valley, sites such as Medellin or Badajoz castle, would be at the top of this hierarchy. Opposing this centralised interpretation, other researchers have proposed a more segmented, cellular model. According to this view, sites like La Mata had the role of stately palaces, surrounded by dependent farms that defined the limits of domains ruled by this emerging aristocracy.

Figure 3. Archaeological plan and hypothetical reconstruction of a) El Manzanillo (according to Rodríguez Díaz et al. 2009) and b) La Mata (according to Rodríguez Díaz et al. 2004). The first can be considered an example of a small farmstead, while the second is an aristocratic residence.

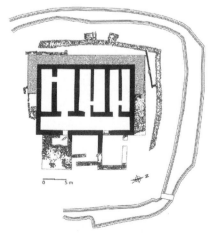

Although the proposal of Almagro is very discerning, the spatial structure of the Medellin territory has been addressed in very general terms. The extension of the area of political influence, main axes of communication and settlements has been identified. However, a more detailed look at the local landscape is necessary in order to define general guidelines for the functioning of the agricultural colonisation process.

In contrast to these heated discussions, the Second Iron Age in the area offers a much smaller documentary base. A hypothesis of a population vacuum in the region during this time is proposed. However, several settlements are known, corresponding to high, easily defendable fortified locations. This is the case of Entrerríos, 18 km from Medellin. It has been assumed that these were almost the only form of human settlement that could be detected during these centuries. However, there is evidence of occupation at that time in the floodplain around Medellin, suggesting that, although weak, small scattered sites could be present in the Second Iron Age landscape. Surveys conducted in other areas in south-western Spain have yielded similar results: although there is a sharp decline in the number of datable finds between the end of the 5th century BC and the 3rd century BC, it seems that some of them survived during this stage. As attention has been focused on the study of the best known settlement patterns, we lack a detailed record of such settlements.

So although there appears to be a process of concentration and contraction of habitat, the degree of organisation and intensity of land use of the landscape immediately before the Roman conquest remains to be assessed. What is clear is that the latter happened in a place with a long tradition of spatial and economic planning.

As in other Peninsular areas, academics have assumed that in many respects the pre-Roman landscape was a kind of blank slate. This is a difficult issue to address, firstly because of the paucity of data on the early stages of the Roman occupation. One of the biggest questions raised by the current project concerns the transformation of the agricultural landscape during this time of crisis. We have to bear in mind that the adverse effect of political and social instability continued at least until the last three decades of the first century BC. This moment (circa 25 BC) corresponds to the foundation of the Roman colony of *Augusta Emerita*, which became the head of the province of Lusitania.

In the case of Medellin the problem is particularly relevant, as historical sources record the foundation of a Roman colony in the early 1st century BC. The impact of this new settlement in the territorial structure has been hardly identified archaeologically. The sample of known settlements of this age is very small, although there is scattered evidence of finds like coin hoards with a Republican chronology. The voluminous record of known settlements of the Imperial times hinders the perception of earlier stages. Nevertheless, the evidence revealed by the survey carried out near the municipality of Valdetorres (about 10 km south-west of Medellin) shows the potential for research on the Roman Republican period in the area (Heras Mora 2009).

FROM PASTURES TO HIGH-YIELD CROPS: A SEVERELY TRANSFORMED LANDSCAPE

The agricultural landscape of the alluvial plains of the Guadiana can be defined as one of the most dynamic environments in the geography of south-western Spain. This is due to the interaction of a series of natural and human factors intricately intertwined. Firstly, there is an important environmental factor:

Figure 4. The agrarian landscape of Medellin today. a) ground levelling for rice cultivation. b) evidence of recent processes of artificial excavation.

the changing shape of the Guadiana. Its riverbed produced countless branches, forming islands and numerous abandoned channels. Many of them are still recognisable despite land levelling and pipe works. Even in the recent past, Guadiana floods caused the reactivation of these watercourses. The influence of the meandering course covers a large area both upstream and downstream. Historically it has been an unhealthy area, due to the abundance of stagnant water. The settlement of Medellin is located precisely at a point where the proximity between elevations, formed by residual reliefs on both sides of the river, forms a narrowing that ensures a stable ford over the year.

Secondly, we must bear in mind the great impact of the agrarian colonisation during the so-called 'Plan Badajoz'. It was implemented along the Guadiana River by Franco's regime from the beginning of the 1950s, and involved the conversion of traditional dry crops and pasture lands into high-yield irrigated crops. Over 9,000 settlers were distributed in new populations. With this great change began a process of ground levelling and excavation of ditches that in some areas profoundly altered the original relief.

We think that the awareness of these problems, far from leading to inactivity, stimulates our curiosity for establishing to what extent these changes have affected the preservation of the archaeological record. We would like to define and map the different degrees of this impact. Eventually, our model will allow us to reconstruct the ancient landscape and will provide tools for the management of archaeological heritage in the area.

METHODOLOGICAL THREADS IN THE MEDELLIN PROJECT

The previous state of the art increased the need for a renewed and systematic approach to the problem of the structure of agrarian landscape and change in settlement patterns in the historical area of the ancient site of Medellin. The strength of this new approach is to undertake the analysis from several complementary disciplinary fields.

ARCHAEOLOGICAL SURVEY

We looked at the distribution of surface evidence, recording not only visible archaeological sites, but also stray finds (Ebert 1992). The surface finds therefore form a continuous variable. The study of the evolution of the sedimentary matrix of finds (from the geological substratum to current land use) helps to give an explanation for their presence. It comprises the identification of small concentrations of sherds that could have been produced by human activity limited in space and time, but also the evidence of extensive agricultural practices like manuring. According to the record obtained in the intensive survey of other Mediterranean landscapes (see for example Wilkinson 1982, 1989, 2000; Bintliff & Snodgrass 1988). The latter can produce a thick carpet of residues, mainly potsherds from the areas where manure accumulates (e.g. landfills, stables and cesspits). The problem is that when the density of fragments derived from these processes is very high, the definition of activity loci is diluted and it becomes necessary to develop methods to distinguish sites from the background noise (Gallant 1986).

The objective of our survey is to implement such methods, in order to establish the nature of the surface record, adapting the modus operandi to the particular traits of each area with a reasonable investment of work. In general terms we followed the survey design implemented on the basis of previous experience developed by our team in other study areas during 2007 and 2008 (see Mayoral et al. 2009 for a detailed description), but new experiments and improvements were introduced. We would like to emphasise that the application of similar procedures in different study areas provides a good opportunity to compare results, and in the long term it will generate a wealth of documentation on surface data in a wide variety of contexts offering many opportunities for future analysis.

Regarding the delimitation of the survey area, the area of historical influence of Medellin was clearly an impossible space for intensive survey. We decided then to use a space well defined physiographically by natural features: the Guadiana River to the south, its tributaries Guadamez and Ortigas to the west and

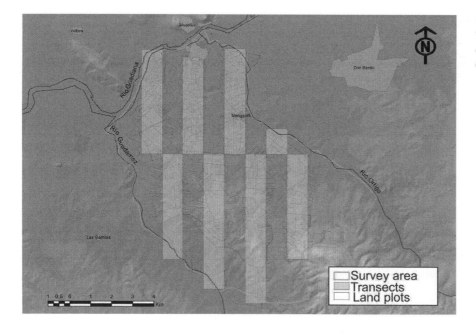

Figure 5. Definition of the survey area. *See also full colour section of this book*

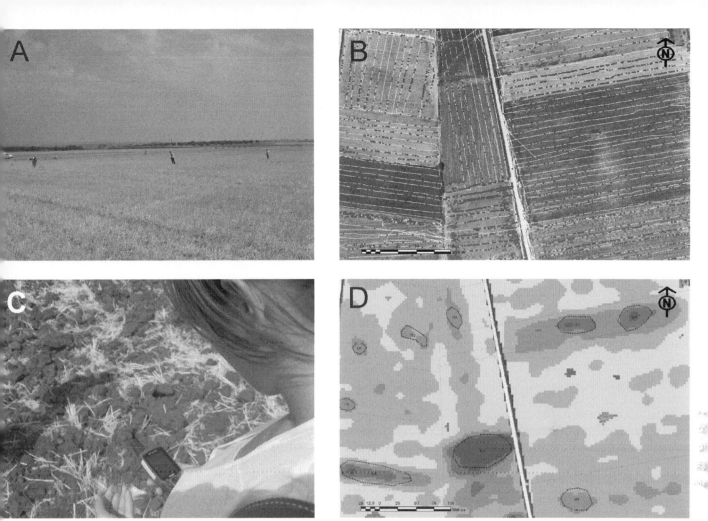

Figure 6. Workflow in the Medellin survey. a) Covering the ground. b) Downloading and representing data in a GIS environment. Clear lines correspond to the fieldwalker's track. Black dots are finds along these tracks. c) Introducing the waypoint of a sherd. d) Density estimation using the kernel method of sherds/square metre. Polygons delimit areas of interest for further examination.

east and the Ortigas Mountains to the north. This area represented a cross section of the Guadiana basin and allowed us the opportunity to explore different landscape units, from the intensively farmed alluvial plain to the rockiest unproductive lands dedicated almost exclusively to pasture. Nevertheless, it was still a very wide area. In order to create a systematic approach, we have outlined seven major non-aligned transects, each one kilometre wide and between two and seven km long, with a north-south orientation. This size was considered easy to handle and able to provide a simultaneous coverage of the different landscape units. The total area comprised by these transects represents 43% of the study area. The implementation of this design is structured in three stages.

The first one is systematic fieldwalking. Land plots are prospected with a spacing of 10 metres. Each field walker was equipped with a GPS receiver that recorded the position of every material of potential archaeological interest located along its track. They consisted mostly of potsherds, building materials (roof tiles, stone blocks, etc.), slag fragments and lithic artefacts. It is considered a first, coarse approach to the spatial distributions of finds. The GPS model we used has an average horizontal typical error of two or three metres. This is not considered a problem, since a very accurate position was not needed. We took into account a homogeneous level of experience of the team members and a similar ability for surface recognition. As a general rule, every team member kept the same GPS during the campaign, so the gathered data would give us the opportunity to assess the inter-observer variability. Fieldwalkers were told to pick up only fragments of pottery rims, bottoms, handles or decorated sherds that are considered of more diagnostic value to obtain chronological references. These samples were labelled with the GPS waypoint number, so it was possible to plot their distribution later. The result we obtained in this stage is a map of global distribution of surface materials in the field. For every landplot we also recorded information about topography, land use, actual vegetation coverage and recent terrain alterations (ditches, irrigation channels, power lines, power buildings etc.). These data were considered relevant in order to interpret differences in surface visibility.

In the second stage, all the information gathered was downloaded onto a GIS. We then analysed the variation in sherd densities. Our calculations take into account several possible solutions, with the aim of producing images that stress local detail and more general views of surface distributions. We must understand that this density estimation corresponds to the overall distribution of recorded waypoints, regardless of its identification as modern bricks, roman roof tiles or prehistoric sherds. Reclassification of density maps provided the delimitation as polygons of discrete zones were we could find potentially archaeological sites or evidence of land use.

All these areas were considered to be of interest and were revisited for closer examination, in order to obtain more detailed evidence. As in the landplot form, information was recorded regarding the characteristics of the terrain and the type of surface finds. We started to work without a normalised procedure for sherd collection in these areas. The reason was that surface evidence in many of them was very homogeneous, consisting of 90% roof tiles and traditional pottery dating from the last 100 years. It was quite easy to recognise their shapes and fabrics in the ethnographic collections. They were in fact the result of contemporary manuring and did not contain any trace of earlier land use. Many of the oldest farmers in the region gave us a detailed explanation of how dung was transported to the field at least until the 1970s, when it was replaced by industrial fertilisers. But on the other hand, in a small number of concentrations we found clear evidence of Protohistoric and Roman occupation. In several cases the abundance and diversity of surface sherds suggested a clear definition of sites. Here polygons defined by density analysis were used as a reference for a systematic gridded collection.

However, there were many occasions when we lacked an undeniable criterion for identifying a site. The problem was then: do we perceive this ancient material just because we are focusing our attention on an area of higher-density modern finds, or are the concentrations of finds really produced by the overlapping of different chronology scatters? Faced with this question, we realised that we were being too selective in our sherd collection and we were lacking a clear, quantitative as well as qualitative characterisation of pottery dispersed in the fields. In order to test the magnitude of this bias, in the 2010 campaign we implemented a collection method inspired by the off-site recording of the Boeotia Survey (Bintliff, Howard & Snodgrass 2007). Now sampling units were 50x100m rectangles adapted to the limits of land plots, and everything identified by 10m spaced fieldwalkers was collected, counted and weighed. This system dramatically increased the time invested in studying small areas and the number of gathered sherds. But at the same time it provided a completely new image of the off-site record, revealing in some cases that ancient pottery was not so spatially localised.

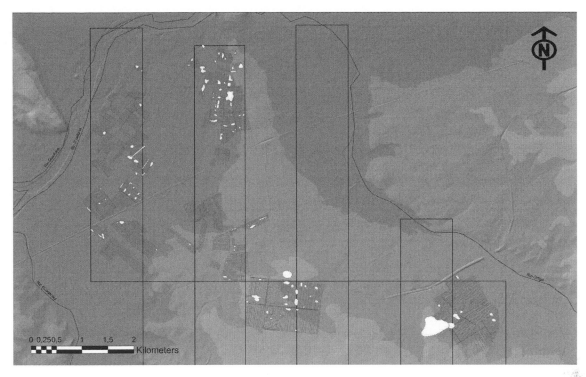

Figure 7. Overall picture of survey areas of 2009 campaign. Gray lines represent GPS tracks. White polygons are areas of interest. *See also full colour section in this book*

We are still beginning to understand the potential of this new strategy, but it seems clear that it will be an irreplaceable tool to study the discarded sherds in a Roman town like Metellinum. The present urban nucleus of Medellin is very small, and we have the opportunity of surveying extensive areas around the ancient site. The main problem here is, as we pointed out before, the impact of irrigation works of the last 50 years.

On the positive side, we note that it has been possible to correct the location of previously known sites. The existing records for most of them were limited to a pair of coordinates and a place name. Now, we have information about their size, a diagnostic about their current state of preservation and new information about chronology. We have also isolated 74 new sherd concentrations, half of which were considered as potential sites or places of interest for later analysis. Obviously the application of intensive methods has increased the number of known sites. In the future we hope to test to what extent the original distribution map has been modified.

GEOARCHAEOLOGICAL RESEARCH IN THE GUADIANA ALLUVIAL PLAIN

Obviously, survey results may contain serious deficiencies and interpretation problems if we do not bear in mind geomorphological evolution. That was the second main methodological aim of the project. In this sense our first objective was to obtain an assessment of recent geomorphological processes, quanti-

Figure 8. Exploration of gravel quarries in the study area. The circle in the right image indicates the find of a Roman roof tile.

fying the intervention of geodynamics (slopes, changing alluvial plain, aeolian deposits) and alterations induced by human intervention (e.g. introduction of irrigation and ground levelling for crops like rice), in order to evaluate its implications in surface visibility and preservation of archaeological finds. This can be both a tool for interpreting and a resource for planning future field work. A GIS-based territorial analysis procedure has been designed, in order to speed up a crossed evaluation of variables corresponding to the characterisation and dynamics of natural environment and human factors.

The first step was a preliminary exploration. All gravel pits and other artificial cross sections in the area were located and described. This work was quite helpful during the process of morphogenetic characterisation and elaboration of a detailed cartography of geomorphologic units. This mapping process relied on a number of initial reference data from several sources and formats, both digital and analogical:

- Aerial photographs of several photogrammetric flights, used as a support for photo-interpretation work.
- Geological Maps of the Spanish Geological Survey (IGME) in vector format (scale 1: 50,000).

- Geomorphological analogical maps, at the same scale, provided by the former survey.
- Orthophotographs of the study area from 2005. Their spatial resolution is 50 cm/pixel.
- Digital Elevation models obtained from several sources. The most accurate and recent was provided by the National Orthophotographic Plan (PNOA), and is based on a photogrammetric restitution that offers one elevation point every five metres. Our idea for the future is to develop the comparison of this model with others obtained from earlier photogrammetric flights (1956, 1983), in order to assess relief changes affecting the surface record and to determine recent morphogenetic evolution. These images are also an invaluable testimony of change in the Guadiana landscape before and after the implementation of the massive irrigation plan in the 1960s.

The main tools used were ArcGIS ® (ESRI ™), GRASS and gvSIG. Thanks to these programmes it was quite easy to superimpose and integrate all available digital cartography within a geographical context defined by the UTM coordinate system and datum referred to European Datum 1950 (ED50).

Regarding the relief modelling in the study area, we can identify two distinct morphodynamic systems. On the one hand, the fluvial system comprises the accumulation of fluvial deposits linked primarily to the river Guadiana. In our study area, the river flows across a terrace level 3.4 metres above the active channel, reaching widths of 4-5 km. The shape of the active riverbed is close to meandering. Numerous channels draw a braided river network, active in episodic events during flood periods. The main course is wide and composed of secondary channels that interweave to form islands. In the section that runs between the hills of Medellin castle and the Enfrente mountains the channel is narrower, which, as we said

Figure 9. Geomorphologic map of the survey area.

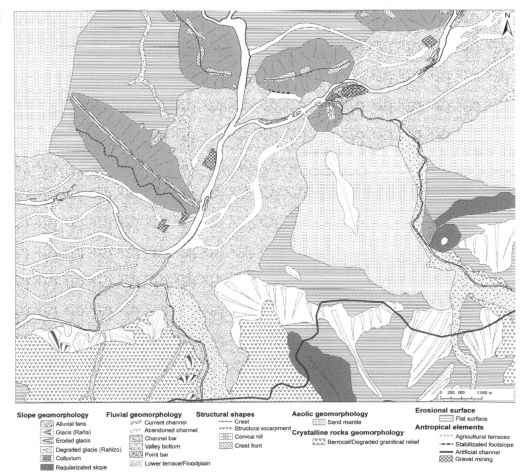

Slope geomorphology
- Alluvial fans
- Glacis (Raña)
- Eroded glacis
- Degraded glacis (Rañizo)
- Colluvium
- Regularizated slope

Fluvial geomorphology
- Current channel
- Abandoned channel
- Channel bar
- Valley bottom
- Point bar
- Lower terrace/Floodplain

Structural shapes
- Crest
- Structural escarpment
- Conical hill
- Crest front

Aeolic geomorphology
- Sand mantle

Crystalline rocks geomorphology
- Berrocal/Degraded granitical relief

Erosional surface
- Flat surface

Antropical elements
- Agricultural terraces
- Stabilizated footslope
- Artificial channel
- Gravel mining

in a previous section, is a factor of considerable strategic value as a crossing point. Other geomorphological units in this system are those related to fluvial modelling: channel bars, floodplain/lower terraces, point bars and valley bottoms.

The mapped terraces show extensive deposits linked to the Guadiana channel. The evolution of the river network has formed three terrace levels, represented in the study area by the most recent one. This is mostly composed of gravel and quartzite ridges. At the top of this level there is up to one metre of sand interleaved with gravels with channelled bases. The abandoned channels are currently part of the major riverbed of the Guadiana, which recovers its banks in times of flood. The alluvial gravels are composed of quartzite with a sandy matrix and intercalated sandy levels. All these channels are built up by similar deposits, with increasing silt content in the shallow parts where there would have been a slower flow rate. Finally, deposits associated with valley bottoms are poorly incised by the river network. Their lithology corresponds to gravel, sand and silty clay. The genesis of these deposits corresponds to river systems influenced by slope and aeolian inputs. One of the main objectives of the geomorphological work in the framework of the project is to define the timing of the alluvial fills, particularly the contributions from the slopes and the large Aeolian sand deposits that fossilise them.

The study area shows several geomorphological units related to slope dynamics, shapes and existing deposits. The major colluvium deposits are associated with cuestas. Regularised slopes and flattened surfaces are maintained in dynamic equilibrium with processes of rill or gully erosion. Geomorphodynamic processes in slopes generate different geomorphological units: conical hills (like Medellin Castle), crests, front crests and structural escarpments are linked to forms of structural origin. Forms associated with slope modelling processes are: alluvial fans, glacis composed by conglomerates (poorly cemented), mainly of quartzites, sand and clay, and sediments corresponding to braided alluvial gravel systems, eroded glacis, degraded glacis and compounds of quartzite gravels in a silty clay matrix. These formations are interpreted as deposits from the remobilisation of previous glacis.

There are present some aeolian mantles or deposits that occupy large areas along the river, on the margins of fluvial deposits. Consisting of fine-grained sands, they generate plains by wind aggradation. These plains have been strongly impacted by farming. Surface archaeological evidence includes several First Iron Age and Roman sites. Lastly, rocky outcrops are located in areas of granite substratum. They appear as rounded massive bodies where the regolith is removed.

We must also consider, as we have seen before, an active anthropogenic system transforming the surrounding area of Medellin. Several gravel quarries are located in the point bars and channel bars. Nevertheless, the most aggressive impact corresponds to agricultural terraces, structures for slope stabilisation and irrigation channels. Sometimes these structures are the only guides to reconstructing the original topography.

DEFINING A SEQUENCE AND CREATING A DYNAMIC MODEL

In order to build a geomorphological sequence, we are working towards establishing chronological references. Sometimes they are provided by finds which appear in the sections of gravel pits. Nevertheless, we still need to establish absolute dated benchmarks, so we take samples from sand deposits using the technique of Optically Stimulated Luminescence (Walker 2005). Analysis will be performed in the Dat-

Figure 10. Workflow of a suitability model for the delimitation of instability scenarios in the survey area of Medellin.

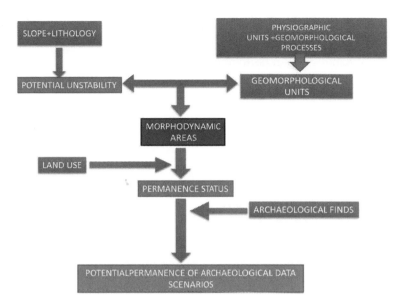

ing and Radiochemical Laboratory of the Universidad Autónoma de Madrid. The OSL dating will provide information about the last time that the sediment was exposed to sunlight. It will provide us with an absolute chronology of the sampled deposit, and should be a valuable tool for the establishment of the stratigraphic sequence. Finally, the dynamics that shape these units and processes will be included in a model of potential instability of the different types of terrain according to factors such as slope, lithology and land use.

This stage of the research is focused on the identification of relationships between the evolution of the natural environment and the human management of the landscape, and on the current condition of the archaeological record. With the aid of GIS tools, we have implemented a multi-criteria analysis based on the integration of all available spatial data. Figure 10 shows the workflow of this cross evaluation model.

In this methodological design two main input sources are defined, in order to obtain a map of morphodynamic areas. They are defined as zones with a variable susceptibility to damage of the archaeological record caused by surface processes. Variables considered for the elaboration of this first interpretative cartography are, on the one hand, slope and lithology, and on the other hand, geomorphological units as they have been defined above. They were obtained from a physiographic analysis and from the assignment to each one of predominant geomorphological processes.

The input layer for land use, measured from the point of view of its trends to change, allows a new, synthetic view of the territory to be produced. A different degree of permanence is then assigned to the discrete areas. It refers to different degrees in which the archaeological remains could have been preserved in their original positions. We cross this zoning with our previously known archaeological sites, and with the new survey results, assessing spatial variation according to chronological and functional indicators of data. In the final stage of the model this allows us to establish what we call permanency scenarios of archaeological data.

Our objective is to obtain an assessment of the degree of permanence, defining a gradient whose variation through space will help to delimit the most affected areas by recent processes and those in which the surface record can be considered more reliable.

PRELIMINARY RESULTS AND FUTURE DEVELOPMENT

Regarding the archaeological survey, it is still too early to offer an evaluation of the finds. Nevertheless we can show some preliminary results. During the two weeks of the survey campaign a first exploration of several transects was made in order to assess the diversity and complexity of the surface record around Medellin. The final area covered was around 700 ha. This represents 18% of the total transect area, in twelve real work days, so we felt we reached a good rhythm of work.

Although unfinished, classification of the finds offers some interesting insights. It is clear that there was intensive occupation during the First Iron Age over the entire basin, from the banks of the Guadiana River to the hills of the margins. Colonisation by small scattered farms goes beyond the optimal agricultural soils and it could be clear evidence of high pressure on the land, and possibly also surplus production by the Medellin community. It is more difficult to interpret the small sample of Second Iron Age sherds in lowland areas, which may correspond to small scattered settlements. With regard to the geoarchaeological study, the detailed mapping that we have produced provides a cornerstone for the development of the proposed suitability model. It will also serve to guide future survey works and to select the best locations for core sampling.

Figure 11. Integration of previously published information and current survey results on Protohistoric and Roman finds.

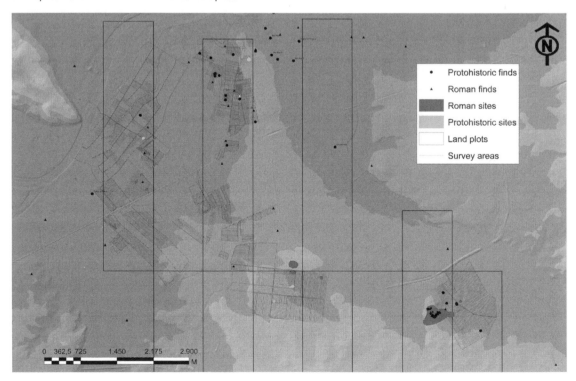

Beyond the specific achievements of the historical study on the territory of Medellin, we believe that the proposed work system may be useful for research on severely altered alluvial landscapes. At the same time, the use of standardised recording procedures in different types of environments opens up the possibility of interesting comparisons that will help to characterise the changing face of surface evidence across the Mediterranean. This may be the basis for flexible methods to monitor the future of this heritage, anticipating the effect of ongoing processes and incoming threats.

ACKNOWLEDGEMENTS

This work has been developed in the framework of the research project: 'The evolution of an agrarian landscape: the Medellin territory from protohistoric to Roman times' (ref. PRI08B050), funded by the Ministry of Economy, Trade and Innovation of the Autonomous Government of Extremadura. We would like to thank Pedro Hidalgo and the cartographic services of the Autonomous Government of Extremadura for providing the thematic cartography which served as the basis for the territorial analysis.

REFERENCES

Almagro Gorbea, M. 1977. *El bronce final y el período orientalizante en Extremadura.* Consejo Superior de Investigaciones Científicas, Instituto Español de Prehistoria, Madrid.

Almagro Gorbea, M. (ed.) 2009. *La necrópolis de Medellín, IV. Interpretación de la necrópolis, V. El marco histórico de Medellín-Conisturgis,* Madrid, Real Academia de la Historia.

Almagro Gorbea, M. & Martín Bravo, A. 1994. Medellín 91. La ladera norte del Cerro del Castillo. In Almagro Gorbea, M. & Martín Bravo, A. M. (eds.) *Castros y Oppida en Extremadura.*Complutum, Extra 4, 77-128.

Alonso, C., Gracia, F.J., Ménanteau, L., Ojeda, R., Benavente, J. & Martínez-del-Pozo, J.A. 2003. Paleogeographie de l'anse de Bolonia (Tarifa, Espagne) à l'époque romaine. In E. Fouache (ed.), *The Mediterranean World. Environment and History.* Elsevier SAS, Paris, 407- 417.

Arteaga, O. & G. Hoffmann. 1999. Dialéctica del proceso natural y sociohistórico en las costas mediterráneas de Andalucía. *Revista Atlántica-Mediterránea de Prehistoria y Arqueología Social* 2, 13-121.

Bintliff, J. 2005. Human impact, land-use history, and the surface archaeological record: A case study from Greece. *Geoarchaeology* 20 (2), 135-147.

Bintliff, J.L., Howard, P. & Snodgrass, A.M. 2007. *Testing the hinterland: the work of the Boeotia Survey (1989-1991) in the southern approaches to the city of Thespiai.* England McDonald Institute of Archeological Research, Cambridge.

Bintliff, J.L. & Snodgrass, A.M. 1988. Off-site pottery distributions: a regional and interregional perspective. *Current Anthropology* 29 (3), 506-513.

Cerrillo Martín De Cáceres, E. 1988. La aplicación de las teorías de lugar central al territorio romano de Augusta Emérita. *Arqueología espacial* 12, 197-204.

Dabrio, C., Zazo, C., Goy, J.L., Sierro, F., Borja, F., Lario, J. González, J.A. & Flores, J.A. 2000. Depositional history of estuarine infill during the last postglacial transgression (Gulf of Cadiz, Southern Spain), *Marine Geology* 162 (2-4), 381-404.

Del Amo y De La Hera, M. 1973. Estudio preliminar sobre al romanización en el término de Medellín (Badajoz), *Noticiario Arqueológico Hispánico* 2, 56-131.

Del Amo y De La Hera, M. 1982. El teatro romano de Medellín, Badajoz. El teatro en la Hispania romana. *Mérida 1980,* 317-324.

Ebert, J.I. 1992. *Distributional Archaeology.* University of New Mexico Press, Albuquerque.

Fernández Corrales, J.M. 1986. Consideraciones para la distribución y delimitación de las ciudades romanas, *Norba, Revista de Arte, Geografía e Historia* 7, 173-176.

Fernández Corrales, J.M. & Cerrillo Martín De Cáceres, E. 1980. Contribución al estudio del asentamiento romano en Extremadura: Análisis espacial aplicado al S. Trujillo. *Norba, Revista de Historia*, 1, 157-176.

Gallant, T.W. 1986. 'Background Noise' and Site Definition: a Contribution to Survey Methodology, *Journal of Field Archaeology* 13 (4), 403-418.

Haba, S. 1998. *Medellín romano. La Colonia Metellinensis y su territorio*. Diputación Provincial de Badajoz, Badajoz.

Heras Mora, F.J. 2009. Fundaciones militares en el origen de la ciudad lusitana: nuevos datos para la reflexión. In P. Mateos Cruz, S. Celestino Pérez and A. Pizzo (eds.), *Santuarios, oppida y ciudades: arquitectura sacra en el origen y desarrollo urbano del Mediterráneo Occidental*. Mérida, Instituto de Arqueología de Mérida y Consorcio de la Ciudad Monumental de Mérida, 299-309.

Howard, A.J., Brown, A.G., Carey, C.J., Challis, K., Cooper L.P., Kincey, M. & Toms, P. 2008. Archaeological resource modelling in temperate river valleys: a case study from the Trent Valley, UK. *Antiquity* 82, 1040-1054.

Jiménez Ávila, F.J. & Haba Quirós, M.S. 1995. Materiales tartésicos del solar de Portaceli (Medellín, Badajoz). *Complutum* 6, 235-244.

Jiménez Ávila, J., Ortega Blanco, J. & López-Guerra, A. 2002. El poblado de 'El Chaparral' (Aljucén) y el asentamiento del Hierro Antiguo en la Comarca de Mérida. *Mérida, Excavaciones Arqueológicas, Memoria 8*, 457-486.

Mayoral Herrera, V., Cerrillo Cuenca, E. & Celestino Pérez, S. 2009. Métodos de prospección arqueológica intensiva en el marco de un proyecto regional: el caso de la comarca de La Serena (Badajoz). *Trabajos de Prehistoria* 66, 7-25.

Rodríguez Díaz, A. (ed.) 2004. *El edificio protohistórico de 'La Mata' (Campanario, Badajoz) y su estudio territorial* Universidad de Extremadura, Cáceres.

Rodríguez Díaz, A., Chautón Pérez, H. & Duque Espino, D. 2006. Paisajes rurales protohistóricos en el Guadiana Medio: Los Caños (Zafra, Badajoz). *Revista Portuguesa de Arqueologia* 9, 71-113.

Rodríguez Díaz, A., Duque Espino, D. & Pavón Soldevilla, I. (eds.) 2009. *El caserío de cerro Manzanillo (Villar de Rena, Badajoz) y la colonización agraria orientalizante en el Guadiana Medio*. Junta de Extremadura, Consejería de Cultura, Mérida.

van Andel, T.H., E. Zangger & Demitrack, A. 1990. Land-use and soil erosion in prehistoric and historical Greece. *Journal of Field Archaeology* 17, 379-96.

Vita-Finzi, C. 1969. *The Mediterranean Valleys: Geological changes in historical times*. Cambridge University Press. Cambridge.

Walker, M. 2005. *Quaternary Dating Methods*. Wiley, Chicester.

Wilkinson, K., Gerrard, C., Aguilera, I., Bailiff, I. & Pope, R. 2005. Prehistoric and Historic Landscape Change in Aragón, Spain: Some Results from the Moncayo Archaeological Survey. *Journal of Field Archaeology* 18 (1), 31-54.

Wilkinson, T.J. 1982. The definition of ancient manured zones by extensive sherd-sampling techniques. *Journal of Field Archaeology* 9, 323-333.

1.7 Talking ruins: The legacy of baroque garden design in Manor Parks of Estonia

Authors

Sulev Nurme[1], Nele Nutt[1], Mart Hiob[1] and Daniel Baldwin Hess[2]

1. Landscape Planning Department, Tallinn University of Technology, Tartu College, Tartu, Estonia.
2. School of Architecture and Planning, University at Buffalo, State University of New York, Buffalo, NY, United States of America

Contact: sulev@artes.ee

ABSTRACT

The late 19th-century and early 20th-century 'grand era' of manor parks in Estonia coincides with a period when English gardening ideas dominated Europe. What is less recognised, however, is that manors in Estonia possess formal French-inspired gardens dating from the mid-18th century (the introduction of Baroque design in Estonia was delayed). Today, about 600 complete manor ensembles remain, retaining distinctive structural characteristics which date from the 18th-19th centuries. It is quite typical that in old parks of Estonia Baroque and English garden styles have merged, giving them a unique and original character. This research reports on archival study, field investigation and map analyses of 45 protected manor parks in Estonia. The analysis suggests that, despite the relatively short period (ca. 1730-1770), formal Baroque gardening was the dominant style practised in Estonia. The movement had a significant influence on local garden design, and on landscape planning more broadly. The Baroque elements in manor lands include formal geometric spaces, axial connections between landscape and buildings, orchestrated vistas and tree-lined roadways. Within the Baroque garden, formal plantings, pathways and water features were arranged in classical configurations. Finding physical traces of Baroque artefacts today is difficult because many manor parks were destructed during the Soviet era in the latter half of the 20th century. Nevertheless, archival materials and present-day visits to garden ruins in manor parks suggest that formal Baroque gardens dating from mid 18th-century manor lands were vivid and sophisticated ensembles of formal terrain, tree allées, sculptural elements and finely orchestrated water elements.

KEYWORDS

landscape design, park planning, manor parks, Baroque garden design, Estonia

INTRODUCTION

The Baroque garden design movement has given to mankind some of the most splendid and grandiose examples of spatial arrangement in the built and natural environment. For example, the legendary park at Versailles near Paris ranks amongst the world's greatest achievements in garden design. However, after the rise of ideals of equality one of the key ideologies of the French monarchy – formal Baroque design – fell out of favour during the 18th century. As the popularity of Baroque design waned in Western Europe, however, formal garden design continued to be practised in Estonian manor parks during 19th century by local German-influenced gentry.

At the beginning of the 20th century, there were 2,017 manors in Estonia (Rosenberg 1994). Today, about half this number survives, and approximately 400 manor parks are protected as natural or heritage areas. These protected manors are preserved (Sinijärv 2008) and they have been visited by experts who have conducted dendrological inventories (Sinijärv et al. 2007). For the most part, the manors and manor parks display 19th-century design characteristics of English landscape parks. Ideas governing manor park design, and the cultural features evident in manor lands, originate from two places. First, manor park design was imported to Estonia from northern and central Germany (Maiste 2005). Therefore, parallels with Germany's contemporary developments – the most famous English-style park being the one in Wörlitz – are useful for understanding the movement that inspired Estonian garden design (Rolf 2007). Second, local Estonian heritage is reflected in manor park design, celebrating local history and local culture. Features of Estonian origin in manor parks are especially evident from the late 19th century and early 20th century, the most splendid period of local manor culture, when existing manors were reconstructed and new manors were established. Shortly after, in 1919, manors were abolished in Estonia.

The late 19th-century and early 20th-century 'grand era' of manor parks in Estonia coincides with a period when English gardening ideas dominated Europe. Surprisingly, however, more than one-third of Estonian manor parks display traits of formal design. There were manor parks established in the 17th century, but unfortunately they are poorly documented and they have practically disappeared today. The major influence of the Baroque style arrived relatively late to Estonia, delayed by the Great Northern War and economic hardship in its aftermath. In one of the earliest examples of Baroque garden design in Estonia, Czar Peter I established Kadrioru park in formal Baroque style near Tallinn in 1718. In the 1740s and 1750s, various manor parks were founded in Estonia and many established formal garden elements (Hein 2007), while at the same time in Western Europe the era of formal Baroque park design came to an end (Turner 2005).

Although there are about 400 relatively well-preserved manor parks in Estonia, most appear today as park ruins. Twentieth-century events in Estonia – including World War I, World War II and the Soviet occupation – caused great losses within the parks as well as poor maintenance of manor land.

Now, to properly preserve the natural environments of manors, radical restoration efforts are needed. However, such restoration works face a number of challenges. For instance, it is often difficult to know whether formal garden elements, which appear to possess Baroque characteristics, are actually authentic

Baroque artefacts or are instead late 19th-century additions to the landscape. To distinguish between the two, it is helpful to identify which features characterise original Estonian Baroque-style gardens and to assess whether or not these features are still in evidence, even in a state of ruin, today. Determining the authenticity of garden elements that appear to date from the Baroque period is challenging for two key reasons. First, the original manor park plans and detailed design documents for manor projects are seldom available for study. In their absence, researchers usually rely on contemporary land-use plans. Secondly, the Baroque elements within manor landscapes are generally fragmented and in poor condition. These two challenges are interrelated, because without original plans it is difficult to identify the original elements of composition.

In this article, we provide a detailed study of Baroque elements of manor parks in Estonia, focusing on various elements of the built and natural environments, including spatial structure, design, characteristics and distinctive features. The research employs archival study, field investigation, and map analyses of 45 protected manor parks in Estonia (Heringas 2009). The objective of the research is to identify the formal, Baroque garden elements and develop trends about spatial construction and the relationship between manor landscapes and their surroundings. In most cases, due to a lack of primary research material, it is impossible to draw conclusions about single artefacts such as sculptures, vases, staircases, or pergolas. Instead, we focus on larger trends and broad design themes. In addition, the research provides an opportunity to better understand the evolution of landscape design in Estonia and the influence of manor landscape planning.

More broadly, this research situates the Baroque gardening movement in manor landscapes as a unique phenomenon in Estonian cultural history. Despite the relatively short period (ca. 1730-1770) that formal Baroque gardening was the dominant style practised in Estonia, it has had a significant impact on local garden design and landscape planning.

AN OVERVIEW OF ELEMENTS AND STRUCTURE OF HISTORIC ESTONIAN MANORS

The territory of Estonia was conquered by German knights during the 13th century. Gradually, a system of manors was developed, whereby large agricultural estates accounted for the majority of agricultural production. From the 17th century onward (and possibly earlier but no evidence remains), the manor centres, with economic and administrative functions, started to flourish as important sites of garden design. Manor owners established elaborate parks near the main manor buildings for their private enjoyment. Until the 19th century, manor parks remained almost the only form of garden design in Estonia.

In the design of manor parks, the most important model was formal Baroque gardening as developed to maturity in France during the 17th and 18th centuries. Thereafter, English-style landscape gardening was favoured in Europe. In Estonia, both styles were influential.

In a typical Estonian manor, a Baroque park space is formed by the connection of the front yard with the main building ensemble, or *cour d'honneur*, on the central axis (see figs. 1, 3). An entrance road provides access to the front yard. The largest part of the manor centre, or backyard, lay behind the main building. The structures are characterised by geometric order and well-defined forms of plants and plantations. Although there is a focus on physical order, the spatial structure of the park in some manors is not symmetrical nor does the central axis focus on the main building (Maiste 2005).

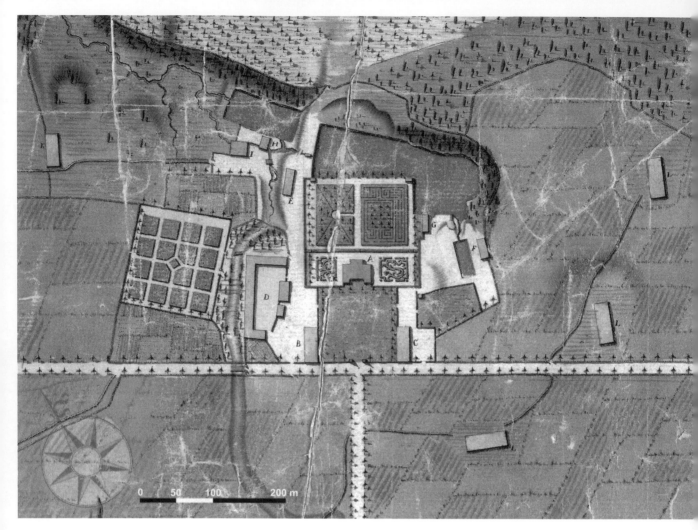

Figure 1. Plan of Palmse manor centre (1753). Source: Pahlen, G.F. 1753. Plan der Hoflage von dem Guthe Palms. *See also full colour section in this book*

The Baroque front yard of an Estonian manor complex is characteristically a spacious area, featuring a circular entrance road from the main gate to the main building entry. An open front yard provided opportunities for imposing views of the front façade; similarly, the view outward from the manor house windows, stairs and balconies focuses on the formality of the landscape design and its central axis. The front yards are usually among the best-preserved parts of the manor ensembles, having maintained their structure and visual and functional connections to the landscape. The largest part of a manor park is typically the backyard, with a formal garden and an adjacent landscape park. The design of these spaces was carefully planned. The backyard was typically divided symmetrically into smaller geometrical parts. It can be assumed that the backyards of Estonian manors, in the immediate vicinity of the main buildings, were more exclusively designed; typical surviving elements of backyards are allées of tree, terraces, water features and park boundary systems, such as stone walls.

INTERNATIONAL INFLUENCES ON GARDEN DESIGN IN ESTONIA

The oldest preserved manor landscapes in Estonia date from the second half of the 18th century – when Estonia was recovering from war and plague – during an important time for building and reconstructing manors (Maiste 2005). During this period, local garden design tended to follow one of two design philosophies. In the first, garden ensembles were created according to the above-mentioned Baroque principles of classical French formal design. This is evidenced by original landscape-planning documents produced in Estonia during the 18th century. The most famous is the 1753 plan of Palmse manor (see figs. 1, 2). In addition, there is evidence that classical French gardening literature was used by local garden designers in Estonia. These works – including André Mollet's *Le Jardin de plaisir* (printed in Stockholm in 1651) and Claude Mollet's *Théâtre des plans et jardinages* (printed in Paris in 1652) – were included in the library of the owner of Anija manor, Jacob Stael von Holstein (Hein 2007). The existence of newly-established Ba-

Figure 2. Palmse manor centre (1753).
Source: Plan of Manor Palmse 1753.

Figure 3. Schematic map of spatial composition of Vasta manor centre. Source: Nurme 2007. *See also full colour section in this book*

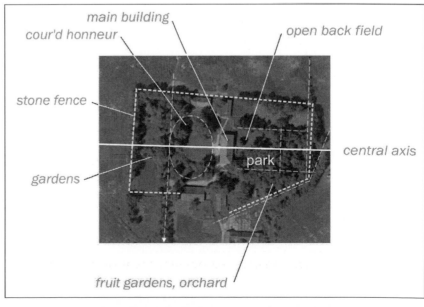

roque gardens in Estonia was confirmed by contemporaneous travellers. For example, the well-known architect Johann Wilhelm Krause produced a number of sketches in the 1790s that clearly depict formal design principles and even single Baroque garden elements in manors in northern Latvia, which at the time formed, together with southern Estonia, the province of Livonia (Janelis 2009).

The second gardening method – an English-style landscaped park – spread throughout Europe during 18th century. In 1785, *Theorie der Gartenkunst* by Christian Cajus Lorenz Hirschfeld was published, which significantly influenced the design of Baltic German gardening (Nutt 2008). Various manor landscapes founded or reconstructed in Estonia during the last quarter of the 18th century and the early years of the 19th centuries, such as Vatla, Aaspere and Õisu, are Baroque in structure, however landscape elements, including winding paths, irregular ponds and varied terrain, are formed in typical English 'picturesque' landscape design.

In fact, the English gardening style was dominant in virtually all new manor landscapes established in Estonia after 1770 (Hein 2007); the pre-eminence of this style gave rise to several beautiful landscaped parks in the 19th century. Nevertheless, the formal style was still dominant in older manor centres, probably because manor centres were already set in 18th-century landscape design and favoured the symmetric relations of the buildings and the park typical of Baroque layout (Maiste 2005). Moreover, the landscape parks surrounding the manor centres had matured to their best by the mid-19th century, and the desire and will to radically rearrange them was understandably weak.

A study of existing plans, drawings and postcards suggests that the designers of manor gardens in Estonia were often more conservative – drawing inspiration from formal, classical structure – than landscape designers elsewhere in Europe. This claim is supported by the built form of several parks created in the mid-19th century; for example, the general design principles evident in manor gardens in Raikküla, Hummuli, or Purila, where the spatial configuration of park elements, especially those closest to manor buildings, has been inspired by the ideas of formal Baroque design. A unique trait from the second half of the 19th century is a mixture of both styles, which is evident in Estonia in late 19th century and early 20th century manor gardens (e.g. Taagepera) or reconstructed manor landscapes, e.g. in Kärstna or Olustvere.

There are several explanations for the popularity of formal Baroque gardening in Estonia. The use of regular *cour d'honneur* as late as in the 19th century cannot be explained by the late arrival of original ideas to Estonia. On the contrary, the idea of 'freely flowing nature' used in Germany in one of the first great English style parks in Wörlitz (Gerhard & Erfurth 2000) was almost simultaneously applied in Estonia in Vana-Vigala manor in 1766, when '*Der Englische Garten*' was constructed (Hein 2007).

In addition, the use of formal Baroque garden elements in Estonian manors may be attributed to the introduction of techniques by international experts. For example, many Baltic Germans had family ties with building masters from Germany and, to a lesser extent, from Russia, Sweden and the Netherlands. For instance, the complex of Hiiu-Suuremõisa was planned by Swedish-French engineer Joseph Gabriel Destain (Särg 2006), Sagadi has been attributed to French-Italian-Russian architect Bartolomeo Francesco Rastrelli (Maiste 1983) and the largest Baroque-style park in Estonia, Kadriorg, was designed by the Italian architect Niccolò Michetti (Kuuskemaa 1985). The relationship between these designers and manors in Estonia demonstrate the great international mobility of landscape architects in the 18th century.

Although there are many examples of trained landscape-design professionals who planned manor gardens, the majority were laid out by the manor owners themselves, and the results reflect their knowledge, taste and views. For example, for a manor envisioned as a villa to be used as refuge from city life, an owner's garden design may have promoted peace and tranquillity (see Ackerman 1993 for a thorough analysis of villas and gardens). These ideas connect the local park design to Western European ideals (Kuuskemaa 1985).

A detailed review of spaces within manor parks protected by the National Heritage Board of Estonia (Heringas 2009; Vaine 2009; Mihkelson 2010) reveals evidence of formal Baroque spatial construction in 150 of 293 manor gardens from the final decades of the 19th century (National Heritage Board of Estonia 2009). Certainly, not all sites date from the 18th century as they are partly a result of the later designs which illustrate the vitality of formal design. At the same time, we often see mixed-era design, especially in parks reconstructed at the end of the 19th century, where formal Baroque structures, English-style planting systems and historical details intertwine (Nurme 2009).

MANOR STRUCTURE AND ELEMENTS

The formal Baroque garden is a distinct element of the manor landscape due to its compact nature and integration – both visual and structural – with the built and natural composition, formed from carefully-chosen axial relationships. Due to the axial structures, manor parks are visible and often dominant in the cultural landscape. The ensemble core, formal garden and landscape elements that are compositionally connected within a typical manor can produce a dramatic visual impact. For example, in Suure-Lähtru, the length of the main road and viewshed along the central north-south axis of the park is 1,200 metres. From the main road, perpendicular intersecting side roads emanate east and west, which in turn provide views of 1,400 metres (Nurme 2009).

Usually, contemporary circulation systems in manor landscapes are focused on roadways established during the grand era of Estonian manors. Therefore, the roads approaching the manor centre from the outskirts are in most cases similar to the original planned circulation system, which makes it possible to observe the park in the landscape from the perspectives that the designers originally planned.

Tree allées line roadways that lead to focal points in the landscape; in addition, tree allées form the boundaries of components of the landscape, delineating the border, for example, of the formal garden from the landscaped park (see fig. 4). Usually, design motifs within this landscape have been preserved only in a fragmented fashion and therefore they are less readable today. However, there is evidence of topiary cuttings, which are a key feature of a formal garden. The study of parks in situ gives valuable information about 'invisible' elements (Järvela 2009); for example, a geo-radar technique has been used to detect buried pathways (Artes Terrae 2010).

In Estonian manor landscapes, low dry-stone walls or higher mortar stone walls often serve as boundaries. Usually, the landscaped park was separated from other sections by walls and gates. In many places, such walls have been preserved, along with occasional gateposts and gate structures.

Water features, including ponds and fountains, were carefully designed, using natural characteristics of the landscape, to be integral features of the garden. For example, a formal garden could include rectangular ponds, circular islands, or a pond system connected with canals (see fig. 5), e.g. in Elistvere (Map of Elistvere manor 1825) and Õisu (Maiste 2008; Map of Õisu manor 1908).

Figure 4 .Luke manor park and tree allées (2008). Photo by S. Nurme, Autumn 2008.

Figure 5. Õisu manor park and canal (2008). Source: Photo by S. Nurme, Autumn 2008.

Terracing the land was an important technique of Baroque garden design, however terraces divided with structural support walls – such as those in Luunja park (Map of Luunja manor 1827) – are quite rare. Most of the original terraces were formed from sloping sections of garden. On one hand this is an indication of Scandinavian influence, and on the other hand it shows relatively mature formal garden design. Stone walls make the garden boundaries more rigid and unnatural, while grass-covered slopes suggest less control and greater organicism.

Engravings, photographs and postcards depicting the former milieu of Estonian manors suggests that, at least during the second half of the 19th century, garden design techniques produced rich, vivid environments. The landscapes in the images depict picturesque views of wooden bridges, pavilions, sculptures and flowerbeds (Nurme 2009), suggesting that much of what people admired in European formal Baroque parks was evident in Estonian manor parks.

Unfortunately, finding physical traces of Baroque artefacts today is difficult because there was much destruction of the cultural heritage of manor parks during the Soviet era in the latter half of the 20th century. As a result of short-sighted practices and a lack of cultural awareness, many manor centres were subdivided into smaller plots, used as construction sites, or abandoned and laid waste. Therefore, today, there is unfortunately little hope of uncovering additional examples of Baroque artefacts in what appear today to be clumps of old trees surrounded by undergrowth that mark the old manor gardens and landscaped parks.

Based upon the compositional features of preserved manor parks and historical documentation of destroyed manor parks, we suggest that manor parks dating from the second half of the 18th century possess classical Baroque garden features, and such features are evident even today, more than a century after they were first established. The rise of manor culture after the Great Northern War enabled the creation of elaborate manor estates, which give distinctiveness to local landscapes. Road networks on manor lands, which unified the manor ensemble together with the orchestrated views of the landscape, gave shape to the manor land, thereby giving shape to the local Estonian landscape which is still visible today.

CONCLUSION

Formal Baroque gardens in Estonia (created between ca. 1730 and 1780), in their purest form, were based on classical Baroque garden design. Due to its late rise compared to Western Europe, the Baroque structures remained an essential part in the design of Estonian manor parks throughout the 19th and 20th centuries. Therefore, regularity in garden design was never fully forgotten, which is evident in the landscape plans of 19th-century manor centres and may be observed in the parks today. It is difficult to determine how many Baroque gardens in Estonia are authentic, dating from the mid-18th century, and which were rebuilt at a later time using French garden design inspiration. As a result, our research allows us to describe the general spatial-design characteristics of a Baroque garden but we cannot fully articulate the detailed formal design when original garden design documents are not available.

Unfortunately, a lack of reliable archival material and a lack of opportunities to view preserved elements in gardens today prevent us from better describing the Baroque gardening period in Estonia. However, many manor lands today exhibit the essential values of Baroque gardens, and this provides opportunities to experience the elements of formal garden design that is still evident in the Estonian countryside more than 250 years after the gardens were estbalished.

A formal Baroque garden was intended to sparkle like the contemporaneous music of Händel. Such gardens, characterised by grandeur and dramatic spaces linking manor centres with other manor features, such as a landscaped park, formed memorable views into the distance. Formal terrain, tree allées forming enclosing 'pillars' and finely orchestrated water elements contributed to the sophisticated ensembles. If a visitor still senses surprise, amazement, playfulness and joy when visiting an unreconstructed park – despite destructive physical transformations of historic landscapes during past centuries – then it is surely an authentic Baroque garden and its uplifting atmosphere prevails.

ACKNOWLEDGEMENTS

An earlier version of this research was presented at the First Landscape Archaeology Conference at the VU University in Amsterdam, sponsored by the European Geosciences Union, the Cultural Heritage Agency and the Royal Academy of Sciences in the Netherlands in January 2010. When this article was written, Daniel B. Hess was a Fulbright Scholar at Tallinn University of Technology, Tartu College, Estonia.

REFERENCES

Ackerman, J.S. 1993. *The Villa: Form and Ideology of Country Houses*. Princeton University Press, Princeton, NJ.

Artes Terrae, Ltd., Nurme, S., Kaare, E. 2010. *The archaeological research of location of historical roads in front yard* [Ajaloolise rondeeli arheoloogiline uuring]. *The reconstruction project of Räpina manor park* [Räpina mõisapargi rekonstrueerimisprojekt], no. 08KP10, Tartu, Estonia.

de Jong, E. 2000. *Nature and Art. Dutch Garden and Landscape Architecture 1650-1740*. University of Pennsylvania Press, Philadelphia, PA.

Hein, A. 2007. Garden and Time. Guidelines from older history of Estonian garden art [Aed ja aeg. Piirjooni eesti aiakunsti vanemast ajaloost]. In Sinijärv, U., Konsa, S., Lootus, K. & Abner, O. (eds.) *Estonian parks I* [Eesti pargid I]. Varrak, Tallinn, Estonia.

Heringas, D. 2009. *Preserved Elements of Estonian Baroque Gardens* [Eesti 18.-19. sajandi regulaarpark. Pargiruumi säilivus]. Master's thesis, Tallinn University of Technology, Tartu College, Tartu, Estonia.

Janelis, I.M. 2009. *Manor Gardens and Parks of Latvia*. Neptuns, Riga, Latvia.

Järvela, G. 2009. *Development and Application of a Research Methodology for the Road Network of Historical Park* [Ajaloolise pargi teedevõrgu uurimise metoodika väljatöötamine ja rakendamine]. Master's thesis, Estonian University of Life Sciences, Tartu, Estonia.

Kuuskemaa, J. 1985. *Baroque Kadriorg* [Barokne Kadriorg]. Perioodika, Tallinn, Estonia.

Maiste, J. 1983. *Architecture and Planning of Manor Ensembles of Northern-Estonia from the Middle of the 18th century to 1917*. Central Institute of Art Research, Moscow, Russia.

Maiste, J. 2005. *Manorial Architecture in Estonia* [Eestimaa mõisad]. Kunst, Tallinn, Estonia.

Maiste, J. 2008. Õisu Manor and Park 1845-2008 [Õisu mõis ja park 1845-2008]. Historical Survey, Artes Terrae Ltd., Tartu, Estonia.

Map of Elistvere manor [Charte von denen zu dem Guthe Ellistfer und dessen Hoflage Johannishoff gehörigen Heuschlaege]. 1825. National Archives of Estonia (EAA), f. 1691, n. 1, s. 195, Tartu, Estonia.

Map of Luunja manor [Charte von dem Hofs-Lande zu privat Guthe Lunia]. 1827. National Archives of Estonia (EAA), f. 1442, n. 1, s. 281, Tartu, Estonia.

Map of Õisu manor [Generalcoupon des Gutes Euseküll belegen im Kreise Fellin und Kirchspiele Paistel]. 1908 copy of 1860 map, 1908. National Archives of Estonia (EAA), f. 3724, n. 5, s. 2768, Tartu, Estonia.

Map of Vasta manor [Special Karte sämmtlisher Ländereien des im Gouv. Estland, Kreise Wierland und Maholmschen Kirchspiel belegenen Privatgutes Waschel mit Unnuks]. 1881. National Archives of Estonia (EAA), f. 3724, n. 4, s. 1810., Tartu, Estonia.

Mihkelson, H. 2010. *Estonian Regular Park in the 18th-19th Centuries: Relationships between Park and Landscape* [Eesti 18.-19. sajandi regulaarpark. Pargi ja maastiku suhe]. Master's thesis, Tallinn University of Technology, Tartu College, Tartu, Estonia.

National Heritage Board of Estonia. 2009. www.muinas.ee (accessed 4 February 2009).

Nurme, S. 2007. Estonian baroque garden inventory field works methodology (Eestimaa baroksete mõisaparkide välitööde metoodika). Tartu

Nurme, S. 2009. The Old Park. Voyage of Discovery to the Periphery of Baroque. [Vana park. Avastusretk baroki ääremaile.] In Külvik, M. & Maiste, J. (eds.), *Park is Paradise in Art and Nature* [Park on paradiis looduses ja kunstis], Estonian University of Life Sciences, Tartu, Estonia.

Nutt, N. 2008. Design of Estonian Manor Parks in Their Times of Glory [Eesti ajalooliste mõisaparkide kujundus nende hiilgeajal] In Nutt, N., Maiste, J., Nurme, S., Karro, K. & Sinijärv, U. *Restoration of Parks* [Parkide restaureerimine]. Tallinn University of Technology, Tartu College, Tartu, Estonia.

Pahlen, G. F. 1753. Plan der Hoflage von dem Guthe Palms. National Archives of Estonia (EAA) f.1690, n.1, s.34

Plan of Manor Palmse [Plan der Hoflage von dem Guthe Palms]. 1753. National Archives of Estonia (EAA), f. 1690, n. 1, s. 3417, Tallinn, Estonia.

Rosenberg, T. 1994. *Estonian Manors* [Eesti mõisad]. Olion, Tallinn, Estonia.

Särg, A. 2006. *Manors and manor owners of Hiiumaa* [Hiiumaa mõisad ja mõisnikud]. Argo, Tallinn, Estonia.

Sinijärv, U. 2008. Parks as Protected Sites [Pargid kui kaitsealad]. In Nutt, N., Maiste, J., Nurme, S., Karro, K. & Sinijärv, U. *Restoration of Parks* [Parkide restaureerimine]. Tallinn University of Technology, Tartu College, Tartu, Estonia.

Sinijärv, U., Konsa, S., Lootus, K. & Abner, O. 2007. *Estonian Parks 1* [Eesti pargid 1]. Varrak, Tallinn, Estonia.

Stavehagen, W.S. 1887. Album. Ehstländischer Ansichten gezeichnet und herausgeben von Wilhelm Siegfried Stavehagen in Mitau in Stahl gestochen und gedruckt von G.G. Lange in Darmstadt. Mitau

Toman, R. 2007. *European Garden Design from Classical Antiquity to the Present Day*. Könemann, Bonn, Germany.

Turner, T. 2005. Garden History. Philosophy and Design 2000 BC – 2000 AD. Routledge, London, UK.

Ulbrich, B.G. & H. Erfurth. 2000. Wörlitz. Ein Bildband in deutscher und englischer Sprache. Mitteldeutscher Verlag, Dessau, Germany.

Vaine, J. 2009. *Compositon of Estonian Baroque Gardens* [Eesti 18.-19. sajandi regulaarpark. Pargiruumi kompositsioon]. Master's thesis, Tallinn University of Technology, Tartu College, Tartu, Estonia.

1.8 Configuring the landscape: Roman mining in the *conventus Asturum* (NW *Hispania*)

Authors

Guillermo S. Reher[1], Lourdes López-Merino[2] , F. Javier Sánchez-Palencia[1] and J. Antonio López Sáez[2]

1. Social Structure and Territory-Landscape Archaeology research group, Institute of History (CCHS, CSIC), Madrid, Spain
2. Archaeobiology research group, Archaeology and Social Processes, Institute of History (CCHS, CSIC), Madrid, Spain

Contact: guillermo.reher@cchs.csic.es

ABSTRACT

The *Conventus Iuridicus Asturum* (mainly, though not only, modern Asturias and León provinces in Spain) was created after the Cantabrian Wars carried out by Augustus himself, which finished in 19 BC. Even though the NW quadrant of the Iberian Peninsula was rich in gold, the *C. Asturum* concentrated the greatest deposits in the western ends of both Asturias and León. The exploitation of gold was a strategic need for Augustus' new Imperial coin, the *aureus*. Those areas with rich pre-Roman goldwork were systematically prospected and mined during the first two centuries of the Christian Era. As a consequence, local populations were subjected to a special form of imperialist policy directed at ensuring the maximum output of minerals. Settlement form and function was radically changed and a tributary system was put in place, thereby changing local society completely.

This policy had a major effect on the landscape in two ways: the mines brought about important geomorphological changes, and the territorial policy changed the rural exploitation of the area with a new landscape management and an increasing importance of cereal cultivation. In this paper these changes are brought forth as what they are: a measure of the impact that Roman gold mining had on the landscape.

KEYWORDS

Roman gold mining; geomorphology; palaeoenvironment; pollen analysis; rural exploitation

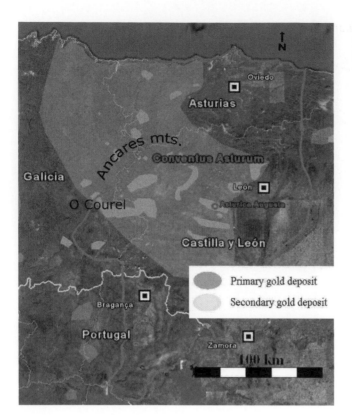

Figure 1: Location of the *Conventus Asturum* and its gold deposits. *See also full colour section in this book*

Primary gold deposit

Secondary gold deposit

GEOMORPHOLOGICAL TRANSFORMATIONS AND FOREST EVOLUTION

In 19 BC the Cantabrian Wars ended. At that moment Roman *Asturia* was configured as the *Conventus Iuridicus Asturum* (see Figure 1), thereby becoming incorporated into the Empire. In that area gold had already been exploited at the artisan level (Fernández-Posse de Arnáiz et al. 2004; Sánchez-Palencia Ramos & Fernández-Posse de Arnáiz 1998), generating a rich catalogue of goldwork from the Iron Age (García Vuelta 2007; Montero Ruiz & Rovira Llorens 1991; Diputación Provincial de Lugo, 1996; Perea Caveda & Sánchez-Palencia Ramos 1995). Augustus had designed a new monetary system which consolidated the *aureus* and the *denarius* as the gold and silver standards, exhibiting the strength of the imperial treasury (Crawford 1985, 258-260). In order to mint the *aurei* the amount of gold needed was multiplied, no doubt spurring a massive survey of the sites throughout the region. As a result, a great number of mostly open-cast mines were opened in the 1st century AD.

Geomorphological transformations of the landscape

Three decades of thorough research on Roman gold mining (Domergue 1986; Sánchez-Palencia Ramos 2000; Sánchez-Palencia Ramos et al. 1996) has allowed a better understanding of Pliny the Elder's description of the techniques used in the peninsular NW: the *aurum arrugiae* (*Nat.* 33.71-72). A hydraulic network of channels (*corrugi*) supplied with water the deposits (*piscinae* or *stagna*) which sat at the edge of the mining front. As the front advanced upwards – or receded back up the mountain – the network had to be redesigned at higher altitudes, thus allowing the exploitation to move from the valleys up the mountain.

The pureness of gold of these sites is low by today's standards. The amount of earth that needed to be removed to reach the richer levels impressed Pliny (see Table 1). The most spectacular method used was described as the *ruina montium* (*Nat.* 33.73). Overall, so much soil was dislodged due to gold mining that the lands of Hispania "advanced into the sea" (*Nat.* 33.76). Drawing on Pliny, two geomorphological transformations are to be expected as results of gold mining; a) the absence of earth due to the mining extraction, which could use the technique of converging furrows or the *ruina montium*; b) the landfill created by the tailings which form cones and artificially accelerates the sedimentation of certain areas.

Table 1. Gold pureness in the NW of *Hispania* (Pérez García et al. 2000, 226).

Deposits	Earth dislodged (m3)	Gold obtained (t)	Gold pureness (mg/m3)
Quaternary fluvial deposits	73	7,3	100
Quaternary moraines and residual placers	12	1,2	100
Pliocene fluvial deposits	20	1,8	90
Miocene fluvial fans	203	10,2	50
Total placers (Neogene+Quaternary)	308	20	67
Primary deposits (late Hercynian)	290	170	600

These geomorphological changes are clearly visible using stereoscopic aerial photography or satellite imagery, techniques which are necessary for an adequate interpretation of these structures (see Figure 2). On the ground they are also visible. At Las Médulas, the largest gold mine of the region, the *ruina montium* altered the morphology of 1100 ha through the extraction of ore in 500 ha and the deposition of tailings

Figure 2: Geomorphological alterations brought about by mining in Las Médulas (León).

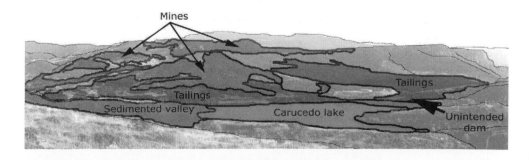

in another 600 ha. The immense volume of displaced earth that suffocated the valley of Carucedo led to the accidental creation of Carucedo lake – stagnant remains of a long-forgotten activity – and the ensuing sedimentation upstream. Today, that little vale enjoys an unusually flat valley bottom of more than 100 ha.

Forest evolution

Even though there is a cliché regarding the need for large-scale logging for the mining activity on behalf of the 'arid and sterile mountains' of *Hispania* (Plin. *Nat*. 33.67), the fact is that the techniques used had no such requirements. Pliny refers to wood being used only in the following ways:

1. Timber used for shoring in galleries or pits, when the technique is *aurum canalicium* (*Nat*. 33.68), which is rare in the region (an example near Porto, Domergue 1987, 524-526). Some beams have been found in mining contexts, such as the Boinás site in the Sierra de Begega (Villa Valdés 1998, 170-175).
2. Fuel needed to roast the ore which has been ground and washed in primary deposits, or else to melt the remaining slag (*Nat*. 33.69). Slag mounds, however, are rare in primary gold-mining sites of the region.
3. The heather (Domergue & Hérail 1978, 285-290) used to retain the gold in the washing channels or *agogae*, which was afterwards burned in order to extract from the ashes the gold particles which had been caught in the branches (*Nat*. 33.76). Pliny himself, however, said that the *arrugiae* technique using water did not require a melting down of gold, for it was freed by the washing process itself (*Nat*. 33.77).

Mining, nonetheless, did have an effect on forests, though perhaps not due directly to mining activity, which can be observed in the pollen sequences analysed. Forests, according to these palaeoenvironmental studies, follow their own evolution. In the pre-Roman period (which is attributed data from approximately 150 BC in all our palynological records) we see a deforestation tendency of upper mountain birch forests in the Ancares mountains (Muñoz Sobrino et al. 1995, 1996, 1997), clearly related to the increase of farming activity. In the Courel mountains this same tendency occurs later, and mainly affecting oak forests, which would be transformed into shrub (Aira Rodríguez 1986; Santos et al. 2000).

In the Roman period (see Figure 3), the oak forests of both the Ancares and Courel mountains suffered greatly from the extension of crop and tree cultivation. Farming pressure and mining activity broke the natural forest cycle in those areas, in contrast with the continuity we can see in the upper Porma river basin (Fombella Blanco et al. 1998; García Antón et al. 1997; Muñoz Sobrino et al. 2003). After the mining activity ceased, the deforestation trend would only increase, possibly echoing steadily growing farming and demographic pressure on mountain areas. The receding forests therefore follow a trend which is not directly affected by mining activity.

From the Iron Age, forests in this region lost ground because of human activity. The cultivation of crops and trees did most of the damage to the Holocene forests remaining, regardless of whether there was or was not gold mining carried out by the Romans.

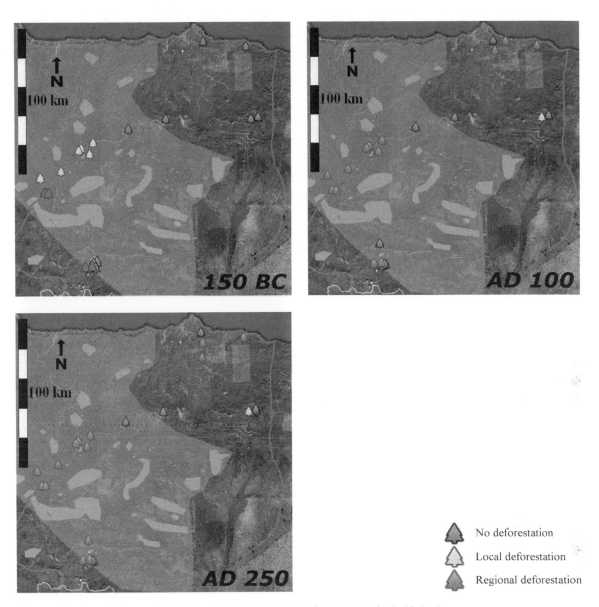

Figure 3: Forest evolution as attested by pollen records. *See also full colour section in this book*

No deforestation
Local deforestation
Regional deforestation

THE 'SIDE-EFFECTS' OF MINING: THE PROVINCIALISATION AND EXPLOITATION OF ROMAN *ASTURIA*

The mining techniques used in *Asturia* required an extensive control of a large territory and the population that inhabited it, which would have to provide the workforce necessary to exploit the mine, build and maintain the hydraulic network, and provide the food and tools necessary. In order to achieve this, the Empire sent its imperial bureaucrats (the *procuratores metallorum* and their *officina*), and used the technical expertise of the army. By establishing the *civitas* system, the local population was for the first time

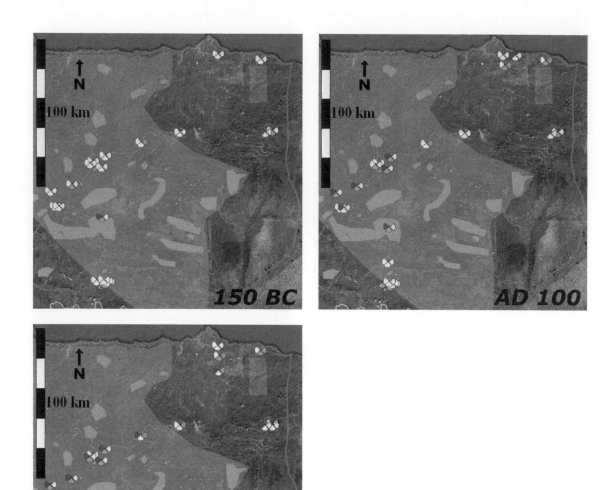

Figure 4: Absence/presence of cultivated species as attested by pollen
records. *See also full colour section in this book*

subjected to a tributary system which, more often than not, entailed labour at the mines. Gold, therefore, brought to the region a very specific and thorough form of imperialism.

The territory was completely reorganised, with new settlement patterns and a tendency to develop, as far as possible in Antiquity, a partial specialisation of roles within it. In Las Médulas, certain settlements housed foundries, while others had a purely farming vocation. There are cases where the only role was the maintenance of the hydraulic network, compelling small groups to live in otherwise uninhabitable areas (Sánchez-Palencia Ramos 2000, 270-271). These specialisation trends were completely alien to the pre-Roman society, and they bear witness to the integrated exploitation strategy imposed by Rome in this area.

The environmental effect of the new territorial and social articulation is important. The presence of cultivated species, including cereals – mainly wheat and barley – and the staple Roman fruit-bearing trees chestnut and walnut bear witness to the impact that the Empire had on how the land was exploited (see Figure 4).

Cereal

Cereal pollen tends to travel very little, which is a problem given that most samples used here are taken at high altitudes. Their presence in a sample usually attests local cultivation, or else in very small amounts, which can be interpreted as 'regional presence'. This low proportion is interpreted here as positive presence of the crop. An overview of the palynological results shows that many areas affected by mining show growing cereal cultivation in their environs, clear proof that settlement patterns favoured nearness to the mines – and high pastures – in detriment to a previous preference for more apt locations.

Chestnut

With chestnut trees interpretation is more complicated. *Castanea sativa* pollen was associated directly with human exploitation (Behre 1990; Conedera et al, 2004). This is, however, challenged by the attested presence of relic chestnut forests (Krebs et al. 2004), which compels further studies on local microevolution models in order to distinguish natural from anthropic presence. In our case, however, the clear association between Roman presence and abundant chestnut groves seems to favour their interpretation as serving the purposes of local exploitation. Their cultivation, however, is much greater in the mountain areas surrounding mining exploitation (Courel and Ancares), appearing more rarely in other mountain areas or lowlands.

Walnut

Juglans regia trees, however, do not follow the footsteps of chestnut trees. Their presence increases slowly with time, and they favour those areas not affected by mining. In many ways, their evolution resembles that of cereal cultivation, but enjoying the moment of maximum impact during the later Roman Empire.

EL CASTRELÍN AND ORELLÁN AS LOCAL EXAMPLES

El Castrelín and Orellán are two well known settlements around Las Médulas (recently López-Merino et al, 2010). The first is a pre-Roman hill fort and the second is a Roman-era settlement with a certain metallurgical role within the mining territory. The difference in pollen presence belonging to the three cultivated species mentioned before is clear (see Figure 5). El Castrelín, though it did exploit its territory with very similar types of vegetation, did not do so as thoroughly as Orellán would in Roman times, when the landscape around it was deeply transformed by the mine and a much greater settlement density. Indeed, the pre-Roman settlement has 84 % tree pollen – of which none is cultivated – in contrast to the 40% found at Orellán – of which 51% is cultivated chestnut and walnut trees.

These two settlements also have comparable zooarchaeological studies which have shown a much greater use of ovicaprids in the pre-Roman El Castrelín than in the Roman Orellán where, however, the pieces found are much more often butchered pieces (Sánchez-Palencia Ramos 2000, 275). This phenom-

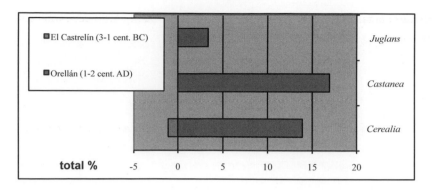

Figure 5: Pollen proportions recovered in both settlements (from Sánchez-Palencia Ramos 2000, 273).

enon suggests local importation of consumption meat, thus illustrating the previously mentioned specialisation trend of settlements during the mining period.

CONCLUSION: THE IMPACT OF MINING ON THE NATURAL LANDSCAPE

This table summarises the effect of mining on the landscape:

Geomorphology	Environment
Direct effects	*Direct effects*
Rasing of the landscape	Devastation of exploitation areas
Tailings accumulation	
Indirect effects	*Indirect effects*
Alteration of the sedimentation regime	Expansion of cereals and chestnuts
	Deforestation

In this paper many pollen samples have been reviewed in order to assess the effect that Roman gold mining had on a particular region of Hispania. Results have, however, shown very little effect directly attributable to the devastation which this activity caused on the landscape. The nature of the sampling techniques available and the location of these sequences in the high mountains make it impossible to measure this. Notwithstanding, the new territorial articulation which mining stimulated throughout the land proved to be a much more powerful transformation force. It is debatable whether there was a significant demographic increase under the Romans in this area, but there is no doubt that the number of settlements, and how thoroughly the landscape was exploited, saw a great increase. Deforestation was a consequence of this increased agricultural activity.

The end of the mining activity (2nd-3rd century AD transition) did not entail an abandonment which would have meant a return to a pre-Roman status. There seems to be, in fact, a further intensification of this exploitation, with the generalisation of the walnut cultivation as an example. Society, and how it made use of the land, had changed for good.

ACKNOWLEDGEMENTS

This paper is generated by the scientific network programme CONSOLIDER – *Research Programme on Technologies for the Conservation and Valorisation of Cultural Heritage/Programa de Investigación para la conservación y revalorización del Patrimonio Cultural (TCP)* (CSD2007-0058; www.heritage-technoscience.com), funded by the Spanish Science and Innovation Ministry (MiCInn). It draws on research carried out under the *Paisajes culturales y naturales del Bierzo: Geoarqueología, Paleoambiente y Paleobiología (BierzoRVN)* (PIF 06-055) research project funded by the CSIC, the collaboration agreement between the Junta de Castilla y León and the CSIC on ancient mining areas of the Castilla y León region (CyL-IA-24.044.0006.07). Some results were generated within the projects: *Geoarqueologia y organizacion del territorio en zonas mineras del occidente de la Meseta Norte (METALA)*, funded by the Junta de Castilla y León (CSI07/03), *Formas de ocupacion rural en el cuadrante noroccidental de la Peninsula Iberica. Transicion y desarrollo entre epocas prerromana y romana (TERRITORIA)*, funded by the Spanish Education and Science Ministry (HUM2004-04010-C02-01 y 02), and *Formación y disolución de la civitas en el Noroeste peninsular. Relaciones sociales y territorios (CIVITAS)*, funded by the Spanish Science and Technology Ministry (HAR2008-06018-C03-01).

REFERENCES

Aira Rodríguez, M.J. 1986. *Contribución al estudio de los suelos fósiles de montaña y antropógenos de Galicia*. Facultade de Bioloxía. Santiago de Compostela, Universidade de Santiago de Compostela.

Behre, K.E. 1990. Some reflections on anthropogenic indicators and the record of prehistoric occupation phases in pollen diagrams from the Near East. In Bottema, S., Entjes-Nieborg, F. & van Zeist, W. (eds.) *Man's role in the shaping of the eastern Mediterranean landscape*. Rotterdam, Balkema.

Conedera, M., Krebs, P., Tinner, W., Pradella, M. & Torriani, D. 2004. The cultivation of Castanea sativa (Mill.) in Europe, from its origin to its diffusion on a continental scale. *Vegetation History and Archaeobotany* 13, 161-179.

Crawford, M.H. 1985. *Coinage and Money under the Roman Republic. Italy and the Mediterranean Economy*. London, University of California Press.

Diputación Provincial de Lugo, 1996. *El Oro y la Orfebrería Prehistórica de Galicia*. Lugo, Museo Provincial de Lugo.

Domergue, C. 1986. Dix-huit ans de recherché (1968-1986) sur les mines d'or romaines du nord-ouest de la Peninsule Iberique. *Actas del I Congreso Internacional Astorga Romana*. Astorga, Ayuntamiento de Astorga.

Domergue, C. 1987. *Catalogue des mines et des fonderies antiques de la Péninsule Ibérique*, Madrid, Boccard.

Domergue, C. & Hérail, G. 1978. *Mines d'or romaines d'Espagne. Le district de la Valduerna (León). Étude géomorphologique et archéologique*. Toulouse, Publications de l'Université de Toulouse-Le Mirail.

Fernández-Posse de Arnáiz, M.D., Sastre Prats, I. & Sánchez-Palencia Ramos, F.J. 2004. Oro y organización social en las comunidades castreñas del Noroeste de la Península Ibérica. In Perea Caveda, A., Montero Ruiz, I. & García Vuelta, Ó. (eds.) *Tecnología del oro antiguo: Europa y América*. Madrid, CSIC.

Fombella Blanco, M.A., Andrade Olalla, A., Puente García, E., Penas Merino, A., Alonso Herrero, E., Matías Rodríguez, R. & García-Rovés Fernández, E. 1998. Primeros resultados sobre la dinámica de la vegetación en la turbera del Puerto de San Isidro (León). In Fombella Blanco, M.A., Fernández González, D. & Valencia Barrera, R.M. (eds.) *Palinología: Diversidad y Aplicaciones*. León, Universidad de León.

García Antón, M., Franco Múgica, F., Maldonado, J., Morla Juaristi, C. & Sainz Ollero, H. 1997. New data concerning the evolution of the vegetation in Lillo pinewood (Leon, Spain). *Journal of Biogeography* 24, 929-934.

García Vuelta, Ó. 2007. *Orfebrería Castreña del Museo Arqueológico Nacional*. Madrid, Museo Arqueológico Nacional.

Krebs, P., Conedera, M., Pradella, M., Torriani, D., Felber, M. & Tinner, W. 2004. Quaternary refugia of the sweet chestnut (Castanea sativa Mill.): an extended palynological approach. *Vegetation History and Archaeobotany* 13, 285-285.

López-Merino, L., Peña-Chocarro, L., Ruiz-Alonso, M., López-Sáez, A. & Sánchez-Palencia, F.J. 2010 Beyond nature: The management of a productive cultural landscape in Las Médulas area (El Bierzo, León, Spain) during pre-Roman and Roman times. Plant Biosystems 144 (4), 909-923.

Montero Ruiz, I. & Rovira Llorens, S. 1991. El oro y sus aleaciones en la orfebrería prerromana. *Archivo Español de Arqueología* 64, 7-21.

Muñoz Sobrino, C., Ramil-Rego, P. & Rodríguez Guitián, M. 1997. Upland vegetation in the north-west Iberian peninsula after the last glaciation: Forest history and deforestation dynamics. *Vegetation History and Archaeobotany* 6, 215-233.

Muñoz Sobrino, C., Ramil Rego, P. & Gómez-Orellana, L. .2003. La vegetación postglaciar en la vertiente meridional del macizo del Mampodre (Sector central de la Cordillera Cantábrica). *Polen* 13, 31-44.

Muñoz Sobrino, C., Ramil Rego, P. & Rodríguez Guitián, M. 1996. Análisis polínico de dos sondeos realizados en la turbera de A Cespedosa (Sierra de Ancares, Lugo). In Ruiz Zapata, M.B., Martín Arroyo, T., Valdeolmillos Rodríguez, A., Dorado Valiño, M., Gil García, M.J. & Andrade Olalla, A. (eds.) *Estudios Palinológicos, XI Simposio de Palinología (A.P.L.E.)*. Alcalá de Henares, Universidad de Alcalá de Henares.

Muñoz Sobrino, C., Rodríguez Guitián, M. & Ramil Rego, P. 1995. Cambios en la cubierta vegetal durante el Pleistoceno y el Holoceno en la Sierra de Ancares (NW Ibérico). In Aleixandre Camposy, T. & Pérez González, A. (eds.) *Reconstrucción de paleoambientes y cambios climáticos durante el Cuaternario*. Madrid, CSIC.

Perea Caveda, A. & Sánchez-Palencia Ramos, F.J. 1995. *Arqueología del oro astur. Orfebrería y minería*. Oviedo, Caja de Asturias.

Pérez García, L.C., Sánchez-Palencia Ramos, F.J. & Torres Ruiz, J. 2000. Tertiary and Quaternary alluvial gold deposits of Northwest Spain and Roman mining (NW of Duero and Bierzo Basins). *Journal of Geochemical Exploration* 71, 225-240.

Sánchez-Palencia Ramos, F.J. (ed.) 2000. *Las Médulas (León). Un paisaje cultural en la Asturia Augustana*. León, Instituto Leonés de Cultura.

Sánchez-Palencia Ramos, F.J., Álvarez González, Y. & López González, L.F. 1996. La minería aurífera en Gallaecia. *El Oro y la Orfebrería Prehistórica de Galicia*. Lugo, Museo Provincial de Lugo (Diputación Provincial de Lugo).

Sánchez-Palencia Ramos, F.J. & Fernández-Posse de Arnáiz, M.D. 1998. El beneficio del oro por las comunidades prerromanas del Noroeste peninsular. In Delibes de Castro, G. (ed.) *Minerales y metales en la Prehistoria Reciente. Algunos testimonios de su explotación y laboreo en la Península Ibérica*. Valladolid, Universidad de Valladolid.

Santos, L., Vidal Romani, J.R. & Jalut, G. 2000. History of vegetation during the Holocene in the Courel and Queixa Sierras, Galicia, northwest Iberian Peninsula. *Journal of Quaternary Science* 15, 621-632.

Villa Valdés, Á. 1998. Estudio arqueológico del complejo minero romano de Boinás, Belmonte de Miranda (Asturias). *Boletín Geológico y Minero* 109, 169-178.

1.9 English town commons and changing landscapes

Author

Nicky Smith

English Heritage, London, United Kingdom
Contact: nicky.smith@english-heritage.org.uk

ABSTRACT

English Heritage has recently completed a project which investigated the archaeological content of English town commons. Commons in urban areas are under pressure and because their historic element is not understood it is unprotected. This paper examines the results of the project and argues that town commons should be recognised as a valid historical entity and a valued part of the modern urban environment. This is an essential first step towards successful informed conservation. It also promotes the view that landscape archaeology is about the fabric of the land as created and modified over a long period, in which our own activities are part of that continuum. This reflects the changing nature of archaeology as a discipline that is increasingly concerned with public understanding, along with land management and conservation.

KEYWORDS

urban, town, archaeology, earthwork, common

INTRODUCTION

A common is an area of land, in private or public ownership, over which rights of common exist. Right of common has been defined as 'a right, which one or more persons may have, to take or use some portion of that which another man's soil naturally produces' (from *Halsbury's Laws of England* [1991], quoted by

Clayden 2003, 10). There are six main rights of common: pasture (the right to graze animals); pannage (the right to feed pigs on fallen acorns and beech mast); estovers (the right to collect small wood, furze and bracken); turbary (the right to cut turf or peat); piscary (the right to fish) and common in the soil (the right to take sand, gravel, stone or minerals). Rights of common were usually held either by all householders of a town, just burgage holders (holders of freehold property), or freemen (possessing citizenship of the town). Over time, they tended to become restricted to senior members of the town corporation or to wealthier townsfolk.

English town commons have been largely disregarded by historians and archaeologists, even though their wildlife and recreational value has been recognised. Typically, they have no Conservation Plans and their historic environment content is unknown, and therefore delivers no conservation benefits. Those that have survived, despite urban expansion and other threats to their existence, are regarded locally as important places, 'green lungs' in cities and havens for wildlife. Although it is rarely recognised, they are also a reservoir of archaeological remains.

In 2009, English Heritage completed a project which investigated the archaeological content and Historic Environment value of English town and urban commons. An initial desk-based assessment revealed that nearly 320 English towns possessed commons at some time and that some level of survival still existed in at least 50 places. Site reconnaissance visits were made to these places, to record visible archaeological remains and, to assess their potential for more detailed work. At sites with the highest archaeological potential detailed topographical surveys of earthworks (mounds and ditches) and other visible features were carried out. Differential Global Positioning (GPS) survey was the method used. The research described here assesses the extent and nature of archaeological survival in this unique category of urban landscape.

ACKNOWLEDGEMENTS

The project was carried out by Mark Bowden, Graham Brown and the author. Contributions were made by many other individuals, and these are gratefully acknowledged. Particular thanks are due to David Field and Deborah Cunliffe.

The antiquity of English town commons has long been appreciated: 'When an American visitor asked Freeman, the Regius Professor of Modern History at Oxford, to show him the most ancient monument in Oxford, Freeman walked him out to Port Meadow' (Hoskins & Stamp 1963, 12). Despite this, the origins of town commons are lost in obscurity. Port Meadow, Oxford, is mentioned in the Domesday Book (1086), and the common meadow of Wilton, Wiltshire, is mentioned in a charter of King Edgar (959-975) (Haslam 1984, 127-8). However, it is possible that these are the first documentary records of an even older agricultural practice. Several charters claiming to establish common rights survive from the 13th century, but they merely confirm rights and practices which had probably existed for hundreds, if not thousands, of years. Sudbury's commons, for example, 'given' to the burgesses (inhabitants of the borough with full municipal rights) by Richard de Clare in 1262, had the same acreage as the land belonging to the burgesses and St Gregory's Church in 1086, suggesting that de Clare's grant confirmed a practice already established by the 11th century (French 2000, 177, n38). Clues to the early origins of urban commons perhaps lie in

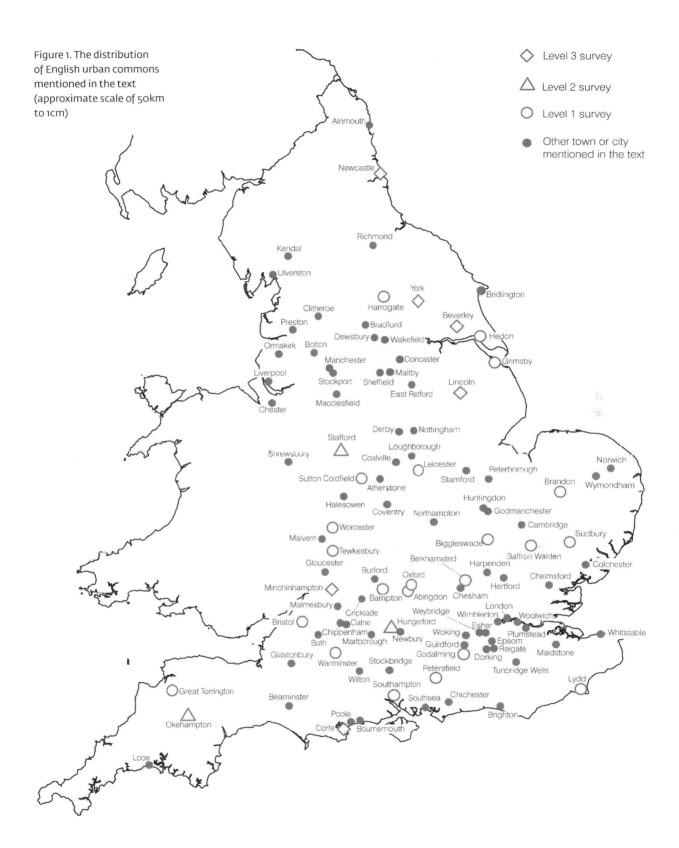

Figure 1. The distribution of English urban commons mentioned in the text (approximate scale of 50km to 1cm)

Level 3 survey
Level 2 survey
Level 1 survey
Other town or city mentioned in the text

the natural landscape. The Greensands (a variety of sandstone), with their thin acidic soils, are of low agricultural potential, hence their use in historic times as commons. Limited evidence from environmental archaeology suggests that these geological zones were already developing as areas of rough grazing in the Bronze Age (Field 1998, 313-14; Graham et al. 2004).

DEFINITIONS

The meaning of 'town' and 'urban' status is not universally agreed. Criteria commonly applied include: significant concentration of population; specialist economic function; the possession of trading rights; sophisticated political form; complex social structure and influence beyond the immediate boundaries of the settlement (e.g. Clark & Slack 1976, 4-5; Beresford 1998, 127-8). Some authorities stress that functions other than the agricultural should predominate (e.g. Everitt 1974, 29; Laughton & Dyer 1999, 26). An interesting facet of this last criterion is that the requirement to demonstrate that a place did not have an agricultural basis in order to qualify as a town tends to make the urban historian underplay, overlook or ignore the town common.

PRE-URBAN LANDSCAPES

Town commons provide rich sources of archaeological remains because of their relatively benign traditional land use, which preserves the physical evidence of the past. Minchinhampton Common, Gloucestershire, has perhaps the earliest surviving earthwork on an urban common, a probable Neolithic burial mound known as 'Whitfield's Tump' (Smith 2002, 13-14). On Corfe Common, Dorset, eight well-preserved Bronze Age burial mounds, part of a more extensive linear cemetery, are prominent landmarks (Fletcher 2003, 8-9, 15). Similarly, on Petersfield Heath Common, Hampshire, there is a cemetery of 21 Bronze Age burial mounds, each standing up to 2m high. Westwood Common, Beverley, has rare Iron Age burial mounds, to which the project has added previously unrecorded examples (Pearson & Pollington 2004, 15-18, 48).

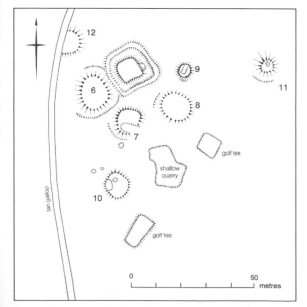

Figure 2. Rare Iron Age square burial mounds, surveyed on Westwood Common, Beverley (© English Heritage).

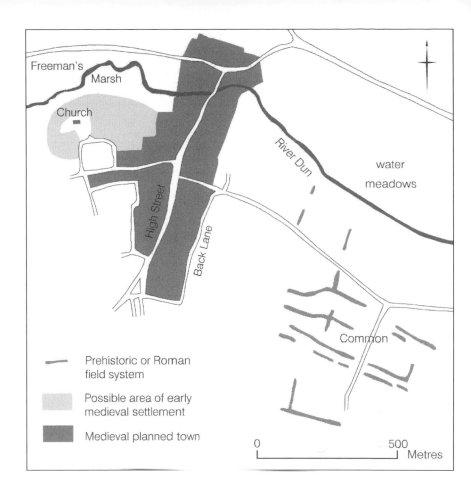

Figure 3. The layout of Hungerford town mirrors the alignment of a probable prehistoric or Roman field system surviving as earthworks on the town common (© English Heritage).

Extensive prehistoric remains are seen on other town commons. Biggleswade Common retains earth-works of Bronze Age burial mounds, a trackway, a field system and prehistoric or Roman-British settlement enclosures. There is also a probable Roman villa or temple complex. Crop-marks in arable fields adjacent to the common show that these are the elements of a wider prehistoric landscape, preserved through common usage from the medieval period onwards. Ironically, while cropmarks have been noted on aerial photographs and recorded, the earthwork remains surviving on the common had largely been overlooked. In some cases the presence of early remains can be seen to influence the layout of later urban space. On Hungerford Common the alignment of a probable prehistoric or Roman field system is followed by the general alignment of the common and the medieval layout of the town itself (Newsome 2005, 14). On Lincoln's South Common the Roman, Ermine Street was traced as a short stretch of agger (bank) and more of its route is delimited by a linear arrangement of small quarries resulting from robbed road materials. At right angles to this, fragments of an extensive field system are preserved. Interestingly the boundary of a medieval hospital, The Malandry, shares the same alignment. The implication is that South Common preserves not only the remains of a major Roman road but also a contemporary field system, the alignment of which, as at Hungerford, continued to have significance into the medieval period.

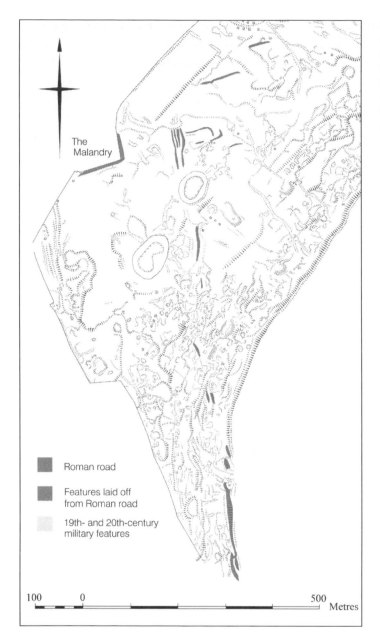

The Malandry

Roman road

Features laid off
from Roman road

19th- and 20th-century
military features

100   0   500
                                      Metres

Figure 4. On the South Common, Lincoln, the route of the Roman Ermine Street is delimited by earthworks. At right angles to this are fragments of an extensive field system and the boundary of a medieval hospital, The Malandry (© English Heritage). *See also full colour section in this book*

TOWNS AND AGRICULTURE

Prior to the Industrial Revolution of the 19th century, the separation between English towns and the countryside was less marked than it is today. Most 17th- and early 18th-century towns had small populations, occupied limited geographical areas and were immersed in the agrarian life of their rural surroundings (French 2000, 171). All historic English towns probably had one or more commons attached to them at some time because townspeople needed pasture for draft animals. These commons served as a municipal pound in which cows could graze securely and in which horses and mares could be rested overnight when not employed during the day. Although archaeological remains associated with grazing are few, cattle pounds or 'pinfolds', for the temporary accommodation of stray animals occasionally survive. These were a constituent of medieval manors and early boroughs (towns with corporations), but examples existing today date from the 17th to 19th centuries. On Hampstead Heath, London, a circular brick-walled enclosure, with supports for its gate fashioned from the jaw bone of a whale, was built in 1787 to replace a pound removed by a man presented at the Manorial Court (Willmott Dobbie 1979, 3, 11, 33, 36). Stock ponds are also found, such as prominent examples on Westwood Common, Beverley, probably dating to the 19th century in their present form, though as established watering holes they could have origins in the medieval period. Longevity of use is evident from the deeply incised sections of track-way leading towards several of them (Pearson & Pollington 2004, 28-9).

Cultivation, above any other agricultural use, has left the most tangible evidence for agriculture on town commons. They were cultivated in times when food supplies could not be relied upon, during wars and other national emergencies such as crop failure. One key period was during the Revolutionary and Napoleonic wars against France (1793-1815). Cultivation earthworks of this period are typified by 'narrow' ridge-and-furrow, produced by ploughing with heavy horses, leaving parallel straight ridges alternating with furrows no more than 5m apart. The Hob Moor, York, was cultivated during this period and narrow ridge-and-furrow can still be seen across large parts of it. Kendal Fell, also cultivated at this time, has earth and stone boundaries, 1m or 2m wide and up to 0.6m high, delimiting an area of narrow ridge-and-furrow, with furrows 2m to 3m apart. Not all ridge-and-furrow found on urban commons represents short-term cultivation. In some instances its presence, or that of field boundaries, indicates that commons have expanded beyond their original boundaries or that their boundaries have been otherwise altered to incorporate former arable fields. On Stafford Common, for example, narrow ridge-and-furrow, field boundary

Figure 5. Ridge-and-furrow cultivation remains surviving on Stafford Common indicate the late addition of former arable lands to the Common (RAF aerial photography 1954 ©English Heritage, NMR).

banks and strip-lynchets (terraces between areas of ploughed land) represent former arable land. In this case it was granted to the householders and commoners of the town in 1801 to compensate for the enclosure of their historic grazing lands. The Pitchcroft, Worcester, similarly comprises two historically different elements, one of which, 'The Moor' (containing narrow ridge–and-furrow) was arable land added to the common in *c*1775 (Hodgetts 2003, 25).

A hay crop was an important product of many urban commons and the period of the year for which grazing was permitted reveals a common's use for this purpose. Traditionally, grazing was prohibited between Candlemas (2 February) and Lammas Day (12 August or 1 August after 1762). Since grazing rights precluded the construction of fences on commons, hay meadows were usually divided into strips marked by posts or stones. Stones surviving on Cricklade Common are probable hay apportionment markers.

The agricultural functions of town commons have declined over time and the advent of motorised transport, agricultural mechanisation and intensive farming systems has put an end to the traditional uses of most. By 1876, only 130 out of 2,300 ratepayers entitled to graze cattle on Stafford Common exercised their rights (*VCH Staffs VI* 1979, 210). Similarly, grazing of Bristol's Downs was in decline by the later 19th century, and by 1872 only 300-400 sheep were being turned out to pasture from a permitted number of at least 1,882. Grazing on Clifton Down ceased completely in the mid-late 19th century, while on Durdham Down it stopped in the 1920s.

ECONOMIC PRESSURES

Economic incentives to 'improve' common pastures always were high, particularly prior to their appropriation by enclosure. Early examples of such works were recorded on Figham Common, Beverley, where there is a network of small drainage channels, some of which may date from as early as the 13th century (*VCH Yorks VI* 1989, 217; Pollington & Pearson 2004, 3, 8). Loss of common land to enclosure between the 16th and 19th centuries was dramatic in the well-populated south and east of England where the pressure for agricultural 'improvement' was greatest (Short & Winter 1999, 616-617) and, even in less favourable areas, reclamation and improvement of 'wastes' gathered pace in the century before 1850. Drainage systems were laid out across many town commons and surface channels, embankments, sluices and machines to remove water all survive as archaeological features. Subtle earthworks of underground drainage systems consisting of straight, shallow linear depressions leading to rivers can also be seen on some town commons, such as Southampton and Saffron Walden. Such schemes continued into the 20th century, being funded by the Ministry of Agriculture and Fisheries in the 1920s to alleviate unemployment, and using prisoner-of-war labour during First and Second World Wars (Bowers 1998, 71).

The archaeology of urban commons is never more closely inter-linked with the history of towns than when considering extractive industry, which has left most urban commons scarred by quarry pits and mounds. Urban commons have long been convenient sources of building materials. Easy availability of these materials fuelled town expansion, threatening the very survival of the commons from which they had been extracted. In Shrewsbury, Shropshire, the town common, named 'The Quarry', was almost certainly the source of materials for the town's buildings. The same is true of small pits on The Pitchcroft, Worcester where, in 1770, a brickworks was established specifically for the building of a new hospital. Dips and mounds at either end of Southampton Common also result from clay digging, where a brickmaker's house with a kiln was built in 1712 (Southampton City Council 2007, 14-15). The extraction and processing of limestone on Kendal Fell was similarly inextricably linked with the creation of the modern town and the single surviving limekiln is now a Scheduled Ancient Monument (no. 34994) (Elsworth 2005, 11, 13, 29-32).

As a general rule early mining and quarrying was a piecemeal and small-scale process. It left remains such as those on Minchinhampton Common, Gloucestershire, where hundreds of small pits and mounds scattered across part of the common may be linked to 12th-century documentary references to inhabitants paying rent for the privilege of quarrying stone (Smith 2002, 34-35). On Kendal Fell, Cumbria, where widespread earthworks of shallow quarrying also remain, the common right of 'stone-getting' was important enough to be preserved following enclosure in 1767 (Elsworth 2005, 11-12). Irregular depressions, c1-2m deep, close to the eastern edge of Westwood Common, Beverley, were probably dug by townsfolk to obtain clay and chalk for building, or other purposes, and could date from the medieval period. These early pits were eclipsed in scale by the Limekiln Pits and two unnamed quarries in the south-eastern corner of the common, which cover over 12ha (Pearson & Pollington 2004, 28).

During the 18th and 19th centuries the demand for rock and lime accelerated rapidly. Hardcore was needed for turnpike roads and railway lines, stone was required for walls erected under Enclosure Acts and lime for soil improvement. Deeper and larger depressions on town commons indicate co-ordinated and organised quarrying suggestive, although not necessarily proof of, this later date. The type of material sought and the geological composition of the area is reflected in the nature of the remains, with excavations for clay and other surface materials tending to be shallow and spread widely, whereas deep excavations were made to reach good quality stone and coal. Linear quarries, or rows of small pits following mineral seams also occur. On Minchinhampton Common, a limestone quarry cut in two by an 18th-century road and hollowed trackways may be the 'Rode Quarry' mentioned in 1516 (Russett 1991, 612). Here larger excavations associated with hollowed track-ways represent 19th- and 20th-century quarrying, the extent and scale of which proved extremely damaging to pasture and to the landscape of the common (Smith 2002, 33).

Figure 6. Small-scale quarry pits visible on Port Meadow, Oxford, as seasonal flooding subsides. These are typical features found on town commons throughout England.

As towns developed into industrial centres, urban commons became convenient 'empty' corridors for the routes of new canals, roads and railways. One example of many was the East Coast main line railway, constructed across Biggleswade Common in 1850, cutting it in two. Material for such road, rail and canal building projects was often extracted from the commons themselves. Hungerford Common, for example, has quarries that probably relate to the construction of the canal and railway nearby (Newsome 2005, 21).

URBANISATION AND TOWN COMMONS

To a large extent the landscape archaeology of an urban common mirrors the economic and social circumstances of its adjacent town. Fluctuating fortunes of the town, changes in its demographic profile, changes in industrial activity and leisure interests, all have an impact upon the archaeological footprint found on the urban common. During the 19th century, one of the most remarkable social changes in Britain was the congregation of the majority of the population into large cities and the creation of the totally 'urban' environment which followed (Mellor 1976, 1, 109). As urban populations swelled and towns spread beyond their ancient boundaries, many of their commons were lost to building development, while others became 'urbanised', encircled by new streets and buildings. In Grimsby, the construction of the railway and the building of a new dock heralded a dramatic increase in the town's population that prompted housing development on the ancient East Marsh common (Gillett 1970, 213). In other towns, such as Dewsbury, influxes of immigrant workers led to illegal settlement on the commons. In some cases of urban expansion the layout of a former common is reflected in the street layout. In Nottingham, for example, the principle routes follow tracks and paths across the common (e.g. Carter 1983; Hoskins 1988, 224-228).

Piecemeal encroachment on town commons has frequently been permitted for the sites of public facilities. Recurring types are workhouses, isolation hospitals, prisons, gallows, schools, sports centres, cemeteries and mortuary chapels; indeed, the presence of any of these, or a barracks, in a modern town may be the best clue as to the location of former common land. Many of these institutions date to the late 19th century, although there are earlier examples, such as St Leonard's leper hospital at Sudbury, Suffolk, which was built in the 14th century and later used for the old and infirm. The hospital lay about a mile to the north of the town on the boundary of what became known as North Meadow Common (Hodson 1891, 268-274). Other examples are the lunatic asylum, built on the northern edge of Chapel Fields, Norwich, in 1712, and infectious disease hospitals, as at Lincoln (Brown 2003, 12).

During times of popular dissent people congregated on urban commons, so it was no accident these were often the places chosen as the sites for penal institutions and gallows. Wormwood Scrubs Prison, Lambeth, London, was built on common land using prisoner labour, in the 1880s. The west side of York's Knavesmire was the site of a gallows named the York 'Tyburn' in imitation of the famous London gallows of the same name. Crowds flocked to see the spectacle of executions there from the early 16th century onwards, including the execution of the most notorious thief-highwayman, Dick Turpin – in 1739. Southampton Common, Hampshire, was also the site of a gallows, and executions were carried out on Brandon Hill, Bristol, which was said to be a haunt of thieves.

CHANGING SOCIAL VALUES AND THE TRANSFORMATION OF URBAN COMMON LANDSCAPES

A requirement for provision of open spaces for public recreation in increasingly overcrowded and un-sanitary towns emerged from the mid-19th century onwards. It arose from the misconception that disease was caused by air pollution, and open spaces were intended to provide reservoirs of wholesome air to purify the blood of the town citizens (Mellor 1976, 110). One effect of this was the transformation of large numbers of town commons into public parks or 'urban open spaces'. The deliberate laying out of parks and walks for public recreation in England was an idea with origins in the 17th century (see, for instance, Elliott 2000, 145) and the earliest recorded public park on an English urban common is at Moorfields, London, where in 1607 the moor was filled in, the ground level raised and two walks, bordered by walls and trees were laid out (Lambert 1921, 87). The activities that could take place in such spaces were limited to the sedate and the polite -manicured 'walks' were a means by which, through the 18th century, 'an increasingly urbanized society sought to retain contact with a retreating rural world' (Borsay 1986, 132). Ironically, this was happening at the time that towns were shedding genuine connections with rural life by enclosing their commons.

As town populations grew, urban commons became the sites of pioneering schemes designed to bring clean water supplies into towns. One such scheme took spring water from King's Meads, Hertford, to London via an aqueduct named the 'New River'. This remarkable aqueduct was begun in 1609 and originally extended for 39 miles (Page 1993, 137-138).

During the 19th century the governing bodies of towns began to show increasing concern over facilities for the health and recreation of citizens, partly due to compulsion (Mellor 1976, 110). An Act of Parliament (1866) compelled local authorities to provide sanitary inspectors and allowed central government to insist upon the provision of sewers and a good water supply (Woodward 1962, 463-465). Land for such works could be taken without agreement and was often appropriated from town commons. In 1868, for example, the East London Waterworks Company issued a notice to the commoners of Tottenham regarding land it was compulsorily acquiring (London Metropolitan Archives: ACC/1016/485). On Minchinhampton Common, an irregular stone-revetted platform is a disused reservoir built by the Stroud Water

Figure 7. The 'New River', a remarkable aqueduct 39 miles long, was built in 1609 to carry spring water from King's Meads, Hertford's town common, to London.

Figure 8. King's Meads, Hertford, under flood. This ancient common, as those elsewhere, provided a convenient corridor for urban infra-structure, in this case a fly-over carrying the A10 dual carriageway.

Company before 1922 (Russett 1991, 66) and a further reservoir built there is still in use. Water is also collected on large upland commons such as Wardle Common, Greater Manchester and Whitworth and Trough Common, Lancashire, where the water authority (West Pennine Water Board) has the right to collect water (Aitchison et al. 2000, CL165-166). On other urban commons deep wells were sunk. In Southampton, Hampshire, the city's population had outgrown its water supply by the 19th century and so an artesian well (in which a bore-hole is cut perpendicular to a synclinal fold of the underlying rock strata) was constructed on the common to meet the increased demand. Despite being sunk 420m deep the required flow of water never appeared, but the cover of the shaft can still be seen today.

Improved utilities were introduced into towns during the late 19th and early 20th centuries, when numerous agreements were entered into for the erection of telegraph poles, electricity cables, sub-stations and the digging of gas and water mains across urban commons. The benefits of these are still enjoyed by townspeople today. Less savoury utilities have also found a place on town commons, which frequently serve as a depository for sewerage and other waste. Lincoln's West Common and the Pitchcroft, Worcester, for example, have both been used as town rubbish tips, while sewerage purification works were established on King's Meads, Hertford; King's Marsh, Sudbury, Suffolk; Biggleswade Common, Bedfordshire; and Earlswood Common, Reigate, Surrey.

THE PRESENT AND FUTURE

Town and urban commons may be viewed as historic and ancient landscapes in terms not only of their intrinsic value as relatively undisturbed resources for the study of ancient ecological systems and environmental change, but also as reservoirs of visible and sub-surface archaeological remains. In the case of Port Meadow, Oxford, extensive archaeological remains with five distinct phases of activity, discovered by aerial photography, topographical survey and excavation include at least six Bronze Age burial mounds and three groups of middle Iron Age farmsteads accompanied by ditched paddocks. Analysis of the archaeology of the natural landscape by earth scientists has provided further insights into the processes of natural change which prompted these settlements. A higher water table in the Iron Age than in the preceding period supported grassland communities similar to those which exist on Port Meadow today and evidence for winter water-logging, suggests that the settlements may have been seasonally occupied for summer grazing. Corroborative evidence was found in the form of plants of disturbed ground, characteristic of rural settlements and farmyards, that are no longer represented on the Meadow (Lambrick & McDonald 1985, 100).

Despite urban commons being converted into 'people's parks', the change is by no means universal. Port Meadow and Wolvercote Common, Oxford, are ecologically unique in their region because of the survival of their hay meadows. Their common status has protected them from ploughing, draining and the application of chemical fertiliser, allowing the continued evolution of grassland communities which developed through traditional management by farmers exercising ancient common rights. This ecological significance led to designation of these and adjacent smaller commons as Sites of Special Scientific Interest (SSSIs) in 1952.

Because so many town commons are on valley floors they are prone to flooding, and the incidence of such flooding will vary naturally with climate change. However, panic measures to prevent or control flooding may be damaging in themselves. At the same time it is anticipated that 'historic open spaces in urban areas will play an important role in ameliorating the effects of a hotter climate' (English Heritage 2008, 12). Biodiversity Action Plans are now in place for several areas, which include town commons such as Staines Moor, Surrey, and there are numerous local authority countryside management projects tackling habitat restoration.

The recognition of town commons as a valid historical entity and a valued part of the modern urban environment is a fundamental first step towards successful informed conservation. An important consideration for the future is maintaining the character of town commons as a different sort of urban open space, distinct from parks and public gardens. The fact that most are no longer working as agricultural commons should not mean that they are treated as urban parks. Their legal situation has been changed by the passing of the Commons Act 2006 which enables more sustainable management through 'commons councils', bringing together commoners and landowners to regulate grazing and other activities, and reinforcing existing protections against abuse, encroachment and unauthorised development. The act places much stress on the natural diversity of commons and their value as wildlife habitats, with much less emphasis on their historical and archaeological value; nevertheless, the historic environment should benefit.

METHODOLOGY

The research methods used give an overview of the historic environment resource on English town commons. Desk-based work, combined with field reconnaissance, proved effective in locating and mapping archaeological remains in these landscapes. However, there is potential for survival in places which were not identified during the desk-based work and this study does not pretend to present a total picture of the surviving archaeological remains.

REFERENCES

Aitchison, J., Crowther, K., Ashby, M. & Redgrave, L. 2000. *The Common Lands of England: A biological survey*. Aberystwyth: University of Wales Rural Surveys Research Unit.

Beresford, M.W. 1998. *History on the Ground* (revised edition). Stroud: Sutton.

Borsay, P. 1986.The rise of the promenade: The social and cultural use of space in the English provincial town c 1660-1800. *British Journal for Eighteenth-Century Studies* 9, 125-40.

Bowers, J. 1998. Inter-war land drainage and policy in England and Wales. *Agricultural History Review* 46 (1), 64-80.

Brown, G. 2003. *An Earthwork Survey and Investigation of Lincoln, West Common (Archaeological Investigation Report Series AI/11/2003)*. Swindon: English Heritage.

Carter, H. 1983. *An Introduction to Urban Historical Geography*. London: Edward Arnold.

Clark, P. & Slack, P. 1976. *English Towns in Transition 1500-1700*. London: Oxford University Press.

Clayden, P. 2003. *Our Common Land: The law and history of commons and village greens*. Henley-on-Thames: Open Spaces Society.

Elliott, P.A. 2000. Derby Arboretum (1840): The first specially designed public park in Britain. *Midland History* 26, 144-76.

Elsworth, D.W. 2005. *Kendal Fell, Kendal, Cumbria: Conservation Plan (Draft)*. Lancaster: Oxford Archaeology North.

English Heritage. 2008. *Climate Change and the Historic Environment*. London: Centre for Sustainable Heritage, University College London with English Heritage.

Everitt, A. 1974. The Banburys of England. *Urban History Yearbook* 1, 28-38.

Field, D. 1998. Round barrows and the harmonious landscape: Placing early Bronze Age burial monuments in south-east England. *Oxford Journal of Archaeology* 17 (3), 309-26.

Fletcher, M. 2003. *Corfe Common, Purbeck, Dorset (Archaeological Investigation Report Series AI/28/2003)*. Swindon: English Heritage.

French, H. 2000. Urban agriculture, commons and commoners in the seventeenth and eighteenth centuries: The case of Sudbury, Suffolk. *Agricultural History Review* 48 (2), 171-199.

Gillet, E.A. 1970. *History of Grimsby*. Oxford: Oxford University Press.

Graham, D., Graham, A. & Wiltshire, P. 2004. Investigation of a Bronze Age mound on Thursley Common. *Surrey Archaeological Collections* 91, 151-166.

Haslam, J. 1984. The towns of Wiltshire. In Haslam, J. (ed.) *Anglo-Saxon Towns in Southern England*. Chichester: Phillimore, 87-148.

Hodgetts, C. 2003. *Worcester's Riverside Parks*. Worcester City Council.

Hodson, W.W. 1891. John Colney's or St Leonard's Hospital for Lepers at Sudbury. *Suffolk Institute of Archaeology & Natural History* 7, 268-74.

Hoskins, W.G. 1988. *Making of the English Landscape*. (First pub 1955, revised edition with introduction and commentary by C.C. Taylor). London: Hodder & Stoughton.

Hoskins, W.G. & Stamp, L.D. 1963. *The Common Lands of England and Wales*. London: Collins.

Lambert, F. 1971. Some recent excavations in London. *Archaeologia* 71, 55-112.

Lambrick, G. & McDonald, A. 1985. The archaeology and ecology of Port Meadow and Wolvercote Common, Oxford. In Lambrick, G. (ed.) *Archaeology and Nature Conservation*. Oxford: Oxford University Department for External Studies, 95-109.

Laughton, J. & Dyer, C. 1999. Small towns in the east and west midlands in the later Middle Ages: A comparison. *Midland History* 24, 24-52.

London Metropolitan Archives: ACC/1016/485 Notice issued to the commoners of Tottenham by the East London Waterworks Company, 1868 (part of a collection of papers from Couchmans, Surveyors and Valuers, Tottenham, Middlesex).

Mellor, H.E. 1976. *Leisure and the Changing City 1870-1914*. London: Routledge & Kegan Paul.

Newsome, S. 2005. *Hungerford Common, Freeman's Marsh and Environs (EH Aerial Survey & Investigation Special Project AER/5/2005)*. Swindon: English Heritage.

Page, F.M. 1993. *A History of Hertford* (2nd edition). Hertford: Hertford Town Council.

Pearson, T. & Pollington, M. 2004. *Westwood Common, Beverley: An archaeological survey (Archaeological Investigation Report Series AI/25/2004)*. Swindon: English Heritage

Pollington, M. & Pearson, T. 2004. *Figham Common, Beverley: An archaeological survey (Archaeological Investigation Report Series AI/23/2004)*. Swindon: English Heritage.

Russett, V. 1991. *The Minchinhampton Commons, Gloucestershire: The National Trust Archaeological Survey*. Gloucester: Gloucestershire County Council.

Short, C. & Winter, M. 1999. The problem of common land: Towards stakeholder governance. *Journal of Environmental Planning & Management* 42 (5), 613-630.

Smith, N.A. 2002. *Minchinhampton Common: An archaeological survey of the earthwork remains (Archaeological Investigation Report Series AI/12/2002)*. Swindon: English Heritage.

Southampton City Council. 2007. *Southampton Common Green Flag Management Plan 2007-2010*. Southampton.

VCH Staffs VI. 1979. *The Victoria History of the County of Staffordshire*, Vol VI. Oxford: Oxford University Press.

VCH Yorks VI. 1989. *The Victoria History of the County of York: East Riding*, Vol VI. Oxford: Oxford University Press.

Willmott Dobbie, B.M. 1979. *Pounds or Pinfolds, and Lockups*. Bath: University of Bath.

Woodward, L. 1962. *The Age of Reform 1815-1870* (The Oxford History of England 13). Oxford: Clarendon Press.

1.10 From feature fetish to a landscape perspective: A change of perception in the research of pingo scars in the late Pleistocene landscape in the Northern Netherlands

Author
Inger Woltinge

Groningen Institute of Archaeology, University of Groningen, Groningen, The Netherlands
Contact: i.woltinge@rug.nl

ABSTRACT

In the northern part of the Netherlands a number of frost mound remnants are known. Some of these have been interpreted as pingo scars which have historically been regarded as good habitation locations for hunter-gatherers because of their relatively high position in the landscape and their easy access to water. The existence of a direct association between the higher ridges, known as ramparts, of the pingo scars and finds from early prehistory has been a common idea in archaeology in the northern provinces of the Netherlands. Indisputably, pingo scars provide an excellent base for environmental reconstruction as they have been filled with organic sediments from the early Holocene onwards and have proven to be an effective pollen trap, making them indispensable for local palynological research.

In a recent study the connection between archaeological remains, pingo scars and the corresponding landscape was investigated, resulting in an unexpected outcome. The historically assumed association between the pingo scars as isolated features and human occupation thereof seems to be nonexistent when researched on a metadata scale. It was concluded that other parameters lie at the basis of this assumption. The human occupation and its interaction with the landscape seem to be based on much larger environmental elements than these specific features themselves.

KEYWORDS

pingo scars, Pleistocene, Stone Age, landscape archaeology, Netherlands

INTRODUCTION TO PINGO SCARS: DEFINITIONS AND DATES

Frost mounds in general and pingos in particular, are ice-pushed mounds growing in periglacial conditions, such as those at the end of the Weichselian in Europe. When the soil cover of a pingo is pushed upwards due to the growth of the ice lens underneath it, the top of the mound will tear at a certain height and the material will start to slide down (gelifluction), which leads to the formation of a rampart. The height at which this happens depends on the thickness of the soil cover or overburden. After a while, the top of the ice core will be exposed and can subsequently melt from the radiation of the sun. Eventually, this leads to the melting of the greater part (or the whole) of the ice lens. The mound collapses and a lake surrounded by rampart, also known as pingo scar, is formed (amongst others, Mackay 1972; Watson 1972; Mackay 1979; De Gans 1988).

It is generally assumed that pingo scars in the Netherlands are remains of the Weichselian Ice Age. The exact dates of their formation are hard to obtain because organic material from underneath ramparts is scarce. Accepting that they were formed during the Weichselian glacial places their age at a maximum between 25,000 and 22,000 BP, at which time the Netherlands were part of the continuous permafrost zone. The erosion of the pingos into pingo scars probably took place when the continuous permafrost degraded between 22,000 and 17,000 BP (according to OSL dates on the Lutterzand section from Bateman & Van Huissteden 1999, 282). Weichselian pingo scars in the Netherlands collapsed due to a temperature rise at the start of the Bølling rather than as a result of reaching their maximum height. Lacustrine deposition in pingo scars therefore roughly coincides with the Bølling period and started around 14,700 BP at the very earliest (De Gans 1988, 307). Precise dating of these events is difficult, as the first registration in pingo scars depends on local hydrological conditions and possibly the moment at which the ice lens was formed. Preliminary results of pollen analyses of the infill of a pingo near Twijzel and Jistrum (Frys-

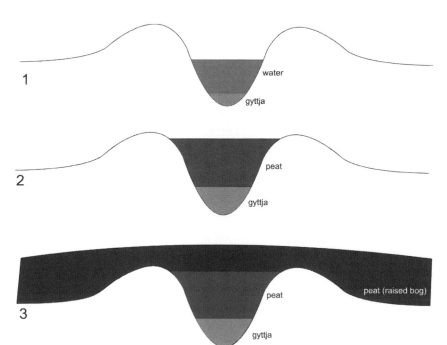

Figure 1. Three stages of pingo scar development after pingo collapse. Stage 1: the pingo becomes a water-filled depression in which gyttja settles. Stage 2: peat grows, following the ground-water level; and Stage 3: a raised bog covers the entire feature.

lân) demonstrate that filling of that particular feature starts late in the Younger Dryas (dated to 10,950 to 10,150 BP) rather than in the Bølling (De Kort 2010). In general, after pingo collapse, there are roughly three stages of pingo scar development (fig. 1).

PREREQUISITES FOR RESEARCH INTO THE RELATIONSHIP OF MAN AND FEATURE

The association between pingo scars and archaeological finds is based on the present-day idea of usefulness of these features to prehistoric man. Reasons for visiting pingo scars in prehistory can be twofold at least, the most frequently named being the presence of water. Apart from being of use to man himself, the water in the pingo scar may have been an incentive for animals to visit the spot, making it easier for hunters to find the animals. The second reason may be the (slightly) raised rampart, which may have been a dry location for a viewpoint and camping. The different uses of a pingo scar will not have been evident at all stages of its development. When the accumulated gyttja has reached the water level and peat starts growing, the amount of open water will diminish. At this stage of development, water could still be originating from the swampy conditions of the inside of the pingo scar and as long as the peat does not completely fill the depression, the spot could still be usable for shelter or water supply. The rampart could also still be used as a camp site. All these uses disappear in the third stage, when the pingo scar is fully overgrown with a raised bog. In this stage, it is most likely indistinguishable from the surrounding landscape, which will be a marshy area by then. The most important reason for assuming a significant relationship between prehistoric man and pingo scars is that of the raised rampart in combination with the vicinity of water. As the exact dating of the infill of pingo scars is problematic and, therefore, conditions for research of this hypothesis are unfavourable, the focus will be on the association of archaeological finds and the slightly higher points at which the ramparts will have been placed in the landscape. In order to be able to make specific statements on the association between ramparts and finds, the width of a pingo scar's rampart was calculated.

SCOPE OF A PINGO SCAR

In this paragraph, the width of the rampart is calculated in order to examine the scope of a pingo scar. The definition of 'scope' in this respect is the area that can be seen as part of the pingo scar, consisting of the depression and the rampart. Examining the scope of a pingo scar is necessary in order to be able to make statements about which finds can be considered as being on the rampart and therefore in direct association with the feature. Finds beyond the width of the rampart can, in the strictest sense, not be considered to be in association with the rampart.

Using a formula for this exercise may be regarded as an effort to be too precise, while prehistoric people obviously did not use any formula for making decisions as to where they were going to leave their mark, but calculating the width of the rampart is a way of getting a better grasp of what constituted the feature 'pingo scar' in prehistoric times. Another important reason for calculating the rampart's width rather than measuring it in the field or on maps lies in the fact that ramparts have often been flattened and as such are not visible in the field. On contour maps like the 'Actuele Hoogtebestand Nederland'

(AHN: Digital Elevation Model), their visibility is also limited due to the fact that most of them are located on the banks of (fossil) stream valleys, often making the distinction between the riverbank and rampart difficult.

If, for simplicity the shape of a pingo is considered a cylinder, the formula that can be used for calculating the width of the rampart (b) is

$$[\,b = \sqrt{(r^2 + V/\pi d)} - r\,]$$

in which r is the radius of the depression, V is the volume of rampart material and d is the average height of the rampart (fig. 2).

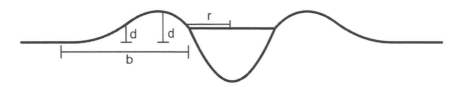

Figure 2. Visual explanation of the variables used in the formula for calculating the scope of a rampart.

The formula is based on the calculation of the volume of the rampart material (V) as the volume of a cylinder with radius b+r minus the volume of a cylinder with radius r.

To fill out this formula the radius of the depression (r), the average height of the rampart (d) and the volume of material constituting the rampart (V) are needed. The radius of the depression (r) is practically always known, because the diameter of the depression is often still visible in the field or on AHN maps. The average height of the rampart (d) is more problematic. The height could be inferred from the AHN data or field measurements, but as said before, most ramparts are not intact and therefore, their height is not the same as it would have been in prehistoric times. The location of most pingo scars on the edges of stream valleys, making them slightly asymmetrical, also means that the rampart is not entirely visible on maps or in the field, making educated guesses to the stretch of a rampart very hard. Assessing the volume of material constituting the rampart (V) is even more problematic. The ice lens in the growing pingo will also have been partially below the surface, making the depression slightly larger than the pushed up soil material (overburden) could fill. Soil will also have slid into the depression on the thawing of the ice, leaving less material on the rampart. Finally, some of the sediment could have been blown away while the pingo was growing. Based on these three assumptions, calculating the volume of the infill of the pingo scar will thus generate a number that is too high for the volume of rampart material. A number or formula for the thickness of the soil cover of the ice lens must be found in order to calculate the volume of the rampart material.

OVERBURDEN AND RAMPART VOLUME

In literature, there is little mention of heights of ramparts or overburden (the original soil material that is pushed up by the ice core). The general assumption, when looking at figures in publications seems to be that the overburden is a couple of metres thick, at most. A more detailed written overview will be published in the author's PhD thesis. Some modern-day active pingos known from literature and their characteristics are listed in Table 1.

Table 1. Characteristics of seven modern-day active pingos and two pingo scars from literature: 1) Ross et al. 2005, 131-135; 2) Mackay 1972, 12; 3) French & Dutkiewicz 1976, 215-216; 4) Pissart 1988, 282-283; 5) Waltham 2003, 214; 6) Watson 1976, 79; 7) De Gans 1988, 301.

Name	Diameter (m)	Height (m)	Overburden (m)	Maximum height rampart (m)	Width rampart (m)
Active pingos					
Riverbed[1]	90 and 50	12	1	-	-
Innerhytte[1]	400 and 200	28	1	-	-
McKinley Bay[2]	90	9	1,5	-	-
Banks Island[3]	45	3,4	1,5	-	-
Yakutsk[4]	100	15	4	-	-
Yakutia[4]	100	13	3	-	-
Ibyuk[5]	300	50	15	-	-
Pingo scars					
Pingo L, Cledlyn[6]	135	-	-	6	65
East Greenland[7]	105	-	-	12	65-70

CALCULATING RAMPARTS

As stated above, the variables needed to calculate the width of a rampart are the radius of the depression, the average height of the rampart and the volume of rampart material. The first is easily measured in the field or on maps; the second is nearly impossible to measure and is therefore better inferred from modern day active pingos. The third variable can be calculated using the information of modern-day pingos in Table 1. The maximum volume of the rampart material can be calculated by using the general formula for the volume of a sphere cap:

$$[V = 1/6\pi \cdot h \, (3r^2 + h^2)]$$

Here V is the volume of the sphere cap, r the radius and h the height of the sphere cap (fig. 3). In order to calculate the volume of material available for the rampart, this formula has to be filled out twice: once with h_1 as the height of the entire pingo and once with h_2 as the height of the ice lens (pingo height minus overburden, Table 1). The latter result has to be subtracted from the former and the number thus found is the volume of available rampart material.

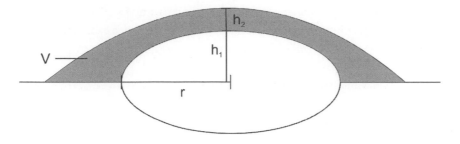

Figure 3. Visual explanation of the variables used in the formula to calculate the volume (V) of material pushed up by the ice lens. When the pingo disintegrates and becomes a pingo scar, the volume of material will make up the rampart.

The width of hypothetical ramparts was calculated based on the present-day pingos in Table 1 if they were to become pingo scars. This was done in order to achieve a number for the maximum rampart width based on the overburden, which is the available soil material that was pushed up by the ice. The width of the rampart was calculated for three different average heights of the rampart: 0.5, 1 and 2 m. These numbers were used because a rampart will generally be relatively wide and not very high: the material slides down while the ice core is already exposed and is fanned out rather than heaped at the foot of the ice core, much like material sliding down from mountains. Ramparts in the Netherlands are mentioned to have maximum heights of less than 1.5m (Maarleveld 1976, 58), and are seldom reported to exceed that height in Scandinavia (Svensson 1976, 35). The highest rampart reported in Wales is 7m, whereas the highest known Belgian rampart is 5m (Maarleveld 1976, 61). Average heights for the entire rampart will obviously be lower than the maximum height of the top of the rampart and therefore the calculations performed use a maximum *average* height of 2m.

Because the pingos of Riverbed and Innerhytte are asymmetrical, the average diameter (and thus radius) was used. With 15m the Ibyuk pingo has a very high overburden and is probably the highest in the world (Pissart 1988). The depth at which the ice core usually starts is roughly between 2 and 10m below ground surface (Castel & Rappol 1992, 131; Jorgenson & Osterkamp 2005, 2102), making 2 to 10m the accepted thickness for overburden. Most of the overburdens seem to be closer to the lower estimate than the higher one (see figures in, for instance, Maarleveld & Van Toorn 1955, 348; Castel & Rappol 1992, 130; Ross et al. 2005, 133-135; Pissart 1988, 282-283).

After calculating the volume of soil material (overburden) that would make up the rampart for the pingos in Table 1, the formula [$b = \sqrt{(r^2 + V/\pi d)} - r$] could be used for these pingos, calculating the maximum width of their ramparts at an average rampart height of 0.5, 1 and 2m respectively. The results of this exercise are listed in Table 2. It is obvious that in this survey, the Ibyuk pingo is an outlier at almost ten times as wide a rampart can stretch compared to any other pingo used for this calculation. It is a large pingo in all respects, with a diameter of 300m and a height of 50m. Even though it is an extraordinary pingo, it was included in the calculations, because a few of the pingo scars in the Netherlands have a depression of (almost) 300m, and not using it might be biasing the outcome of the calculations. It has to be mentioned however that none of the pingos in the research area used in this paper are that large. When the characteristics for the individual pingos were calculated, the average width or stretch of a rampart was calculated for the average rampart heights of 0.5, 1 and 2m. These widths are 102, 63 and 37m respectively, which means that a very slight rampart height of 0.5m the distance the rampart material stretches from the edge of the depression of the pingo scar is 102m at most. Finds beyond that distance can, in the strictest sense, not be considered to be in association with the pingo scar or its rampart.

Table 2. Results of the calculations of the scope of a rampart of Arctic pingos mentioned in Table 1. The width of the ramparts is given for an average height of the rampart of 0.5m, 1 and 2m.

Name	Diameter (m)	Overburden (m)	b (m) at d=0.5	b (m) at d=1	b (m)at d=2
Riverbed	90	1	19	11	6
Innerhytte	400	1	64	35	18
McKinley Bay	90	1,5	27	15	8
Banks Island	45	1,5	13	7	4
Yakutsk	100	4	65	39	22
Yakutia	100	3	52	30	17
Ibyuk	300	15	472	303	187
Averages	160	3,9	102	63	37

NEARNESS

The average scope of a pingo's rampart was calculated in order to have a number that could be used as a rule of thumb for what should be considered as part of a pingo scar. By calculating the average maximum stretch of a rampart, a tentative distance from the depression's edge can be quantified, rather than saying something is 'near' or 'close' to the pingo scar. In the next paragraph, the actual scope of a rampart will be discussed and a finer definition for 'in association' will be suggested.

Performing such a suggestively exact calculation was considered necessary for a number of reasons, the first being that simply measuring ramparts on maps will give insufficient data. Most ramparts have been disturbed or flattened and will not represent the rampart's width as it must have been in prehistory. A second problem is the result of the location of pingo scars on the banks of (former) river systems and in stream valleys. On contour maps made with GIS, it is often hard or impossible to distinguish (part of) the rampart from the river bank or sand ridges they lie adjacent to. When in literature pingo scars are mentioned as being nearby archaeological sites, there is no way of knowing what is considered to be 'nearby' by the authors, or if a (palaeo) water source was responsible for the choice of location. Even setting aside the fact that there may be an underlying topography the prehistoric people used (that we need to study in order to get a better understanding of the landscape they moved in), we need to get a clearer picture of

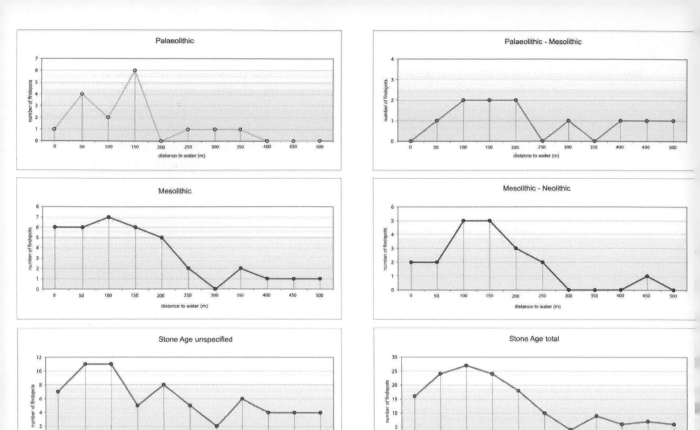

Figure 4. These graphs show the distance of finds from Palaeolithic through to the Neolithic to the water's edge. The bottom left graph shows finds from the Stone Age that have not been attributed to a specific period. The bottom right graph show the total of all finds from all Stone Age periods. Adapted from Slootweg 2008, 46.

what is 'close' and what constitutes as part of the pingo scar. Based on the characteristics found in literature, the maximum average scope or width of a pingo scar's rampart was calculated to be about 100m, but would often be much less as one very large example was included in the calculations (Table 2). Any finds beyond that distance will have to be seen as being outside the rampart and therefore, not directly associated with the pingo scar. A recent study into the location of prehistoric finds in regards to water sources has shown however that most finds concentrate in an area between 100 to 250m from the water's edge (fig. 4).

QUANTIFYING NEARNESS

Using the outcome of the calculations of the average width of a rampart for quantification of association is done using statistics. The results obtained from calculating the radius of ramparts were used as a starting point for a statistical test of several areas in Friesland.

In Archis, a Dutch database of the Cultural Heritage Agency, archaeological finds and monuments can be plotted on different maps and, when plotting them on the geomorphological map, the link between pingo scars and archaeology seems to be missing. A simple statistical test was performed using such maps based on the geomorphological map and registered archaeological finds. The aim of the test was to determine whether or not there is a significant difference between the amount of finds reported in or directly adjacent to pingo scars and the amount of finds not associated directly with pingo scars.

The null hypothesis (H_0) is that there is no statistically significant difference between the number of finds near pingo scars and finds in other locations. The alternative hypothesis (H_1) is that there is a significant difference. Because the H_1 is expected to occur in one way only (namely, the significant difference will be in the association of pingo scars with archaeological finds, not the area outside the pingos), the test was performed one-tailed. In statistics, there are two classes of mistakes possible as a result of an either too strict or too broad significance level. An α mistake means the null hypothesis is rejected although it is correct, whereas a mistake of the β kind means the null hypothesis is wrongly accepted (Siegel & Castellan 1988, 8-9). The significance level in this case should be the stricter .01 level to correct for mistakes of the β kind, because of the diversity in sampling due to the provenance of the data: coming from an open source such as ARCHIS, chances for differences in methods of collecting and sampling are large.

Several geomorphological maps of areas with a lot of known pingo scars were created. The areas chosen have the highest density of pingo scars in Friesland (and indeed in the Netherlands) and in-

Figure 5. Example of one of the maps made in Archis, showing pingo scars as water-filled depressions on the geomorphological map. Archaeological finds present in the Archis database are shown on the map as red dots. One square on the map represents one hectare (100 x 100m). Created in Archis, Rijksdienst voor Cultureel Erfgoed. *See also full colour section in this book*

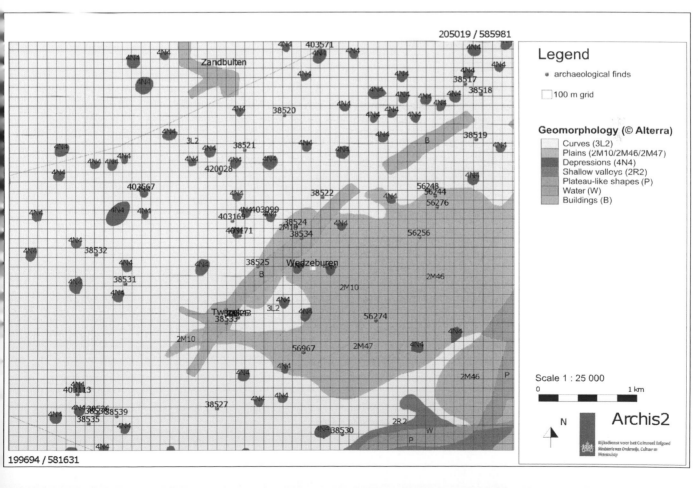

clude the environs of Twijzelerheide, Waskemeer, Siegerswoude, Surhuisterveen, Bergum, Wijnjewoude, Drogeham and Ureterp. The locations of registered finds were plotted onto these maps in Archis (fig. 5).

The criterion used in deciding whether a find was in association with a pingo scar or not was that it lay within 102 metres of the edge of the pingo scar. This number is the maximum average width of ramparts of the pingos in Table 1 at a rampart height of 0.5m. The smallest height was chosen for this exercise, as it gives the biggest area for a rampart. This way, the chances of a find being within the width of a rampart and thus being in association with a pingo scar were greatest. Using a higher rampart in the calculations would lead to a smaller area of that rampart, thus diminishing the chances of archaeological material being found within the range of the rampart. By using the slightest height, the conditions for finding an association between pingo scars and archaeological finds were at its most favourable.

Each map represented an area of 23 km². Per map, the total area occupied by pingo scars including the average rampart width of 102m was calculated. $\prod r^2$ was used to calculate the area of each individual pingo scar. To ensure the best chances of an association between pingo scars and archaeological finds being revealed, the depression within the rampart of each of these pingo scars was taken out of the equation. The reason for that is to be found in a research bias: all of the pingo scars visible on the soil maps have a water-bearing depression. These depressions will therefore positively never be included in field work, as coring in them is time consuming and difficult (one has to use a boat or wait for a severe winter to freeze the water into ice strong enough to carry people). Including these water features in the area of the pingo scars could lead to the conclusion that there are no finds from these areas, which in the end could then lead to the conclusion that prehistoric people avoided these areas, because nothing will be found in them. That would however be the result of a lack of work of modern researchers, not of negative favouritism on the prehistoric people's part. The area that was included in the statistical analyses is therefore made up of the ramparts of all the pingo scars on the maps. As stated before, these have been calculated to be 102m wide on average at maximum.

The sum of all ramparts per map together is used as the area belonging to the pingo scars. The strongest statistical test for analysing 2x2 contingency tables is Fisher's Exact (Siegel & Castellan 1988, 95-104). This test was used in all cases. The results are displayed in Table 3. The p values range from .17 to .78, which means the relationship is highly insignificant (the significance level was .01), in other words, the distribution of finds has a great chance of being random rather than showing a specific preference for pingo scars. It therefore seems reasonable to accept the H_0 in all cases, which means that based on statistical analysis, there is no reason for assuming a significant difference between the number of archaeological finds on ramparts of pingo scars and those outside these features.

There are some comments to be made on the source of the data used in this statistical exercise. The Archis database is filled out by a large number of people over a long period of time and therefore, not all finds in the database have the same level of detail. Not all records in the database will really be archaeological finds either: some may only be secondary indicators such as charcoal or unworked flint. Besides that, not all material found will be entered in the database, like single finds by amateur archaeologists wanting to keep their find locations secret. However, it is the only national archaeological database we have and errors were deemed as possible on the one side as on the other. For instance, the chance that a find associated with a pingo scar is not a primary indicator is thought to be just as large as the chance that a find in the non-pingo area is not a primary indicator. Therefore, the outcomes of the statistical tests keep their validity, even if not all variables are known to the same detail.

Table 3. Results of Fisher's Exact test (one-tailed) of probability for eight areas with a high density of pingo scars in Friesland.

Twijzelerheide	km²	N finds	Total
Area outside pingos	21	25	46
Total pingo area	2	4	6
Total	23	29	52
Result Fisher's exact	p = .45		

Bergum	km²	N finds	Total
Area outside pingos	21	3	24
Total pingo area	2	0	2
Total	23	3	26
Result Fisher's exact	p = .78		

Drogeham	km²	N finds	Total
Area outside pingos	22	16	38
Total pingo area	1	2	3
Total	23	18	41
Result Fisher's exact	p = .41		

Wijnjewoude	km²	N finds	Total
Area outside pingos	22	23	45
Total pingo area	1	2	3
Total	23	25	48
Result Fisher's exact	p = .53		

Waskemeer	km²	N finds	Total
Area outside pingos	21	17	38
Total pingo area	2	3	5
Total	23	20	43
Result Fisher's exact	p = .43		

Surhuisterveen	km²	N finds	Total
Area outside pingos	22	12	34
Total pingo area	1	0	1
Total	23	12	35
Result Fisher's exact	p = .66		

Siegerswoude	km²	N finds	Total
Area outside pingos	22	29	51
Total pingo area	1	3	4
Total	23	32	55
Result Fisher's exact	p = .44		

Ureterp	km²	N finds	Total
Area outside pingos	22	19	41
Total pingo area	1	4	5
Total	23	23	46
Result Fisher's exact	p = .17		

The above table shows that all areas under investigation yield no significant statistical result for the association of archaeological finds and pingo scars. This means that the variation seen in the distribution of archaeological finds in these areas is the result of sampling, measuring and analytical errors, not of actual variation in the archaeological record. In other words, this does not imply that there is any statistical basis to assume prehistoric man favoured pingo scars over other spots in the landscape. It does however give a hint that the idea of pingo scars as a preferable location may be more of a modern-day archaeological provenance than an etic prehistoric view.

PINGO SCARS AND ARCHAEOLOGY: ASSOCIATION OR ASSUMPTION?

With all prerequisites in place for research into the relationship between prehistoric man and pingo scars, research into the origin of this idea can commence. After conducting a literature study and talking to international scholars in both archaeology and geology, the pre-emptive conclusion had to be that the idea of the existence of this relationship is limited to the Netherlands. The suspicion arose that the emphasis put on pingo scars as important prehistoric features is a modern artefact rather than a prehistoric one. Academic archaeological publications of excavations of ramparts or infills of pingo scars are unknown. Pingo scars are mentioned in results of prospective research in gray literature, but none of those have been excavated. The archaeological material presented in these reports is hardly ever unequivocally connected to the pingo scars, and often concerns secondary indicators.

The assumed association between pingo scars as isolated features in a prehistoric landscape and man's use thereof seems highly contestable. Another hypothesis is that pingo scars as isolated elements within a landscape had no traceable special attraction for prehistoric people and the reasons for choosing a certain location for a certain task has to be found elsewhere. The hypothesis that there is no direct association between pingo scars and archaeology could only be tested adequately if there was a clear definition of the surface and ground cover of a pingo scar and its rampart. A literature survey was performed to gather data on the characteristics of pingos and pingo scars. These characteristics were then used as a basis for defining what can be seen as 'part of a pingo scar'. To avoid discussions about whether or not something should be considered in association with a pingo scar, these characteristics where also used as a basis for a statistical test. The aim of the test was to see whether there was a significant difference between archaeological finds from within the 'pingo scar area' and those outside that area.

A BROADER VIEW

There is no doubt that valuable source pingo scars act as geological elements and as an archive of ecological evidence in earlier times (See for instance Bakker 2003; Hoek 1997; Mook-Kamps & Bottema 1987; De Gans & Cleveringa 1981). Their value as archaeological archives however remains to be seen. Based on the results of this study there are little unambiguous associations between pingo scars and archaeological sites. Most finds associated with pingo scars come from disturbed ramparts or are surface finds in the vicinity of pingo scars. From inside pingo scars, hardly any material is known. There may be a research bias here due to the fact that most pingo scars these days are lakes or completely filled with peat. Research on areas deeper within the depression will be difficult and expensive and therefore not often undertaken. In the cases where pingo scars have been dug out for other than archaeological reasons, finds are seldom mentioned either.

Most artefacts associated with pingo scars are flint tools or débitage. From the periods in which the pingo scars may have been used in prehistoric times (mostly Late Palaeolithic and Early Mesolithic), we often find flint to be the only finds category available. As with all sites above, the flint assemblage is usually made up of surface finds. This assemblage is sometimes combined with artefacts from excavations, though that is virtually never the case in pingo scar research. Most finds in areas around pingo scars have been moved either vertically or horizontally, or both, which makes these find spots unsuitable for spatial analysis. The information that can be gathered from the flint assemblages is in most cases limited to a typological overview of the site, with no direct indications for further use of the environment by prehistoric man.

Pingo scars have a definite value as a research area that is well-defined, both in terminology and physical appearance. Its enclosed structure, combined with organic infill, makes it a potentially undisturbed catchment area for pollen and for dating sediments. If archaeological organic materials such as bone, wood or fibres are present inside pingo scar infill, they have a good chance of being preserved. Even though all these conditions for pingo scars are good to be of value for archaeology, the fact is that there is no reason to assume a particular prehistoric preference for pingo scars. The idea that these small ponds were favoured as a water source and a location on the transition between higher, drier grounds and wet areas seems to be based on their visibility in the current landscape rather than their place in that of prehistoric times. Besides the considerable amount of pingo scars located on the Drents-Friese till plateau, pingo scars in the Netherlands are practically always situated on the slopes of stream valleys or river banks. Most of this relief was formed in the Saalian and is invisible in the current landscape. However, the original relief from that period can be traced on contour maps made with the 'Actueel Hoogtebestand Nederland' (AHN: Digital Elevation Model) as a basis. It seems far more likely that river systems as a whole had a strong attraction for prehistoric people. To focus archaeological research on the small parts that are still visible of that prehistoric landscape is neglecting the landscape in its entirety.

In conclusion, these Ice Age features can only be valuable from an archaeological point of view if we see them for what they are: indicators of a now mostly disappeared landscape, capable of inferences to that landscape on a very local scale. Clues for specific prehistoric use or indeed a preference for pingo scar locations have not been found, and are not expected to be unearthed. Rather, this preference is found in modern research questions and as such is a far likelier artefact of these times than of prehistoric origin.

ACKNOWLEDGEMENTS

Dr A.H. de Vries of the research group Molecular Dynamics (Faculty of Mathematics and Natural Sciences) of the University of Groningen is thanked for his help on the formulas and for reading a first draft of this paper. Comments by Dr S. Bohncke and especially Prof. Dr J. Vandenberghe, both of the Faculty of Earth and Life Sciences of the VU University Amsterdam, greatly improved the paper.

REFERENCES

Bakker, R. 2003. *The emergence of agriculture on the Drenthe Plateau: A palaeobotanical study supported by high-resolution 14C dating*. Archäologische Berichte 16, Bonn.

Bateman, M.D. & J. van Huissteden. 1999. The timing of last-glacial periglacial and aeolian events, Twente, eastern Netherlands. *Journal of Quaternary Science* 14 (3), 277-283.

Castel, I.I.Y. & M. Rappol. 1992. Het Weichselien – Drenthe in de ijstijd. In M. Rappol (ed.) *In de bodem van Drenthe – Geologische gids met excursies*. Lingua Terrae, Amsterdam.

French, H.M. & L. Dutkiewicz. 1976. Pingos and pingo-like forms, Banks Island, Western Canadian Arctic. *Biuletyn Peryglacjalny* 26, 211-222.

Gans, W. de. 1988. Pingo Scars and Their Identification. In: M.J. Clark (ed.) *Advances in Periglacial Geomorphology*. John Wiley & Sons, Chicester, 299-322.

Gans, W. de & P. Cleveringa. 1981. Stratigraphy, palynology and radiocarbon dating of middle and late Weichselian deposits in the Drentse Aa valley system. *Geologie & Mijnbouw* 60, 373-384.

Hoek, W.Z. 1997. *Palaeogeography of lateglacial vegetations: aspects of lateglacial and early holocene vegetation, abiotic landscape, and climate in the Netherlands*. Koninklijk Nederlands Aardrijkskundig Genootschap, Utrecht.

Jorgenson, M.T. & T.E. Osterkamp. 2005. Response of boreal ecosystems to varying modes of permafrost degradation. *Canadian Journal of Forest Research* 35. NRC Press, Vancouver, 2100-2111.

Kort, I. de. 2010. *The Neolithic in a Pollen diagram from the Opperkooten-pingo, Friesland*. Master Thesis. Department of Palaeoclimatology and Geomorphology, Faculty of Earth and Life Sciences VU University, Amsterdam.

Maarleveld, G.C. 1976. Periglacial phenomena and the mean annual temperature during the last glacial time in The Netherlands. *Biuletyn Peryglacjalny* 26, 57-77.

Maarleveld, G.C. & J.C. van den Toorn. 1955. Pseudo-sölle in Noord-Nederland. *Tijdschrift van het Koninklijk Nederlandsch Aardrijkskundig Genootschap*, Tweede Reeks, Deel LXXII, no. 4, 344-360.

Mackay, J.R. 1972. The World of Underground Ice. *Annals of the Association of American Geographers* 62 (1), 1-62.

Mackay, J.R. 1979. Pingos of the Tuktoyaktuk peninsula area, Northwest territories. *Géographie Physique et Quaternaire* 33 (1), 3-61.

Mook-Kamps, E. & S. Bottema. 1987. Palynological investigations in the Northern Netherlands. *Palaeohistoria* 29, 169-172.

Pissart, A. 1988. Pingos: An Overview of the present State of Knowledge. In M.J. Clark (ed.) *Advances in Periglacial Geomorphology*. John Wiley & Sons, Chicester, 279-297.

Raemaekers, D.C.M. 2005. Boren in een pingoruïne bij Augustinusga. In G. de Langen & F. Veenman, *Archeologische kroniek van Fryslân over 2003 en 2004*. De Vrije Fries 85, 203-204.

Ross, N., C. Harris, H.H. Christiansen & P.J. Brabham. 2005. Ground penetrating radar investigations of open system pingos, Adventdalen, Svalbard. *Norsk Geografisk Tidsskrift* 59, 129-138.

Siegel, S. & N.J. Castellan. 1988. *Non-parametric Statistics for the Behavioral Sciences*. McGraw-Hill, New York, 2nd edition.

Slootweg, E. 2008. *Het Actueel Hoogtebestand Nederland als bron voor het verleden. Analyse van locatiegegevens van Paleolithische en Mesolithische vindplaatsen in het Pleistoceen Noordenveld op basis van het AHN*. Master thesis, RACM, Amersfoort.

Steenbeek, P., P. Cleveringa, & W. de Gans. 1981. Terreinvormen in Friesland uit de laatste ijstijd. *It Beaken* 43 (1), 249-271.

Svensson, H. 1976. Pingo problems in Scandinavian countries. *Biuletyn Peryglacjalny* 26, 33-40.

Waltham, T. 2003. Pingos of Tuk. *Geology Today* 19 (6), 212-215.

Watson, E. 1972. Pingos of Cardiganshire and the Latest Ice Limit. *Nature* 236, 343-344.

Improving temporal, chronological and transformational frameworks

2.1 Pre-industrial Charcoal Production in southern Brandenburg and its impact on the environment

Authors

Horst Rösler[1], Eberhard Bönisch[1], Franz Schopper[1], Thomas Raab[2] and Alexandra Raab[2]

1. Brandenburgisches Landesamt für Denkmalpflege und Archäologisches Landesmuseum, Zossen, Germany
2. Brandenburgische Technische Universität, Cottbus, Germany
Contact: franz.schopper@bldam.de

ABSTRACT

Due to modern lignite mining in southern Brandenburg and northern Saxony (East Germany), entire landscapes are being destroyed. In the area of the lignite extraction, the BLDAM (Brandenburgisches Landesamt für Denkmalpflege und Archäologisches Landesmuseum) concurrently carries out large-scale archaeological surveys and excavations to study and document evidence of past land use by prehistoric and historic cultures. On the area of the Jänschwalder Heide (Lower Lusatia, southern Brandenburg) one of the largest archaeologically investigated charcoal production areas in Germany was discovered, demonstrating the great intensity of energy production in historical times. The charcoal was probably used in the nearby ironworks of Peitz, where bog iron ore was smelted since 1567. Meanwhile, remnants of more than 400 charcoal hearths are excavated. To charge those piles, large areas had to be cleared, which certainly had major consequences for the environment and the character of the landscape. At least for a while, the vegetation was completely absent on the deforested areas, which were used as farmland although the soils are very sandy and poor in nutrients. Wind-blown sediments covering the charcoal pile relics prove that clearing and agricultural use has induced aeolian soil erosion and the remobilisation of Quaternary sands.

One of the main aims of the ongoing investigation is to build up a chronological framework of the former charcoal production. These findings have to be correlated with the major phases of the landscape dynamics, which are documented by the relics of soil erosive landforms, human-induced aeolian sediments, and buried soils.

KEYWORDS

lignite mining, charcoal burning, Lower Lusatia, land-use history, anthropogenic impact

INTRODUCTION

Opencast lignite mining results in the total destruction of cultural landscapes and even small towns. Therefore, over the past years systematic archaeological research has been carried out in the opencast pits in Lower Lusatia (southern Brandenburg, Germany), prescribed by the regulations of the Brandenburgisches Denkmalschutzgesetz (BbgDschG). For the opencast pit Jänschwalde, it is expected that during the year 2010, an area of approximately 200 ha will be utilised (4 km length of the opencast pit and 500 m width of the excavated stripe). However, the large-scale impact of lignite extraction offers the opportunity for archaeologists to study landscape and settlement history as a whole instead of recording single findings and find spots.

During the archaeological investigations in the apron of the opencast mine Jänschwalde, presumably the largest charcoal production area in Central Europe, was detected (Rösler 2008). Meanwhile the remnants of more than 400 upright circular kilns have been prospected and excavated. The number of charcoal kilns suggests an immense impact on the environment caused by charcoal burning and related activities like logging.

In former times, charcoal burning was a widely distributed practice, carried out within forested parts of Europe (Groenewoudt 2005). Charburner was a respectable profession due to the importance of charcoal for energy supply and aggregates in pre-industrial production facilities like glass kilns, brickworks and iron works. This is also proven by the family names 'Kohler' or 'Köhler', which are common in Germany. In Central Europe, charcoal was mainly produced in the wooded low mountain ranges like the Erz Mountains, the Black Forest, the Vosges, the Jura Mountains, the Thuringian Forest, the Upper Palatinate Forest and the Bavarian Forest (e.g. Rösler 2008). Investigations by Groenewoudt (2005) show that charcoal burning was also carried out in the eastern part of the Netherlands, although the investigations in Jänschwalde suggest that former charcoal production in Lower Lusatia exceeds the charcoal production in the low mountain ranges. However, historical charcoal burning is not as well investigated as in other regions like the Netherlands (Groenewoudt 2005) or the Black Forest (Ludemann 2009).

In Brandenburg, the demand for charcoal resulted in a considerable development of charcoal burning. The earliest evidence of targeted charcoal burning in larger quantities in Lower Lusatia derives from the opencast pit Welzow-Süd. On the Wolkenberg, charcoal piles were found at the individual smelting furnace sites of a Germanic smelter centre (Lipsdorf 2001). The remains of pit kilns (German: *Grubenmeiler*) document the use of the oldest technique of traditional charcoal production (Lipsdorf 2001; Spazier 1999). From the 16th century onwards, charcoal hearths (German: *Platzmeiler*), were used. Charcoal burning for industrial use, particularly in smelters, is also known from several sites in Lower Lusatia, e.g. at Lauchhammerschlag near Altdöbern, where charcoal was produced for the iron work Lauchhammer, built in 1725 (Lipsdorf 2001).

First of all, this paper gives a résumé of the current state of research on historical charcoal burning in the Jänschwalde area (Lower Lusatia) which is mainly based on archaeological survey. Furthermore, since 2010 the Research Group 'Anthropogenic Landscape Development and Palaeoenvironmental Research', the Chair of Geopedology and Landscape Development, the International Graduate School (all BTU Cottbus) and the BLDAM have jointly studied the environmental consequences of charcoal burning. The main objectives are to investigate the spatial extent of the affected area, to compile a chronology for charcoal burning, to study the effects of charcoal burning on the landscape (e.g. reactivation of aeolian dynamics, soil erosion, deforestation) and on soil physics and chemistry in matters of soil productivity. Therefore the research concept comprises four different scientific approaches: 1) an archaeological approach, 2) a pedological-geomorphological approach, 3) an archival research and 4) a GIS-based reconstruction of past environmental conditions and anthropogenic induced landscape change.

STUDY AREA AND GEOGRAPHIC SETTING

The study area Jänschwalde (opencast pit; 51°47′31N′′, 14°32′23′′E) is situated c. 10 km northeast of Cottbus (Lower Lusatia, southern Brandenburg, Germany) (fig. 1). Opencast mining started during the 1970s and the affected area is c. 6015 ha. The investigated charcoal production area lies within the opencast pit,

Figure 1. Location of the study area opencast pit Jänschwalde (Lower Lusatia, southern Brandenburg), situated c. 5 km northeast of Peitz.

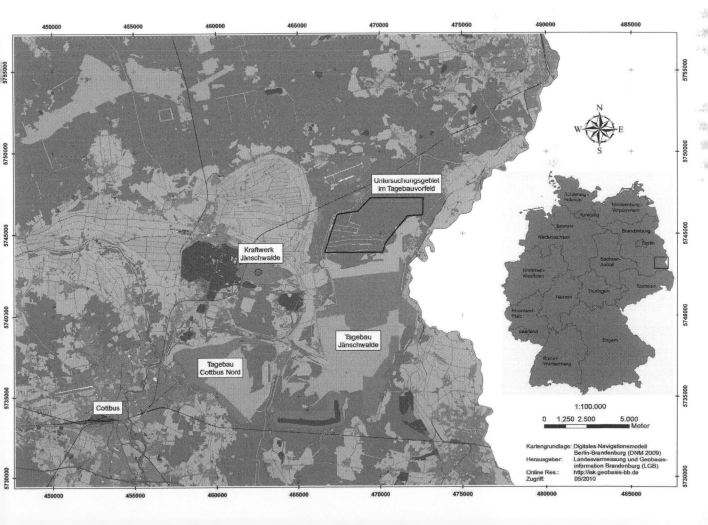

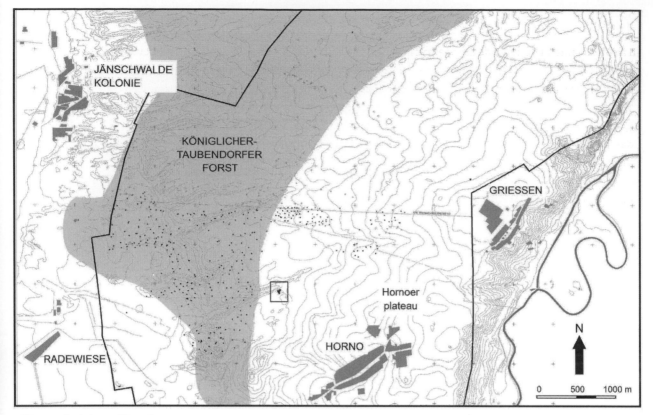

● excavated charcoal hearths
○ prospected charcoal hearths

Figure 2. Overview of the location of prospected and excavated charcoal piles in the opencast pit Jänschwalde (state: 01.04.2010). The small rectangle marks the location of the large charcoal hearths assembly shown in Figure 4.

east of Peitz (fig. 1). In former times, this area was named 'Königlicher-Taubendorfer Forst' or 'Tauerscher Forst' and was in possession of the royal family (fig. 2).

Generally, Lower Lusatia is characterised by a continental climate with an annual mean air temperature between 8 and 9 °C and the temperature amplitude is about 18 °C. The annual precipitation sum ranges between 510 and 610 mm (Scholz 1962).

The landscape was formed by Pleistocene glaciations. The opencast pit Jänschwalde lies on the terrain in the border area of Old and Young drift (Lippstreu et al. 1994; Nowel 1995). The Hornoer plateau (up to 110 m a.s.l., fig. 2) is a relic of the ground moraine deposit of the Saale-complex (Cepek et al. 1994). It is a gentle undulated plain with elevations up to 110 m a.s.l. divided by basins at about 75 m a.s.l (Lippstreu et al. 1994). The near subsurface of the Hornoer plateau is built up by subglacial tills (i.e. ground moraine) as well as sandur and fluvio-glacial deposits, which were altered by periglacial geomorphological processes during the Weichselian. The plateau is in some extent bordered by the River Neiße in the east, and in the west by the broad plain of the Weichselian sandur, the so-called Taubendorfer Sander (Lippstreu et al. 1994). Directly to the south of the Hornoer plateau adjoins to the Baruther Urstromtal, which is filled up with meltwater deposits deriving from the Late Weichselian continental ice sheet (Brandenburger Stadium) (Lippstreu et al. 1994; Nowel 1995).

In addition, inland dune fields are present in the study area. Sedimentological, stratigraphical and pedological investigations were carried out in the dune field area in the apron of the opencast pit Jänschwalde, at the western border of the Düringsheide between the River Malxe in the west and the southern cape of the Hornoer plateau in the east (Poppschütz 2001). Postglacial landscape history studies in the region along the River Malxe between the villages Grötsch and Heinersbrück were carried out in connection with archaeological investigations (Stapel 2000a, b, c; Bittmann 2000). Research on anthropogenic impact on the landscape during (pre-)historic times were for example conducted on V-shaped erosion valleys of the Hornoer plateau (opencast pit Jänschwalde) and on late medieval ridge and furrow in the opencast pit Cottbus-Nord (Woithe & Rösler 2001; Geldermacher et al. 2003; Woithe 2003; Bönisch 2005).

The dominant soils on the sandy-loamy ground moraines are cambisols, luvisols and podsols. The mainly poor soils are partly truncated by soil erosion. Buried soils are present under colluvial deposits on the slopes or below sand dunes. In depressions, wet lowlands and on the foot of the dunes iron- and iron-humus horizons may occur (Woithe 2003).

METHODS AND TECHNIQUES

The ongoing archaeological investigation aims to survey the cultural landscape as a whole. The systematic research comprises three stages of investigation: 1) prospection by site inspection and aerial photo analyses, 2) sondages (in stripes and in grids) and test trenches carried out manually and/or with an excavator, 3) open area excavations of selected areas which are especially likely to produce good results or to answer specific questions. The study area is surveyed with differential GPS (Global Positioning System). Additionally, airborne laser scanning maps, courtesy of the Vattenfall Europe Mining Group, are used for orientation, topographic information and mapping. For absolute age determination with radiocarbon and dendrochronological dating, charcoal samples are collected from buried agricultural soil horizons and the remnants of selected charcoal kilns.

ARCHAEOLOGICAL BACKGROUND OF THE OPENCAST PIT JÄNSCHWALDE

The excavations in the apron of the opencast pit Jänschwalde are yielding plenty of findings (fig. 3). Archaeological evidence of prehistoric settlement in the study area includes numerous Mesolithic chipping floors. For the Neolithic Period and the Early Bronze Age, burial grounds were found with arrowheads and flint dirks as grave goods (Rösler 2001). Furthermore, Bronze Age post buildings with granaries for cereals as well as graves were detected. Wells were found, which served as water supply on the Hornoer plateau (Bönisch 2004). In addition, a Germanic village from the 3rd and 4th century was excavated. The ground plans of the buildings, the wells, a cereal mill and much more, complete the picture of the way of living of the Germanic people in Lower Lusatia. Particularly interesting is the discovery of a forge for the production of ornaments, which was present within the settlement. In addition, *fibulae* were found, which were part of the period costume of the entombed dead persons on the graveyard nearby (Schultz 2008).

Figure 3. Aerial photograph of the apron of the opencast pit Jänschwalde with archaeological longitudinal sections and ground plans of charcoal piles (photo: H. Rösler). *See also full colour section in this book*

RESULTS AND DISCUSSION

Archaeological features of charcoal burning in the area of the opencast pit Jänschwalde

On the area of the opencast pit Jänschwalde, up to now, more than 400 remnants of circular upright hearths have been documented by prospection and excavation. The location and spatial distribution of the kiln sites is shown in Figure 2, which displays preliminary results. The majority of the charcoal piles are present in the former 'Königlicher-Taubendorfer Forst'.

The remnants of the charcoal kilns are characterised by black, charcoal-bearing layers or by circular to oval surrounding or interrupted ditches, filled with charcoal. The ground plans of selected charcoal hearths are shown in Figure 4. The circular ditches were dug to extract the surrounding soils and substrates to cover and seal the stack. Following the lightning of the stack and the carbonisation and cooling down processes, the charcoal piles were opened with pokers and the ditches were backfilled with charcoal remains. The relicts of the charcoal hearths are clearly distinct as black circles present in the light sandy substrates (figs. 5, 6).

In addition, there are diverse finds of pits and post-settings, both inside and outside the circles. Particularly, single circular and trough-shaped pits are present with *c.* 1.5 to 1.9 m in diameter, which were

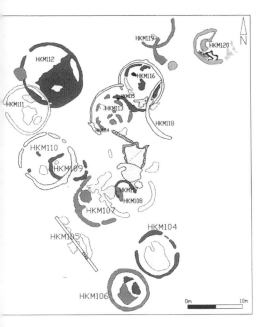

Figure 4. Large charcoal hearths assembly in the apron of the opencast pit Jänschwalde. The different circular ground plans are clearly distinct in the surrounding substrate by gray-coloured features and charcoal filling (drawing: M. Pingel). For the location of the large charcoal hearths assembly, see Figure 2.

Figure 5. Two overlapping ground plans of charcoal piles proving the multiple-shift usage of a charcoal burning site (photo: R. Piskorski).

Figure 6. Excavation of a charcoal pile (find spot 10, charcoal pile 2) buried under a 1m thick aeolian sediment. The covering by the dune sands indicates the remobilisation of Quaternary sands as a consequence of clearing for charcoal production (photo: H. Rösler).

filled with charcoal and sometimes with tar. These pits were found in different orientations directly in front of the circular ditches. Probably, these pits were used as pits to light the fire (German: *Zündfeuergruben*) or as pits to collect the tar (German: *Teerauffanggruben*). The overlapping ground plans of charcoal piles (see figs. 4, 5) prove the multiple-shift usage of a charcoal kiln site.

Concerning the age of the charcoal kilns, it is assumed that they have a medieval to modern age as suggested by the condition of the remains (Lipsdorf 2001). To date, only one charcoal kiln is absolutely dated by dendrochronological age determination. It was determined on a charcoal piece (pine) derived from the ditch of charcoal hearth 2 (find spot 10, fig. 6) dating to the year 1850 AD (Lab.-Nr. C44165), which fits well with the assumed age. It is presumed that not all the charcoal kilns are contemporaneous, but rather date from several centuries. Therefore, for further research more absolute age determinations are required.

The inner diameters of the charcoal kilns range from 3 to 20 m and they can be subdivided into three size classes: (a) small (3-8 m), (b) medium (8-14 m) and (c) large (14-20 m). The classification is preliminary based on the evaluation of one hundred ground plans. The majority of the charcoal hearths are large, with an inner diameter of up to 20 m. This result points at a production of charcoal in large quantities for industrial use. However, this has to be substantiated by further research. In contrast, the charcoal produced in the smaller hearths was possibly used by individuals for domestic application or by smaller craft producers.

Most probably the larger part of the charcoal produced in the study area was taken to the former ironwork Peitz nearby (Lipsdorf 2001). This smelter existed from the 16th century until 1858 (Lipsdorf 2001). Furthermore, there is evidence that the ironwork Peitz operated its own charcoal burning site, situated in the 'Königlicher Taubendorfer Forst', north-east of the ironwork. It is a fact that the ironwork was supplied with charcoal from the Großen Peitzer Heide. Prior to 1600, oak wood was used for charcoal production, and pine wood later on (Reichmuth 1986). So far, it is not clear if and to what extent the charcoal burning sites south-east of Peitz were also operated by the ironwork. At Peitz bog iron ore, a typical raw material of the lowlands in Brandenburg, was used for iron production. Today, the former mining sites are not visible in the landscape, because they are levelled by natural processes or filled.

For charcoal burning in upright circular kilns the site selection is crucial (Lipsdorf 2001). The considerations include the location (e.g. substrate, inclination of the ground surface) and the availability of wood and water. Concerning the location, a level surface and a loamy-sandy substrate are needed for the aeration of the stack during the charring process. In this respect, the charcoal burning area in the open-cast pit Jänschwalde offers ideal conditions. The investigations of charcoal hearths near Horno showed that they were situated on the slopes of the Hornoer plateau, which are dissected by V-shaped valleys caused by erosion. This suggests that hillside situations were preferred, probably because of the natural windbreak since the control of the oxygenation was the major difficulty. The charcoal hearths were mainly present on the flat footslope between 65 and 70 m a.s.l., extending over a linear line of 1.3 km length. The inclination of 20 to 30 cm from the centre of the stacks was ideal for the drainage of condensation water and tar and guaranteed more efficient oxygenation. The charcoal hearths sites were used repeatedly since a new setting meant both additional work and expense (Lipsdorf 2001). A further important precondition for the charcoal site selection was the availability of wood and the proximity to the wood resource. Therefore, prior to the beginning of charcoal burning, the presence of extended woodland is inferred. Investigation of the charcoal sites on the slopes of the Hornoer plateau showed that the majority of the kilns were situated at the rim of the V-shaped erosion valleys and close to brooks, but the proximity of the charcoal kiln sites to water supplies presumably played a secondary role in the site selection (Lipsdorf 2001).

Though the site selection for charcoal burning is important, as described in detail above, at this stage the distribution of the charcoal hearths seems to be more related to the historic land tenure than to distinct natural landscape units. This question is part of the ongoing research.

The effects of charcoal production on the environment in the area of the opencast pit Jänschwalde

Large quantities of wood are required for charcoal production. The computational model for a rough calculation of the woodland consumption for one charcoal kiln is shown in Figure 7. It is based on data from current forest management in Brandenburg (Frommhold 2010). Accordingly, a woodland consumption of *c.* two hectares per stack is calculated, demonstrating the intensive utilisation of the woodland area. Actually, the calculated numbers are extrapolations, and a transfer of this value to the entire charcoal pile area is only possible with reservations. A more exact calculation of the woodland consumption is an essential part of the ongoing research. Nevertheless, charcoal production and related activities must inevitably have had tremendous consequences on the environment.

One of these consequences was the remobilisation of Quaternary sands, which was initiated by forest clearance. Quaternary sands are mainly present in the west of the study area, as Taubendorfer Sander and south of the Hornoer Plateau in the Baruther Urstomtal, which maybe the source of the relocated sands. The causal connection of clearing for charcoal production and wind erosion is for example proven by the find of a charcoal pile (1850 AD, Lab.-Nr. C44165, dendrochronological age) buried below a *c.* 1m thick aeolian sediment cover (Fig. 6). Furthermore, in the study area former ploughing horizons (fAp-

Figure 7. Computational model for the woodland consumption of one charcoal pile. For the rough calculation the geometric shape of a truncated cone is used. The model is based on data for pine from current forest management in Brandenburg (Frommhold 2010). The most frequent ground plans found have an average inner diameter of 14m, therefore this diameter is used for calculation.

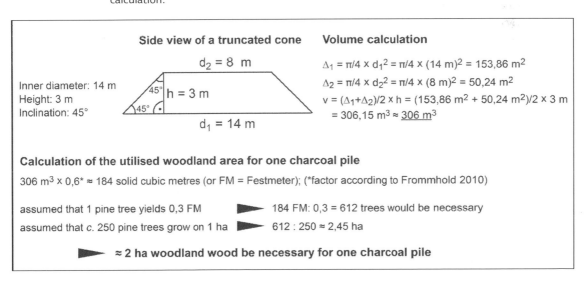

Side view of a truncated cone

Inner diameter: 14 m
Height: 3 m
Inclination: 45°

$d_2 = 8$ m
$45°$ $h = 3$ m
$45°$
$d_1 = 14$ m

Volume calculation

$A_1 = \pi/4 \times d_1^2 = \pi/4 \times (14\,m)^2 = 153{,}86\,m^2$
$A_2 = \pi/4 \times d_2^2 = \pi/4 \times (8\,m)^2 = 50{,}24\,m^2$
$v = (A_1 + A_2)/2 \times h = (153{,}86\,m^2 + 50{,}24\,m^2)/2 \times 3\,m$
$= 306{,}15\,m^3 \approx \underline{306\,m^3}$

Calculation of the utilised woodland area for one charcoal pile

306 m³ x 0,6* ≈ 184 solid cubic metres (or FM = Festmeter); (*factor according to Frommhold 2010)

assumed that 1 pine tree yields 0,3 FM ▶ 184 FM: 0,3 = 612 trees would be necessary

assumed that *c.* 250 pine trees grow on 1 ha ▶ 612 : 250 ≈ 2,45 ha

▶ ≈ 2 ha woodland wood be necessary for one charcoal pile

Figure 8. Buried agricultural soil (fAp-horizon) covered with an up to 150 cm thick aeolian sand cover (photo: A. Raab).

horizons) buried by dune sediments are present. Radiocarbon dating (^{14}C-AMS) of charcoal particles from the fAp-horizon of the cross-section shown in Figure 8 resulted in an age of 970±44 a BP (989 AD-1162 AD, 2 Sigma, Erl-15502). Moreover, in the opencast pit Cottbus-Nord, ridge and furrow were found, also covered by aeolian sand. Based on ceramics found in the buried topsoil, the agricultural use is dated to the 15th to 16th century (Geldermacher et al. 2003). In conclusion, the investigations demonstrate a highly dynamic landscape caused by anthropogenic impact during the past centuries.

DISCUSSION, CONCLUSIONS AND PROSPECTS

The opencast pit Jänschwalde is probably the largest charcoal production area in Central Europe. Primarily, the excavations in the apron of the opencast pit Jänschwalde provide evidence for the use of upright circular kilns for charcoal burning. The large number, in total more the 400 charcoal hearths, and especially the occurrence of big charcoal kilns situated in the former 'Königlicher-Taubendorfer Forst' hints at the connection between industrial charcoal production and the ironwork Peitz. The ongoing excavations in the opencast pits south of Peitz supply further evidence of the interrelation with the ironwork. For example, in the opencast pit Cottbus-Nord, the stream that drove the water wheels and therefore the machines of the Hammerwerk (water-powered drop forge) is currently being investigated. This artificial channel of the River Spree is a brilliant engineering achievement.

Charcoal burning on such a large scale inevitably caused damage to the environment. One of the unintended results was the remobilisation of Quaternary sands by wind erosion, initiated by forest clearance. The sands went on to cover former agricultural soils and charcoal burning sites. This attests to large-scale man-induced landscape change during past centuries. Besides the implications for the soils and landscape dynamics, by forest clearance, by charcoal burning and by the incorporation of charcoal fragments in the former top soils, the carbon cycle was also affected, at least on a regional scale. Finally, deforestation certainly had an impact on the water balance, which has not yet been investigated.

First of all, one of the main targets concerning further research in the opencast pit Jänschwalde is to establish a chronology for the land use history, especially for charcoal burning, based on absolute age determinations by dendrochronological and radiocarbon dating. Kiln site anthracology (analysis of wood charcoal) could supply complementary information on forest vegetation and woodland history.

With the continuation of the opencast pit Jänschwalde and passing the village of Grießen in a northerly direction, the archaeological work will concentrate on the Hornoer plateau and the adjacent western boundary areas. The time pressure caused by the mining activity in the apron of the opencast pit affords an effective strategy for the archaeological survey and for the accompanying geomorphological and pedological investigations. Concerning the latter, the issues are ideal for the application of modern, rapid and low-cost handheld techniques of soils and sediment analysis, e.g. handheld X-ray fluorescence (XRF) analyses. Finally, the results from the studies combined with the historic map analysis are combined and evaluated using a Geographic Information System (GIS) to reconstruct the dimension of charcoal burning and its impact on the environment.

In conclusion, the opencast pits in Lower Lusatia offer the outstanding opportunity for a comprehensive land use reconstruction. The large dimension of the area affected by lignite mining and the numerous outcrop situations caused by the lignite extraction provide an extraordinary insight into a complete landscape unit comprising geology, geomorphology, pedology and archaeology. This has created the opportunity for interdisciplinary cooperation between archaeologists, geographers, soil scientists and palaeobotanists, to the benefit of both the historical and natural sciences.

REFERENCES

Bittmann, F. 2000. Pollenanalytische Untersuchungen zur Landschaftsgeschichte des Malxetals. *Ausgrabungen im Niederlausitzer Braunkohlenrevier* 1999, 25-28.

Bönisch, E. 2004. Häuser, Speicherplätze und Siedlungsmuster. Neues zu Siedlungen der Lausitzer Kultur links der Neiße. *Biblioteka Archeologii Srodkowego Nadodrza* 2, 91-120.

Bönisch, E. 2005. Frühe anthropogene Veränderungen der Landschaft der Niederlausitz. *Change, Internationales Symposium Landschaft und Energie*, Tagungsband, 79-86.

Cepek, A.G., Hellwig, D. & W. Nowel. 1994. Zur Gliederung des Saale-Komplexes im Niederlausitzer Braunkohlenrevier. *Brandenburger Geowissenschaftliche Beiträge* 1 (1), 43-83.

Geldmacher, K., Woithe, F. & H. Rösler. 2003. Die Dokumentation von Bodendenkmalen und Archivböden im Niederlausitzer Braunkohlenrevier. *Petermanns Geographische Mitteilungen* 3, 44-49.

Frommhold, H. 2010.: Lecture notes. http://www6.fh-eberswalde.de/forst/forstnutzung/ifem/homepage/dokumente/vorlesung/pdf/3.pdf

Groenewoudt, B. 2005. Charcoal burning and landscape dynamics in the Early Medieval Netherlands. *Ruralia VI, Arts and crafts in Medieval rural environment*, 327-337.

Lipsdorf, J. 2001. Köhler über der Kohle. Ausgrabungen von Holzkohlemeilern am Tagebau Jänschwalde. *Ausgrabungen im Niederlausitzer Braunkohlenrevier* 2000, 213-223.

Lippstreu, L., Hermsdorf, N., Sonntag, A. & H.U. Thieke. 1994. Zur Gliederung der quartären Sedimentabfolgen im Niederlausitzer Braunkohlentagebau Jänschwalde und in seinem Umfeld – Ein Beitrag zur Gliederung der Saale-Kaltzeit in Brandenburg. *Brandenburgische Geowissenschaftliche Beiträge* 1, 15-35

Ludemann, T. 2009. Past fuel wood exploitation and natural forest vegetation in the Black Forest, the Vosges and neighbouring regions in western Central Europe, Palaeogeography, Palaeoclimatology Palaeoecology, doi:10.1016/j.palaeo.2009.09.013

Nowel, W. 1995. Geologische Übersichtskarte des Niederlausitzer Braunkohlereviers. Lausitzer Braunkohleaktiengesellschaft Senftenberg (Hg), 3. Auflage 1995.

Poppschötz, R. 2001. Beobachtungen zur Dünenentwicklung östlich von Heinersbrück, *Ausgrabungen im Niederlausitzer Braunkohlenrevier* 2000, 43-54.

Reichmuth, G. 1986. Die Produktion im ehemaligen Eisenhüttenwerk Peitz. Geschichte und Gegenwart im Bezirk Cottbus, 103-112.

Rösler, H. 2001. Gräber der Schnurkeramik an der Hornoer Hochfläche. *Ausgrabungen im Niederlausitzer Braunkohlenrevier* 2000, 111-119.

Rösler, H. 2008. Zur Köhlerei für das Eisenhüttenwerk Peitz in Brandenburg. *Archäologie in Deutschland* 3, 36-38.

Scholz, E. 1962. *Die naturräumliche Gliederung Brandenburgs*. Potsdam.

Schulz, D. 2008. Verbrannt und zugeweht. Germanische Gräber bei Jänschwalde. *Ausgrabungen im Niederlausitzer Braunkohlenrevier* 2007, 177-187.

Spazier, I. 1999. Neue Ergebnisse aus dem germanischen Eisenverhüttungszentrum Wolkenberg. *Ausgrabungen im Niederlausitzer Braunkohlerevier* 1998, 97-103.

Stapel, B. 2000a. Als die Rentiere von Weißagk nach Horno zogen..., *Ausgrabungen im Niederlausitzer Braunkohlenrevier* 1999, 9-16.

Stapel, B. 2000b. Landschaftsgeschichte im Malxetal, *Ausgrabungen im Niederlausitzer Braunkohlenrevier* 1999, 17-24.

Stapel, B. 2000c. Die ersten Bauern an der Malxe, *Ausgrabungen im Niederlausitzer Braunkohlenrevier* 1999, 33-37.

Woithe, F. & H. Rösler. 2001. Bodenkundliche Untersuchungen überdünter Wölbäcker in den Fluren von Merzdorf und Dissenchen, Tagebauvorfeld Cottbus-Nord. *Ausgrabungen im Niederlausitzer Braunkohlenrevier* 2000, 197-202.

Woithe, F. 2003. *Untersuchungen zur postglazialen Landschaftsentwicklung in der Niederlausitz*. Dissertation, Universität Kiel.

2.2 Landscape transformations in North Coastal Etruria

Authors

Marinella Pasquinucci and Simonetta Menchelli

Dipartimento Scienze Storiche del Mondo Antico, Pisa University, Pisa, Italy
Contact: pasquinucci@sta.unipi.it

ABSTRACT

This paper concerns North coastal Tuscany (ancient *Etruria*), Italy. Multidisciplinary diachronic research provides evidence of palaeo-environmental changes, of water and risk management practices and of rural and urban landscapes both in the coastal district and the hinterland. From north to south the littoral is articulated in three sections: the Luni – Livorno shoreline, which prograded westwards from the 2nd-1st cent. BC up to about 1830; the Livorno terrace and Livorno-Castiglioncello coastal strip, which are rocky and stable; the Vada – Cecina shoreline, which is low and stable. In the Luni – Livorno district, the coastal and hydrologic evolution strongly affected the sea- and river ports. Three main critical phases are identified, dated to the early 5th century BC, the late Republican- early imperial period and late Antiquity. The cities Pisa, Volterra, Lucca and Luni are examined in their changing landscapes.

KEYWORDS

north coastal Tuscany; Etruria; coastal progradation; Magra, Serchio, Arno, Cecina rivers; Portus Pisanus

TERRITORY AND MULTIDISCIPLINARY RESEARCH

In this paper we study north coastal *Etruria*, focusing on the littoral area from the Magra to the Cecina rivers and the hinterland with the main rivers lower valleys (Magra, Serchio, Arno, Cecina) (figs.1, 3). From the 6th to the early 2nd centuries BC the northern part of this territory was dominated by Pisa, the south-

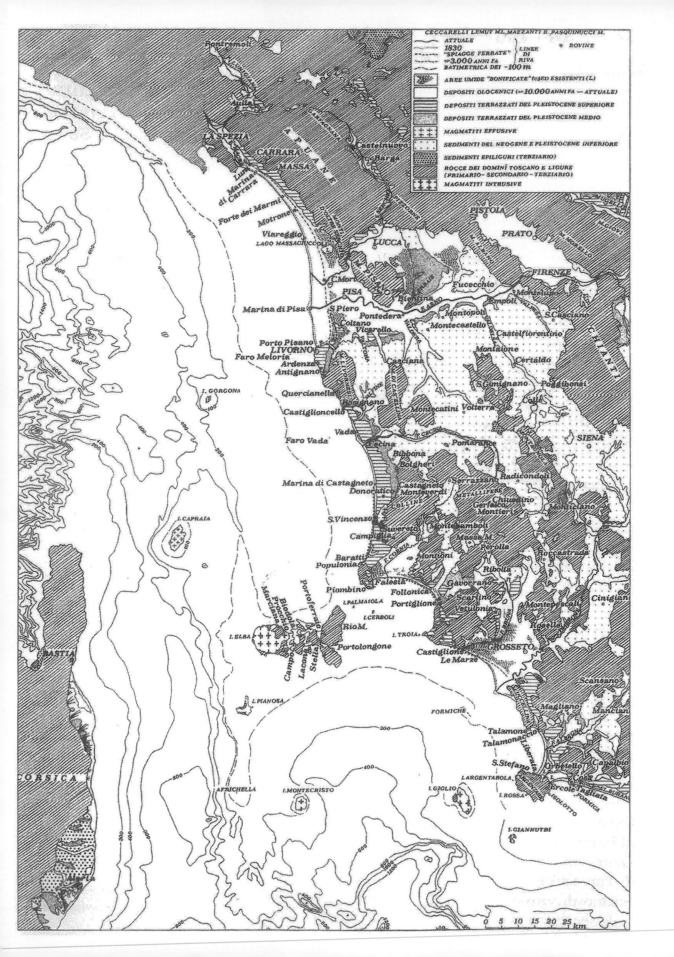

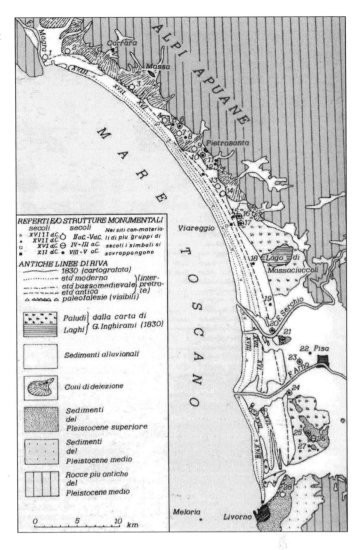

Right
Figure 2. The Luni-Livorno shoreline (after Mazzanti 2003).

ern by Volterra. In the early decades of the 2nd century and the late decades of the 1st century BC the establishment of colonies changed the settlement patterns and had a strong impact on the rural landscapes and the coastal progradation (fig.2).

In this area and its hinterland we run an intensive 'total archaeology' research project (Darvill 2001, 36). Diagnostic techniques include geomorphology, palaeogeography, remote sensing, geophysical surveys, archaeological research (including intensive surveys and monitoring of surveyed areas, stratigraphic excavations, underwater archaeology), archaeometric and archaeological studies of finds (metals and pottery), bioarchaeology, the study of ancient and medieval epigraphy, literary sources and toponyms and historical cartography.

The archaeology dates from the late Bronze Age up to the early Medieval period, with our main focus on the late Etruscan, Roman and Late Roman period. At the time of writing the research was still in progress: in this paper we present just part of the results, a few *tesserae* of a larger mosaic. In order to pro-

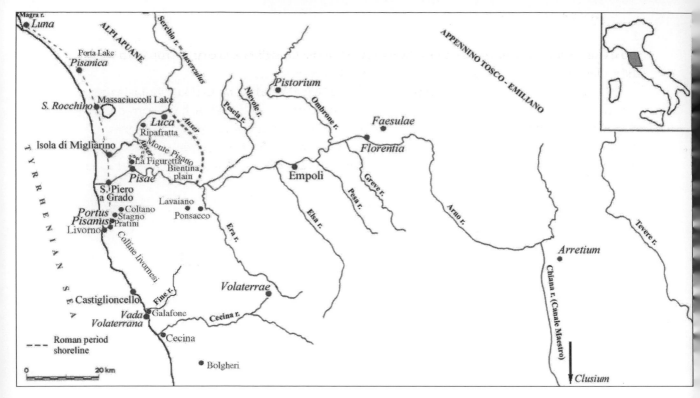

Figure 3. North Etruria: sites and rivers quoted in the text (by Giulia Picchi, Pisa).

vide data for temporal, chronological and transformational frameworks, we will focus on three subjects: evidence for Bronze and Iron Age coastal landscapes, the evolution of the Luni-Pisa littoral and the ancient ports of Luni and *Portus Pisanus*, the main cities in their diachronic rural landscapes.

THE NORTH ETRUSCAN LITTORAL: BRONZE AND IRON AGE LANDSCAPES

The north Etruscan littoral is articulated in three sections (Pasquinucci et al. 2001) (fig. 1):

- The Luni-Livorno shoreline, characterised by strong progradation from the 2nd-1st centuries BC to around 1830 AD;
- The Livorno terrace and Livorno-Castiglioncello coastal strip, which is rocky and stable;
- The Vada-Cecina shoreline, which is low and stable.

The Luni-Livorno shoreline extends from the Magra river mouth (immediately west of the high and rocky Punta Bianca promontory) to the low and rocky Livorno terrace. It is about 63 km long. In this district the main rivers are the Magra in the north and the Serchio and Arno in the south (figs. 1, 2, 3). In the hinterland are Quaternary plains, the Apuan Alps and the Monte Pisano. The Pisa plain is an alluvial-delta plain formed by the Arno river. A branch of the Serchio (Roman *Auser*) river descending from the north flowed into the Arno at Pisa, as documented by literary and archaeological evidence (Pasquinucci 2003; Camilli

2004; Camilli 2005 (fig. 3). The Livorno terrace and the Livorno-Castiglioncello coastal strip, both rocky and stable, are bordered in the hinterland by the Colline Livornesi (fig. 1).

The Vada-Cecina shoreline (figs. 1, 3) is low and stable. In the hinterland the Cecina is the main river.

In this district some evidence of the coastal environment in the Bronze and Iron Ages is provided by archaeological research. In the Late Bronze-early Iron Age a vast settlement was constructed north-east of Livorno (Stagno, figs. 1, 3) along the shore of a brackish lagoon and on a platform resting on piles driven into the muddy bottom of the lagoon (Zanini 1997: fig. 3 Stagno-Pratini). The platform was made of tree trunks, mostly of elm, more rarely of oak and only in one case of ash; the interstices were filled with bundles of branches lashed together. Abundant traces of leguminous plants, tree fruit and mostly *vitis vinifera* (*sylvestris* and *sativa*) provide evidence at least of selective gathering of certain species combined with

Figure 4. Coltano and *Portus Pisanus* area shown in the lower Arno Valley in a detail of the geomorphological map of Mazzanti (1994). The numbers identify geomorphological units (Mazzanti, 1994) and the colored symbols identify archaeological sites (Pasquinucci, 1994) dated to prehistory (black), archaic and classical period (red), middle ages (violet), and modern times (green). *See also full colour section in this book*

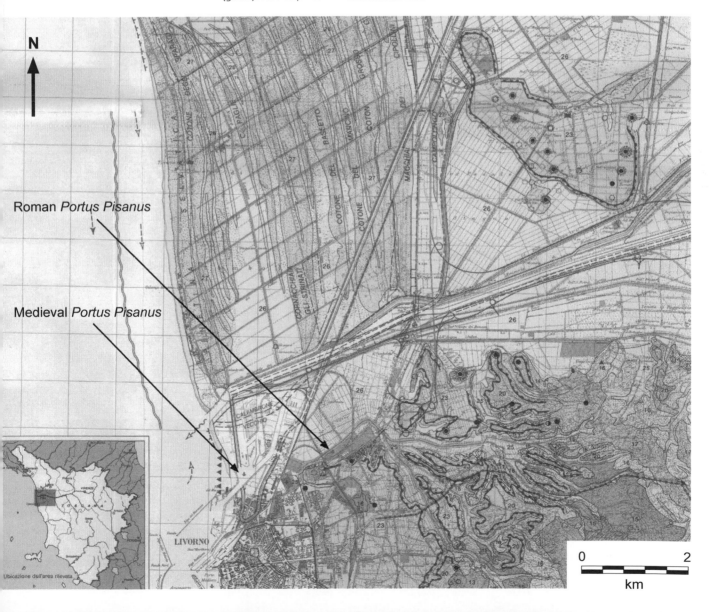

the earliest forms of their cultivation. Domesticated animals were raised (cattle, swine, mostly sheep and goats); hunting (deer, ducks, seagulls) and fishing (sea mullet) were other important activities.

In the Bronze and Iron Ages coastal salt marshes are indirectly documented by archaeological evidence of briquetage salt production (Weller 1998, 282-283) in the ancient coastal strip at Isola di Coltano (south of Pisa, figs. 3, 4) and at Galafone (South of Castiglioncello, fig. 3). At Isola di Coltano surveys and excavations identified a village dated to the middle and final Bronze Age (1600-1200 BC) located on the banks of a coastal lagoon which repeatedly submerged the area. No evidence of huts or waste foods was found. We can infer that the excavated village was a 'manufacturing', not a dwelling one (Pasquinucci & Menchelli 1997, 2002). The anthropic layers were formed by mounds of fragmented coarse vessels (more than 10,000 fragments were collected) and by a few fireplaces near which numerous firedogs were found. A charcoal sample from one of these fireplaces is dated to 1686-1538 BC (C14 analysis by the Centre for Isotope Research at Groningen). The lagoon floods are documented by yellow silt deposits containing molluscs (mostly *Cerastoderma edule*) covering the anthropic layers. These molluscs are characteristic of a low brackish marine habitat, typical of a lagoon. They belonged to various biologic cycles as the identification of different dimension individuals has shown. Their valves were closed: it provides evidence they died *in loco*.

Concerning the chronology of the site, the lower calcrete layers contain vessels dated back to the Middle Bronze age (1600 BC: this datum wholly agrees with the mentioned C14 analysis); some Final Bronze Age items (1200 BC) have been found in the superficial levels. Therefore the village appears to have been in existence for four centuries: the site was submerged by the lagoon at least four times, but it was always reoccupied by people involved in salt making, as shown by the morphological and technical continuity of the ceramics (Di Fraia & Secoli 2002). The frequent reoccupation of the village proves that the site was highly suitable for such economic activity. The selection of the site must have been influenced by the vicinity of the lagoon, the availability of fuel and clay, the proximity to roads and the access to coastal and inland lagoon/river navigation systems, which made it possible to distribute the salt.

In the early Iron Age, the briquetage technique is documented at Galafone, North of Vada (fig. 3). Large mounds of fragmented coarse vessels were identified by intensive surveys, together with layers of ashes and some parallelepiped firedogs, similar to the Coltano ones (Pasquinucci et al. 2002).

It is to be noted that in the Roman and Medieval periods salt production is documented in this same area, but different techniques were used, evidently connected with a changed environment. In fact, in the early 5th century AD the production of sea salt in the Vada area is noted by the poet *Rutilius Namatianus* (*de red*. 1, 475 ff.), who described the saltpans (*salsa palus*) in the area of *Vada Volaterrana* (present Vada: figs.1, 3). According to his description, the seawater entered the marshes through canals, and a drainage ditch irrigated the water basins. During the summer, locks were closed in order to break the communication between the sea and the marsh; the seawater evaporated in the pools, depositing the salt. In the Middle Ages, salt production at Vada is documented by archival evidence from 754 AD up to 1237 (Ceccarelli Lemut 2000; Collavini 2010, 35).

In the Iron Age, evidence of coastal landscape transformation is provided by archaeological research at Vada (figs.1, 3). A hut village located on the coastal palaeodunes, dated to the 9th century BC, was submerged by a coastal lagoon (C14 analysis by the Centre for Isotope Research at Groningen). Its remains (wooden posts, pieces of clay daub, coarse pottery) were covered by various layers of sand containing fossils (mostly *Cerastoderma edule*, *Abra alba* and *Gastrana fragilis*) typical of a lagoon environment sedimentation. This event can be referred to a phase of high sea level, probably followed by a relative drop which caused the drying of the lagoon area. The site was abandoned until the 1st century AD, when a quarter of the Roman settlement *Vada Volaterrana* was built in this area (Pasquinucci et al. 2002).

According to our research, the probably discontinuous Late Iron Age (8th century BC) shoreline from the Magra river mouth to Livorno (fig. 2) was almost stable up to the 2nd century BC. In this timespan a cool and relatively damp climate phase (dated to the 9th-3rd century BC) was followed by a warm phase up to the 4th century AD (Pinna 1996, 121-124). From the 2nd-1st centuries BC up to about 1830 AD the coastline prograded 7 km westwards, mostly in correspondence with the Arno mouth (fig. 2), as a result of a marked increase in alluvial sedimentation (Pasquinucci et al. 2001). In a period characterised both by sea level rise (Lambeck et al. 2004) and by the absence of drastic climatic changes, the sedimentation was most probably due to anthropic causes. Among these, the main elements were the construction of new towns and settlements, deforestation and increased agricultural and manufacturing activities connected with the establishment of colonies in the early 2nd century BC (*Luca*/Lucca and *Luna*/Luni) and in the late 1st century BC (triumviral/Augustan colonies at Luni, Lucca, Pisa, Volterra, Firenze, Arezzo), in particular with the organisation of their territories (*centuriatio* and land allotments) (Ciampoltrini 1981; Ciampoltrini 2004; Pasquinucci & Menchelli 1999; Pasquinucci & Menchelli 2003) (see below and fig. 3).

This phenomenon came to an end around 1830 AD, when the Arno-Serchio rivers were diverted away from the sea and into the lagoon as part of the systematic land reclamations pursued by the Lorena (Barsanti Rombai 1986, 47). The river sediments filled the low-lying lagoon, but the cessation of sediment deposition in the river mouth led to the erosion of the Arno delta, which began in the late 19th century (fig.2). All these phenomena point out the close relationship between the fluvial sediment transport and beach growth.

As for the rivers, remote sensing, geomorphologic and palaeogeographic research, geophysical investigations, archaeological surveys and excavations provide evidence of an extremely complex network of palaeorivers, of Etruscan, Roman and Medieval embankments, channels and ditches (Mazzanti 1994; see below). The Greek and Latin sources mention only the main ones in the district (*Macra, Arnus, Auser, Caecina*) and the *Fossae Papirianae*, a Roman artificial canal documented in the *Tabula Peutingeriana* North of Pisa. Medieval sources mention various rivers and river branches (Magra, Arno, *Auser, Auserculus, Tubra* flowing across the Pisa plain, Cecina).

The hydrogeological evolution of the Pisa-Lucca territory and the long term human actions aimed at optimising the rivers regime were perceived by the ancients, as documented in the early Imperial period by Strabo, after his sources (5.2.5, 222C):

> Pisa is situated between, and at the very confluence of two rivers, the Arnus and the Auser, of which the former runs from Arretium, with great quantities of waters (not all in one stream, but divided into three streams), and the latter from the Apennine Mountains, and when they unite and form one stream they heave one another up so high by their mutual resistance that two persons standing on the opposite banks cannot even see each other; and hence, necessarily, voyages inland from the sea are difficult to make; the length of the voyage is about twenty stadia. And the following fable is told: when these rivers first began to flow down from the mountains, and their course was being hindered by the natives for fear that they would unite in one stream and deluge the country, the rivers promised not to deluge it and kept their pledge... (translated by H.L. Jones, Loeb Classical Library, 1949).

The present courses of the Arno and Serchio rivers result from complex natural transformations and anthropic actions taken over the centuries, ranging from the straightening of river segments and the draining of stagnant waters to the construction of embankments, dikes, *casse di colmata* and the building of canals to channel flood waters (Ceccarelli Lemut et al. 1994).

Of peculiar interest is the complex history of the Serchio (*Auser/Auserculus*/Serchio) river (figs. 1, 3), flowing from the Apennines to the south and the south-west. In Roman times, north-east of Lucca its course divided into two branches. One of them flowed across the Bientina plain into the Arno river, the other westwards across the Stretta di Ripafratta to the coastal plain, where it divided into two branches: one of them (*Auserculus*) mounded into the sea, the other southwards reached Pisa and flowed into the Arno (fig. 3). Recent field researches in the district provide evidence of the flooding frequency of the Arno/Serchio river system. Most probably climatic fluctuations triggered these events (Pasquinucci 2008b; Pasquinucci & Menchelli 2009).

Starting from 400-450 AD and up to the late 8th century, a new humid-cold phase took place (Pinna 1996, 125). Concerning this period, ancient authors describe exceptional rainfalls which caused catastrophic floods. In particular *Paulus Diaconus* mentions the *Diluvium* (the deluge), referring to 589 AD (Paulus Diaconus, *Historia Langobardorum* III, 23; Dall'Aglio 1997).

This climatic crisis explains the origin of the tradition concerning the miracle attributed to Saint Frediano, bishop of Lucca between 560 and 588 AD. In the 6th century AD floods and large marshes affected the plains through which the Serchio flowed. In order to reclaim the soil, actions were taken that were handed on by the medieval hagiography as a miracle attributed to Frediano (Gregorius Magnus, *Dialogi* 3.9; Ceccarelli Lemut & Garzella 2005, 11; Mazzanti 1994). According to this medieval text, the Saint di-

Figure 5. The *Portus Pisanus* late republican seabed (photo by Stefano Genovesi, Pisa).

Page 187 >
Figure 6. Luni in its changing landscape (courtesy of Monica Bini, Pisa).

verted the course of the Serchio river using a rake and prevented the frequent river floods. Most probably the intervention was actually carried out on the initiative of bishop Frediano, who possibly superintended the work too. Actually, in late Antiquity and early Middle Ages the bishops frequently practised not only the religious functions pertaining to their *status*, but also the civil ones that were no longer guaranteed due to the collapsing state structures.

From the Middle Ages up to the Renaissance and in the centuries of vast land reclamation projects which continued up to and even after World War II, many interventions on the Serchio and the Arno courses are documented. For example, in the Middle Ages the Comune di Pisa and later the Medici intervened in the management of the lower Arno and Serchio basins, by cutting a few meanders off these rivers and straightening the Arno's terminal segment (Ceccarelli Lemut et al. 1994).

TWO ANCIENT PORTS IN THEIR CHANGING ENVIRONMENT: *LUNI* AND *PORTUS PISANUS*

The coastal and hydrological evolution was particularly impressive in the Luni district and in the Pisa territory (*ager Pisanus*) and heavily affected the north Etruscan ports and calls network. The Luni (*Luna*) plain underwent major landscape changes over the last 3000 years (figs. 1, 3). A geomorphologic and archaeological project is studying the palaeogeography of the Luni plain, where the coastline shifted southward and westward; the project is also addressing the long debated question of the location of the *Luna* port(s) (Bini et al. 2009 a, b, c; cf. Bernieri et al. 1983; Delano Smith 1986). Since a few centu-

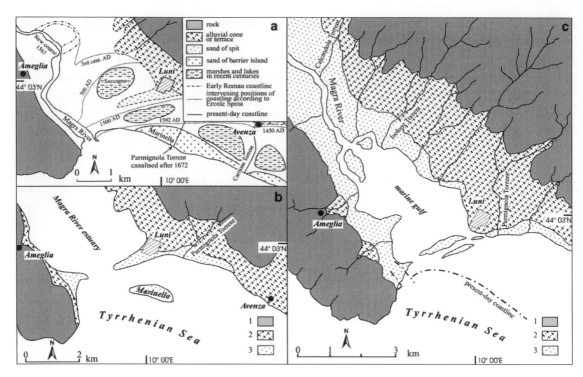

Figure 7. Palaeogeographic scenarios of the Luni territory (after Bini et al. 2009b)

ries before the foundation of *Luna*, the coastal plain had been characterised by a complex architecture of swamps and marshes, limited by dune ridges and fluvial sand bars. The positions of these landforms shifted, mainly depending on the spatial relationship between the coastline and the rivers mouths (Bini et al. 2009 a, b).

Luna (Luni), a Roman colony, was founded in 177 BC (Liv. 41.13.5) in a newly conquered territory in the coastal strip about 3 km south of the present Magra river, between the mouths of the Magra and the Parmignola stream (Bini et al. 2009 a, b) (figs. 6, 7). In Roman times the first was north of *Luna* (Loc. Bocceda, Sarzana: Gervasini 2007, 160). The Parmignola is likely to have changed its position after Roman times, gradually migrating from west to east; it currently flows 1 km east of *Luna* (Bini et al. 2009 a, b).

The new town was founded on an alluvial fan on the eastern side of the marine gulf into which the river flowed (Durante 2010; Bini et al. 2009c). In the territory, the *centuriatio* provided the drainage and road grid. The Seccagna, north-west of Luna, between the town and the Magra, was a vast shallow basin traditionally interpreted as the port of the colony. This was progressively filled by peat, which was rich in Roman pottery fragments, and by dark alluvial clay (Bini et al. 2009b; Durante 2001). In the early 6th century the *Itinerarium maritimum* mentions *Lune, fluvius Macra*, providing evidence that the *Luna* port was perceived in connection with the Magra river mouth. The presence of structures interpreted either as docks or as shoreline reinforcements is not confirmed by archaeological research.

As for Pisa, in Antiquity the city and its territory had a well integrated network of sea- and river-ports (Pasquinucci 2003, 2007). The main port was situated NNE of present Livorno. It is denominated *Portus Pisanus* by the *Itinerarium Maritimum* (501) in the early 6th century AD. A harbour station named

Labro by Cicero (*ad Quintum fr.*, 2.5) was probably located in the area (cf. current hydronym Calambrone: Pasquinucci 2003) (fig. 4).

The *Itinerarium* provides the accurate position of *Portus Pisanus*: the harbour was located 23 *milia* from *Vada Volaterrana* and 9 from the Arno river mouth in those times (*Pisae fluvius*). In the 18th and 19th centuries AD conspicuous remains of the Roman harbour settlement were still standing (Targioni Tozzetti 1768; Banti 1943). In this area (figs. 1, 4) the low and rocky ancient coastline (current 'Gronda dei Lupi') was set up south-west/north-east, in the same direction of the prevailing wind (Libeccio). The sea storms were damped down by the wave refraction near the coast and by the Meloria shoals, located 10 km south-west of the Gronda in the offing.

Thanks to its favourable geomorphologic peculiarities, this area north-north-east of Livorno was frequented by ships at least since the late 7th-early 6th century BC to the 6th century AD, as is shown by palaeogeographic and archaeological research. Recent excavations have brought to light evidence of a seabed (fig. 5, see below) and a port settlement served by an aqueduct originating from the nearby hills (Pasquinucci & Menchelli 2010).

The first provides evidence of an area that was navigated by appropriate crafts loading and unloading wares at least since the late 7th-early 6th century BC. In the 4th-3rd century BC stone blocks and posts were set up either to reinforce the shoreline or to provide a dock. The stretch of water was progressively and rapidly silted up by alternate sand and *posidonia* (seagrass) layers since the mid-2nd century BC; in the late 1st century BC activities connected with navigation could no longer be performed in this area and were therefore shifted westwards (Pasquinucci 2003, 2004). In the port settlement (2nd-6th century AD) *horrea* and a necropolis were excavated (Pasquinucci & Menchelli 2010).

In the 5th century AD *Rutilius Namatianus* landed at *Portus Pisanus*, both well sheltered (*de red.* 1, 559: *puppibus meis fida in statione locatis*), busy and rich (*portum quem fama frequentat Pisarum emporio divitiisque maris*). In the hinterland, the nearby hills were covered by woods where *Rutilius* and his companions went hunting boars. The same woods supplied several amphora, brick and tiles and coarse pottery workshops with fuel (Cherubini et al. 2006).

The silting chronology matches the above-mentioned data, which indicates that the north Etruscan coast progradation was a consequence of the late Republican colonisation impact on the hinterland and possibly of a phase characterised by intense rainfalls. The Imperial, Late Roman and Medieval *Portus Pisanus* were located one after the other westwards of the Roman Republican one, confirming the coastal progradation (fig. 4; see also fig. 2).

THE CITIES AND THEIR TERRITORIES: LANDSCAPES IN A DIACHRONIC PERSPECTIVE

In Etruscan times Pisa (*Pisae*) and Volterra (*Velathri, Volaterrae*) were the main cities in north-western Etruria. Their territories respectively included the Magra-Fine rivers and the Fine river- Bolgheri coastal strips and large hinterlands (figs. 1, 3). Volterra was conquered by the Romans in the first half of the 3rd century BC. Pisa became *civitas foederata* in the mid-3rd century BC, and in the 2nd century BC it was a Roman military base during the wars against the *Ligures*. In 180 BC Pisa granted part of its northern territory for the establishment of the Latin colony *Luca* (Lucca). In 177 BC, when the *Ligures* were defeated in north-west Tuscany, the Roman colony *Luna* (Luni) was established (Coarelli 1985-1987; Ciampoltrini 2004).

Between 42 (*Philippi* battle) and 31 BC (*Actium* battle), or after *Actium*, the cities and most plains in north Etruria were remodelled by veterans' colonisation (Ciampoltrini 1981, 2007, 14; Keppie 1983) (fig. 3). The *centuriationes* and the related drainage systems carried out in this period are largely preserved, since they were shaped according to the local geomorphology and hydrology by highly skilled *mensores*. In some areas of earlier colonisation (e.g. *Luca*: 180 BC) the orientation was slightly changed, most probably in order to match the hydrological situation.

In north-western Etruria the triumviral-Augustan colonisation was marked by economic growth in farming, manufacturing and trade (Pasquinucci & Menchelli 1999; Pasquinucci & Menchelli 2003) and had a strong impact on landscape and territory. The landscape remained substantially the same until the mid-6th century (see below).

For what concerns Pisa, recently the toponym (Greek: *Pisa/Pisai/Peisa/Peisai*; Latin: *Pisae*) has been interpreted as Indo-European, meaning a site rich in water, both stagnant and flowing (Dini 1994). In the late 7th century BC the Etruscan town originated in a few settlements separated by major and minor streams, at the confluence of the river *Auser* (an ancient branch of the Serchio) into the Arno (Strab. 5.2.5; Plin. *N.H.*, 3.50; Rut. Nam. 1.566; Schol. Ptol. 3.1.4.) on the right bank of the latter (fig. 2). Pisa was 20 *stadia* far from the seacoast according to Strabo (5.2.5) and his sources (therefore in the late republican-early imperial period). The site was an important crossroad, where the north Tyrrhenian coastal route intersected the route following the Arno and Serchio rivers banks and linking the riverine ports of call. The rivers were a natural defence, a resource and a risk. Both the *Arnus* and the *Auser* were still busy waterways in late Antiquity: in the 6th century *Theodericus* took actions against the fishermen who used to hamper navigation of both rivers by weirs (Cass. *Var.*, 5, 17, 6; 20, 3: referring to 523-526 AD).

The distribution of the archaeological finds demonstrates that the pre-Roman settlement was scattered on several low mounds formed by previously deposited fluvial sediments and separated by river branches and streams. Excavations in the city centre (via S. Apollonia) provide evidence both of a low mound (embanked by means of vertical poles and horizontal planks) and of an adjoining humid zone that was filled by the deposition of peat layers and by the accidental or deliberate accumulation of large amounts of pottery fragments, bones and vegetal elements that consolidated the soil (Corretti & Vaggioli 2003, 57). This evidence is dated from the 7th to the early decades of the 5th century BC, when a silt and clay layer deposited by a river flood covered the previous phases. In the same period (around 480 BC) a flood damaged a large part of the town area (Bruni 1998, 137, 198: evidence in piazza Duomo, piazza Dante). Here, as elsewhere in the town, the streams progressively silted up after the 4th-3rd centuries BC. In the early imperial age the differences in level were filled by residual brick, tile and pottery fragments, in order to provide a suitable urban soil (Corretti & Vaggioli 2003, 60.). In the ancient town suburb (northwest periphery of present Pisa) evidence of the late-Republican to late antique landscape is provided by a few shipwrecks sunken in the *Auser* by floods, one of which is dated around 10 AD (Camilli 2004, 2005; Benvenuti et al. 2006; Martinelli & Pignatelli 2008; cf. Leveau 2008). The wrecks are dated from the second half of the 2nd century BC to the late 6th-early 7th AD.

In the territory, archaeological research provides evidence of dramatic hydrological instability. East of Pisa a flood destroyed an Etruscan settlement at the turn of the 6th to 5th century BC (Pasquinucci et al. 2008, 41-74.) (fig. 3). Etruscan drainage works in the plain east of Pisa are demonstrated by the presence of a 5th-century reclamation channel (Bonamici 1990, 115; Pasquinucci 1994, 189; Maggiani 1990; Pasquinucci 2008a) (fig. 3), which is the clue to Etruscan drainage systems constructed in a period characterised by hydrological instability.

In the last decades of the 1st century BC the colonia *Opsequens Iulia Pisana* was established and a vast *centuriatio* was constructed in the *Pisae* plain. This had a dramatic impact on the town and its territory. The *centuriae* were bounded by a largely surviving grid of roads, ditches and channels. The channels are a characteristic feature of the landscape in the southern part of the plain, providing the area with the necessary drainage. Literary, epigraphic and archaeological sources show a landscape greatly moulded by anthropic activities, where the natural resources were exploited within well organised enterprises. As usual, the colonisation implied deforestation, tillage of previously uncultivated areas, development of the road network, construction of drainage systems.

Hydraulic works in the Arno valley must have been very demanding because of drainage troubles mostly due to plain subsidence (Pasquinucci et al. 2001). Drainage canals dated 50 BC-50 AD have been identified in the area north-east of *Pisae* (Bonamici 1989). The *fossae docariae* quoted by medieval archive sources in various sites along the Arno left bank can be connected with the Roman hydraulic works (Ceccarelli Lemut et al. 1994, 420).

Along the *limites*, farmsteads were set up (Pasquinucci & Menchelli 2003). The countryside reorganisation improved the agricultural production, mainly grain and wine crops (Plin. *N.H.* 18, 86-87; 18,109: *siligo*, *alica*; Plin. *N.H.* 18, 109: *Pariana uva*). The north-Etruscan production of wine amphoras (Graeco-italic, Dressel 1, Dressel 2-4, Spello, Forlimpopoli and Empoli types) confirms the importance of the pisanvolaterran viticulture (Cherubini et al. 2006). Olive oil production is archaeologically documented by equipment (an oil press and decanting tanks) excavated in local *villae* (Pasquinucci & Menchelli 2003) and by bioarchaeological data (Motta 1997; Mariotti-Lippi et al. 2006, 2007).

Woods were very important in both landscape and economy of North coastal Etruria: e.g. in the late Republican-early Imperial period the timber was used *in loco* for ship construction and was exported to Rome as building material (Strabo 5.2.5). The Pisan stone-quarries, located in the Monte Pisano slopes, strongly affected the natural landscape (Strabo 5.2.5). The local limestone has been identified in Roman buildings in *Pisae* and its territory.

The Pisa district was characterised by pottery production. River banks and woods were intensively exploited for the raw material (clay and fuel) necessary for the pottery production that included *terra sigillata*, bricks and tiles, wine amphoras, coarse vessels and *dolia*. In the Augustan period the ceramic activities reached their peak with the *terra sigillata* vessels: it was the main phenomenon in the north Etruscan economy and one of the most important in the history of Roman pottery (Menchelli et al. 2001).

Population pressure, soil erosion caused by ploughing and woodcutting increased the fluvial sediment transport and caused remarkable changes in the north Etruscan coastal landscape (see above and Grove & Rackham 2001). In the Late Roman period (5th-6th century AD) a progressive degradation and depopulation is documented in the Pisa territory, but in different ways and times according to the geographical areas and the socio-economic contexts. Already during the 3rd century AD, in the Eastern Pisa plain (Lavaiano area) a spreading marsh caused some early imperial farmsteads to be abandoned (Pasquinucci et al. 1997, 241).

In many sectors of the Pisa plain the small and middle-sized farmsteads continued in use up to the end of the 5th century. The rural settlement patterns appear to document a still remarkable agricultural production; the imported vessels and amphoras found in the farmsteads and villas provide evidence they were producing not only for subsistence but also a surplus to be traded. The 'Empoli' type amphora proves that the local wine production continued conspicuously at least up to the end of the 5th century.

Economic activities such as shipbuilding (cfr. Claudian, *Bell. Gild* 483) and vessel ceramic production are documented all through Late Antiquity (Cherubini et al. 2006).

In the 6th century the ecological and economic situation started to change. The small/middle-size farmsteads disappeared; a large villa at Massaciuccoli (figs. 1, 3) was still active but was occupied by a Pieve (Parish), which became the centre of the church's management of the territory (Ciampoltrini 1994). The *saltus* landscape spread, mainly in the inner valleys and mountainous areas where forestry and sheep-breeding were always prevalent. Hunting and fishing (e.g. Cassiodorus, *Var.* 5.17 and 20) were important economic activities.

During the Goth-Byzantine war (535-556 AD) destruction, epidemics and famine likely affected the district. Lacking State control, the hydraulic works necessary for the plain drainage were abandoned and wide marshes started spreading in areas which had undergone *centuriatio*. Many Medieval documents mention toponyms referring to forests (e.g. *Selva, Travalda*) and marshes (e.g. *Stagno, Putrido*) (Ceccarelli Lemut 1994, 415) in areas which were intensively occupied by farmsteads and villas in Roman times. In the late 6th century *Pisae* still kept an important role, in the context of the Byzantine sea routes which focused just on the strategic areas that were necessary for the Empire's defence. At the beginning of the 7th century the Lombards conquered the coastal strip: the Late Roman cultural landscape changed into the Medieval one.

A new town, the Latin colony *Luca* (Lucca), was founded in 180 BC (Liv. 40.43.1) on the left bank of the *Auser* (Serchio) river (figs. 1, 3). As elsewhere, the river was a natural defence, a resource and a risk. The eastern segment of the northern city walls is curving in plan, most probably because it ran parallel to the *Auser* (a palaeochannel is identified in the aerial photographs of the area). This portion of the fortified perimeter was particularly at risk, being close to the river at the time. Archaeological research provides evidence that here a portion of the outer face of the city walls was mended after damage produced by a flood in Triumviral – early Augustan times. A flood dated between 10 and 20 AD is documented in the eastern suburb of the town (Ciampoltrini 2007).

The *Luca* 2nd-century *centuriatio* reclaimed land in the *Auser* plain and was oriented north-south. Evidence of several floods dated to the 2nd and 1st century BC is provided by archaeological research in the territory (Ciampoltrini 2007). Possibly because of such hydrological instability the later *centuriatio* dated to Augustan times was oriented north-north-east, like the town plan. A few streams were canalised. In-mid imperial times, at least the marginal areas of the plain started becoming marshy: on the eastern plain a *decumanus* underwent several episodes of maintenance before it was replaced by a wooden viaduct in the mid-imperial period (Ciampoltrini 2007, 42). In the Middle Ages the *centuriatio* survived several ecological crises in many areas of the *ager Lucensis*, mainly south and east of the town.

RESULTS AND FURTHER RESEARCH

Our multidisciplinary diachronic research provides sound evidence for temporal, chronological and transformational frameworks in north-western *Etruria*. We will further develop this intensive 'total archaeology' project in the next years, in order to outline the palaeogeography, the settlements patterns and the changing landscapes in the district.

REFERENCES

Antonioli, F., O. Girotti, S. Improta, M. F. Nisi, C. Puglisi & V. Verrubbi. 2000. Nuovi dati sulla trasgressione marina olocenica nella pianura versiliese. Proceedings of the Meeting 'Le Pianure', Ferrara, November 1999, Regione Emilia Romagna, 214-218.

Banti, L. 1943. Pisa.*Memorie della Pontificia Accademia Romana di Archeologia* 3 (6), 67-141.

Barsanti, D. & L. Rombai. 1986. *La 'guerra delle acque in Toscana'*. Storia delle bonifiche dai Medici alla Riforma agraria, Florence.

Benvenuti, M., M. Mariotti-Lippi, P. Pallecchi & M. Sagri. 2006. *Late Holocene catastrophic floods in the terminal Arno river (Pisa, Central Italy) from the story of a Roman riverine harbour*, The Holocene, SAGE Publications online, 863-876.

Bernieri, A., T. Mannoni & L. Mannoni. 1983. *Il porto di Marina di Carrara*. Storia e attualità, Genova.

Bini, M., A. Chelli, P. Federici, M. Pappalardo & F. Biagioni. 2009a. Environmental features of the Magra River lower plain (NW Italy) in Roman times. In De Dapper, M., Vermeulen, F., Desprez, S. & Taelman, D. (eds.). Ol' Man River: Geo-archaeological Aspects of Rivers and River Plains. *ARGU (Archaeological Reports Ghent University)* 36, 111-126.

Bini, M., A. Chelli, A.M. Durante, L. Gervasini & M. Pappalardo. 2009b. Geoarchaeological sea-level proxies from a silted up harbour: a case study of the Roman colony of Luni (northern Tyrrhenian Sea, Italy). *Quaternary International* 206, 147-157.

Bini, M., H. Brueckner, A. Chelli & M. Pappalardo. 2009c. New Chronological constrains on the Holocene palaeo-geography of the Luni coastal plain (Eastern Liguria, Italy) Geoitalia. *Epitome* 3, 267.

Bonamici, M. 1989. Contributo a Pisa arcaica, Atti Secondo Congresso Internazionale etrusco Firenze 1985, Roma, 1135-1147.

Bonamici, M. 1990. L'epoca etrusca: dall'età del Ferro alla romanizzazione, in San Giuliano Terme. *La storia , il territorio Pisa*, 97-124.

Bruni, S. 1998. *Pisa Etrusca*, Milano, Longanesi.

Camilli, A. 2004. Le strutture 'portuali' dello scavo di Pisa-San Rossore. In Zevi, A.G. & Turchetti R. (eds.), *Le strutture dei Porti e degli approdi antichi, Atti II Seminario Progetto ANSER* (Interreg IIIB MEDOCC), Ostia antica 2004, Soveria Manelli, 67-86.

Camilli, A. 2005. Il contesto delle navi antiche di Pisa. Un breve punto della situazione, in Fastionline 2005, 1-7, www.fastionline.org/docs/2005-31.pdf.

Ceccarelli Lemut, M.L. 2000. Vada: le attività produttive., In Regoli, E. & Terrenato, N. (eds.), *Guida al Museo Archeologico di Rosignano M.mo*, Siena, Nuova immagine, 154-155.

Ceccarelli Lemut, M.L. & G. Garzella. 2005. Tipologia, funzioni e connotati istituzionali. In Ceccarelli Lemut, M.L. & Garzella, G. 2005 (eds.), *Terre nuove nel Valdarno pisano medievale*, Pisa, Pacini Editore, 9-47.

Ceccarelli Lemut, M.L. R. Mazzanti & P. Morelli. 1994. Il contributo delle fonti storiche alla conoscenza della geomorfologia. In Mazzanti, R. (ed.) *La pianura di Pisa e i rilievi contermini*, 401-429.

Cherubini, L., A. Del Rio & S. Menchelli. 2006. Paesaggi della produzione: attività agricole e manifatturiere nel territorio pisano-volterrano in età romana. In Menchelli, S. & Pasquinucci, M. (eds.), *Atti convegno Internazionale Territorio e produzioni ceramiche*, Pisa, Edizioni PLUS Pisa University Press, 69-76.

Ciampoltrini, G. 1981. Note sulla colonizzazione augustea nell'Etruria settentrionale. *Studi Classici e Orientali* 31, 41-55.

Ciampoltrini, G. 1994. Gli ozi dei Venulei. Considerazioni sulle 'Terme' di Massaciuccoli. *Prospettiva* 73-74, 119-130.

Ciampoltrini, G. 2004. La seconda fase della guerra: dall'attacco a Pisa alla presa del Ballista (193-179 a.C.). In De Marinis, R.C. & Spadea, G. (eds.), *I Liguri. Un antico popolo auropeo tra Alpi e Mediterraneo*. Milano, Skira Editore, 306-307.

Ciampoltrini, G. (ed.) 2007. *Ad limitem. Paesaggi d'età romana nello scavo degli Orti del San Francesco in Lucca*, Lucca, I Segni dell'Auser.

Coarelli, F. 1985-1987. La fondazione di Luni. Problemi storici ed archeologici, in Centro Studi Lunensi. *Quaderni* 10, 17-36.

Collavini, S. M. 2010. *Rosignano Marittimo il Medioevo. Ambiente, economia e società.* Livorno, Debatte.

Corretti, A. & M.A. Vaggioli. 2003. Pisa, via S. Apollonia: secoli di contatti mediterranei. In Tangheroni, M. (ed.), Pisa e il Mediterraneo. Uomini, merci, idee dagli etruschi ai Medici. Milano, Skira Editore, 57-63.

Dall'Aglio, P.L. 1997. Il Diluvium di Paolo Diacono e le modificazioni ambientali tardoantiche: un problema di método. Ocnus 5, 97-104.

Darvill, T. 2001. Traditions of landscape archaeology in Britain: issues of time and scale. In Darvill, T. & Gojda, M. (eds.) *One Land, Many Landscapes*, 33-45.

De Dapper, M., F. Vermeulen, S. Deprez, D. Taelman (eds.) 2009. *Ol' Man River. Geo-archaeological Aspects of Rivers and River Plains.* Academia Press, Gent.

Delano Smith, C. 1986. Changing Environment and Roman Landscape: the ager Lunensis, Papers British School at Rome 56, 125-140.

Di Fraia, T. & L. Secoli. 2002. Il sito di Isola di Coltano. In Negroni Catacchio, N. (eds.), *Preistoria e Protostoria in Etruria*. Quinto incontro di Studi, Milano, 79-93.

Dini, P.U. year. Sul toponimo Pisa in una prospettiva indoeuropea. *AION-Linguistica*16, 283-316.

Durante, A.M. 2010. *Città antica di Luna. Lavori in corso,2*, Genova, Fratelli Frilli Editore.

Fabiani, F. 2006. *'stratam antiquam que est per paludes et boscos...' Viabilità romana tra Pisa e Luni*. Pisa, Edizioni PLUS Pisa University Press.

Gervasini, L. 2007. La linea del Magra: un territorio fra la seconda età del Ferro e la romanizzazione. In De Marinis, R.C. & Spadea, G. (eds.), *Ancora su I Liguri. Un antico popolo europeo tra Alpi e Mediterraneo*, Genova, De Ferrari, 159-167.

Grove, A. & T. Rackham. 2001. *The Nature of Mediterranean Europe. An Ecological History*. New Haven/London, Yale University Press.

Keppie, L. 1983. *Colonization and veteran settlement in Italy, 47-14 B.C.* London/Rome, British School at Rome.

Lambeck, K., F. Antonioli, A. Purcella & S. Silenzi. 2004. Sea level change along the Italian coast for the past 10.000 yrs. *Quaternary Science Review* 22, 309-318.

Leveau, Ph. 2008. Les inondations du Tibre à Rome: politiques publiques et variations climatiques à l'èpoque romaine. In Hermon, E. (ed.), *Vers une gestion intégrée de l'eau dans l'Empire Romain*, Roma, «L'Erma» di Bretschneider, 137-146.

Maggiani, A. 1990. La situazione archeologica nell'Etruria settentrionale nel V sec. a.C. In *Crise et trasformation de sociètès archaiques de l'Italie antique au Ve siècle*, av. J.-C., Actes de la Table Ronde Rome 1987, Roma.

Mariotti-Lippi, M., C. Bellini, C. Trinci, M. Benvenuti, P. Pallecchi & M. Sagri. 2006. Pollen analysis of the ship site of Pisa San Rossore, Tuscany, Italy: the implications for catastrophic hydrological events and climatic change during the Late Holocene. *Vegetation History and Archaeobotany* 15, 1-13.

Mariotti-Lippi, M., M. Guido, B.I. Menozzi, C. Bellini & C. Montanari. 2007. The Massaciuccoli Holocene pollen sequence and the vegetation history of the coastal plains by the Mar Ligure (Tuscany and Liguria, Italy). *Vegetation History and Archaeobotany* 16, 267-277.

Martinelli, N. & O. Pignatelli. 2008. Datazione assoluta di alcuni relitti dal contesto delle navi di Pisa. Risultati preliminari delle indagini dendrocronologiche e radiometriche col 14C(*), *Gradus* 3 (2), 69-78.

Mazzanti, R. (ed.) 1994. La pianura di Pisa e i rilievi contermini. Roma, Società geografica Italiana.

Mazzanti, R. 2003. Le conoscenze sullo sviluppo del litorale toscano nel Quaternario. In Tangheroni, M. (ed.), Pisa e il Mediterraneo. Uomini, merci, idee dagli etruschi ai Medici. Milano, Skira Editore, 333-338.

Menchelli, S. 2004. Ateian sigillata and import-export activities in North-Etruria. In Poblome, J., Talloen, P., Brulet, R. & Waelkens, M. (eds.), *Early Italian Sigillata. The chronogical framework and trade patterns* BAbesch Supplement 10, 271-277.

Menchelli, S., C. Capelli, A. Del Rio, M. Pasquinucci, V. Thiron-Merle & M. Picon. 2001. Ateliers de céramiques sigillées de l'Etrurie septentrionale maritime: données archéologiques et archéométriques. *Rei Cretariae Romanae Fautorum Acta* 37, 89-105.

Motta, L. 1997. I paesaggi di Volterra nel tardo-antico. *Archeologia Medievale* 24, 245-276.

Pasquinucci, M. 1994. Il popolamento dall'età del Ferro al tardo-antico. In Mazzanti, R. (ed.) La pianura di Pisa e i rilievi contermini. Roma, Società geografica Italiana, 183-204.

Pasquinucci, M. 2003. Pisa e i suoi porti in età etrusca e romana. In Tangheroni, M. (ed.), Pisa e il Mediterraneo. Uomini, merci, idee dagli etruschi ai Medici. Milano, Skira Editore, 93-97.

Pasquinucci, M. 2004. Paleogeografia costiera, porti e approdi in Toscana. In De Maria, L. & Turchetti, R. (eds.), Evoluciòn paleoambiental de los puertos y fondeaderos antiguos en el Mediterráneo occidental, Atti del I Seminario Progetto ANSER (Interreg IIIB MEDOCC), El patrimonio arqueológico submarino y los puertos antiguos, Alicante (November 2003), Roma, Soveria Mannelli, Rubbettino Editore, 69-73.

Pasquinucci, M. 2007. I porti di Pisa e di Volterra. Breve nota a Strabone 5.2.5, 222C. Athenaeum 95, 677-684.

Pasquinucci, M. 2008a. Paesaggi modellati dalla natura e dall'uomo:dalla preistoria alla fine dell'antichità. In Pasquinucci, M. & Ceccarelli Lemut, M.L. (eds.), Il territorio nell'età antica. Un'eredità di lungo periodo. San Giuliano, Pisa, Edizioni ETS. 7-27.

Pasquinucci, M. 2008b. Water Management practices and risk management in North Etruria (Archaic period to Late Antiquity): a few remarks. In Hermon, E. (ed.), Vers une gestion intégrée de l'eau dans l'histoire environnementale: savoirs traditionnels et pratiques modernes, Roma, «L'Erma» di Bretschneider, 147-156.

Pasquinucci, M., A. Del Rio & S. Menchelli. 2002. Terra e acque nell'Etruria nord-occidentale. In Negroni Catacchio, N. (ed.), *Preistoria e protostoria d'Etruria. Paesaggi d'acque. Ricerche e scavi*, Atti V incontro studi, Milano, Edizioni Centro Studi Preistoria e Archeologia, 51-61.

Pasquinucci, M., S. Mecucci & P. Morelli. 1997. Territorio e popolamento tra i fiumi Arno, Cascina ed Era: ricerche archeologiche, topografiche e archivistiche. In Atti I Congresso SAMI Pisa, Firenze, Edizioni all'Insegna del Giglio.

Pasquinucci, M. & S. Menchelli. 1997. Isola di Coltano (Coltano, Pisa). In Zanini, A. (ed.), *Dal Bronzo al Ferro. Il II Millennio a. C. nella Toscana centro-occidentale* Pisa, Pacini Edirore, 49-53.

Pasquinucci, M. & S. Menchelli. 1999. The landscape and economy of the territories of Pisae and Volaterrae (coastal North Etruria). Journal Roman Archaeology 12 (1), 122-141.

Pasquinucci, M. & Menchelli, S. 2002. The Isola di Coltano Bronze Age Village and the Salt Production in North Coastal Tuscany (Italy). In Weller, O. (ed.), Archéologie du sel. Techniques et sociétés dans la Pré- et la Protohistoire européenne. Rahden/Westfalen, Verlag Marie Leidorf GmbH., 177-182.

Pasquinucci, M. & S. Menchelli. 2003. Insediamenti e strutture rurali negli agri Pisanus e Volaterranus. Journal Roman Topography 12, 137-152.

Pasquinucci, M. & S. Menchelli. 2009. Variazioni climatiche nella Toscana nord-occidentale: indagini multidisciplinari e prime riflessioni. In Hermon, E. (ed.), *Société et climats dans l'Empire romain. Pur une perspective historique et systémique de la gestion des ressources en eau dans l'Empire romain*. Napoli, Editoriale Scientifica, 377-388.

Pasquinucci, M. & S. Menchelli. 2010. Il sistema portuale di Pisa: dinamiche costiere, import-export, interazioni economiche e culturali (VII sec. a.C.-I sec. d.C.). *Bollettino di Archeologia online, Numero Speciale*, www.beniculturali.it/bao

Pasquinucci, M., S. Menchelli & N. Leone. 2008. Paesaggi antichi nella bassa valle dell'Arno: il caso dell'insediamento pluristratificato in loc. Le Melorie di Ponsacco (PI). In Ciampoltrini, G. (ed.), La Valdera romana fra Pisa e Volterra, Pisa, Pacini Editore, 41-74.

Pasquinucci, M., S. Menchelli, M. Marchisio, R. Mazzanti & L. D'Onofrio. 2001. Coastal Archaeology in North Etruria. Geomorphologic, archaeological, archive, magnetometric and geoelectrical researches. *Revue d'Archéométrie* 25, 187-201.

Pinna, M. 1996. *Le variazioni del clima*. Milano, Franco Angeli Editore.

Stefaniuk, L., C. Roumieux, Ch. Morhange & M. Pasquinucci. 2007. Dynamiques environnementales du complexe deltaïque Arno/Calambrone à l'Holocène récent et localisations des ports de Pise/Livourne. Archéometrie '07, Colloque du G.M.P.C.A., Aix-en-Provence.

Targioni Tozzetti, G. 1768. Relazioni d'alcuni viaggi fatti in diverse parti della Toscana, II. Firenze, Stamperia Granducale.

Weller, O. 1998. L'exploitation du sel: techniques et implications dans le neolitique europeen. In Atti XIII Congresso UISPP, Forli 1996, vol. 3, 281-287.

Zanini, A. (ed.) 1997. Dal Bronzo al Ferro. Il II millennio a.C. nella Toscana centro-settentrionale Pisa, Pacini Editore.

2.3 Can the period of Dolmens construction be seen in the pollen record? Pollen analytical investigations of Holocene settlement and vegetation history in the Westensee area, Schleswig-Holstein, Germany

Authors

Mykola Sadovnik[1, 2], H.-R. Bork[2, 3], M.-J. Nadeau[1,4] and O. Nelle[1, 2]

1. Palaeoecology Research group, Institute for Ecosystem Research, Christian-Albrechts-Universität zu Kiel, Kiel, Germany.
2. Graduate School 'Human Development in Landscapes' Christian-Albrechts-Universität zu Kiel, Kiel, Germany.
3. Ecosystem Development Research Group, Institute for Ecosystem Research, Christian-Albrechts-Universität zu Kiel, Kiel, Germany.
4. Leibniz-Laboratory for Radiometric Dating and Isotope Research, Christian-Albrechts-Universität zu Kiel, Kiel, Germany

Contact: msadovnik@gshdl.uni-kiel.de

ABSTRACT

This study focuses on the high-resolution reconstruction of land use and forest history and the changes of vegetation connected to the erection and use of megalithic graves at Krähenberg in Schleswig-Holstein, Northern Germany. Pollen analysis of a peat core from a small mire directly neighbouring the graves was performed in connection with the analysis of archaeological data. A chronological framework is provided by ^{14}C AMS radiocarbon dating. Analysis of known archaeological records using GIS-technique provides information of the intensity and time periods of human activities in the study area. The megaliths in close vicinity of the investigated mire present a clearly visible evidence of anthropogenic use during the middle Neolithic. The pollen diagram shows the vegetation development of the study site from the end of the Atlantic, Subboreal and early Subatlantic period. A small forest opening is suggested around 3500 BC, possibly in connection with the construction of the megaliths, but there is no strong evidence of considerable woodland clearances. The archaeological data indicate that human impact in the area took place during the Neolithic, the Bronze Age and Iron Age, which is corroborated by the pollen record, suggesting that human impact in the study area occurred periodically from the end of the Atlantic period, with an increasing intensity during the Bronze Age.

KEY WORDS

Pollen analysis, Neolithic, megalithic graves, human impact, Schleswig-Holstein

INTRODUCTION

In the prehistory of northern Europe megalithic graves belong to the most remarkable and mysterious structures. The time of their construction, as well as their function and role in the development of human culture are intensely discussed topics not only in archaeology, but also in the natural sciences, dealing with the impacts of human activities on the landscape. The boulders pertaining to the megalithic structures in the study area potentially yield information on their geological age and origin, but they do not provide evidence of when or by whom the structures were erected nor their possible function. In archaeology, the question of dating megalithic structures to a particular Neolithic cultural period can be done by characteristic finds (Müller 1997), which are potentially related to different cultures.

Difficulties of such dating are mainly caused by the phenomenon of reusing the megaliths and its surrounding area during thousands of years. Even if the megaliths possessed a particular primary function, this could have been changed several times during the Neolithic and Bronze Age periods (Steinmann 2009). At present, using archaeological and radiocarbon dating methods (Klassen 2001; Persson & Sjögren 1995), the construction time of northern German and Scandinavian megalithic graves is estimated between the late Early Neolithic period until the early Middle Neolithic period. These monuments were probably erected during a very short time period (Schuldt 1976). Forests in the surroundings have been opened possibly in connection with the construction of the megaliths (Andersen 1992). After their construction the monuments were used as graves by people associated with the TRB culture (Hoika 1990). The megaliths become a part of the landscape as objects and indicators of human influences on the landscape.

Pollen analysis has shown that small mires represent valuable archives for the reconstruction of vegetation changes in the immediate surrounding area (Behre & Kučan 1986; Kühl 1998; Prentice 1985; Rickert 2001) and are thus particularly suitable to track small-scale human impact on the landscape. Changes of vegetation resulting from human impact, including the erection of megalithic graves, are often reflected in the pollen records by the occurrence of pollen grains of anthropogenic indicators. In this study, we investigate whether the erection and use of megalithic graves is reflected in the pollen diagram. By combining the pollen record with archaeological data, we are attempting to identify whether and when local woodland clearance took place as a possible result of the construction and use of the megalithic graves.

MATERIALS AND METHODS

Study area

The research site 'Krähenberg' (Crows Hill) is located in northern Germany within the municipality of Westensee, Schleswig-Holstein, south-west of the city of Kiel. The area belongs to the Westensee and mo-

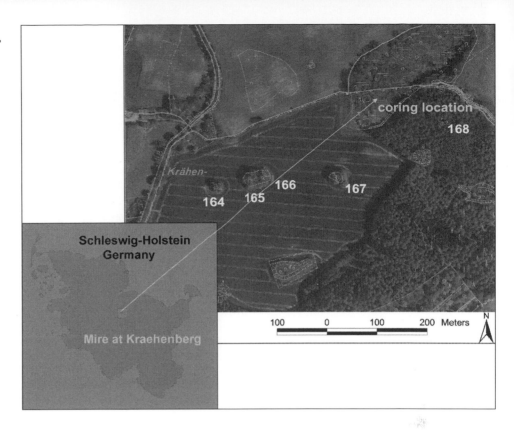

Figure 1. Study site Krähenberger mire 54°15'40.15" N, 9°54'10.05"E (pollen coring location: arrow) and five megalithic graves (164–168) after Sprockhoff (map Digital Ortho Photo 1:5000, Land Surveying Office Schleswig-Holstein ® 2007).

raine region (Meynen & Schmithüsen 1962). The landscape was formed by the Weichselian glaciation spreading from the Baltic Sea almost to the middle of Schleswig-Holstein. The climate can be classified as temperate humid, the mean annual precipitation ranges from 750 to 850mm and the mean annual temperature is 8.5 °C. The dominating soils in the area are cambisols on sand and clay or luvisols on boulder clay (Kielmann 1996). Today, woodlands are dominated by beech (*Fagus sylvatica*) or planted conifers. Historically old woodland is found in the investigation area. The wooded area is separated from agricultural fields by hedge rows. Small land depressions (kettle holes) are numerous due to Late Glacial dead ice relicts. They developed into lakes and were later transformed into small mires (Dierssen 2005). Due to the usually continuous deposition of sediments, these old, small (1-8 ha) mires are especially valuable for a high-resolution pollen analytical reconstruction of local vegetation and landscape history (Rickert 2006). The Krähenberg mire (Crows Hill mire) is located in the Westensee region about 2 km south from the Lake Westensee (54°15'40.15" N, 9°54'10.05"E). The mire surface covers an area of 3.45 ha and was separated in the past by the hedge bank. At a distance of approximately 100m of the mire, five megalithic graves exist. The graves 164-168 (Sprockhoff 1966) are arranged in a more or less straight line and nowadays are surrounded by arable fields (fig.1).

GIS-analyses of the archaeological records and geomagnetic surveys

In a periphery of 5 km diameter of the study site the known archaeological records of all prehistoric time periods were analysed using GIS-techniques. The morphology of the area adjacent to the graves was investigated using airborne laser altimetry data (Land Survey Office Schleswig-Holstein® 2009). Additionally, geomagnetic surveys at this site have been done.

Peat coring and pollen analysis

A 560 cm long peat core (KRM) was retrieved in July 2009 from the centre of the western part of the Krähenberger mire using the high-precision rod-operated Usinger piston corer (Mingram et al. 2007). The extracted undisturbed 1m-long cores are 80mm in diameter. Stratigraphic features were recorded in the field. Cores were cut longitudinally, thus two halves are available in half plastic tubes for sampling.

The uppermost 2.24m of the core was sampled in 2cm intervals. Samples were processed following standard laboratory techniques (Fægri & Iversen 1989; Moore 1991), and microscopically analysed with 400x and 1000x magnification by using the pollen reference collection of the Palaeoecology research group at the Institute for Ecosystem Research, University of Kiel, as well as the pollen atlas by Beug (2004). Glycerine was used as the embedding medium to prepare slides. A minimum of 500 arboreal pollen grains were counted in each sample. Microcharcoal fragments bigger than 10 μm were counted to reflect fire events (Tinner & Hu 2003). The pollen diagram was constructed using the program TGView© Version 2.1 (Grimm 1994). Percentage calculations of pollen taxa and types are based on the terrestrial pollen sum. Additionally, the AP/NAP ratio was calculated. For that, pollen grains of *Corylus* (hazel) were excluded from the AP and calculated as NAP (Overbeck 1975). Pollen grains of wetland plants, Cyperaceae and aquatic plants are excluded from the terrestrial pollen sum, as well as spores of cryptogams. The pollen diagram is composed as follows, from left to right: tree taxa indicating long distant pollen transport (*Pinus*), trees and shrubs of the local and regional vegetation, plants of the heath family (Ericales), upland herbs, cereals, anthropogenic indicators, cryptogams, wetland and aquatic plants, non-pollen palynomorphs and microcharcoal fragments (> 10 μm).

^{14}C AMS dating and radiocarbon calibration

Five ^{14}C measurements with the accelerator mass spectrometry system AMS were provided by the Leibniz-Laboratory for Radiometric Dating and Stable Isotope Research, University of Kiel (Grootes et al. 2004). Sediment samples were checked under the microscope and an appropriate amount of material was selected for dating. Selected material was then extracted with 1% HCl, 1% NaOH at 60°C and again 1% HCl alkali residue. The combustion to CO_2 was performed in a closed quartz tube together with CuO and silver wool at 900 °C. The sample CO_2 was reduced at 600 °C with H_2 over about 2 mg of Fe powder as catalyst, and the resulting carbon/iron mixture was pressed into a pellet in the target holder. Conventional ^{14}C ages were calculated according to Stuiver & Polach (1977) with a $\delta^{13}C$ correction for isotopic fractionation based on the $^{13}C/^{12}C$ ratio measured by the AMS-system simultaneously with the $^{14}C/^{12}C$ ratio. 'Calibrated' or calendar ages were calculated using OxCal v4.1.6 (Bronk Ramsey 2009), data set IntCal09 (Reimer et al. 2009).

RESULTS AND DISCUSSION

Archaeological evidence for prehistoric landscape use

In a periphery of 5 km of the study site there are about 250 located places of discovery from all archaeological periods. Archaeological records from the Palaeolithic and Mesolithic are represented by three records of stone tools including a tranchet axe. In the area of Krähenberg several megalithic structures and Neolithic findings are recorded: 17 megalithic graves, 9 earth graves, 34 findings of silex-axe, silex-chisel, four records of silex-knifes and seven stone axes. Altogether 58 archaeological records in the surrounding of Krähenberg date back to the Neolithic (fig. 2). The graves at Krähenberg have been described by Sprockhoff as three extended dolmens and two passage graves (Sprockhoff 1966, 164-168), but have not been investigated in detail or dated radiometrically. Dating of similar structures in Schleswig-Holstein and in Northern Germany (Fansa 2000; Baldia 1995/2009; Midgley 1992) yielded dates of early to middle Neolithic age. Based on the types of the found artifacts, especially the silex-axes (fig. 3), the assumed human impact in the study area can be put mainly in the time of Middle Neolithic (3500-3300/3200 BC). Therefore, it is suggested that the graves at Krähenberg might date back to the same period in Northern Germany, i.e. the Funnel Beaker (TBK) culture (4100/4000-2800/2700 BC).

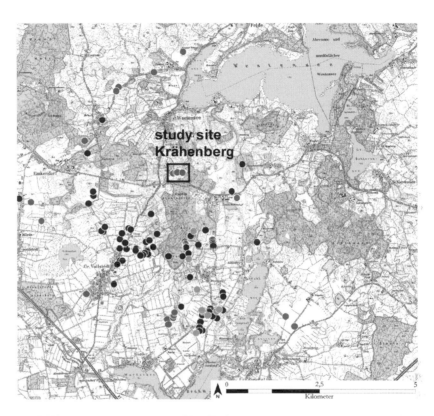

Figure 2. Archaeological records from the Neolithic (4100-1800/1700 BC) around the study site: megalithic graves (dark gray), earth graves (light gray), and findings (black).

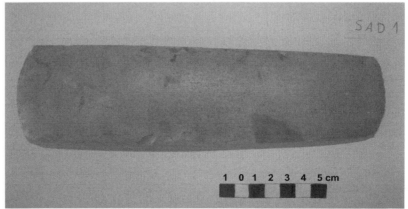

Figure 3. Silex-axe (TBK-culture) from the study site.

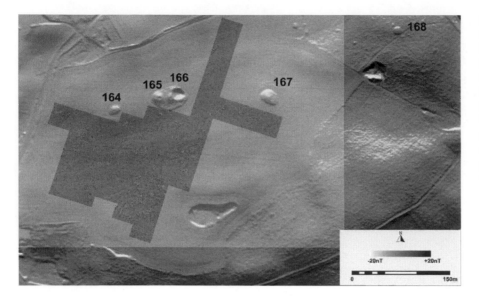

Figure 4. Digital terrestrial model based on airborne laser altimetry LIDAR-data (Land Surveying Office Schleswig-Holstein ® 2009) and magnetogram of the study site (graves: Sprockhoff 164-168).

Studies of the early Bronze Age (Aner et al. 2005) show in contrast to the archaeological recordings of the Archaeological State agency of Schleswig-Holstein (only 15 records) continuous and intensive human activities in the area, especially in the surroundings of the study site. The Bronze-age finds are usually devices and decoration made of bronze or also made of amber. The Iron Age however is particularly represented by ceramic findings and the remains of urn grave fields.

The megaliths at Krähenberg represent the first, clearly visible evidence of the landscape use by people during the Middle Neolithic. The magnetogram shows unknown anomalies and remains of the mound fill and of the stone kerbs surrounding the graves (fig. 4). The two rows are 11m apart from each other, run over a distance of ca.100m along the graves. The accurate parallelism of the rows within an otherwise rather unimpaired range south of the graves are clearly of anthropogenic origin. They can be interpreted as prehistoric construction structures (Sadovnik et al. 2010). However, clarity can only be brought by an excavation.

Pollen analysis and radiocarbon calibration

The pollen record (fig. 5) shows that the upper peat layer of the mire was removed by historical peat cutting in the 18th century (Hedemann-Heespen 1906). The pollen of Beech (*Fagus*) as a dominant tree species in natural woodland communities since the migration period in Northern Germany (Schmitz 1951; Aletsee1959; Wiethold 1998; Wiethold & Lütjens 2001) is not sufficiently present in the upper part of the core. However, the periods since the Neolithic to Bronze Age are fully presented in the pollen record, without any indication for a hiatus, thus providing a high resolution record of the Neolithic and Bronze Age. This is supported by the five ^{14}C AMS datings.

Nine local pollen assemblages zones (LPAZ) were identified according to the pollen assemblages (Birks 1986; Hedberg 1972a, b) KRM-1 to KRM-9 (table 1). Ages of the zone boundaries and samples age were modelled with OxCal v4.1.6 (Bronk Ramsey 2009), based on the five ^{14}C AMS samples from peat (table 2). AMS measurements covered a period from about 4000 cal BC until 361-204 cal BC.

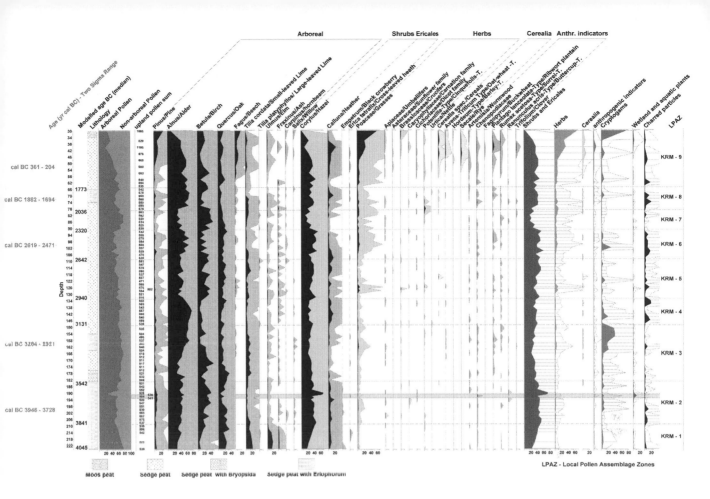

Figure 5. High resolution pollen diagram KRM 0.30 – 2.24 m (simplified, only selected pollen types/taxa). *See also full colour section in this book*

The lowermost zone KRM-1, from 4045 BC to 3841 BC, represents the transition from the Atlantic to the Subboreal period. At the beginning of this period the study area was covered by deciduous forests, dominated by oak (*Quercus*) with high amounts of lime (*Tilia*) and elm (*Ulmus*). Decrease in *Ulmus* and *Tilia* pollen around 3841 cal BC, at the depth of 2.08m, is interpreted as the 'elm decline'. The chronology of these phenomena is not yet fixed to an absolute date, but a number of conventional and AMS datings assemble it around 3800 cal BC in Schleswig-Holstein (Nelle & Dörfler 2008). The change in the pollen spectra of arboreal pollen and increase in pollen of *Corylus* indicate some human impact on the landscape. At this time, the change from the Mesolithic culture of hunters and gatherers to Neolithic economic systems with agriculture and permanent settlements occurred in Northern Germany (Behre 2008). Change of burial rites also took place at this time with the beginning of the period of megalithic graves.

Further evidence of an anthropogenic effect on the vegetation composition is seen in the pollen record from 3841 to 3542 BC, in zone KRM-2. At the depth from 1.90m to 1.94m (3646-3525 cal BC) the diagram shows a succession of the occurrence of grasses, Ericales and a marked peak of *Corylus*. A decrease of oak and lime in the pollen diagram and an increase of wetland and aquatic plants can be interpreted as a short but significant, moderate opening of the landscape, which might have been related to the con-

Table 1. Local pollen assemblage zones of the pollen record KRM (0.30 – 2.22 m depth).

LPAZ/AGE	ZONE	Depth (m)	Characteristics
KRM-9 from 1773 BC	Alnus-Corylus-Poaceae Zone	0.63 – 0.30	Increase of grasses, cereals and anthropogenic indicators, decrease of Tilia (0.30 cm end of core)
KRM-8 (2036-1773 BC)	Alnus-Corylus Zone	0.79 – 0.63	Decrease of arboreal pollen, increase of anthropogenic indicators and charcoal
KRM-7 (2320-2036 BC)	Betula-Quercus-Tilia Zone (forest regeneration)	0.91 – 0.79	Decrease of Alnus,Corylusand Ericales, increase of Betula and Tilia
KRM-6 (2642-2320 BC)	Alnus-Corylus Zone	1.09 – 0.91	Increase of grasses and Plantago lanceolata-type, increase of Corylus and charcoal particles
KRM-5 (2940-2642 BC)	Betula-Corylus Zone	1.33 – 1.09	Decrease of Alnus, increase of Betula and Corylus, increase of Tilia, small peak of grasses
KRM-4 (3131-2940 BC)	Alnus-Calluna Zone	1.49 – 1.33	Increase of Alnus and Calluna, increase of charcoal particles, slight decrease of Corylus decrease of Tilia
KRM-3 (3542-3131 BC)	Betula-Quercus-Tilia Zone (forest regeneration)	1.83 – 1.49	Increase of Tilia,and increase Quercus in lower part of zone, increase of grasses, very low charcoal value
KRM-2 (3841-3542 BC)	Alnus-Betula-Corylus Zone	2.09 – 1.83	Increase of Pinus-pollen, Corylus-peak, slight increase of grasses, decrease of Tilia, decrease of Quercus in upper part of zone
KRM-1 (4045-3841BC)	Tilia-Ulmus-Quercus Zone	2.24 - 2.09	Decrease of Tilia, elm-decline, transition from the Atlantic to the Subboreal 3841 BC

struction of megaliths at Krähenberg. A – probably short – opening of the forest might have resulted in an increase of surface runoff and spring activity, which promoted the wetness of the mire. The peak of *Corylus* is especially remarkable. At the moment, it is not yet possible to tell whether small anthropogenic gaps in the forest were already present before the construction of megaliths, or whether this peak might be related to some canopy opening during and after the construction. *Corylus* produces especially high numbers of pollen after disturbance (cutting) and has a relatively high light demand for its growing. This peak will be checked for reproducibility with a second core in the future. The increase of long-distance

Table 2. ¹⁴C-AMS dating of the core KRM.

Lab. ID	Sample	Depth (m)	Fraction	Corrected pMC†	δ13C(‰)‡	Conventional Age (14C yr BP)	2–σ range(s) cal yr BC (probability 95,4 %)
KIA40648	KRM-52	0.52	Peat, alkali residue, 4,3 mg C	75,87 ± 0,20	-29,60 ± 0,16	2219 ± 21	378 – 341 (17,2 %) 326 – 204 (78,2 %)
KIA40649	KRM-72	0.72	Peat, alkali residue, 4,7 mg C	64,95 ± 0,22	-27,21 ± 0,37	3466 ± 27	1882 – 1735 (89.7 %) 1714 – 1694 (5.7 %)
KIA40631	KRM-100	1.00	Peat, alkali residue, 5,2 mg C	60,62 ± 0,22	-27,86 ± 0,26	4020 ± 30	2619 – 2607 (2.9 %) 2599 – 2593 (1.0 %) 2586 – 2471 (91.6 %) 2770 – 2763 (1.0 %)
KIA40632	KRM-160	1.60	Peat, alkali residue, 3,1 mg C	57,71 ± 0,19	-27,96 ± 0,28	4416 ± 25	3264 – 3241 (4,8 %)
			Peat, humic acid, 4,9 mg C	59,12 ± 0,21	-29,53 ± 0,16		3105 – 2921 (90,6 %)
KIA40633	KRM-200	2.00	Peat, alkali residue, 4,2 mg C	53,45 ± 0,20	-25,41 ± 0,25	5032 ± 31	3946 – 3760 (90.6 %) 3741 – 3728 (1.9 %) 3726 – 3714 (2,9 %)

transported pollen of *Pinus* is also a sign of a temporal opening of the woodlands around the mire, during a comparatively short period of time.

For the Zone KRM-3 from 3542 to 3131 cal BC the increase of grass pollen as well as a slight increase of other anthropogenic indicators like *Plantago*-Type and *Rumex*-Type can be recorded. At this time the increase of lime and oak pollen percentages indicates a regeneration of forest ecosystems. However, this phenomenon can be interpreted as a sign for using the tree leaves to provide fodder (Andersen 1990). Pollarded trees tend to produce more flowers (and pollen) in the upper parts of the crown. Rickert (2006) discusses this use of Neolithic forests to exist locally and for a short period, not longer than 500 years. The landscape in the surrounding of the megaliths at that time might have been a semi open pasture. The openings in the forested landscape must have been small and local. The microcharcoal amounts are low. Very few pollen grains of the *Cerealia*-Type were found in the pollen record during this period. Thus, we do not assume arable fields to be present close to the mire.

For the zone KRM-4, from 3131 to 2940 cal BC, a rise of *Alnus* (alder, growing on wet sites) up to 70% was observed. The curves of *Quercus*, *Tilia* and *Betula* (birch) show very low values. *Corylus* decreases, while *Calluna* (heath) increases. Charred particles increase during this pollen zone, which indicates more fire activity, which is most likely connected to human activities.

Pollen zone KRM-5, from 2940 to 2642 cal BC, is characterised by elevated frequencies of grasses,

Corylus and ruderal plants. While *Alnus* pollen decreases, *Betula* increases. A regeneration of the forest in the older part of the zone is reflected by a slight increase of *Tilia* and *Quercus*. This zone corresponds to the end of the TBK cultural period in Northern Germany (2800/2700 BC) and the beginning of the Single Grave period.

In zone KRM-6, from 2642 to 2320 cal BC, the pollen record shows again an increase of *Alnus* and *Poaceae* (wild grass) pollen as well as elevated frequencies of charred particles. This is paralleled by high values of *Corylus* and an increase of anthropogenic indicators, but there is no indication of cereal cultivation. A moderate intensity of human impact can be assumed for this period. During the following zone 7 (2320 to 2036 cal BC) the regeneration of forests (increase of *Tilia*) together with the reduction of pollen of the anthropogenic indicators, *Poaceae* and *Calluna* can be observed in the diagram. From 2036 to 1773 cal BC (KRM-8), arboreal pollen decrease, while grass pollen grains increase, as well as microcharcoal and anthropogenic indicators. At the end of the late Neolithic period (c. 1800 cal BC) the stratigraphic boundary at the depth of 0.78m is indicated with a change of the peat composition of the mire from sedge peat with *Bryopsida* to sedge peat with *Eriophorum*. Sequence calibration of AMS-data of the core KRM shows that in this period the peat growth of the mire decreased dramatically. The age for this period was modelled from 1898 to 1773 cal BC and corresponds to the beginning of the Early Bronze Age in Northern Germany. Clear evidences of land use with arable fields and cereal farming could be detected only for the Bronze Age (pollen zone KRM-9) with an increase of cereal pollen grains.

CONCLUSIONS

Though there is no clear signal of woodland clearance, the forest was opened slightly in a period between 3560-3512 BC (by modelled age using OxCal v4.1.6) in the immediate surroundings of the megaliths. We assume that the forest was opened in connection with the construction of the megaliths c. 3500 BC. After this short, moderate opening, the surrounding landscape was used as a semi-open pasture, but with no grain cultivation. Forest regeneration took place thereafter over a period of c. 500 years.

A further forest regeneration around the megaliths after their construction permits conclusions over the role of megalithic graves in the landscape of Middle Neolithic in Schleswig-Holstein. As is the case for current investigations of the megaliths of Altmark (Demnick et al. 2008), the graves at Krähenberg lie at the crow mountain in an isolated manner from settlement and arable fields areas in the time of the TBK-culture. After the construction of the megalithic graves the site probably might have been used as a cult and ritual place, but the graves were present in a wooded landscape. Clear evidence of land use with arable fields and cereal cultivation was recorded only for the Bronze Age.

Forest composition changed with a succession from *Quercus* to *Betula* to *Alnus* periodically. The reasons and processes of this phenomenon are yet to be understood.

ACKNOWLEDGEMENTS

We would like to thank Hartmut Usinger and Aiko Huckauf for essential support with coring, and Ingmar Unkel, Andrey Mitusov, Johannes Müller and Susann Stolze for their help and useful discussions. We are also grateful to the forest management of the manor Deutsch-Nienhof, particularly to the landowner Sven von Hedemann-Heespen, for the friendly permission to carry out field and archive research on the territory of Deutsch-Nienhof. We are grateful to the Leibniz-Laboratory for Radiometric Dating and Stable Isotope Research (University of Kiel, Germany) for providing the AMS measurements and to Georg Schafferer, (Institute of Prehistoric and Protohistoric Archaeology, University of Kiel) for providing the geomagnetic survey. The paper benefitted from reviews by Dr S.J.P. Bohncke (VU University Amsterdam, the Netherlands) and an anonymous reviewer.

REFERENCES

Aletsee, L. 1959, Zur Geschichte der Moore und Wälder des nördlichen Holsteins. *Nova Acta Leopoldina Neue Folge* 139 (21), 51 S.

Aner, E., Kersten, K. & K.-H. Willroth. 2005. Kreis Rendsburg-Eckernförde (südlich des Nord-Ostsee-Kanals) und die kreisfreien Städte Kiel und Neumünster. Bearbeitet von Karl Kersten und Karl-Heinz Willroth mit einem Beitrag von Bernd Zich. *Die Funde der älteren Bronzezeit des nordischen Kreises in Dänemark, Schleswig-Holstein und Niedersachsen* 19. Neumünster.

Andersen, S.Th. 1992. Early and Middle Neolithic agriculture in Denmark: Pollen spectra from soils in burial mounds of the Funnel Beaker Culture. *Journal of European Archaeology* 1, 153-180.

Andersen, S.Th. 1990. Changes in agricultural practices in the Holocene indicated in a pollen diagram from a small hollow in Denmark. In: Birks, H.H., Birks, H.J.B., Kaland, P.E., & Moe D. (eds.) *The Cultural Landscape – Past, Present, Future*, 395-407, Cambridge, Cambridge University Press

Baldia, M.O. 1995/2009. A Spatial Analysis of Megalithic Tombs. Vol. 1-2. Ph. D. Dissertation. Southern Methodist University. http://www.comp-archaeology.org/06nonmeg.htm

Behre, K.-E. 1981. The interpretation of anthropogenic indicators in pollen diagrams. *Pollen et Spores* 23, 225-245.

Behre, K.-E. & D. Kučan. 1986. Die Reflektion archäologisch bekannter Siedlungen in Pollendiagrammen verschiedener Entfernung. Beispiele aus der Siedlungskammer Flögeln, Nordwestdeutschland. In , K.-E. Behre (ed.), *Athropogenic indicators in pollen Diagramms*. Rotterdam/Boston, A.A. Balkema, 95-114.

Behre, K.-E. 2008. *Landschaftsgeschichte Norddeutschlands, Umwelt und Siedlung von der Steinzeit bis zur Gegenwart*. 300 S., Neumünster, Wachholtz.

Beug, H.-J. 2004. *Leitfaden der Pollenbestimmung für Mitteleuropa und angrenzende Gebiete*. München, F. Pfeil Verlag.

Birks, H.J.B. 1986. *Numerical zonation, comparision and correlation of quaternary pollen stratigraphical data. Handbook of Holocene Palaeoecology and Palaeohydrology*. J. Wiley & Sons, Chichester, 734-744.

Demnick, D., Diers, S., Fritsch, B., Müller, J., Bork, H.-R. 2009. Bestimmend für die Raumnutzung-Grosssteingräber der Altmark. *Archäologie in Deutschland* 4, 34-40.

Dierssen, K. 2005. Vegetation of Schleswig-Holstein. In *Excursion Guide Dedicated to Leopoldina-Meetings, Change of the Earth Surface in Past Century*. Ecology Centre, CAU Kiel.

Fægri, K. & J. Iversen. 1989. *Textbook of Pollen Analysis* 4. Wiley, Chichester.·

Grimm, E.C. 1994. *Tilia Graph 2.1*. Illinois State Museum, Springfield.

Grootes, P.M., Nadeau, M.-J., Rieck, A., 2004. ¹⁴*C-AMS at the Leibniz-Labor: Radiometric dating and isotope research. Nuclear Instruments and Methods in Physics Research*. Section B 223-224, 55-61.

Hedberg, H.D. (ed.) 1972a. *Introduction to an International Guide to Stratigraphic Classification, Terminology and Usage*. Report 7a – Boreas 1: 199-211.

Hedberg, H.D. (ed.) 1972b. *Summary of an International Guide to Stratigraphic Classification, Terminology and Usage*. Report 7b – Boreas 1: 213-239.

Hedemann-Heespen, P. 1906. *Geschichte der adeligen Güter Deutsch-Nienhof und Pohlsee*. Bänder I-III.

Hoika, J. 1990. Megalithic graves in the Funnel Beaker culture of Schleswig-Holstein. *Przeglad archeologiczny* 37, 53-119.

Klassen, L. 2001. Frühes Kupfer im Norden. Untersuchungen zu Chronologie, Herkunft und Bedeutung der Kupfer-funde der Nordgruppe der Trichterbecherkultur. *Jysk Arkæologisk Selskab* 36, Moesgård.

Kühl, N. 1998. Pollenanalytische Untersuchungen zur Vegetations- und Siedlungsgeschichte in einem Kesselmoor bei Drangstedt, Ldkr. Cuxhaven. *Probleme der Küstenforschung im südlichen Nordseegebiet* 25, 303-324.

Fansa, M. 2000. *Großsteingräber zwischen Weser und Ems*. Isensee Verlag, Oldenburg.

Klassen, L. 2001. Frühes Kupfer im Norden. Untersuchungen zu Chronologie, Herkunft und Bedeutung der Kupfer-funde der Nordgruppe der Trichterbecherkultur. *Jysk Arkæologisk Selskab* 36, Moesgård.

Kielmann, K. 1996. Deduction of Soil Maps and Ecological Data from Reich soil assessment for the Community Westensee. Diploma Thesis, Christian-Albrechts University, Kiel.

Lang, G. 1994. *European quaternary vegetation history. Methods and results*. Gustav Fischer publishing, Jena/Stuttgart/New York.

Land Survey Office Schleswig-Holstein®. 2005. Digital topographic maps in scale 1:5,000 DTK 5 and ATKIS® Digital Orthophotos DOP 5 in scale 1:5,000.

Land Survey Office Schleswig-Holstein®. 2009: ATKIS®-DGM1.

Moore, P.H., Webb, J.A. & M.E. Collinson. 1991. *Pollen Analysis*. Oxford, Blackwell Scientific Publications.

Midgley, M.S. 1992. *TRB-culture. The First Farmers of the North European Plain*. Edinburgh, Edinburgh University Press.

Mingram, J., Negendank, J.F.W., Brauer, A., Berger, D., Hendrich, A., Köhler, M. & H. Usinger. 2007. Long cores from small lakes – recovering up to 100 m long lake sediment sequences with a high-precision rod-operated piston corer (Usinger-corer). *Journal of Paleolimnology* 37, 517-528.

Meynen, E. & J. Schmithüsen. 1962. *Handbuch der naturräumlichen Gliederung Deutschlands*. Bundesamt für Raumord-nung und Landeskunde, Bonn.

Müller, J. 1997. Zur absolutchronologischen Datierung der europäischen Megalithik. In B. Fritsch, M. Maute, I. Matuschik, J. Müller, C. Wolf Hrsg., *Tradition und Innovation*. Espelkamp, Festschrift Ch. Strahm.

Nelle, O. & W. Dörfler. 2008. A summary of the Late- and Post-glacial vegetation history of Schleswig-Holstein. In Jürgen Dengler, Christian Dolnik, Michael Trepel (eds.), *Flora, Vegetation, and Nature Conservation from Schleswig-Holstein to South America – Festschrift for Klaus Dierßen on Occasion of his 60th Birthday* 65, 45-68, Selb-stverlag, Kiel.

Persson, P. & G. Sjögren. 1995. Radiocarbon and the chronology of Scandinavian megalithic graves. *Journal of European Archaeology* 3 (2), 59-88.

Prentice, I.C. 1985. Pollen Representation, Source Area and Basin Size: Toward a Unified Theory of Pollen Analysis. *Quaternary Research* 23, 76-86.

Reimer, P.J. et al. 2009. IntCal09 and Marine09 radiocarbon age calibration curves, 0–50,000 years cal BP. *Radio-carbon* 51 (4), 1111-50.

Rickert, B.-H. 2001. Untersuchungen zur Entwicklungsgeschichte und rezenten Vegetation ausgewählter Kleinst-moore im nördlichen Schleswig-Holstein. In: *Mitteilungen der Arbeitsgemeinschaft Geobotanik in Schleswig-Holstein und Hamburg* 60, Kiel.

Rickert, B.-H. 2006. Kleinstmoore als Archive für räumlich hoch auflösende landschaftsgeschichtliche Untersu-chungen – Fallstudien aus Schleswig-Holstein. *EcoSys. Beiträge zur Ökosystemforschung* 45, 173.

Sadovnik, M., Schafferer, G., Mischka, C., Nelle, O. in press. *Pollendiagramm und Magnetogramm: Eine Verknüpfung von paläobotanischen und archäologischen Methoden in der neolithischen Forschung am Krähenberg, Westensee. Die Archäologischen Nachrichten aus Schleswig-Holstein 2010*. Archäologischen Landesamt und der Archäologis-chen Gesellschaft Schleswig-Holstein, Wachholtz Verlag.

Schmitz, H. 1951. Die Zeitstellung der Buchenausbreitung in Schleswig-Holstein. *Forstwissenschaftliches Centralblatt* 70 (4), 193-203.

Schuldt, E. 1976. Die mecklenburgischen Megalithgräber. *Ausgrabungen und Funde* 21, 54-58.

Sprockhoff, E. 1966. *Atlas der Megalithgräber Deutschlands*. Habelt Verlag, Bonn.

Spek, T., Groenman-VanWaateringe, W., Kooistra, M., Bakker, L. 2003. Formation and land-use history of Celtic fields in north-west Europe. An interdisciplinary case study at Zeijen, the Netherlands. *European Journal of Archaeology* 6 (2), 141-173.

Steinmann, C. 2009. Großsteingräber in Mecklenburg-Vorpommen-wichtige Orte mit Bestattungen. *Archäologie in Deutschland* 4, 30-34.

Stuiver, M. & H.A. Polach. 1977. Discussion: Reporting of 14C Data. *Radiocarbon* 19 (3), 355-363.

Tinner, W. & Hu, F.S. 2003. Size parameters, size-class distribution and area–number relationship of microscopic charcoal: relevance for fire reconstruction. *The Holocene* 13, 499-505.

Wiethold, J. 1998. Studien zur jüngeren postglazialen Vegetations- und Siedlungsgeschichte im östlichen Schleswig-Holstein (mit einem Beitrag von H. Erlenkeuser). *Universitätsforschungen zur prähistorischen Archäologie* 45, Habelt (Bonn).

Wiethold, J. & I. Lütjens. 2001. Paläo-ökologische Untersuchungen an jahresgeschichteten Sedimenten aus dem Belauer See, Kr. Plön, Schleswig-Holstein. Ergebnisse zur Vegetations- und Siedlungsgeschichte des westlichen Ostholsteins von der vorrömischen Eisenzeit bis zum hohen Mittelalter. *Regensburger Beiträge zur Prähistorischen Archäologie* 7, 239-257.

2.4 Geo- and Landscape archaeological investigations in south-western Lazio (Italy): An approach for the identification of man-made landscape transformation processes in the hinterland of Rome

Authors

Michael Teichmann and Hans-Rudolf Bork

Graduate School 'Human Development in Landscapes', University of Kiel, Kiel, Germany
Contact: mteichmann@gshdl.uni-kiel.de

ABSTRACT

The landscape of south-western Lazio is studied in a combined geo- and settlement archaeological approach, focusing on the changes caused by humans in the Roman Republican and Imperial Age. First results of a brief, non invasive assessment of the nature and extent of man-made landscape alterations in a part of the suburban hinterland of Rome are presented and the methodological frame of the approach is outlined.

KEYWORDS

geoarchaeology, Settlement Archaeology, GIS, Republican and Imperial Rome, Central Italy, Lazio

INTRODUCTION

The relation of man to his surrounding environment and the diachronic, dynamic changes of settlement patterns on a regional scale are key aspects for the understanding of past landscapes. These aspects are central issues on the research agenda of a project focusing on the Roman Republican and Imperial Era in central Italy, for which geoarchaeological field research is combined with GIS based analysis of settlement site preference factors.

The study area is situated in south-western Lazio, the immediate hinterland of Rome (fig.1). A geo-archaeological overview survey was conducted in 2009, mainly focusing on the area between the Alban

Figure 1. Overview study area with geoarchaeological and archaeological sites entered so far. *See also full colour section in this book*

Sites held in the GIS so far
● Geoarchaeological Sites
○ Archaeological Sites
▒ Study area geoarchaeology

0 5 10   20 Kilometres

N

hills and the Astura River, therefore the northern part of the overall study area, which reaches from Rome and Ostia to Terracina and from the Tyrrhenian Sea to the Alban Hills and Lepini an Ausoni Mountains. This paper introduces this particular aspect of the project outlining the chosen approach and presenting first results.

SCIENTIFIC OBJECTIVES

The main objective of the geoarchaeological research was to gain an overview on quality and quantity of human-induced landscape transformation processes in the past within the study area. While all stages of human action may have affected the formation of erosional deposits in one way or the other, the particular focus was on the Period of Roman Antiquity, for which the settlement patterns and settlement dynamics are studied as part of the wider project.

The type and the quantity of erosion may be used as an indicator to questions as: Did deforestation take place? Was the land used for pasture or farming? Was the use of the landscape sustainable or did the land use deteriorate the former conditions significantly? Erosion could have been caused by heavy rainfall events, which may have affected the past population and land use. So, are there any indications for heavy rainfall events?

We did not expect to find definite answers to all posed questions in a single campaign, but the aim was to get an idea of general trends and to evaluate the potential for further in-depth research.

Similar phenomena of landscape transformation processes in younger periods were regarded as relevant in respect to the primary research objectives as well: In case of significant landscape transformations in post-Roman times, the archaeological record might be buried or eroded and would therefore affect the distribution of recordable archaeological sites, which are used as a basis for settlement pattern analysis. Archaeologists are aware of the effects of post-depositional processes on the archaeological record and its visibility, but it is difficult to assess the actual role of these factors, without the application of geoarchaeological research. The topography of Roman Times and the changes since then may be reconstructed by geoarchaeological research, too.

Research questions linked to this aspect were: How significant were post-Roman landscape alterations? Do we find areas, where we have to be particular aware of 'modern' landscape transformation processes?

As a starting point, it was assumed that the area was heavily exploited in Pre-Roman/Protohistoric Times as well as in the Roman Period with a decline of exploitation from the Late Antiquity onwards. A further peak of exploitation was expected for Recent and Sub-recent Times.

HISTORICAL AND ARCHAEOLOGICAL BACKGROUND

The latter assumption was made due to the history of the study area, which is described briefly as the awareness of historical developments and of the ideological concept of the Roman suburban area are crucial factors for the understanding of its formation and transformation.

Within the study area numerous Protohistoric centres were situated. Well known places like Alba Longa, Ardea, Aricia, Lanuvium and Lavinium are to be mentioned as a few examples. In the course of the Roman expansion, the former pattern was succeeded by different stages of intensive rural and urban settlement in Roman Republican and Imperial Times when the area was part of the heartland of Roman Italy. The wider *suburbium*, the hinterland of Rome, whose exact definition and extent is controversially discussed (Mayer 2005, 16-18; Witcher 2005, 191), was linked in numerous ways to the faith of the metropolis: Economically, the immediate surrounding was very suitable to grow food, that could not be transported over long distances, like vegetables, and to supply the growing *urbs* with meat (in particular poultry) and staples. With the growth of the Empire and its power to draw on resources from remote provinces and in respect to the exemption of direct taxes for Roman citizens from 167 BC onwards, the economical freedom for the rural population will have been increased and may have enabled an even more differentiated economical use of the suburban landscape (Kolendo 1994; Moorley 1996). Production was specialised that even roses were grown for economical purposes and sold to the capital (Moorley 1996, 87). The growth of Rome and the related demographic implications will have been affected by and will have had major effects on the demography of the wider hinterland (Witcher 2005).

In addition, the *suburbium* played an important social function for society (Mayer 2005; Adams 2008). In particular from Late Republican Times onwards, the elites withdrew from Rome for recreational purposes to the countryside. Villas used for these purposes often maintained a productive complex, but yields were not the primary concern in contrast to remote agricultural production estates (Mayer 2005, 33). Emperors chose, among other places, the Alban hills, and rural Latium or seaside resorts to escape the summer heat, stimulating the nobility to imitate this custom. The imperial villas of Castel Gandolfo, Tor Paterno, Anzio and the so-called Villa of the Quntilii close to the Via Appia can be mentioned as examples. Summing up all these factors, a complex mesh of motivations will have been relevant for the organisation of space in the hinterland of Rome.

METHODS

The theoretical concept follows the four dimensional ecosystem analysis, developed by Bork, which considers all temporal and spatial factors and processes, which lead to the creation of a soil deposit (Bork 2006, 18-22; compare for a similar methodology Bruckner & Vött 2008).

An empirical research strategy was applied to find suitable geoarchaeological sites. The majority of roads outside built-up areas were driven along to locate road cuts and building sites, where soil exposures had been created. The position of potential sites was recorded by GPS. Soil profiles orientated parallel to the direction of soil deposition (that is in the direction of the slope inclination) were generally preferred.

Figure 2. Studied geoarchaeological sites, mentioned in the text. (Source: Google Earth, modified).

In a second stage the stratigraphy of soil exposures with the highest potential was documented (photo, sketch, description) as well as the artifacts, which were found within the sections (fig. 2). Soil samples were taken for further chemical lab analysis. The stratigraphy itself provides evidence for the relative chronology of the different layers to each other, while artefacts contained in the layers give a *terminus post quem* (maximum age: no older than) for the deposition of the respective layer. In case of absence of archaeological artifacts (in our case pottery, brick and tile fragments), radiocarbon dating and/or OSL dating would have been conducted. For the presented cases we were fortunate enough to encounter sufficient datable archaeological material.

The survey was non-invasive as we did not undertake excavations ourselves, which was of advantage in respect to funds and permissions. A disadvantage was that we had no control over the distribution of the recordable sites, which could have been distributed over the different particular landscape zones and topographic situations and be chosen in respect to the vicinity of dated archaeological sites otherwise. The stratigraphy was reconstructed and analysed to explain the factors leading to its formation.

RESULTS AND INTERPRETATION

Landscape transformation probably dating to antiquity

In the course of the enlargement of the modern Via Laurentina, one of the radial roads connecting Rome to south-western Lazio, soil profiles were exposed by construction work (fig. 3). Close to the Via Laurentina (road kilometre sign 13), a layer (M2) with a thickness of approximately 30 to 40cm contained several brick and pottery fragments (indicated by black triangles in the profile), a further layer (M1), contained a single find. While most of the fragmented finds were not datable, a few diagnostic sherds date most probably to Roman Imperial Age, given the shape of the vessel, the matrix and the degree of fusion (figs. 4a-c). One of the sherds was probably part of an amphora. All of them were coarse ware, which is on the one hand in respect to its shape typical for the Roman Period, but was on the other hand produced over longer

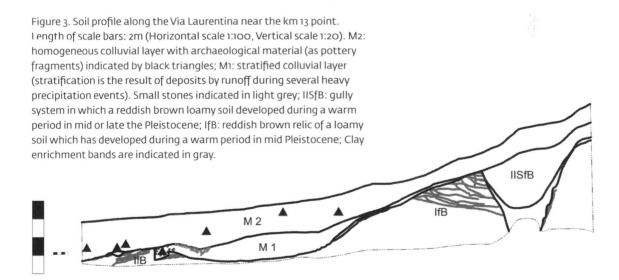

Figure 3. Soil profile along the Via Laurentina near the km 13 point. Length of scale bars: 2m (Horizontal scale 1:100, Vertical scale 1:20). M2: homogeneous colluvial layer with archaeological material (as pottery fragments) indicated by black triangles; M1: stratified colluvial layer (stratification is the result of deposits by runoff during several heavy precipitation events). Small stones indicated in light grey; IISfB: gully system in which a reddish brown loamy soil developed during a warm period in mid or late the Pleistocene; IfB: reddish brown relic of a loamy soil which has developed during a warm period in mid Pleistocene; Clay enrichment bands are indicated in gray.

Figure 4. Pottery fragments from the profile along the Via Laurentina (km 13).

Figure 5. Geoarchaeological site 'Via Laurentina' (km 13) in archaeological context. 158, 160: Settlement sites interpreted as villas by De Rossi, 159 secondary path, 161: Reconstructed course of ancient road, 162: Necropolis with 'Cappuccina tombs'. (Source: De Rossi 1967, map – modified).

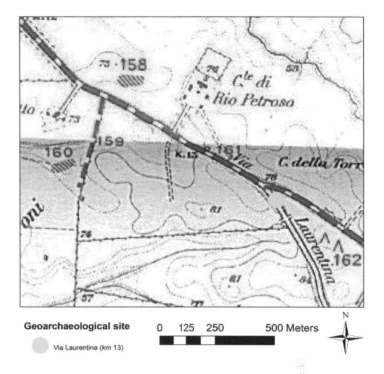

time spans in a similar fashion as coarse ware shapes are mainly defined by their function. No archaeological finds in the respective layers did point to a period later than Roman Antiquity. The soil was compact.

The absence of younger diagnostic finds and the compactness of the deposited material both indicate that the mentioned layer was most probably deposited in the Roman Imperial Age or soon after. The layer might be the result of intensive sheet erosion and deposition during Roman Imperial Age in a period of low vegetation cover density, probably caused by intensive agricultural land use further uphill. Soil erosion during the Roman Imperial Age or soon after had lowered the upper slope area by 20 to 30cm and raised the lower slope area by 30 to 40cm. Soil fertility was most probably reduced significantly during that period as a result of the removal of the ploughing horizon which contained most organic matter.

The interpretation of the geoarchaeological evidence is supported by archaeological evidence. While the archaeological record is fragmentary by its nature, the stage of research is fair for the area under study, as it was partly treated in volumes of the *Forma Italiae* publication series, whose monographs assimilate the style of an archaeological inventory, written in the Italian tradition of ancient topographical research.

Sites close to the Via Laurentina (at the km 13 site) comprise (fig. 5):

1. The ancient road, situated close to the modern road as evidenced by *basoli* (road stones), found in close vicinity (De Rossi 1967, 165, site 161).
2. A necropolis consisting of 'Cappuccina tombs' (a simple type of tomb, where the buried tomb was covered with tiles) was situated approximately 300m to the south-east of the site and was probably of some importance in Imperial Times (De Rossi 1967, 165, site 162).

3. Two sites interpreted by De Rossi as 'Roman villas' are located approximately 820m and 920m to the north and north-west of the site. The publication does not give a dating of these sites, but the presence of *vernice nera* pottery and mosaic *tesserae* in black and white (De Rossi 1967, 162, site 158) at one site and the presence of white mosaic *tesserae* and *opus caementitium* (De Rossi 1967, 163-165, site 160) at the other may allow an approximate dating to Late Republican and Imperial Times. The presence of these sites, in particular of the villas, implies from an archaeological perspective a dense cultivation of the area in Roman times.

A geoarchaeological situation, similar to the Via Laurentina evidence could be observed close to the modern suburb of 'Fonte Laurentina', still on the Via Laurentina, but close to the Grande Raccordo Anulare (the motorway ring road around Rome). The soil profile of 'Fonte Laurentina' was situated at the foot of the slope of a hill. A colluvial layer contained body sherds attributable to Roman coarse ware pottery.

The evidence is interpreted as intensive sheet erosion, which has modified the topography significantly during the Republican and Imperial Period: The top and the upper part of the slope of the hill were lowered by intensive sheet erosion. The material was deposited at the foot of the slope. Finally, the land surface was much less inclined compared with Roman Times.

The archaeological context of this site comprises a recently excavated and so far unpublished site, dating probably to the Republican an Imperial Period (due to the presence of *opus reticulatum* and *vernice nera* pottery sherds visible at the site). A Roman tomb (De Rossi 1967, 138, site107) was observed approximately 560m to the south of the site in former times, but already no traces remained as the spot was assessed by De Rossi (1967, 138).

A profile close to the Via Nettunense was washed out recently by drainage water from a road, as a

Figure 6. Profile along the Via Nettunense.

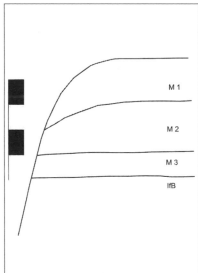

Figure 7. Soil profile along the Via Nettunense (schematic sketch). Length of scale bar: 1m; M3: upper colluvial layer (probably deposited during Modern times as a result of intensive soil erosion upslope); M2: middle colluvial layer with large fragments of a dolium; M1 Ah: Humic horizon which has developed in the lower colluvial layer M1; IfB: relics of a soil.

Figure 8. Dolium fragment from the profile of Via Nettunense.

gabion close by was badly placed. It shows the volcanic geological bedrock with three different layers of colluvial deposits on top (figs. 6, 7). The stratigraphy is formed by yellowish-reddish soil at the bottom, and three further layers. The lower, upper and top layers are colluvial deposits. A humic horizon developed in between the lower and upper layer as an intermediate stratum.

The processes that caused this stratigraphical record are interpreted as follows: The original yellowish-reddish soil that had developed during early and middle Holocene was partly eroded, most probably before the Republican Period. This is proven by the lower colluvial layer. Afterwards a humic horizon developed in the lower colluvial layer under grassland in a period of several decades or a few centuries. A second colluvial layer which contains numerous large fragments of a *dolium* (huge storage vessel) (fig. 8) was deposited probably during Late Republican and Early Imperial Times. Numerous large fragments of the *dolium* were found, indicating that they were moved only a little distance, as can be reasoned due to the weight and the degree of fragmentation. A third colluvial layer was deposited during Modern Times. This profile indicates sheet erosion and soil formation not only for the Late Republican and the Imperial Periods, but already for a preceding period. It points to changing land use patterns for the area situated uphill: Agriculture was probably succeeded by pasture. At least agriculture came to an end and grassland and humus developed, which was once again succeeded by arable use. To our knowledge, no archaeological site is known for the immediate vicinity so far. The *dolium* might indicate the presence of a site close by.

In contrast to these cases of significant landscape modification and manmade landscape impact, a soil profile encountered close to the station of Campoleone Scalo maintained still in-situ remains of the soil that originally developed on this spot during early and middle Holocene. Due to certain characteristics as thick clay enrichment bands, it seems likely that these soils were already in antiquity seasonally water saturated and therefore the range of suitable uses was limited.

Landscape transformation probably dating to recent times

Close to Trigoria Alta finds, datable latest within the last 150 years, but most probably to the last decades were encountered in a canal at the base of a sequence of alluvial layers with a thickness varying from 2m to 3m on a length of several hundred metres. The finds are proof of significant man-made landscape modifications taking place in the recent past in the basin.

The presence of Roman sites was nevertheless documented by De Rossi: Approximately 450m to the south-west surface finds consisting of bricks, roof tiles and sherds, probably dating to the Republican Period, indicate a settlement site (De Rossi 1970, 134, site 437). A further site (De Rossi 1970, 134, site 436) was situated approximately 650m to the south-east. It is described as a scatter of surface material, including *vernice nera* pottery, therefore maybe datable to the Republican Age as well. Finally, a site identified by De Rossi (1970, 134, site 439) as the site of a villa of Imperial date with some wall remains still *in situ* in his days, was situated approximately 650m to the south-west. This site was taken by Nibby, a pioneer of Italian ancient topography, for the site of the Protohistoric settlement of Politorium, but convincing evidence is lacking so far (Nibby 1848-1849, II, 571). In Medieval times a watchtower occupied this strategically fortunate spot (De Rossi 1970, 134).

Reasoning from the geoarchaeological and archaeological evidence, the floodplain has had a much more differentiated topography and a position 2m to 3m below the recent soil surface in Roman Times than today.

Another interesting example is the hill occupied today by Savello castle. Colluvial layers with a thickness up to 110cm were documented in soil profiles along a hollow way leading up Monte Savello, which contained modern 'glazed' pottery, found in the respective strata.

These finds state significant sheet erosion events taking place within the last 150 years, strata of ancient erosion may be covered therefore deeply. Today mainly olives are grown further uphill on agricultural terraces.

It seems that the site, situated in an elevated, strategically fortunate and fertile (volcanic) position, close to the Alban hills, was inhabited by humans at least from the Archaic Age onwards. By some scholars it was identified as the site of the city Apiolae (De Rossi 1970), though this question remains unsolved. For the Roman period, the spot was most probably occupied by a residential complex, while remains of the Medieval fortifications are preserved on top of the hill until today.

CONCLUSION

Pointing out that the results are preliminary, the raised scientific questions can be addressed as follows: Several sites (Via Laurentina, Fonte Laurentina, Via Nettunense), show traces of intensive soil erosion, mostly sheet erosion. These cannot be dated with certainty to Roman antiquity as the archaeological material, used for the dating of the strata, gives only a *terminus post quem* (maximum age). Nevertheless due to the absence of material, which can be dated to younger periods and therefore a supposed homogeneity of Roman material and due to physical characteristics as the compactness of the soil, it is probable that the respective erosion events took place in Roman Times. It seems that the agricultural regime that caused the documented sheet erosion processes was not sustainable and probably therefore conditions deteriorated in the course of time. In the period of the erosion events land use uphill consisted most probably of farming. If there would have been a dense cover of wood, erosion of the documented kind, would not have taken place. So far, no particular evidence for extremely heavy rainfall events was found. Archaeological evidence, provided by the analysis of settlement data for the study area, supports the hypothesis of intensive land use for the respective area and timeslot. Topography was modified since Roman Times significantly in most research areas.

While at numerous sites, which were not explicitly discussed in this paper due to the absence of evidence, all original soil cover was already eroded totally, it was at a site close to Campoleone Scalo still encountered partly in situ. Intensive soil erosion processes, most probably postdating the Roman Period by far, could be documented at two presented sites (Trigoria Alta, Monte Savello), proving major changes of the topography in the course of the last two millennia.

First results reveal a good potential for further research, which seems promising to shed more light on the environmental and archaeological past of south-western Lazio. While the archaeological record is comparably well documented for the study area, geoarchaeological research is so far in its initial state and has a good potential to be explored in further research. Therefore archaeological and geoarchaeological data have to be seen as complementary components of an approach for the holistic understanding of past landscape dynamics. Spatially, research in this particular area contributes to fill a gap in a geoarchaeologically less studied area situated between better studied areas in and around Rome (Volpe & Arnoldus Huyzendveld 2005), close to the Tiber River (Giraudi et al. 2009; Arnoldus Huyzendveld 1995; Arnoldus Huyzendveld 2004; Arnoldus Huyzendveld 2005) and along the Laurentine shore (Bicket et al. 2009) as well as in the Pontine region (Attema 1993, 18; Van Leusen 2002; Van Jolen 2003; Van Leusen & Feiken, 2007).

ACKNOWLEDGEMENTS

The presented research is funded and enabled by the Interdisciplinary Graduate School 'Human Development in Landscapes' at the Christian-Albrechts-University Kiel. The respective PhD thesis is supervised by Prof. F. Rumscheid, Prof. H.-R. Bork and co-supervised by Prof. P. Attema. Some of the GIS data was kindly provided by members of the Pontine Region Project at Groningen, in particular by O. Satijn and L. Alessandri. Prof. G. Olcese, La Sapienza University, Rome, provided kind support for the dating of pottery. The organisers of the First International Landscape Archaeology Conference (LAC2010) are to be thanked for the acceptance of this contribution to the conference and two reviewers are to be thanked for remarks on the text.

REFERENCES

Adams, G.W. 2008. *Rome and the Social Role of Elite Villas in its Suburbs*. Archeopress, Oxford.

Arnoldus Huyzendveld, A. & L. Paroli. 1995. Alcune considerazioni sullo sviluppo storico dell'ansa del Tevere presso Ostia. *Archeologia Laziale* XII, 383-392.

Arnoldus Huyzendveld, A. 2004. L'evoluzione ambientale della Bassa Valle Tiberina in epoca storica. In Serlorenzi, M., B. Amatucci, A. Arnoldus Huyzendveld, A. De Tommasi, H. Di Giuseppe, C. La Rocca, G. Ricci & E. Spagnoli (eds.): *Nuove acquisizione sulla viabilità dell'Agro Portuense. Il rinvenimento di un tratto della via Campana e della via Portuense*. Bollettino della Commissione Archeologica Comunale di Roma.

Arnoldus Huyzendveld, A. 2005. The natural environment of the Agro Portuense. In Keay, S., M. Millett, L. Paroli & K. Strutt (eds.), *Portus: An archaeological Survey of the Port of Imperial Rome*, 14-30. Oxbow Books, Oxford.

Attema, P.A.J. 1993. *An Archaeological Survey in the Pontine Region. A contribution of the early settlement history of south Lazio. 900-100 BC*. Groningen Rijksuniversiteit, Groningen.

Bicket, A.R. H.M. Rendell, A. Claridge, P. Rose & F.S.J. Brown. 2009. Multiscale Geoarchaeological Approaches from the Laurentine Shore, Castelporziano, Lazio, Italy. *Geophysical Research Abstracts* 11.

Bork, H.R. 2006. *Landschaften der Erde unter dem Einfluss des Menschen*. Primus-Verlag, Darmstadt.

Brückner, H., A. Vött. 2008. Geoarchäologie – eine interdisziplinäre Wissenschaft par excellence. In Kulke, E. & H. Popp (eds.), *Umgang mit Risiken. Katastrophen-Destabilisierung-Sicherheit. Tagungsband Deutscher Geographentag 2007 Bayreuth*. Herausgegeben im Auftrag der Deutschen Gesellschaft für Geographie, 181-202. Bayreuth, Berlin.

De Rossi, G.M. 1970. *Apiolae. Forma Italiae Regio 1 Vol. 9*. De Luca Editore, Roma.

De Rossi, G.M. 1967. *Tellenae. Forma Italiae Regio 1 Vol. 4*. De Luca Editore, Roma.

Giraudi, C., Tata, C. & L. Paroli. 2009. Late Holocene evolution of Tiber river delta and geoarchaeology of Claudius and Trajan Harbor, Rome. *Geoarchaeology* 24 (3), 371-382.

Kolendo, J. 1994. Praedia suburbana e loro redditività. In Carlsen, J., P. Ørsted & J.E. Skydsgaard (eds.), Landuse in the Roman Empire, 59-71. 'L'Erma' di Bretschneider, Roma.

Mayer, J.W. 2005. *Imus ad villam, Studien zur Villeggiatur im stadtrömischen Suburbium in der späten Republik und frühen Kaiserzeit*. Geographica Historica 20. Franz Steiner Verlag, Wiesbaden.

Moorley, N. 1996. *Metropolis and hinterland. The city of Rome and the Italian economy 200 B.C.- A.D. 200*. Cambridge University Press, Cambridge.

Nibby, A. 1848-1849. *Analisi storico-topografico-antiquaria della* carta de' dintorni di Roma. Vol. II, 2nd Edition, Tip. delle Belle arti, Roma.

Van Jolen, E. 2003. *Archaeological land evaluation. A reconstruction of the suitability of ancient landscapes for various land uses in Italy focused on the first millennium BC*. Groningen Rijksuniversiteit, Groningen. http://dissertations. ub.rug.nl/faculties/arts/2003/e.van.joolen/.

Van Leusen, M. 2002. *Pattern to Process. Methodological investigations into the formation and interpretation of spatial patterns in archaeological landscapes*. Groningen Rijksuniversiteit, Groningen. http://dissertations.ub.rug.nl/ faculties/arts/2002/p.m.van.leusen/.

Van Leusen, P.M. & H. Feiken. 2007. Geo-archeologie en Landschapsclassificatie in Midden- en Zuid-Italië. *TMA* 37.

Volpe, R. & A. Arnoldus Huyzendveld. 2005. Interpretazione dei dati archeologici nella ricostruzione storica e ambientale del paesaggio suburbano: l'area di Centocelle nel suburbio sudorientale. In Santillo Frizell, B. & A. Klynne (ed.), *Roman villas around the urbs. Interaction with landscape and environment*. Proceedings of a conference at the Swedish Institute in Rome, 17-18 September 2004. www.svenska-institutet-rom.org/villa/.

Witcher, R. 2005. The extended metropolis: Urbs, suburbium and population. *Journal of Roman Archaeology* 18, 120-138.

2.5 The medieval territory of Brussels: A dynamic landscape of urbanisation

Authors

Bram Vannieuwenhuyze[1], Paulo Charruadas[2], Yannick Devos[2] and Luc Vrydaghs[2, 3]

1. Medieval History Research Unit, Katholieke Universiteit Leuven, Leuven, Belgium
2. Centre de Recherches en Archéologie et Patrimoine, Université libre de Bruxelles, Brussels, Belgium.
3. Research Team in Archaeo- and Palaeosciences, Brussels, Belgium

Contact: Bram.Vannieuwenhuyze@arts.kuleuven.be

ABSTRACT

The urbanisation process has a huge impact on both the urban and rural landscape. Not only does it thoroughly modify the urban area, it also has a tremendous impact on the rural hinterland. We propose to take medieval Brussels (Duchy of Brabant) as an example to illustrate this complex issue. According to our different fields of research, a multidisciplinary point of view will be adopted, combining urban history (the study of human urban society), rural history (agricultural developments and rural socio-economic change), historical geography (interaction between medieval people and their spatial environment) and natural sciences (through the archaeopedological and phytolith study of Dark Earth). Firstly, we briefly discuss the essential concepts 'medieval city' and 'medieval urban landscape' and try to apply them to the case of medieval Brussels. Secondly, we address some essential characteristics of landscape transformation, by tackling the major stages of the emergence and development of medieval Brussels and its changing impact on the regional landscape. We argue that the urbanisation process, generally allocated solely to the urban area, is key to understanding landscape transformation of the medieval territory of Brussels.

KEYWORDS

landscape history, urbanisation, urban morphology, archaeopedology, archaeobotany, agriculture

INTRODUCTION

The study of ancient rural landscapes has quite a long tradition of interdisciplinarity, involving archae-ologists, historians, geo-archaeologists, archaeobotanists and archaeozoologists (Rapp & Hill 1998; Wilkinson & Stevens 2003). The study of the medieval urban landscape, however, has mainly been the playground of urban historians, geographers and archaeologists (Heers 1990; Whitehand & Larkham 1992; Verhulst 1999; Lilley 2002; Schofield & Vince 2003). Nevertheless, since the late 1970s and the be-ginning of the 1980s, we have witnessed a growing interest in urban soil. Natural scientists became in-creasingly involved in the study of urban contexts (Macphail 1981; Barles et al. 1999). Nowadays, urban archaeology has also become a truly interdisciplinary research field, incorporating specialists from di-verse research fields. However, most studies are limited to historical city centres and their development, and seldom consider the interplay of the urban environment with its hinterland (Hall & Kenward 1994).

Of course, this issue cannot be tackled without demarcating and clearly defining the topic. How can we define a medieval city and a medieval urban landscape? What does medieval urbanisation en-compass? Many scholars, including historians, archaeologists, soil scientists and ecologists, have already tackled these issues, in either local or general studies (see for instance Lavedan & Hugueney 1974; Heers 1990; Dutour 2003; Heimdahl 2005; Fondrillon 2007). Generally speaking, the parameters of urbanisa-tion, be they medieval or not, are often defined in terms of demography, centrality, density (or concen-tration) and identity. For instance, Clark's historical overview on European urbanisation starts with the following words:

> Since the Middle Ages, Europe has been one of the most urbanised continents on the planet and its cities have stamped their imprint on the European economy, as well as on European social, political, and cultural life. Rarely sprawling mega-cities such as those of present-day Latin America or Asia, mostly compact and coherent, they have been communities characterised by heavy mortality (un-til the twentieth century) and high immigration. Often endowed with administrative functions and political privileges, they have always functioned as commercial and business centres, while religion (up to recent times), education, leisure activity, and a distinctive townscape have helped define ur-ban cultural identity. (Clark 2009).

From a demographic point of view, a city is seen as a concentration of people (Bairoch et al. 1988), regu-larly fed by new immigrants, because of its central and economic functions. Nevertheless, the population structures in a medieval city were very complex. Social, economic and political urban institutions and frameworks were created to organise society and to centralise various functions: trade, industry, cultural manifestations, administration, power, etc. This evolution gave birth to a specific urban mentality. Ac-cording to Reynolds, 'the inhabitants of towns regard themselves, and are regarded by the inhabitants of rural settlements as a different sort of people. However deeply divided they may be among themselves, they tend to be united at least in regarding themselves as different from the rustic simpletons outside.' (Reynolds 1992). Urban identity and the feeling of superiority/otherness were maintained or even defend-ed by administrative, architectural and juridical creations such as city walls, town rights and territorial jurisdictions. Finally, if we consider the built environment, we notice that the built form of a city is often considered as a concentration of buildings of different types (houses, churches, hospitals, castles, walls,

Figure 1. Figured initial dating from the middle of the 15th century, representing the city of Brussels (© Archives of the State, Brussels). From the left to right are shown respectively the entire figured initial, the city view, urban space and finally urban buildings.

etc.) built in different materials (wood, stone, loam, clay, straw, bricks, etc.). Of course, building density is an essential factor which distinguishes the city from the countryside. Medieval citizens themselves portrayed their city in this way, as is clearly shown by the illuminated manuscript initial representing the 15th-century townscape of Brussels (fig. 1). Nearly 84% of the entire image is urbanised, clearly representing a concentration or compilation of buildings (nearly 87% of the urban space, some 72% of the entire image).

Together with the features mentioned above, this kind of image gave birth to a traditionalist vision of the medieval city and urban landscape, namely that of an isolated island, well protected by its walls, inhabited and ruled by a distinct sort of men (citizens). Within these walls a strictly urban landscape would exist, where specific urban functions and activities took place. For a long time historians and archaeologists adopted this paradigm in their studies. However, due to the implementation of other disciplines into the field of urban history, a remarkable change of mindset has been seen over the last few decades. Of course, Despy (1968), Nicholas (1971) and Jourdan-Lombard (1972) have already stressed the importance of considering medieval towns and the countryside as a whole in their pioneering works, but these analyses were often restricted to social, economic or political interaction. In landscape research, this mindset came much later. Today, researchers increasingly consider the hinterland to be an essential part of the urban landscape (see for instance Thoen 1993; Hall & Kenward 1994; Clark & Lepetit 1996; Giles & Dyer 2007; Limberger 2008; Bijsterveld et al. in press). The medieval city can no longer be separated from its surrounding countryside. As a result, the definition of the medieval urban landscape also needs to be reassessed. Was it really that dense, specific and centripetal? In other words, was it really all that urban? In addition, did urbanisation only mean a strengthening of these features?

We want to tackle some of these questions by presenting some results of ongoing research on medieval Brussels. However, it is necessary to stress that this contribution must be considered as one of the first steps within broader research aiming to reconstruct the medieval landscape evolution of the town and its surrounding hinterland. Here we shall try to combine the results of individual research work, bearing in mind that other results will need to be integrated in the future. In particular, we refer to the very recent synthetic view on the Brussels landscape before 1200, published by the archaeological cell of the Ministry of the Brussels Capital Region. It merges the results of successive archaeological campaigns undertaken over the last twenty years with historical data (Degraeve et al. 2010). We have made use of some archaeological findings, but we also look forward to stimulating cross-fertilisation between the various scientific disciplines involved in this research. In this respect, the opinions stated here should be regarded as work in progress.

INTERDISCIPLINARY RESEARCH ON MEDIEVAL BRUSSELS

Brussels is what is called a second generation city (Billen 2000). Unlike for example London or Paris, it does not have Roman origins. The city emerged at the beginning of the 11th century. Following the traditional vision, the nucleus of the latter town lay alongside the river Senne, where a *castrum* and small port were constructed (Henne & Wauters 1845; Des Marez 1935; Bonenfant 1949; Martens 1976). More recent scholars have agreed that the late medieval town evolved from the merging of scattered settlements (Despy 1979 & 1997; Dierkens 1989; Billen 2000; Deligne 2003), although they experienced serious difficulty in locating them precisely and providing a clear chronology for their spatial evolution. From the 13th century onwards, Brussels became an important political, industrial and ecclesiastical centre (Des Marez 1901; Favresse 1932; Lefèvre 1942; Vannieuwenhuyze 2008; de Waha & Charruadas 2009). In the centuries that followed, the town was considered one of the 'capital cities' of the Duchy of Brabant (Martens 1973; Smolar-Meynart 1985). This major role was confirmed in the 15th and 16th centuries with, ultimately, the creation of central political and administrative institutions within the city (Charruadas & Dessouroux 2005).

Reconstructing and understanding the dynamics of medieval urbanisation and the process of landscape change in Brussels presents a challenge. Except for some well-known 'monuments', surviving medieval remains within the former town and its surrounding hinterland are quite scarce. In 1695, Brussels was subjected to an important bombardment which destroyed a large part of the city (Culot et al. 1992). During the 19th and 20th centuries, the urban territory and its environment were affected by massive town planning and (re)construction campaigns that drastically altered the landscape of the former pre-industrial town (Verniers 1934; Apers et al. 1982; Demey 1990, 1992). Nevertheless, the medieval landscape did not entirely disappear, although its relicts have been extensively transformed and are often well concealed. As a result, there is a huge need for appropriate search strategies to discover, analyse and preserve both written and material records. Moreover, new scientific methods to unravel the dynamics of medieval urbanisation are required. To understand the 'landscape of medieval urbanisation', it is not sufficient to compare the viewpoints of various disciplines, but rather to broaden our horizons by integrating different approaches, methods, scales and results.

We have all studied and are currently studying Brussels' medieval landscape from different viewpoints and at different scales. The historical analysis relates to the study of human urban society, with particular emphasis on the interactions between people and space, enabling the medieval town development of Brussels to be understood. This multidisciplinary research is based on an in-depth study of morphological relicts of the medieval urban landscape within the town (e.g. the few remaining medieval monuments, cartographical representations, street patterns, archaeological data) and on medieval texts (e.g. toponyms, laws, property transactions, urban myths), considered to be relicts of human perception of spatial phenomena or evolution (Vannieuwenhuyze 2008). The archaeopedological research focuses on the study of urban soil, in particular the Dark Earth units discovered throughout the actual city centre. These Dark Earth units are dark, humus-rich, non-peaty, strongly melanised and apparently homogeneous layers which have been sealed in the soil. Although they are an important part of the medieval stratigraphy, their significance cannot easily be understood using a traditional archaeological approach (Devos et al. submitted). Archaeopedology (including field study, micromorphology and physico-chemical analysis), combined with archaeobotany and, more specifically, phytolith analysis have been shown to be powerful tools in the identification of agents and in understanding the processes and activities responsible for their formation (Macphail 1981; Devos et al. 2009; Devos et al. 2010; Devos et al. submitted). The integrated study of Brussels' Dark Earth demonstrates its potential to identify ancient activity zones (Devos et al. 2009; Degraeve et al. 2010) and to reconstruct the medieval landscape through the evaluation of the role and interaction of natural and human factors (Devos et al., submitted). Finally, rural history makes it possible to observe the broader landscape of Brussels and study the relationship between the medieval town and countryside. Particular attention has been paid to agricultural change and its reciprocal links with urban development. This approach is not dominated by the paradigm of urban imperialism, but rather by the idea of a social, economic and political continuum between town and countryside (Charruadas 2008).

These different points of view also have repercussions on the use of different scales. For historical analysis, various morphological levels have been selected, going from plots of land in neighbourhoods to the entire city. Human action is studied at each level. To unravel the story hidden within the enigmatic and homogeneous Dark Earth, it is necessary to go from macroscopic field study down to microscopic scale (Macphail 1981; Devos et al. 2009). Archaeobotany combines the analysis of macro-remains (charcoal and seeds) and micro-remains (phytoliths and pollen) originating from several sites within the actual city. Unfortunately, the conservation of pollen and non-carbonised seeds on the higher parts of Brussels, where most excavations have taken place, is not good. More recently, there have been some incursions into the lower parts of Brussels, but we still have to wait for the results. And last but not least, by studying rural history, phenomena in both the town and the countryside have been observed. This broad viewpoint enables us to understand the connections (or the lack thereof), contrasts and adequacy of the gap between what happened in town and in the countryside. These phenomena deal in particular with the economic issue, namely urban demand for rural products, and with the elites who settled both in the city and surrounding area. This approach shows the way in which urban society shaped the rural environment and, conversely, how urban structures influenced rural society.

THE DYNAMIC LANDSCAPE OF MEDIEVAL URBANISATION

By studying how humans changed the medieval urban landscape, some general trends have been discerned. Man clearly had a marked effect on the management of the landscape, especially by organising public works (the creation of the road network, hydrographical infrastructure, city walls, etc.), by constructing buildings, mainly for housing and industrial purposes, and through the establishment of administrative, political and juridical frameworks. Up to now, scholars have taken a particular interest in the evolution of these features. Building campaigns were reconstructed thoroughly by archaeologists and building historians, especially major medieval buildings and infrastructures (Des Marez 1918; De Jonge 1991; Dickstein-Bernard 1995-1996; Bonenfant et al. 1998; Blanquart et al. 2001; Maesschalck & Viaene s.d.), while historians studied in detail the medieval political, administrative and juridical institutions both within and outside the town (Godding 1960; Favresse 1932; Martens 1939; Dickstein-Bernard 1977; Smolar-Meynart 1991).

However important they may be, these and other studies barely reveal the spatial complexity of the medieval urbanisation process. Firstly, it is now commonly accepted that this process cannot be outlined by simply listing the main historical 'events' or reconstructing the evolution of the main sites and buildings. Secondly, although some scholars claim the spatial pattern of medieval Brussels evolved in an organic way (Lavedan & Hugueney 1974), ever more scholars now agree that organic town development did or does not exist (Pinol 2003; Boerefijn 2005). The townscape was constantly changing as a result of top-down actions as well as bottom-up reactions and with successive ups and downs, both at local or regional scale. The construction of two successive stone city walls during late medieval times serves as a good illustrative example. Up to now, a very clear distinction has always been drawn between Brussels' so-called first city wall, built in the 13th century (Demeter 2001), and Brussels' so-called second city wall, built in the second half of the 14th century (Dickstein-Bernard 1995-1996). Both constructions were impressive works, so it is clear that their construction was ordered and financed by the authorities, the Duke of Brabant and the Brussels town council respectively. However, these two building projects should not be considered as isolated and chronologically-defined spatial phenomena, as historians have usually done (and still do). In fact, recent research shows that the Brussels people continuously and systematically expanded and transformed these walls, adapting them in accordance with changing demographic, political, economic and spatial evolutions within the town and the entire territory (Vannieuwenhuyze 2008). In this respect, the distinction between both late medieval city walls seems to originate from a teleological interpretation *post factum*. In summary, these city walls did not suddenly appear nor did they develop organically.

As a result, the already mentioned keywords used to define cities and urban landscapes must be adopted with caution. Urban centrality, density and identity were constantly changing. This was also true for the urban landscape. Recent historical analysis shows that the sole medieval urban landscape was indeed not always that 'urban'. Here, we are restricting ourselves to two concrete examples to illustrate this point. In-depth analysis of the road network and street names has made it clear that different kinds of streets existed during medieval times. Within the town, long and continuous streets juxtaposed narrow alleys. Other streets were created to reach infrastructural works (city walls, canals) or rural estates. From a morphological point of view, these streets were constantly changing. Little alleys (*straatjes*) were created to access the areas within the blocks and thus played an important role in communication at micro-level. They were systematically heightened and paved each time new houses were built. Some still exist, although most have been enclosed by private houses. At a macro-level, Brussels' old artery network came into being in close harmony with the economic potential of the hinterland (Vannieuwenhuyze 2009). Firstly, smaller roads linked the town to the surrounding villages and countryside. At a later stage, a whole new network of artificial arteries was created by the town government, facilitating direct access to the town. In addition, the creation of this artery network clearly reflected urban economic and political imperialism towards the surrounding countryside.

To take a second example, digital analysis of the oldest (but geometrically very reliable) town plan,

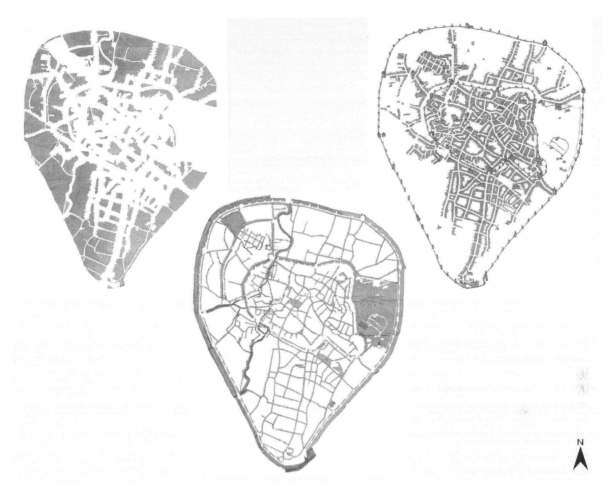

Figure 2. Cartographic analysis of the 16th-century town plan of Brussels by Jacob of Deventer, realized in a scale of approximately 1:8600 (© Royal Library of Belgium, Brussels). From left to right are shown respectively the open spaces (36%), the road network, watercourses and green spaces (34%) and buildings (30%). *See also full colour section in this book*

drawn up by Jacob van Deventer in the middle of the 16th century, shows that the largest part (36%) of urban territory within the city walls consisted of open space. In addition, 17%, 12% and 5% of the territory were respectively allocated to the road network, the watercourses and green elements (trees, bushes and parks), while 'only' 30% consisted of buildings (Vannieuwenhuyze 2008, see also fig. 2). This means that at least until the middle of the 16th century, the urbanisation process did not only imply the creation of different kinds of buildings, as often pointed out by historians. On the contrary, in previous decades and centuries the evolution in open spaces, improvements to the road network and hydrographical changes – whether or not these were natural – were if anything more important. It is not surprising that the latest generation of historians consider the landscape of medieval Brussels to be a collection of expanding and/ or diminishing settlements, separated by open spaces and connected over time by public works such as roads, canals and walls.

The original 'broek-sele' (i.e. Brussels) seems to be one of these settlements, probably of secondary importance and devoted to cattle husbandry. Another small settlement expanded on a nearby hill, the *Coudenberg*, and became the seat of ducal power, perhaps by the 11th century, but certainly from the beginning of the 12th century onwards (Vannieuwenhuyze 2008). Ducal power clearly stimulated the urbanisation process with the foundation of churches, the draining of wetlands, the creation of a port and markets, etc. Meanwhile, other settlements emerged between the Zenne River and the *Coudenberg*. These were often devoted to small-scale trade and/or agriculture, as was the case with the *Oud Korenhuis* (Degraeve et al. 2010). From the 13th century onwards, Brussels town council took over the big public work projects, for instance, by constructing the so-called second city wall, whose perimeter fixed the urban territory *strictu sensu* until the beginning of the 19th century, when it was demolished (Lelarge 2001). During late medieval times and the modern era the urbanisation process was limited to small-scale operations, such as the densification and petrifaction of the urban fabric. Both phenomena have been documented in recently

Figure 3: a) Soil profile of Dark Earth on the site of *Hôtel de Lalaing-Hoogstraeten*; b) Graph showing enhanced phosphorus levels for the Dark Earth units (US 7338 and US 7321); c) Granulometric data showing the high similarity between the units US 7338 and US 7321 and the natural soil (US 7340), suggesting they share the same matrix; d) Thin section micrograph showing phosphorus-rich excrement proving the addition of manure (plain polarised light); e) Thin section micrograph showing a textural pedofeature enabling its identification as former at least temporary unprotected topsoil (plain polarised light); f) Thin section micrograph showing dendritic phytoliths (plain polarised light). *See also full colour section in this book*

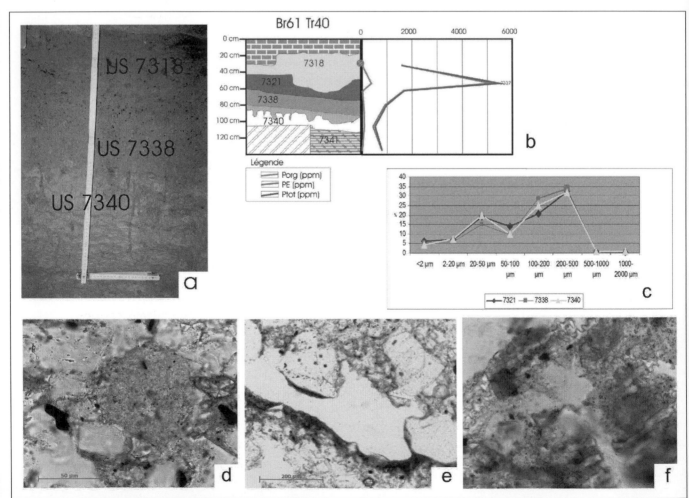

Figure 4. Thin micrograph section of a calcitic biospheroid (e). Its presence proves enhanced earthworm bioturbation due to the addition of soil amendments, site of *rue de Dinant* (crossed polarisers).

conducted research in urban archaeology, showing for instance that private stone or brick buildings only emerged from the 14th century onwards (see for instance Cabuy & Degré 1992; Degré 1995; Diekmann 1997; Nachtergael 1997; Claes 2008).

Of course, these statements give rise to new questions on the evolution of land use within this so-called 'urban' territory. As has already been argued sole analysis of historical records will never be sufficient in tackling this complex issue. Dark Earth constitutes a significant part of Brussels' medieval stratigraphy and has been shown to provide a valuable source of information on this topic. This archaeopedological study reveals that it originates from a complex interplay between human activities and natural factors. Its formation and transformation can be understood as site-specific, involving an ongoing process of accumulation, erosion, decomposition, homogenisation and other types of soil development, which stop once the Dark Earth is sealed. As such, sequences of activities can often be revealed. Among the human activities dating back to the 11th-13th centuries digging, ploughing (fig. 3), manuring (fig. 3) and waste disposal have been identified. The main natural factors are bioturbation (fig. 4), erosion and colluviation (Devos et al. submitted).

The botanical study of the identified agricultural plots is quite difficult. Conservation of pollen and non-carbonised seeds and fruits is very poor. Nonetheless, the few surviving pollen and microfossils confirm the presence of pasture land and crop fields (Court-Picon 2008). Phytolith and charcoal preservation however, is excellent. As phytoliths tend to provide very local information, they are a useful tool confirming the presence of crop fields (Vrydaghs et al. 2007; Devos et al. 2009). Phytolith markers for cultivated crops have been recorded for all the Dark Earth identified as former plough lands (Table 1), demonstrating the cultivation of wheat, oats and barley. Charcoal tends to provide information on the wood people have used. Current data do not contradict pressure on relict forests throughout the 11th-13th centuries, possibly bearing witness to the need for new space (Devos et al. 2007a).

Table 1: Cereals crops identified by phytolith analysis (simplified table).

Excavation site	Units	Triticum	Hordeum	Avena
Treurenberg	115	X	X	X
Lalaing	7338	X	?	X
Vieille Halle aux Blés		?	?	X
Impasse Papier	VD Labour	?	?	X
Pauvres Claires	415	?	?	X
Rue de Dinant	413 a, b & c; 631 a & b	X	X	X

It is clear that due to its complex formation process, Dark Earth should be analysed on an individual basis, meaning that the data only relate to the collection site. Nevertheless, thanks to research frequency, a general pattern has emerged. On most sites, we have noticed that patterns of activity seem to have undergone frequent changes over time. This confirms the hypothesis that Brussels had a dynamic medieval landscape. Taking the *Treurenberg* site as an example, proof of a former stone quarry was discovered. At some point this was filled in and transformed into a crop field, which in turn was sealed off by the construction of the first city wall during the 13th century (Devos et al. 2007b). As far as activities recorded within the medieval urban centre are concerned – namely the space enclosed by this first city wall – archaeologists have noticed a heavy dominance of primary sector activities. Between the 10th and the 13th centuries, traces of crop fields, pasture land, stone quarries and soil extraction pits have been identified (Devos et al. 2007a; Degraeve et al. 2010, see also fig. 5). Apparently, artisan activities only appeared at a later stage, when these areas were transformed into building plots, with a combination of different kinds of buildings and infrastructures (Devos et al. 2007a).

The study of economic and agricultural changes makes it possible to indicate some factors that considerably influenced and altered this dynamic Brussels landscape. As regards the supply of food and cereals, the medieval Brussels region seemed to have been self-sufficient. Unlike, for instance, Flemish cities (Tits-Dieuaide 1975), Brussels did not require massive imports of wheat before the end of the Middle Ages. Between 1100 and 1300, Brussels became a centre for the countryside. At that time the city functioned as the main market attracting rural production surplus (Charruadas 2007a). This is fundamental in understanding the scope of agricultural expansion in the Brussels area.

Regional agricultural growth can be divided roughly into two main phases. From the 11th to the 13th century onwards, the region was characterised by an important clearing process and the emergence of a series of rural settlements in a context of demographic and urban expansion (Verhulst 1990). The agricultural system consisted of cereal growing, associated with intensive, stable-fed cattle breeding. This clearing process appears to have declined sharply around 1250, while demographic growth continued. New farming techniques appeared and clearly took the place of extensive farming. We noticed the regular introduction on fallow land of leguminous plants (peas and beans) and fodder plants such as turnips in particular. These new crops were mainly intended for livestock. Indeed, they enabled their numbers to increase. Thanks to improved and greater transfers of fertility from stall to field (also observed by archaeologists, see above), they resulted in increased production yields on cereal-producing land (Charruadas

Figure 5. Map of Brussels with localisation of crop fields, pasture land, stone quarries and soil extraction pits identified within and outside the first city wall.

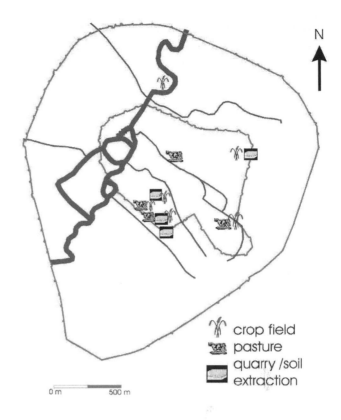

N

% crop field
% pasture
% quarry /soil
extraction

0 m   500 m

2007b; Charruadas 2008). This seems to correlate with the observation that new, initially quite poor soils were put under cereal cultivation by adding considerable quantities of fertilising manure (Devos et al. 2011). This agro-system certainly provided an improved level of supply for both the regional and urban population.

Following the viewpoints of some economists and agronomists (Boserup 1965; Tits-Dieuaide 1981), we can adopt a systemic view to emphasise that the creation of such an advanced farm system was only possible given strong demographic growth. Within the pre-industrial context lacking mechanisation and artificial fertilisers, agricultural growth was closely linked to increased human labour. Hence, we could argue that Brussels agricultural system must have been developed in a context marked by demographic pressure. These changes naturally had a major influence on regional landscape structures, especially land ownership patterns.

In this context, but from a social viewpoint, the economic and political development of the city of Brussels at that time also generated an important effect, with the emergence of a group of urban elite. From the end of the 12th century onwards, former landowners, mainly feudal aristocrats, found themselves in economic distress. Conversely, this new urban elite group, active both within the city and the countryside, was at the height of its wealth in the 13th century. Ownership structures therefore changed considerably with important purchases of land made by these new rulers, who preferred smaller plots which they could exploit by seeking high profitability. This process brought about a marked change in the city's surrounding rural landscape (Charruadas 2008).

SOME PRELIMINARY CONCLUSIONS

In medieval times, the Brussels landscape presented a more important rural component than previously assumed. The digital analysis of 16th-century maps suggested the persistence of open spaces within the 'urban area' *strictu sensu* and integrated archaeopedological and archaeobotanical studies provide direct evidence for open areas such as crop or pasture lands. This open urban landscape appears to have been quite diverse and the result of complex and dynamic processes closely connected to regional development. Written records show that the 12th and 13th centuries were an important turning point. However, our analysis also indicates that the forces driving medieval urbanisation were not only 'urban' (i.e. from the city itself), but also emanated from the countryside. Urban policy and planning also contributed to the urbanisation process. Nevertheless, we can state that this often depended on rural exodus. Data emanating from rural analysis are in fact crucial to understanding the general process.

The combination and comparison of the various approaches show the urbanisation process to be complex and interactive. It can be divided into two phases. The first is a 'light' version of urbanisation dominated by a strong rural component. Strictly-speaking prior to the 13th-century demographic growth affected the rural as much, if not more than, the urban area. This urbanisation phase took place in the Brussels hinterland. Medieval Brussels itself was no more than a fusion of scattered rural settlements. All the signs indicate that the majority of people during this period were scattered throughout the region and took part in the land reclamation movement. Demographic concentration within the city would follow, with the stabilisation of agricultural growth after the 13th century. At this time population increase could no longer find an outlet in the countryside. Consequently, rural exodus on a much wider scale occurred. It generated a greater urbanisation process during the 14th and 15th centuries. Streets and plots became more densely occupied and stone and brick materials gradually started to replace former wooden buildings.

This chronological reconstruction provides a negative answer to the questions addressed at the beginning of this paper. As seems also to have been the case in French towns (Leguay 2009), the medieval urban landscape of Brussels was not that dense, certainly not during the first stages of town development. It is only from the late medieval and early modern era onwards that the urban territory slowly intensified. This means that the concepts of density and centrality paradigmatically used to define cities should be refined. These urban qualities were possibly more imagined or invented than real. In this respect, we could link them to the genesis of an urban identity during late medieval times. According to Howell, 'this was a history of ideology, a history of how cities were imagined, represented, and conceptualized and of how the rights and capacities of urban citizens were defined and legitimated.' (Howell 2000). We hope we have made it clear that interdisciplinary research on landscape evolution and urbanisation can play an essential role in debunking such widespread and often persistent assumptions.

ACKNOWLEDGEMENTS

The Direction of Monuments and Sites of the Brussels Capital Region; The Université libre de Bruxelles and Ghent University; Britt Claes (Musées Royaux d'Art et d'Histoire, Brussels); The Institute for the encouragement of Scientific Research and Innovation of Brussels; The Royal Belgian Institute for Natural Sciences, Brussels; Hugues Doutrelepont and Mona Court-Picon (ROOTS).

REFERENCES

Apers, J., J. Vandenbreeden & L. Van Santvoort. 1982. Chronologisch overzicht van de belangrijkste stedebouwkundige feiten in en rond Brussel, 1780-1982. In *Straten en stenen. Brussel: stadsgroei 1780-1980*. Tentoonstelling ingericht door de Generale Bankmaatschappij in samenwerking met het Sint-Lukasarchief en G. Abeels. 18 November 1982 - 21 January 1983: 10-122. Generale Bankmaatschappij – Sint-Lukasarchief, Brussels.

Bairoch, P., J. Batou & P. Chèvre. 1983. *The population of European cities: data bank and short summary of results: 800-1850*. Droz, Genève.

Barles, S., D. Breysse, A. Guillerme, C. Leyval. 1999. *Le sol urbain*. Anthropos, Paris.

Bijsterveld, A.J., J. Naylor, A. Wilkin & D. Keene. In press. *Dynamic Interactions. Town and Countryside Relations in Northwestern Europe in the Middle Ages.*, Brepols, Turnhout.

Billen, C. 2000. Espace et société. In Billen, C. & Duvosquel, J.M. (eds.), *Bruxelles: 36-139*. Mercatorfonds, Antwerp.

Blanquart, P., S. Demeter, A. De Poorter, C. Massart, S. Modrie, I. Nachtergael & M. Siebrand. 2001. *Autour de la première enceinte. Rond de eerste stadsomwalling*. Ministère de la Région de Bruxelles-Capitale – Musées Royaux d'Art et d'Histoire, Brussels.

Boerefijn, W. 2005. De totstandkoming van de stedelijke vorm. In R. Rutte & H. van Engen, *Stadswording in de Nederlanden. Op zoek naar overzicht: 123-142*.Verloren, Hilversum.

Bonenfant, P. 1949. Une capitale au berceau: Bruxelles. *Annales, Economies, Sociétés, Civilisations* 4, 298-310.

Bonenfant, P.-P., M. Fourny & M. Le Bon. 1998. Fouilles archéologiques à la cathédrale de Bruxelles 1987-1998. Un premier bilan d'ensemble. *Annales de la Société Royale d'Archéologie de Bruxelles* 62, 223-257.

Boserup, E. 1965. *The conditions of Agricultural Growth. The Economics of Agrarian Change under Population Pressure*. Allen & Unwin, London/New York.

Cabuy, Y. & S. Degré. 1992. Intervention place Saint-Géry. Structures médiévales et post-médiévales. *Bulletin du Crédit Communal de Belgique* 182, 13-17.

Charruadas, P. 2007a. Croissance rurale et action seigneuriale aux origines de Bruxelles (Haut Moyen Âge-XIIIe siècle). In Deligne, C. & Billen, C. (eds.), *Voisinages, coexistences, appropriations. Groupes sociaux et territoires urbains*. Actes du Colloque international de Bruxelles (4-6 December 2004): 175-201. Brepols, Turnhout.

Charruadas, P. 2007b. Champs de légumes et jardins de blés. Intensification agricole et innovations culturales autour de Bruxelles au XIIIe siècle. *Histoire et Sociétés Rurales* 28, 11-32.

Charruadas, P. 2008. *Bruxelles et ses campagnes. Croissance économique et actions aristocratiques (haut Moyen Âge-XIIIe siècle)*, unpublished Ph D. thesis, Université libre de Bruxelles, Brussels.

Charruadas, P. & Dessouroux, C. 2005. Histoire d'une capitale: Bruxelles des origines à 1958. In (eds.) Braeken J, Charruadas P., De Kuyper E., Dumont P. & Vander Brugghen, *Région de Bruxelles-Capitale. Bruxelles, 175 ans d'une capitale*, 11-27. Mardaga, Liège.

Charruadas, P. & M. de Waha. 2009. Centralité religieuse et développement urbain: notes sur la fixation du doyenné de Bruxelles aux XIe-XIIe siècles. *Cahiers Bruxellois* 41, 43-72.

Claes, B. 2008. *Archeologisch onderzoek van het Brusselse Arme Klarenklooster*, unpublished report, Ministère de la Région de Bruxelles-Capitale – Musées Royaux d'Art et d'Histoire, Brussels.

Clark, P. 2009. *European Cities and Towns 400-2000*. Oxford University Press, New York.

Clark, P. & B. Lepetit (eds.) 1996. *Capital Cities and their Hinterlands in Early Modern Europe*. Scholar press, Aldershot.

Court-Picon, M. 2008. *Résultats d'analyses polliniques des sediments archéologiques des sites de Treurenberg et Hôtel de Lalaing-Hoogstraeten (Région de Bruxelles-Capitale, Belgique)*, unpublished report, ROOTS, Brussels.

Culot, M., E. Hennaut, M. Demanet & C. Mierop. 1992. *Le bombardement de Bruxelles par Louis XIV et la reconstruction qui s'ensuivit*. Archives d'Architecture Moderne, Brussels.

Degraeve, A., S. Demeter, Y. Devos, S. Modrie & S. Van Bellingen. 2010. Brussel vóór 1200: een archeologische bijdrage. In Dewilde, M., Ervynck, A. & Becuwe, F. (eds.), *Cenulae recens factae. Een huldeboek voor John De Meulemeester*, 141-157. Academia Press, Ghent.

Degré, S. 1995. *Brasserie au quartier Sainte-Catherine*. Ministère de la Région de Bruxelles-Capitale – Musées Royaux d'Art et d'Histoire, Brussels.

De Jonge, K. 1991. Het paleis op de Coudenberg te Brussel in de vijftiende eeuw: de verdwenen hertogelijke residenties in de Zuidelijke Nederlanden in een nieuw licht geplaatst. *Revue Belge d'Archéologie et d'Histoire de l'Art* 60, 5-38.

Deligne, C. 2003. *Bruxelles et sa rivière. Genèse d'un territoire urbain (12e-18e siècle)*, Brepols, Turnhout.

Deligne, F. 2002. *Bruxelles-Treurenberg : résultats anthracologiques*, unpublished report, GIeP, Brussels.

Demeter, S. 2001. La première enceinte, un patrimoine majeur pour Bruxelles. In Blanquart, P., Demeter, S., De Poorter, A., Massart, C., Modrie, S., Nachtergael, I. & Siebrand, M., *Autour de la première enceinte. Rond de eerste stadsomwalling*, 12-28. Ministère de la Région de Bruxelles-Capitale – Musées royaux d'Art et d'Histoire, Brussels.

Demey, T. 1990. *Bruxelles: chronique d'une capitale en chantier. 1: Du voûtement de la Senne à la jonction Nord-Midi*. Legrain, Brussels.

Demey, T. 1992. *Bruxelles: chronique d'une capitale en chantier. 2: De l'Expo '58 au siège de la CEE*. Legrain, Brussels.

Des Marez, G. 1901. *L'organisation du travail à Bruxelles au XVe siècle*. Académie Royale de Belgique, Brussels.

Des Marez, G. 1918. *Guide illustré de Bruxelles*. Touring Club de Belgique-Société Royale, Brussels.

Des Marez, G. 1935. Le développement territorial de Bruxelles au Moyen-Age. Etude de géographie historique urbaine. *Premier Congrès International de Géographie Historique*, 1-90. Brussels.

Despy, G. 1968. Villes et campagnes aux IXe et Xe siècles: l'exemple du pays mosan. *Revue du Nord* 50, 145-168.

Despy, G. 1979. La genèse de la ville. In Stengers, J. (ed.), *Bruxelles. Croissance d'une capital*, 28-39. Mercatorfonds, Antwerp.

Despy, G. 1997. Un dossier mystérieux: les origines de Bruxelles. *Bulletin de l'Académie royale de Belgique (Classe des Lettres)* 8, 241-303.

Devos, Y., L. Vrydaghs, C. Laurent, A. Degraeve & S. Modrie. 2007a. L'anthropisation du paysage bruxellois au 10e-13e siècle. Résultats d'une approche interdisciplinaire. In *On the road again. L'Europe en mouvement. Medieval Europe Paris 2007*. 4e Congrès international d'Archéologie Médiévale et Moderne, Session 7, Archéologies environne-mentales. 3-8 September. Institut National d'Histoire de l'Art, Paris (http://medieval-europe-paris-2007.univ-paris1.fr/Y.Devos%20et%20al..pdf).

Devos, Y., K. Fechner, L. Vrydaghs, A. Degraeve & F. Deligne. 2007b. Contribution of archaeopedology to the palaeoenvironmental reconstruction of (pre-) urban sites at Brussels (Belgium). The example of the Treuren-berg site. In Boschian, G. (ed.), *Proceedings of the Second International Conference on Soils and Archaeology*, Pisa, 12-15 May 2003, 145-151. Società Toscana di Scienze Naturali, Pisa.

Devos, Y., L. Vrydaghs, A. Degraeve & K. Fechner. 2009. An archaeopedological and phytolitarian study of the 'Dark Earth' on the site of Rue de Dinant (Brussels, Belgium). *Catena* 78, 270-284 (http://dx.doi.org/10.1016/j.catena.2009.02.013).

Devos, Y., L. Vrydaghs & S. Modrie. 2010. L'étude des Terres Noires bruxelloises: l'exemple du site de l'hôtel d'Hoogs-traeten (Région Br.). *Archaeologia Mediaevalis* 33, 63-65.

Devos, Y., L. Vrydaghs, K. Fechner, C. Laurent, A. Degraeve & S. Modrie. 2011. Buried Anthropic Soils in the Centre of Brussels (Belgium): Looking for Fields in a (Proto-) urban Context. In Fechner, K., Y. Devos, M. Leopold & J. Völkel (eds.), *Enclosed and buried surfaces as key sources in Archaeology and Pedology. Proceedings of the Session 'From microprobe to spatial analysis – Enclosed and buried surfaces as key sources in Archaeology and Pedology' organised at the European Association of Archaeologists*, 12th Annual Meeting, Krakow-Poland. 19-24 September 2006, Oxford (British Archaeological Reports. International Series).

Devos, Y., L. Vrydaghs, A. Degraeve & S. Modrie. Submitted. Unravelling urban stratigraphy. The study of Brussels' (Belgium) Dark Earth. An archaeopedological perspective, Medieval and Modern Matters.

Dickstein-Bernard, C. 1977. *La gestion financière d'une capitale à ses débuts: Bruxelles, 1334-1467*. Société Royale d'Archéologie de Bruxelles, Brussels.

Dickstein-Bernard, C. 1995-1996. La construction de l'enceinte bruxelloise de 1357. Essai de chronologie des travaux. *Cahiers bruxellois* 35, 91-128.

Diekmann, A. 1997. *Artisanat médiéval et habitat urbain. Rue d'Une Personne et Place de la Vieille-Halle-aux-Blés*. Ministère de la Région de Bruxelles-Capitale – Musées royaux d'Art et d'Histoire, Brussels.

Dierkens, A. 1989. Le haut Moyen Âge. In Smolar-Meynart, A. & Stengers, J. (eds.), *La région de Bruxelles. Des villages d'autres à la ville d'aujourd'hui*, 36-41. Crédit communal de Belgique, Brussels.

Doutrelepont, H. 2009. *Site des Pauvres Claires (Br 100). Déterminations taxonomiques des « micro »s charbons de bois provenant de l'horizon de 'terre noires' US415*, unpublished report, ROOTS, Brussels.

Dutour, T. 2003. *La ville médiévale: origine et triomphe de l'Europe urbaine*. Paris, Odile Jacob.

Favresse, F. 1932. *L'avènement du régime démocratique à Bruxelles pendant le Moyen Age (1306-1423)*. Maurice Lamertin, Brussels.

Fondrillon, M. 2007. *La formation du sol urbain: étude archéologique des terres noires à Tours (4e-12e siècle)*, unpublished PhD-thesis, Université François Rabelais, Tours.

Giles, K. & Ch. Dyer (ed.) 2007. *Town and country in the Middle Ages. Contrasts, Contacts and Interconnections, 1100-1500*. Society for Medieval Archaeology-Maney, Leeds.

Godding, Ph. 1960. *Le droit foncier à Bruxelles au Moyen Age*. Publications de l'Institut de Sociologie Solvay, Brussels.

Hall, A.R. & H.K. Kenward. 1994. Urban-rural connexions: perspectives from environmental archaeology. *Symposia of the Association for Environmental Archaeology* 12, Oxbow, Oxford.

Heers, J. 1990. *La Ville au Moyen Age en Occident: paysages, pouvoirs et conflits*. Fayard, Paris.

Heimdahl, J. 2005. *Urbanised Nature in the Past. Site formation and environmental development in two Swedish towns, AD 1200-1800*, unpublished PhD-thesis, University Stockholm, Stockholm (http://su.diva-portal.org/smash/record.jsf?pid=diva2:197550).

Henne, A. & A. Wauters. 1845. *Histoire de Bruxelles*. Librairie encyclopédique de Perichon, Brussels.

Howell, M.C. 2000. The Spaces of Late Medieval Urbanity. In M. Boone & P. Stabel (eds.), *Shaping Urban Identity in Late Medieval Europe*, 3-23. Garant, Leuven/Apeldoorn.

Jourdan-Lombard, A. 1972. Oppidum et banlieue: sur l'origine et les dimensions du territoire urbain. *Annales. Economies, Sociétés, Civilisations* 27, 373-395.

Lavedan, P. & J. Hugueney. 1974. *L'urbanisme au Moyen Age*. Droz, Genève.

Lefèvre, P. 1942. *L'organisation ecclésiastique de la Ville de Bruxelles au Moyen Age*. Bibliothèque de l'Université catholique de Louvain, Leuven.

Leguay, J.-P. 2009. *Terres urbaines: places, jardins et terres incultes dans la ville au Moyen Age*. Rennes, Presses Universitaires.

Lelarge, A. 2001. *Bruxelles, l'émergence de la ville contemporaine: la démolition du rempart et des fortifications aux XVIIIe et XIXe siècles*. CIVA, Brussels.

Lilley, K.D. 2002. *Urban life in the Middle Ages: 1000-1450*. Palgrave, New York.

Limberger, M. 2008. *Sixteenth-century Antwerp and its Rural Surroundings. Social and Economic Changes in the Hinterland of a Commercial Metropolis (ca. 1450 - ca. 1570)*. Brepols, Turnhout.

Macphail, R. 1981. Soil and botanical studies of the 'Dark Earth'. In M. Jones & G.W. Dimbleby (ed.), *The Environment of Man: the Iron Age to the Anglo-Saxon Period*, 309-331. BAR Ltd., Oxford.

Maesschalck, A. & Viaene, J. s.d.: *Het stadhuis van Brussel*, s.n., Kessel-Lo.

Martens, M. 1953. *L'administration du domaine ducal en Brabant au moyen âge (1250-1406)*. Académie Royale de Belgique (Classe des Lettres), Brussels.

Martens, M. 1973. Bruxelles, capitale de fait sous les Bourguignons. *Westfälische Forschungen* 15, 180-187.

Martens, M. 1976. *Histoire de Bruxelles*. Privat, Toulouse.

Nachtergael, I. 1997. Sauvetage archéologique dans le quartier des Marolles à Bruxelles, rue des Chandeliers, n° 12-16. *Vie Archéologique*, supplément 47, 7-72.

Nicholas, D.M. 1971. *Town and Countryside: Social, Economic and Political Tensions in Fourteenth-Century Flanders*. De Tempel, Bruges.

Pinol, J. (ed.) 2003. *Histoire de l'Europe Urbaine. I. De l'Antiquité au XVIIIe siècle. Genèse des villes européennes*. Seuil, Paris.

Rapp, G. Jr. & C.L. Hill. 1998. *Geoarchaeology, The Earth-Science Approach to Archaeological Interpretation*. Yale University Press, New Haven/London.

Reynolds, S. 1992. The writing of medieval urban history in England. *Theoretische Geschiedenis* 19, 49-50.

Schofield, J. & A. Vince. 2003. *Medieval Towns. The Archaeology of British Towns in their European Setting*. Equinox, London-Oakville.

Smolar-Meynart, A. 1985. Bruxelles : l'élaboration de son image de capitale en politique et en droit au moyen âge. *Bijdragen tot de Geschiedenis* 68, 25-45.

Smolar-Meynart, A. 1991. *La justice ducale du plat pays, des fôrets et des chasses en Brabant (XIIe-XVIe siècle): sénéchal, maître des bois, gruyer, grand veneur*. Société Royale d'Archéologie de Bruxelles, Brussels.

Thoen, E. 1993. The count, the countryside and the economic development of the towns in Flanders from the eleventh to the thirteenth centuries. Some provisional remarks and hypothesis. In E. Aerts, B. Henau, P. Janssens & R. Van Uytven, *Studia Historica Oeconimica. Liber amicorum Herman Van der Wee*, 259-278. Universitaire Pers, Leuven.

Tits-Dieuaide, M.J. 1975. *La formation des prix céréaliers en Brabant et en Flandre au XVe siècle*. Editions de l'Université de Bruxelles – Centre d'Histoire économique et sociale, Brussels.

Tits-Dieuaide, M.J. 1981. L'évolution des techniques agricoles en Flandre et en Brabant du XIVe au XVIe siècle. *Annales. Economies-Sociétés-Civilisations* 36, 362-381.

Vannieuwenhuyze, B. 2007. Le pavage des rues à Bruxelles au Moyen Âge. In Actes des VIIe Congrès de l'Association des Cercles francophones d'Histoire et d'Archéologie de Belgique (AFCHAB) et LIVe Congrès de la Fédération des Cercles d'Archéologie et d'Histoire de Belgique. Congrès d'Ottignies – Louvain-la-Neuve (26-28 August 2004), 299-307. Editions Safran, Brussels.

Vannieuwenhuyze, B. 2008. *Brussel, de ontwikkeling van een middeleeuwse stedelijke ruimte*, unpublished Ph.D thesis, Ghent University, Ghent.

Vannieuwenhuyze, B. 2009. Wegen in beweging. De in- en uitvalswegen van middeleeuws Brussel vóór de 13de eeuw. *Cahiers bruxellois* 41, 7-29.

Vannieuwenhuyze, B. 2010. Wegen in beweging. De in- en uitvalswegen van middeleeuws Brussel van de 13de tot 15de eeuw. *Cahiers bruxellois* 42, 3-32.

Verhulst, A. 1990. *Précis d'histoire rurale de la Belgique*. Editions de l'Université de Bruxelles, Brussels.

Verhulst, A. 1999. *The Rise of Cities in North-West Europe*. Maison des Sciences de l'Homme – Cambridge University Press, s.l.

Verniers, L. 1934. Les Transformations de Bruxelles et l'Urbanisation de sa Banlieue depuis 1795. *Annales de la Société Royale d'Archéologie de Bruxelles* 37, 84-220.

Vrydaghs, L., Devos, Y., Fechner, K. & A. Degraeve.. 2007. Phytolith analysis of ploughed land thin sections. Contribution to the early medieval town development of Brussels (Treurenberg site, Belgium). In Madella, M. & D. Zucol (eds.), *Plants, people and places. Recent studies in phytolith analysis*, 13-27. Oxbow Books, Oxford.

Whitehand, J.W.R. & P.J. Larkham. 1992 (eds.). *Urban landscapes: international perspectives*. Routledge, London.

Wilkinson, K. & C. Stevens. 2003. *Environmental Archaeology. Approaches, techniques and applications*. Tempus Publishing Ltd., Stroud, Gloucestershire.

Linking landscapes of lowlands to mountainous areas

3.1 A qualitative model for the effect of upstream land use on downstream water availability in a western Andean valley, southern Peru

Authors
Ralf Hesse[1] and Jussi Baade[2]

1. State Office for Cultural Heritage Baden-Württemberg, Esslingen am Neckar, Germany
2. Department of Geography, Friedrich Schiller University Jena, Jena, Germany
Contact: ralf.hesse@rps.bwl.de

ABSTRACT

The rise and decline of pre-Columbian cultures in coastal Peru has been the subject of numerous studies. Availability of and access to water have long been recognised as the key issues for the habitability of valley oases in the coastal desert where agriculture depends on seasonal river discharge from the Andes. In general, the reason for cultural changes has often been seen in 'natural disasters' or climatic changes. We propose an alternative qualitative model to explain changes of human-environment interactions in the region. This model focuses on patterns arising from the exploitation of and adaptation to the limited – but on the whole not necessarily declining – resources of water and arable land. Agriculture along the rivers of the Peruvian coastal desert was probably first practised in the wide, gently sloping lowland valley floor areas at the foot of the Andes, where large tracts of land can be irrigated with relatively low expenditure of labour. Subsequent expansion of agriculture necessitated the utilisation of progressively marginal areas along the upstream reaches of the rivers. While the spatially limited valley floor areas can to a certain extent be irrigated with short irrigation canals, irrigation of the steep valley slopes of up-valley areas requires the labour-intensive construction and maintenance of canals and terraced fields in difficult – and arguably less productive – terrain. Diversion of water onto up-valley terraced fields can be expected to have reduced water availability in the lowland valley floor fields. Thus, the adaptation to the constraints of one limited resource (irrigable land) may have led to a suboptimal exploitation of another limited resource (water), leading to an overall decline in agricultural productivity per unit of arable land. Such a feedback between land use and water availability is consistent with archaeological findings such as declining population density and increasing conflict during the Early Intermediate Period.

KEYWORDS

qualitative modelling, Palpa Valley (Peru), Nasca, irrigation, water management, early agricultural land use systems (2800-550 cal BP)

INTRODUCTION

The Palpa Valley, a river oasis at the foot of the Andes in the desert of coastal southern Peru (fig. 1), has been inhabited for at least 3500 years. Settlement was based on irrigation agriculture which by distributing sediment-laden river water to the fields has created thick accumulations of irragic anthrosol (Hesse & Baade 2009). The Palpa Valley is a wide valley floor that is shared by the rivers Rio Palpa and Rio Vizcas and that has an area of approximately 15 km². Because mean annual precipitation is less than 10mm (ONERN 1971), agriculture depended completely on irrigation with river water before the introduction of mechanised pumping. Seasonal river discharge from the Andes therefore is the key factor for agricultural productivity in this and other coastal valleys.

In this environmental context, archaeological investigations document the rise and decline of pre-Columbian cultures (Reindel et al. 2001; Silverman 2002; Schreiber 1999). Both the availability of river water in terms of climatic fluctuations (Shimada et al. 1999) and the access to water in terms of irrigation technology, labour mobilisation and control (e.g. Pozorski & Pozorski 2003) have long been recognised as the key issues for the habitability of valley oases in the coastal desert. Several qualitative models have been suggested to describe and explain observed changes in the archaeological record. A qualitative model is a description of processes, interactions and cause-effect relationships within a system which can be compared with the observed characteristics of that system. It can thus help to improve the understanding of past societies where there is insufficient quantitative data. Taking into account the research history

Figure 1. Map showing the location of the Palpa Valley. The two valleys reaching the Palpa Valley from the east are the valleys of Rio Palpa and Rio Vizcas. *See also full colour section in this book*

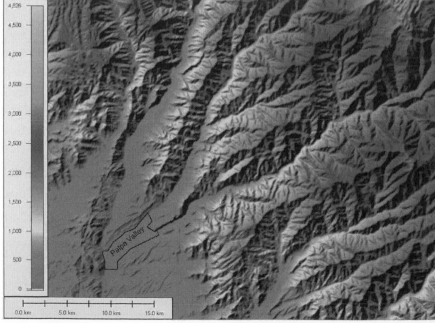

in the study area, the purpose of this paper is to present a qualitative model that focuses on human-environment interactions and feedbacks rather than one-directional cause-effect relationships.

QUALITATIVE MODELS FOR HUMAN-ENVIRONMENT INTERACTIONS

Cultural development in the region is characterised by the rise and decline of the Paracas culture during the Early Horizon (2800-2150 cal. BP) and the Nasca culture during the Early Intermediate Period (2150-1300 cal. BP). A period of cultural decline (1300-950 cal. BP) was followed by renewed cultural elaboration during the subsequent Late Intermediate Period (950-550 cal. BP) (Reindel et al. 2001; Silverman 2002; Unkel et al. 2007). Several models have been proposed to explain changes in the archaeological record of coastal Peru. Most of them assume a strong impact of the environment on cultural development which may be conceptually problematic (cf. Van Buren 2001). The degree to which interactions between humans and their environment are taken into account varies from model to model.

El Niño and natural disasters
The most frequently cited natural phenomenon impacting societies on the coast of Peru are El Niño flood events which may destroy irrigation systems, fields and settlements (Reindel et al. 2001; Silverman 2002; Wells & Noller 1999; Zaro & Alvarez 2005). However, it has also been noted that the disruptive effects of El Niño events are of short duration and may partially be offset by positive effects such as the possibility of planting or herding in the greening desert (e.g. Murphy 1926; Arntz & Fahrbach 1991).

Climatic change
The impact of more gradual environmental changes toward drought and a shifting desert margin may have affected the ecology of the desert and pre-hispanic populations (Eitel et al. 2005). Presently, the balance of evidence does not support this hypothesis (Hesse & Baade 2007).

Agricultural collapse (Irrigation system collapse)
River incision necessitates the upstream shift of irrigation canal intakes and the lowering of canal gradients until diversion of river water onto fields becomes impossible and irrigation agriculture collapses (Moseley 1983). In the Palpa Valley, artificial straightening and narrowing of the Rio Vizcas has resulted in an incision by up to 5m since the middle of the 20th century. Rather than causing the abandonment of fields, this problem has been mitigated by an upstream displacement of four irrigation canal intakes by between 70 and 310m (Hesse 2008).

Environmental degradation
The destruction of riverine forests makes irrigated valley floor areas susceptible to deflation and fluvial erosion, ultimately leading to the abandonment of river oases (Beresford-Jones 2004). While this model has been developed for the Samaca Basin in the lower Ica Valley, it is not clear whether it can also apply to the Palpa Valley, which differs from the Samaca Basin in being much larger and in being topographically protected from deflation.

Figure 2. Abandoned agricultural terraces in the upper valley of the Rio Palpa.

AN ALTERNATIVE MODEL

Common to most previously proposed models is the contention that cultural decline is to a large extent attributed to changes in the physical environment. Therefore, the purpose of the present paper is to explore whether or not the inherent dynamics of cultural landscape evolution in the context of irrigation-based societies can explain the observed changes in the archaeological record. Field observations in the upper valleys of Rio Palpa and Rio Vizcas had documented the presence of abandoned agricultural terraces (fig. 2). Such abandoned terraces cover 7500 km² in Peru (Moseley 1999). Their final abandonment is likely attributable to the early colonial collapse of the indigenous population (Cook 1981). A first appraisal of high-resolution satellite images in Google Earth covering less than 10% of the combined catchment area of Rio Palpa and Rio Vizcas yielded approximately 2 km² of abandoned terraces. These occur at elevations between 1580 and 3570m and occupy both valley bottom and valley slope locations. The lower terraces can be related to (likely seasonal) tributaries draining catchment areas well above 2000m fed by seasonal rainfall. By diverting runoff from tributaries onto irrigated fields, the river discharge at downstream locations is diminished. While water infiltration from irrigated fields may recharge local aquifers and increase base flow, the additional evapotranspiration leads to an overall negative effect on river discharge.

Based on these observations, the following sequence is proposed for the Palpa Valley (fig. 3):

1. Irrigation agriculture begins at least 3500 years ago in the central part of the valley where a wide, gently sloping valley floor can be irrigated with low expenditure of labour (Hesse & Baade 2009).

2. During the Paracas and early Nasca periods of cultural development (Reindel et al. 2001; Silverman 2002), the irrigation system is extended first to the south-eastern and then to the north-western sides of the valley. In the early Nasca period, the irrigation system reaches the limits of the easily irrigated valley floor (Hesse & Baade 2009).

3. Reaching these limits marks a crucial point in cultural landscape evolution: Expansion of agricultural activities now necessitates the utilisation of progressively marginal up-valley areas. Irrigation of steep slopes requires the labour-intensive construction and maintenance of terraced fields and canals in difficult – and arguably less productive – terrain.

4. Diversion of river water onto up-valley terraced fields reduces downstream water availability in the Palpa Valley. A decline in down-valley water availability is thus not necessarily driven by climatic fluctuations. Low and intermediate flows – which are those that can be harnessed for irrigation – are most strongly affected. This is contemporaneous with cultural decline (Silverman 2002), increasing violence and declining population (Schreiber 1999) as well as an up-valley shift of settlement during the middle and late Nasca periods (Reindel et al. 2001).

In this model of human-environment interactions, the adaptation to the constraints of one limited resource (irrigable land) leads to a suboptimal exploitation of another limited resource (water), to an overall decline in agricultural productivity per unit of arable land and to increasing vulnerability.

Given the presently hypothetical character of this model, some words of caution are in order. Until now, the total extent of abandoned up-valley terraces is unknown. More importantly for the validity of the proposed model, there are presently no temporal constraints on the periods of construction and use of these terraces. Furthermore, a quantitative analysis of the potential impact of up-valley water diversion

Figure 3. Qualitative model for human-environment dynamics in the Palpa Valley catchment.

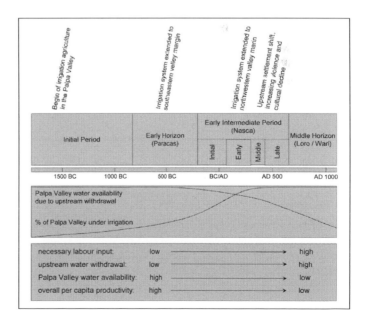

on down-valley water availability has to be performed. Such an analysis will have to take into account the strong interannual variability of river discharge.

CONCLUSIONS

The cultural changes in the archaeological record of river oases in coastal southern Peru can not be confidently attributed to any one of the competing qualitative models for human-environment interactions. Most previously proposed models explicitly or implicitly assume a strong environmental impact which may underestimate the ability of societies to adapt to such changes. The qualitative model presented here is driven by human behaviour in the exploitation of and adaptation to limited – but not necessarily declining – resources. The limiting factor in this case is the spatially constrained nature of irrigable land. The inherent dynamics of cultural landscape evolution between down- and up-valley irrigation agriculture create a pattern in which resource use feeds back onto resource availability. The evolution of resource use can create a situation where the overall efficiency of resource use is reduced, thus leading to a decline in overall productivity. It can qualitatively explain the observed broad changes in the archaeological record of the area without invoking 'natural disasters' or climatic changes.

The proposed model is, however, not meant to replace the previously proposed models. Rather, it is intended as a contribution to the ongoing debate on potential causes for cultural change in pre-hispanic coastal Peru. Comparing environmentally and culturally based models will help to develop new research designs. It may also be of wider applicability in studies regarding the development of irrigation-based societies. It is emphasised that at present this model is a working hypothesis and that further research is necessary, in particular regarding the time of construction and use of up-valley terraces as well as regarding quantitative analysis of the potential impact of upstream water diversion on down-valley water availability.

ACKNOWLEDGEMENTS

Funding for this study was partly provided by the German Research Foundation (DFG) under the grant BA1377/6-1, 6-2. Ronald T. Van Balen and J.C.A. Kolen provided very helpful comments on an earlier version of this paper.

REFERENCES

Arntz, W.E. & E. Fahrbach. 1981. *El Niño. Klimaexperiment der Natur. Physikalische Ursachen und biologische Folgen.* Birkhäuser Verlag, Basel.

Beresford-Jones, D. 2004. *Pre-hispanic Prosopis-human relationships on the south coast of Peru: riparian forests in the context of environmental and cultural trajectories of the lower Ica valley.* Dissertation, University of Cambridge.

Cook, N. 1981. *Demographic Collapse: Indian Peru, 1520-1620.* Cambridge University Press, Cambridge.

Eitel, B., Hecht, S., Mächtle, B., Schukraft, G., Kadereit, A., Wagner, G.A., Kromer, B., Unkel, I. & M. Reindel. 2005. Geoarchaeological evidence from desert loess in the Nazca-Palpa region, southern Peru: palaeoenvironmental changes and their impact on pre-Columbian cultures. *Archaeometry* 47, 137-158.

Hesse, R. 2008. Fluvial dynamics and cultural landscape evolution in the Rio Grande de Nazca drainage basin, southern Peru. BAR *International Series* 1787.

Hesse, R. & J. Baade. 2007. Palaeoenvironmental changes in the Nazca-Palpa region, Southern Peru – alternative interpretations of geoarchaeological evidence: A comment on Eitel et al. (2005). Archaeometry, Vol. 47(1). *Archaeometry* 49 (3), 595-602.

Hesse, R. & J. Baade. 2009. Irrigation agriculture and the sedimentary record in the Palpa Valley, southern Peru. *Catena* 77, 119-129.

Moseley, M.E. 1983. The good old days were better: agrarian collapse and tectonics. *American Anthropologist* 85, 773-799.

Moseley, M.E. 1999. Convergent catastrophe: past patterns and future implications of collateral natural disasters in the Andes. In Oliver-Smith, A. & Hoffman, S.M. (eds.), *The angry earth: disaster in anthropological perspective*, pp. 59-88. Routledge, New York.

Murphy, R.C. 1926. Oceanic and climatic phenomena along the west coast of South America during 1925. *Geographical Review* 16, 26-54.

ONERN. 1971. *Inventario, evaluación y uso racional de los recursos naturales de la costa: cuenca de Rió Grande (Nazca)*, Vol. 1+2.

Pozorski, T. & S. Pozorski. 2003. The impact of the El Niño phenomenon on prehistoric Chimu irrigation systems of the Peruvian coast. In Haas, J. & Dillon, M.O. (eds.), *El Niño in Peru: biology and culture over 10,000 years*. Papers from the VIII annual A. Armour III Spring Symposium, 28-29 May 1999 Chicago. *Fieldiana Botany, New Series* 43, 71-89. Field Museum of Natural History, Chicago.

Reindel, M., Isla Cuadrado, J., Grün, A. & K. Lambers. 2001. Neue Erkenntnisse zu Siedlungen, Bodenzeichnungen und Kultplätzen in Palpa, Süd-Peru: Ergebnisse der Feldkampagne 2000 des Archäologischen Projektes Nasca-Palpa. *Jahresbericht der Schweizerisch-Liechtensteinischen Stiftung für Archäologische Forschungen im Ausland (2000)*, 81-104.

Schreiber, K. 1999. Regional approaches to the study of prehistoric empires. Examples from Ayacucho and Nasca, Peru. In Billman, B.R. & Feinman, G.M. (eds.), *Settlement pattern studies in the Americas. Fifty years since Virú*, 160-171. Smithsonian Institution Press, Washington.

Shimada, I. Schaaf, C.B., Thompson, L.G. & E. Mosley-Thompson. 1999. Cultural impacts of severe droughts in the prehistoric Andes: application of a 1,500-year ice core precipitation record. *World Archaeology* 22 (3), 247-270.

Silverman, H. 2002. *Ancient Nasca settlement and society*. University of Iowa Press, Iowa City.

Unkel, I., Kromer, B., Reindel, M., Wacker, L. & G. Wagner. 2007. A chronology of the pre-Columbian Paracas and Nasca cultures based on AMS ^{14}C dating. *Radiocarbon* 49, 551-564.

Van Buren, M. 2001. The archaeology of El Niño events and other 'natural' disasters. *Journal of Archaeological Method and Theory* 8, 129-149.

Wells, L.E. & J.S. Noller. 1999. Holocene coevolution of the physical landscape and human settlement in the northern coastal Peru. *Geoarchaeology* 14, 755-789.

Zaro, G. & A.U. Alvarez. 2005. Late Chiribaya agriculture and risk management along the arid Andean coast of southern Peru, 1200-1400. *Geoarchaeology* 20, 717-737.

3.2 Connecting lowlands and uplands: An ethno-archaeological approach to transhumant pastoralism in Sardinia (Italy)

Author
Antoine Mientjes

Research Institute for the Cultural Landscape and Urban Environment, VU University Amsterdam, Amsterdam, The Netherlands
Contact: ac.mientjes@gmail.com

ABSTRACT

Historical and geographical studies have stressed the pivotal role of plains (particularly coastal plains), valleys and mountains and their interconnections in Mediterranean history. However, many landscape archaeological studies have tended to focus one-sidedly on the recording of archaeological structures and artefacts in the lowlands. Allegedly, the 'poor' character of material culture and related to this the difficult retrieval of archaeological remains have discouraged landscape archaeologists from studying the mountainous regions of the Mediterranean. This is regrettable, because the relation between plains, valleys and mountains has been a central feature of Mediterranean rural life, especially in regions in which pastoral economies (i.e. sheep herding in particular) have figured prominently. In this article, I will discuss landscape archaeological studies of pastoral economies in the Mediterranean, with particular attention to transhumant patterns of mobility, i.e. seasonal movement of shepherds and their flocks between different regions. I will show that most studies to date have adopted ethno-archaeological approaches in order to aid archaeological interpretations of pastoral economies and their spatial features during prehistoric, classical and medieval times. I will present my ethno-archaeological research on pastoral landscapes in Sardinia, with a focus on how these landscapes have changed over the last 200 years. My contention is that it is possible to study transhumant pastoralism through archaeological methods. In addition, I will show how a landscape archaeological approach in particular can shed new light on the ways in which shepherds inhabited and exploited the countryside.

KEYWORDS

mountains, lowlands, ethno-archaeology, pastoralism, transhumance, pastoral settlements

INTRODUCTION

He tends to linger over the plain, which is the setting for the leading actors of the day, and does not seem eager to approach the high mountains nearby. More than one historian who has never left the towns and their archives would be surprised to discover their existence. And yet how can one ignore these conspicuous actors, the half-wild mountains, where man has taken root like a hardy plant; always semi-deserted, for man is constantly leaving them? How can one ignore them when often their sheer slopes come right down to the sea's edge? The mountain dweller is a type familiar in all Mediterranean literature. According to Homer, the Cretans were even then suspicious of the wild men in their mountains and Telemachus, on his return to Ithaca, describes the Peloponnese as covered with forests where he lived among filthy villagers, 'eaters of acorns'. (Braudel 1972, 29-30)

This quote from Fernand Braudel's work shows that the mountains should occupy an important place in any historical study of the Mediterranean landscape. Seemingly, Braudel accuses his fellow-historians of neglecting the mountains and their influence on the large currents of Mediterranean history. However, his book *La Méditerranée et le Monde Méditerranéen à l'Époque de Philippe II* was first published in 1949. Since then many historical and geographical studies have appeared which have given particular attention to the history, geography and ecology of the Mediterranean mountains and their connection with the lowlands, for example McNeill (1992) and more recently Horden & Purcell's (2000) influential book *The Corrupting Sea: A Study of Mediterranean History*.

By contrast, archaeological studies and especially large-scale regional archaeological surveys in the Mediterranean have focused predominantly on (proto-)urban sites and their rural hinterlands in the inland plains, coastal plains and valleys until the present-day. Classical examples are the South Etruria Survey in the Tiber Valley to the north-east of Rome, and the Biferno Valley project in the region of the Molise in central Italy (Barker 1995; Potter 1979). Moreover, those surveys and other archaeological research projects, which also included mountainous areas, have tended to document predominantly monumental sites such as rural sanctuaries or road networks, which connected urban centres in different regions (e.g. Lloyd 1995; Maaskant-Kleibrink 1987, 1992).

The systematic collection of archaeological artefacts at the surface in mountainous regions, in which particular attention has been also given to so-called *off-site* distributions (e.g. Alcock et al. 1994; Annis et al. 1995; Dommelen 1998; Foley 1981), has been rare to date for two main reasons. Firstly, the rough character of mountainous areas in the Mediterranean makes a field-walking method rather difficult to adopt. The vegetation of scrub and woods makes many mountainous areas difficult to walk, and moreover hinders the visibility of archaeological artefacts at the surface. Secondly, there is a common view among landscape archaeologists that mountainous areas would be characterised by a 'poor' material culture in both the distant and recent past. This is to say that tools and other objects were made from materials directly available and easy to transport, and therefore mostly organic materials such as wood. For instance, Joanita Vroom (1998) has eloquently shown that the rural communities in the Aetolian Mountains in central Greece knew an aceramic tradition during early modern history, and most objects such as dishes were made of wood. She even documented that before the 1950s skins of onions were often used as spoons. Evidently, these kinds of materials do not last very long in archaeological contexts, and rapidly decay at the surface. As a consequence, few large-scale regional archaeological surveys have been set up to date in the Mediterranean mountains, with studies focusing instead on extensive archaeological sites such as hill-top settlements and their geographical distribution.

This state of affairs is regrettable, because many Mediterranean (pre)histories cannot be analysed adequately if the relations between mountains, valleys and plains are not fully considered. Therefore, this article will discuss an ethno-archaeological approach to pastoral landscapes in mountain and lowland regions in Sardinia. Pastoralism, i.e. the chief reliance on herded domestic animals such as sheep, goats and cattle for subsistence as well as market production, has occupied an important place in (early) modern rural economies in the Mediterranean. Distinctive characteristics of Sardinian pastoralism have been transhumant patterns of seasonal mobility and pastoral settlements, and the admittedly ephemeral and perishable nature of much pastoral material culture, both artefactual and architectural. In spite of this latter problem, I will contend that types of pastoral settlements and their geographical settings can inform us about the integration of lowland and upland economies in the past.

ARCHAEOLOGICAL APPROACHES TO ANCIENT AND RECENT PASTORALISM IN THE MEDITERRANEAN

Since about the 1970s, a substantial body of archaeological studies has appeared which has attempted to detect pastoralism and other forms of animal husbandry during Mediterranean (pre)history, such as pigs held on farmsteads. One strand of archaeological research has focused on excavated faunal remains, i.e. animal bones and teeth, and ecofacts such as mineral and botanical residues often present in sheep and goat faeces (e.g. Chang & Koster 1986, 97, 107-9; Payne 1973; Reid 1996). This kind of research has been helpful in determining domestic animal species and identifying animal enclosures. However, a serious problem is the poor preservation of animal bones and teeth and ecological data at archaeological sites in the open air which are not waterlogged. As a result, this type of research has mostly been confined to cave sites such as the Grotta dell'Uzzo (i.e. Uzzo Cave), which is a well-known Mesolithic and Neolithic site in north-west Sicily, used in recent history to pen sheep (Brochier et al. 1992). Moreover, faunal and ecological evidence alone often does go beyond the mere observation of the presence or absence of architectural structures to pen domestic animals. In other words, this evidence does not give us much detail about different strategies in pastoral production and the occupation and exploitation of rural landscapes.

Palynological evidence is a second category of archaeological material, which played an especially prominent role in discussions about the rise of mainly large-scale and transhumant pastoral economies during prehistoric and early historical times (e.g. Halstead 1996). The clearance of large tracts of lowland and upland woods could indicate extensive grazing and the frequent and regular movement of shepherds and their flocks between regions with different altitudes. But instances of large-scale woodland clearances in the Mediterranean are mostly absent from the botanical evidence for periods before the Late Middle Ages (e.g. Edwards et al. 1996; Willis & Bennett 1994). Secondly, the evidence for deforestation does not go beyond suggesting the presence or absence of a considerable role for pastoralism in ancient rural economies. Forest clearance in both lowlands and uplands does not necessarily point to transhumant or other patterns of pastoral mobility.

The outlined problems demonstrate that we need a *different* approach, especially if we want to study distinctive strategies in pastoral production and land use, which connected plains, valleys and mountains in the past. Therefore, a landscape approach seems to offer more potential by studying pastoral structures and their landscape context. This has also been argued by Claudia Chang in her ethno-archaeological study of recent pastoral economies in the Argolid on the Peloponnese and the Pindos Mountains in northern Greece (Chang 1992, 66). Chang's studies are among the best examples to date of archaeological research on pastoral landscapes in the Mediterranean. Her analysis concentrates on pastoral site locations in the total landscape, in which animal enclosures are regarded as the principal structures of pastoral production. However, there are other relevant categories of sites entirely or partly related to pastoralism, including huts, grazing areas, wells, cisterns and springs, trails for animals (including transhumance routes), and barns and bins (e.g. for storage of animal fodder). In the case of the Argolid pastoral settlements are distributed in a circle around the village on the periphery of a low basin and the beginning of the mountains (Chang 1981, 9). Factors in pastoral site location are access to, and control of pasture, climate and the availability of water (Chang 1981, 42).

In the Pindos Mountains Chang's research has observed a different geographical pattern of pastoral settlements and land use (Chang 1992, 1993). In relation to altitude, four environmental zones can be distinguished related to their location and herd pressure on resources, and the relationships with other rural activities, principally cereal and other types of cultivation (Chang 1992, 78-79). In the highest upland zone pastoral settlements are highly dispersed to prevent conflict over grazing land. In the three lower zones pastoral structures are more diverse functionally and are clustered together near natural pasture or cultivated land, which can be used as temporary grazing land after the harvest. Pressure on grazing resources is more intense in these zones, and consequently there is more competition among shepherds and between pastoral and agrarian groups.

However, it is evident that this type of research has not yet fully developed. To date, most landscape archaeological studies of pastoralism have adopted ethno-archaeological approaches, which means the investigation of recent historical and modern material culture aimed at the creation of methodologies useful in the study of the archaeological past. Likewise, my research in Sardinia, which will be discussed below, focused on recent pastoralism (Mientjes 2004, 2008a, 2008b). Here, I want to acknowledge that it is difficult to directly observe distinctive strategies in pastoral production and patterns of pastoral mobility in the archaeological record. Nonetheless, it will be shown that a detailed archaeological analysis of types of pastoral settlements and their geographical locations in plains, valleys and mountains can inform us about functional and social aspects of pastoral practices and seasonal transhumance between regions with different altitudes. This seems a promising approach, in particular when it is used in combination with other types of evidence such as historical maps, land registries, deeds, aerial photographs and ethnographic accounts from local people. Secondly, I will argue that archaeological studies of pastoral landscapes need to highlight the social and political dimensions of herding economies, *contra* Chang, who has predominantly followed a functional and ecological approach (Chang 1992, 76).

Page 253 >
Figure 1. The island of Sardinia (Italy) with the village territories of Fonni and Solarussa in the central mountains (so-called Barbagia) and the northern Campidano Plain, and other zones and places mentioned in the text.

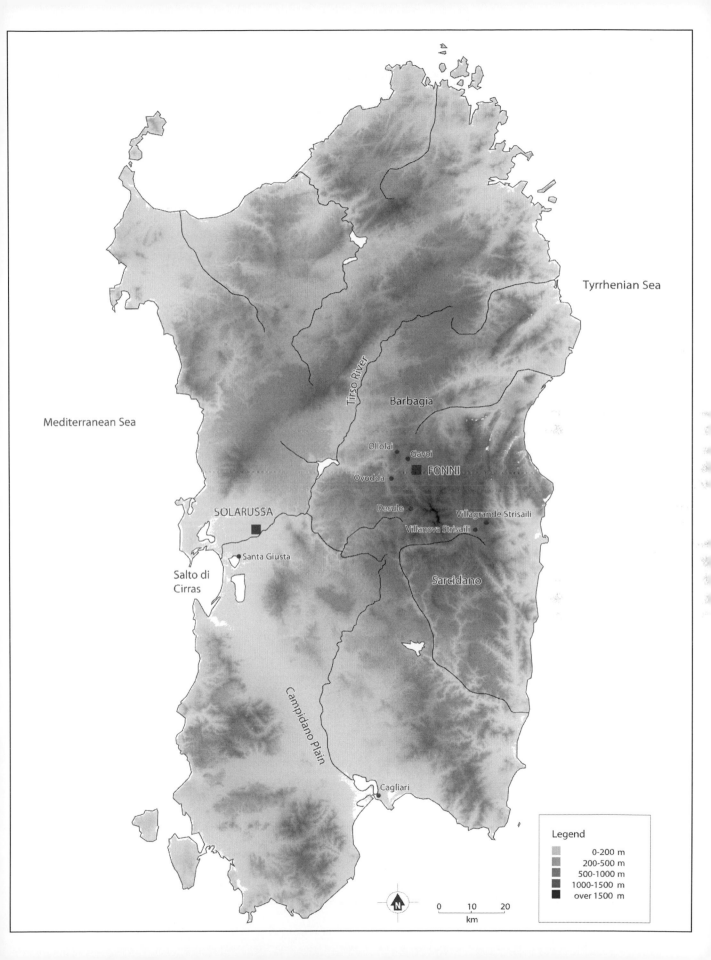

Tyrrhenian Sea

Mediterranean Sea

Tirso River

Barbagia

Ollolai

Gavoi

FONNI

Ovodda

Desulo

Villagrande Strisaili

Villanova Strisaili

SOLARUSSA

Santa Giusta

Salto di
Cirras

Sarcidano

Campidano plain

Cagliari

Legend

0-200 m
200-500 m
500-1000 m
1000-1500 m
over 1500 m

N

0   10   20
km

TRANSHUMANT SHEPHERDS AND THEIR PASTORAL SETTLEMENTS IN SARDINIA

Introduction: the study areas

Sardinia has been renowned for its pastoral economy, i.e. of predominantly sheep herding, during recent history. For example, in 1985 Sardinia had a density of about 105 sheep per square kilometre and 25,000 active shepherds out of a total of 70,000 workers in the agricultural sector generally, and an area of pasture which covers almost half of the entire island (Angioni 1989, 12-13). Moreover, Sardinian sheep are considered among the best milk-producing species in the Mediterranean, and the cheese made from their milk is even sold in the United States.

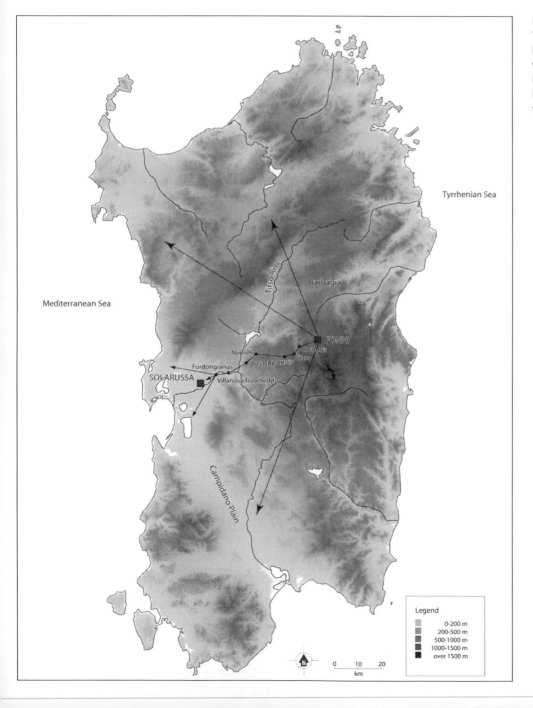

Figure 2. Transhumance routes between Fonni and various lowland areas in Sardinia, which includes the village territory of Solarussa.

Figure 3. Field road with stone field walls on both sides inside the village territory of Fonni (in Sardinian dialects called *utturu*).

Two regions have been studied in detail, i.e. the village territory of Fonni in the central mountains (called Barbagia) and the village territory of Solarussa in the northern Campidano Plain (fig. 1). Both regions were connected by transhumance routes until about the 1970s, when mechanisation (cars and tractors) and agricultural modernisation caused the widespread development of a settled form of pastoral production. In practice, before the 1970s shepherds from Fonni with their flocks of sheep moved twice a year to various lowlands on the island (mainly coastal), which included the village territory of Solarussa (fig. 2). More specifically, the transhumance to the lowlands covered the winter to early spring months (November-April) and the late summer to early autumn months (August-October). In these periods shepherds and their flocks travelled between 80 and 150 kilometres, and various scholars have therefore defined this type of seasonal mobility between Fonni and the (coastal) plains as long-distance transhumance (e.g. Le Lannou 1979, 171-176).

From the historical and ethnographic sources some detailed information has been collected on the transhumant pastoral economy of Fonni and its development during approximately the last 200 years. Archival documents show that shepherds from Fonni were already travelling with their flocks to other village territories and the Campidano Plain during the first half of the nineteenth century (source: State Archive of Nuoro: 'Atti dei notai della tappa di Oliena', years 1820-1840). Statistical data indicate that the community of Fonni counted circa 40,000 sheep in the year 1838 (Angius 1834-1856, 723), a substantial number which suggests a well-established market economy. The actual numbers of sheep have since grown, with an estimated sheep population of 63,317 in the 1980s (E.R.S.A.T. 1987, 6), but the overall pattern indicates continuity in the economy and transhumance patterns. It is not a surprise therefore that Fonni has frequently been considered as one of the classic examples of transhumant pastoral economies in Sardinia, together with communities such as Desulo, Ovodda, Gavoi and Ollolai in the interior mountains (fig. 1).

Shepherds interviewed at Fonni specified that the transhumance route between Fonni and Solarussa passed the villages of Ovodda, Tiana, Austis, Neoneli, Busachi, Fordongianus and Villanova Truscheddu (fig. 2). As a rule, field roads were followed (in Sardinian dialects called *utturu*); they had stone walls on both sides in order to protect the surrounding crop fields against the intrusion of domestic animals (fig. 3). Generally, field roads in village territories were laid out in such a way that shepherds and their flocks did not need to pass the village centres, although local shepherds and farmers used these roads more frequently. It appears that the intermediate communities attempted to control the movement of the transhumant flocks in this way, together with the appointment of country guards. Moreover, shepherds were obliged to announce their journey in advance to the village communities on the route (cf. Caltagirone 1986, 32). However, in spite of this ethnographic information it was impossible to identify transhumance routes between Fonni and the lowlands by means of archaeological methods alone. The combined archaeological, ethnographic and historical evidence demonstrate that transhumance routes have been an integral part of the general rural infrastructure in Sardinia. In other words, roads in the countryside were travelled by a variety of people, and no single route was used exclusively by transhumant shepherds and their flocks. Historically the only example encountered of a field road constructed on purpose for transhumant movements is located in the wooded region of the Sarcidano (fig. 1). Rich landlords originating from Milan constructed this road around World War II to prevent shepherds and their sheep from disturbing their hunting parties in this zone.

As a consequence, other archaeological signatures in the countryside had to be found which testify to ways in which shepherds have occupied, exploited and travelled through the landscape. I will contend

Figure 4. Zone of Monte Novu in the south-eastern part of the village territory of Fonni with the five extensive pastoral settlements indicated.

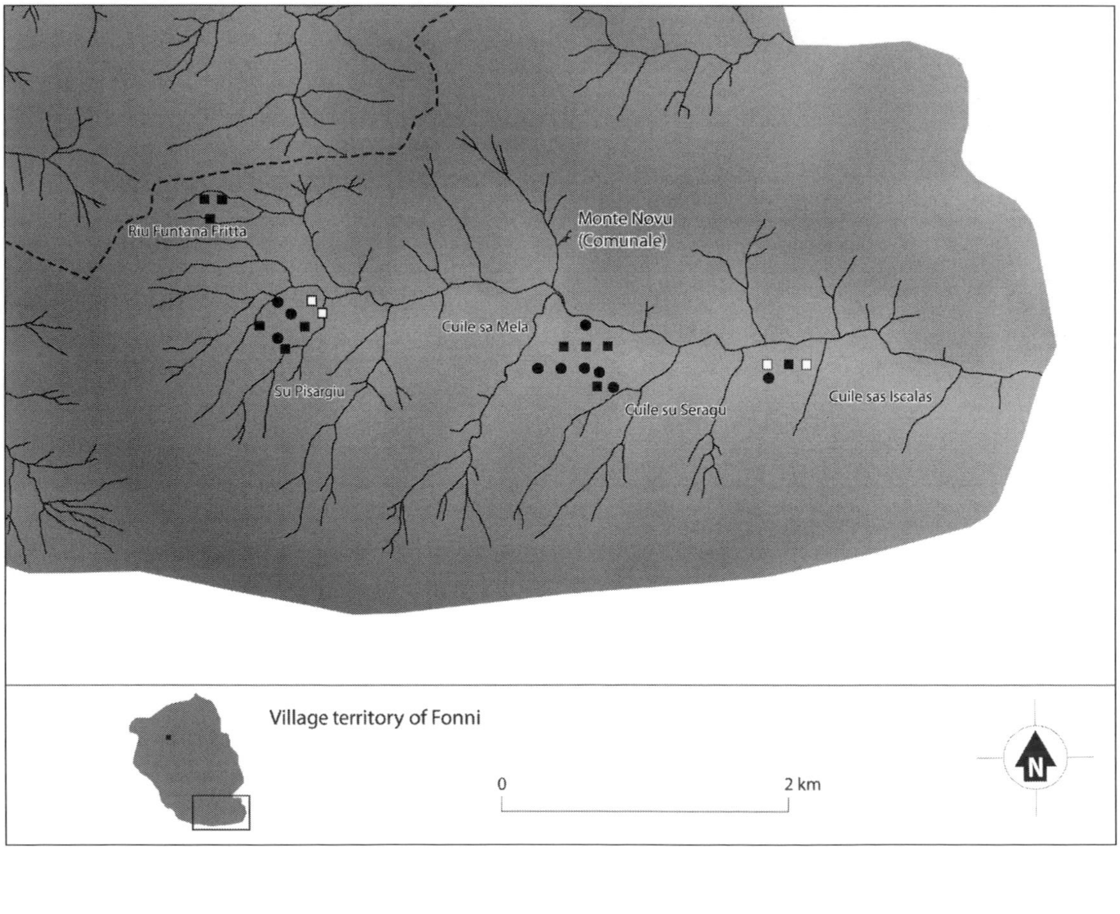

Figure 5. View of the mountainous zone of Monte Novu in the south-eastern part of the village territory of Fonni with *Cuile su Seragu* visible in the centre of the photograph. *See also full colour section in this book*

below that certain types of pastoral settlements at Fonni and Solarussa provide some clues to the transhumant organisation of Sardinian pastoralism during the nineteenth and earlier twentieth century.

Common land and pastoral settlements at Fonni

In the south-eastern part of the village territory of Fonni, five extensive pastoral settlements have been identified in a zone called Monte Novu (literally meaning: 'New Mountain'). The names of these recently abandoned settlements are (from west to east): *Riu Funtana Fritta*, *Su Pisargiu*, *Cuile sa Mela*, *Cuile su Seragu* and *Cuile sas Iscalas* (figs. 4, 5). The five pastoral settlements are characterised by the extensive layout of stone-built shepherds' huts, small sheds, milking pens and other animal enclosures. A good example of the structural features of these pastoral settlements is *Su Pisargiu*, which comprised three stone-built huts, three animal enclosures and two more recently added huts of corrugated iron and wooden beams (fig. 6). As such, the extensive pastoral settlements at Monte Novu significantly deviate from the typical pastoral settlements found in the countryside, which usually consisted of one stone-built shepherd's hut and one or two animal enclosures constructed of wooden branches or cane.

Monte Novu is a markedly different zone within the Fonni territory, in that the land has been communally used for grazing sheep and other domestic animals since the early nineteenth century until the 1990s. The zone covers an area of 3,621 hectares and has a mountainous character with altitudes varying between 1,100 and 1,800 metres above sea level. Locally, Monte Novu is also called Comunale, which can be translated as the 'Commons'. The zone was managed by the village council of Fonni, and a system of common-use rights to land was applied, which in the Sardinian literature is referred to by the term *ademprivi* (e.g. Masia 1992, 18-9; Solmi 1967, 126). The general rule underlying these common rights of use to land was that membership of a rural community was the exclusive condition granting access to certain zones inside a village territory (Meloni 1988, 134). In practice this rule gave the residents the unlimited

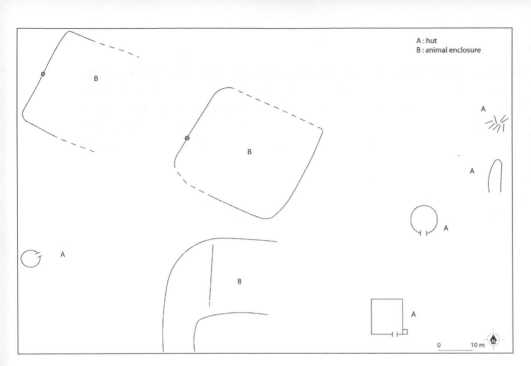

Figure 6. Plan of extensive pastoral settlement of *Su Pisargiu*, which is situated at the bottom of the deeply incised valleys inside the zone of Monte Novu in the south-eastern part of the village territory of Fonni.

A : hut
B : animal enclosure

right to pasture sheep and other domestic animals, to collect acorns as fodder for pigs and wood for fuel and building (Le Lannou 1979, 120). In the case of Monte Novu, shepherds from Fonni had the right to enter and graze their flocks in this zone between the months of April and December, i.e. the period in which they returned from the lowlands.

The Comunale at Fonni has also an interesting history, which is relevant to this argument. Before the nineteenth century the area witnessed severe conflicts between the village communities of Fonni, Villanova Strisaili and Villagrande Strisaili, which are located in the region of the Ogliastra and to which the zone of Monte Novu juridically belonged and still belongs (fig. 1) (Lai 1988, 196-97; Mereu 1978, 236-37). Various historical reasons are mentioned for the disputes between these villages, but the most convincing one appears to be population growth at Fonni, especially during the eighteenth century, and the related increase in sheep numbers. New pasture had to be acquired for the flocks, which had to be sought outside the official territory of Fonni. Between the second half of the seventeenth century and the year 1811, in which a final settlement was attained, a history of violent clashes between the three communities is documented. In the year 1800, for example, a large group of men from Fonni with horses and dogs attacked and murdered shepherds from Villanova Strisaili and Villagrande Strisaili and looted their sheep pens at Monte Novu. The cause of this violent clash was the reoccupation of pastoral settlements in part of Monte Novu, which according to the people of Fonni had been assigned to their community by the regional authorities in the preceding year.

It is important to note that in these inter-village conflicts, claims on land were made according to traditions of use. One of the arguments was that pastoral settlements were always those of shepherds from the villages in the Ogliastra (Mereu 1978, 238). In this context they often claimed customary rights by speaking of 'da tempo immemoriale' signifying that things belonged 'since time immemorial' to their villages (cf. Lai 1988, 193). The crucial notion is that rights to common land had to be constantly established between pastoral households and between shepherds and farmers by use on a daily, seasonal, yearly, lifetime and also generational basis. This enabled access to certain grazing grounds in areas which in theory were open to every member of the local community. The cessation of use by a shepherd or group of shepherds even for a brief period could allow others to take 'possession' of the abandoned structures and disused grazing ground. It is illustrative in this context that certain pastoral settlements were used by several shepherds, but that the one who constructed the pastoral structure had the first right to use it, after which others could follow.

The foregoing discussion suggests that a critical element in establishing rights of use was the actual use of pastoral settlements and the surrounding grazing areas. I therefore argue that the extensive layout of the pastoral settlements at Monte Novu and – most importantly – their enduring material character, can be connected to strategies by shepherds geared to the continuance of access to land within communally-used zones. This conclusion is strengthened by the fact that on average groups of seven, eight or more shepherds used to join their flocks and work together at the extensive pastoral settlements, which was a type of pastoral co-operation described locally as (*semmos*) *a unu* (literally meaning: '(we are) together'). Likewise, this pastoral collaboration can be connected to shepherds' strategies to secure access to pasture in zones which in theory were open to every member of the local community. I will explain the relation between these forms of pastoral co-operation and transhumant pastoralism more thoroughly in the final section below, which describes one particular pastoral settlement at Solarussa. This settlement has been used by shepherds from Fonni or other shepherds from the interior mountains of Sardinia during recent history.

Admittedly, the presence of common land for grazing flocks of sheep and other domestic animals provides no direct proof of transhumant patterns of pastoral mobility between mountains, valleys and plains. However, large-scale and specialised pastoral economies, which are often also transhumant, can only develop when extensive areas of pasture are available and accessible with considerable ease. The outlined system of common-use rights to land provides for this need. The argument therefore corresponds with the current view among many archaeologists and historians that transhumance is connected with the development of large-scale, market-oriented and often state-sponsored forms of specialised pastoral economies in the Mediterranean since the Late Middle Ages (e.g. Delano Smith 1979; Halstead 1987), of which the Mesta in central Spain and the Dogana delle Pecore in the region of Apulia in southern Italy are the best known historical examples (Braudel 1972, 89, 91). These scholars claim that transhumant patterns of pastoral mobility are primarily a function of the search for sufficient pasture in the context of drastically increasing numbers of sheep or other domestic animals. Consequently, they largely reject, in the same way as I do, the ecological perspective in which transhumance is predominantly seen as a highly effective adaptation and therefore almost natural response to the substantial annual fluctuations in the Mediterranean climate at different altitudes (e.g. Barker 1989; Skydsgaard 1974).

THE EXTENSIVE PASTORAL SETTLEMENT AT NURAGHE MEDDARIS, VILLAGE TERRITORY OF SOLARUSSA

Finally, one pastoral settlement at the northern border of the village territory of Solarussa helps to reinforce my line of argument about the connection between transhumant pastoralism and extensive pastoral settlements (fig. 7). This pastoral settlement consists of four stone-built huts and seven animal enclosures (fig. 8), and is located at the edge of a basalt plateau (at about 170 metres above sea level) and moreover

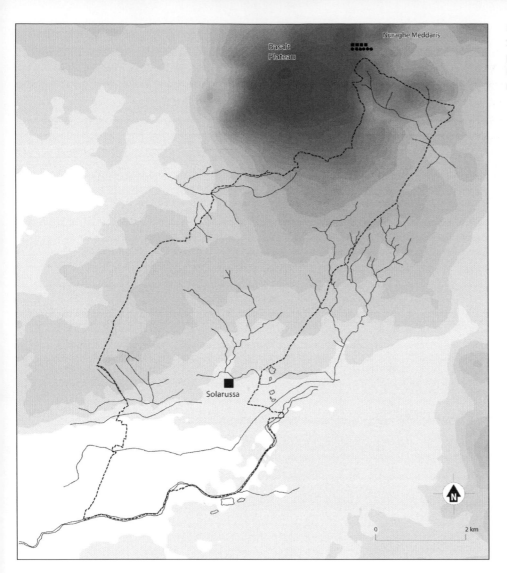

Figure 7. Village territory of Solarussa with the extensive pastoral settlement at Nuraghe Meddaris to the north.

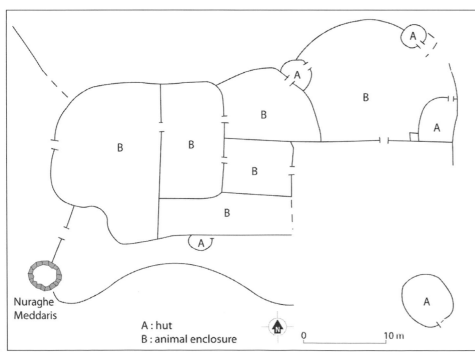

Figure 8. Plan of extensive pastoral settlement at Nuraghe Meddaris, village territory of Solarussa.

next to Nuraghe Meddaris, which is a well-preserved Bronze Age tower (second millennium BC) (fig. 9). Both the extensive layout of this settlement and its location imply occupation by transhumant shepherds and their flocks in the recent past. It is uncertain if these shepherds came from Fonni, but the relatively large amount of pasture acquired by shepherds from Fonni in the territory of Solarussa from the 1940s onwards, according to the cadastral information, is remarkable in this respect (source: State Archive of Oristano: 'catasto nuovo').

Firstly, the peripheral location at the northern edge of Solarussa's territory corresponds with historical and ethnographic information on the seasonal dwelling of transhumant shepherds from the interior mountains in the (coastal) lowlands. Since before World War II, shepherds operated as 'self-sufficient' units – even for example bringing bread provisions for several months from the interior mountains – and tended to seclude themselves from the lowland communities with whom relationships were often strained. Therefore, shepherds were also inclined to occupy the peripheral zones of the village territories in the valleys and (coastal) plains. As such, it is possible to view the spaces in which transhumant shepherds operated in the (coastal) lowlands as 'enclaves', whose defining characteristic was a lack of social integration with the local communities (cf. Caltagirone 1989, 48). Significant in this context are the related and interchangeable words *s'isverrare* ('to hibernate') and *s'istrangiare* ('going among unknown people') to indicate the journey from the familiar mountain villages to lowland areas during winter and early spring (Murru Corriga 1990, 29).

A second reason for the location of pastoral settlements and grazing areas in the outer zones of village territories relates to patterns of landed property. The peripheral areas frequently comprised large estates, which were owned by wealthy nobles and bourgeois landlords. During recent history, leasing lowland estates – mostly of a low soil quality from an agricultural perspective – to groups of transhumant shepherds was a profitable business for large landowners. As a response to expanding markets for sheep cheeses since the late nineteenth century, the value of leases had increased steadily and often reached

Figure 9. View of Nuraghe Meddaris (Bronze Age tower; second millennium BC) in the northern periphery of the village territory of Solarussa.

more than half the value of the total milk production a flock (Angioni 1989, 179; Olla 1969, 30-31). Additionally, the wealthy landlords forced shepherds to introduce the maximum number of sheep an estate could support in terms of nutritious pasture. A good historical example of this practice is found in the zone of Salto di Cirras in the village territory of Santa Giusta, at a short distance to the south-west of Solarussa (fig. 1). It was said that one person from Fonni leased this zone from a rich family residing in the island capital of Cagliari in the decades around World War II, and in turn seasonally sub-let pasture to shepherds from the interior mountains. The area was divided into five separate sectors of which each could sustain about 2,000 sheep. This situation compelled transhumant shepherds to collaborate in large groups in order to deliver the amount of milk demanded by the primary lessee.

The large forms of pastoral co-operation in the lowlands show similarities with the ones observed in the zone of Monte Novu at Fonni. Likewise, in the valleys and plains transhumant shepherds had to collaborate and merge their flocks to secure access to grazing areas in which the type and size of landed properties was the determining factor. Therefore, I contend that the extensive pastoral settlement at Nuraghe Meddaris provides convincing evidence for large forms of pastoral co-operation among transhumant shepherds in the lowlands. Finally, the robust stone construction and enduring character of the huts and animal enclosures at this settlement can perhaps be related to collective strategies among shepherds to make long-term claims on grazing areas for years, decades and possibly even longer periods.

CONCLUDING REMARKS

In an attempt to link mountains, valleys and lowlands in the Mediterranean region, I proposed that landscape archaeological studies of pastoral economies and patterns of mobility offer one promising field of research. Since late medieval times and also partly for Roman Italy there is a wealth of historical (and for modern periods also ethnographic) evidence for large-scale, specialised and often state-sponsored forms of transhumant pastoral economies (e.g. Campbell 1964; Frayn 1984; Pasquinucci 1979). Despite the existence of historical and ethnographic information on the importance of sheep herding and other forms of animal husbandry in many Mediterranean regions, archaeological research into ancient pastoralism has been rare and has encountered many methodological and interpretative problems. Therefore, most archaeological studies of pastoral landscapes have been undertaken within the field of ethno-archaeology with the aim of developing better methods and interpretative models to detect and analyse types of ancient pastoralism. In the context of my own research project in Sardinia I have claimed that settlements of shepherd's huts and animal enclosures and their geographical locations in lowlands and uplands can inform us about the nature of pastoral landscapes, their development, and (transhumant) patterns of pastoral mobility. Crucially, the case studies from Sardinia showed that transhumant pastoral economies are structured by dynamic networks of social and political relationships between shepherds and with outside persons and groups such as wealthy estate owners and village councils. Likewise, the modes of access to grazing areas, which included common-use rights to land, played a pivotal role in the ways in which pastoral landscapes developed and patterns of mobility were conditioned. Surely, the landscape archaeological and (ethno)historical evidence demonstrates that transhumant patterns of pastoral mobility are predominantly the product of particular historical circumstances, instead of a natural response to allegedly 'unchanging' environmental factors in the Mediterranean region.

REFERENCES

Alcock, S.E., J.F. Cherry & J.L. Davis. 1994. Intensive Survey, Agricultural Practice and the Classical Landscape of Greece. In I. Morris (ed.), *Classical Greece: Ancient Histories and Modern Archaeologies (New Directions in Archaeology)*, 137-70. Cambridge University Press, Cambridge.

Angioni, G. 1989. *I Pascoli Erranti: Antropologia del Pastore in Sardegna*. Liguori Editori, Napoli.

Angius, V. 1834-1856. Fonni. In V. Angius & G. Casalis (eds.), *Dizionario Geografico Storico – Statistico – Commerciale degli Stati di S.M. il Re di Sardegna*. G. Maspero, Torino.

Annis, M.B., P. van Dommelen & P. van de Velde. 1995. Rural Settlement and Socio-Political Organisation: the Riu Mannu Survey Project in Sardinia. *Babesch: Bulletin Antieke Beschaving* 70, 133-52.

Barker, G. 1989. The Archaeology of the Italian Shepherd. *Transactions of the Philological Society of Cambridge* 215, 1-19.

Barker, G. (ed.) 1995. *A Mediterranean Valley: Landscape Archaeology and Annales History in the Biferno Valley*. Leicester University Press, London/New York.

Braudel, F. 1972. *The Mediterranean and the Mediterranean World in the Age of Philip II*. Collins, London.

Brochier, J.E., P. Villa, M. Giacomarra & A. Tagliacozzo. 1992. Shepherds and Sediments: Geo-Ethnoarchaeology of Pastoral Sites. *Journal of Anthropological Archaeology* 11, 47-102.

Caltagirone, B. 1986. Lo Studio della Transumanza come Dispositivo di Analisi del Mondo Pastorale. *Etudes Corses* 27, 27-44.

Caltagirone, B. 1989. *Animali Perduti: Abigeato e Scambio Sociale in Barbagia*. Celt Editrice, Cagliari.

Campbell, J.K. 1964. *Honour, Family and Patronage*. Clarendon Press, Oxford.

Chang, C. 1981. *The Archaeology of Contemporary Herding Sites in Didyma, Greece*, Unpublished Ph.D. Dissertation, Anthropology Department, State University of New York at Binghamton, New York.

Chang, C. 1992. Archaeological Landscapes: The Ethnoarchaeology of Pastoral Land Use in the Grevena Province of Northern Greece. In J. Rossignol & L. Wandsnider (eds.), *Space, Time and Archaeological Landscapes*, 65-89. Plenum Press, New York.

Chang, C. 1993. Pastoral Transhumance in the Southern Balkans as a Social Ideology: Ethno-archaeological Research in Northern Greece. *American Anthropologist* 95 (3), 687-703.

Chang, C. & H.A. Koster. 1986. Beyond Bones: Toward an Archaeology of Pastoralism. *Advances in Archaeological Method and Theory* 9, 97-148.

Dommelen, P. van. 1998. *On Colonial Grounds: A Comparative Study of Colonialism and Rural Settlement in First Millennium BC West Central Sardinia*. Faculty of Archaeology, University of Leiden, Leiden.

Delano Smith, C. 1979. *Western Mediterranean Europe: A Historical Geography of Italy, Spain and Southern France since the Neolithic*. Academic Press, London.

Edwards, K.J., P. Halstead & M. Zvelebil. 1996. The Neolithic Transition in the Balkans – Archaeological Perspectives and Palaeoecological Evidence: A Comment on Willis and Bennett. *The Holocene* 6, 120-22.

E.R.S.A.T. (Ente Regionale di Sviluppo e Assistenza Tecnica in Agricoltura). 1987. *Piano di Valorizzazione della Zona di Sviluppo Agro-Pastorale*. Comune di Fonni (Indagine Conoscitiva), Cagliari.

Foley, R. 1981. Off-site Archaeology: An Alternative Approach for the Short-Sited. In I. Hodder, G. Isaac & N. Hammond (eds.), *Pattern of the Past*, 157-83. Cambridge University Press, Cambridge.

Frayn, J.M. 1984. *Sheep-Rearing and the Wool Trade in Italy during the Roman Period*. Redwood Bury Ltd, Trowbridge Wiltshire.

Halstead, P. 1987. Traditional and Ancient Rural Economy in Mediterranean Europe: Plus Ça Change?, *Hesperia, Journal of Hellenic Studies* 107, 77-87.

Halstead, P. 1996. Pastoralism or Household Herding? Problems of Scale and Specialisation in Early Greek Animal Husbandry. In D.T. Kenneth (ed.), *Zooarchaeology: New Approaches and Theory*, World Archaeology 28 (1), 20-42.

Horden, P. & N. Purcell. 2000. *The Corrupting Sea: A Study of Mediterranean History*. Blackwell, Oxford.

Lai, F. 1988. Contestazioni Territoriali e Comunità in Sardegna tra la Fine dell'700 e la Prima Metà dell'800. *Quaderni Bolotanesi* 14, 191-204.

Le Lannou, M. 1979. *Pastori e Contadini di Sardegna*. Edizioni della Torre, Cagliari.

Lloyd, J. 1995. Roman Towns and Territories (c. 80 BC - AD 600). In Barker, G. (ed.), *A Mediterranean Valley: Landscape Archaeology and Annales History in the Biferno Valley*, 213-53. Leicester University Press, London/New York.

Maaskant-Kleibrink, M. 1987. *Settlement Excavations at Borgo Le Ferriere 'Satricum', volume I: The Campaigns 1979, 1980 and 1981*. Forsten, Groningen.

Maaskant-Kleibrink, M. 1992. *Settlement Excavations at Borgo Le Ferriere 'Satricum', volume II: The Campaigns 1983, 1985 and 1987*. Forsten, Groningen.

Masia, M. 1992. *Il Controllo sull'Uso della Terra: Analisi Socio-Giuridica sugli Usi Civici in Sardegna*. Cuec, Cagliari.

McNiell, J.R. 1992. *The Mountains of the Mediterranean World: An Environmental History*. Cambridge University Press, Cambridge.

Meloni, B. 1988. Forme di Mobilità ed Economia Locale in Centro Sardegna. *Quaderni Bolotanesi* 14, 127-141.

Mereu, A. 1978. *Fonni Resistenziale nella Barbagia di Ollolai e nella Storia dell'Isola*. Tip. Solinas, Nuoro.

Mientjes, A.C. 2004. Modern Pastoral Landscapes on the Island of Sardinia (Italy): Recent Pastoral Practices in Local versus Macro-Economic and Macro-Political Contexts. *Archaeological Dialogues* 11 (1), 161-90.

Mientjes, A.C. 2008(a). *Paesaggi Pastorali. Studio Etnoarcheologico sul Pastoralismo in Sardegna*. Cuec, Cagliari.

Mientjes, A.C. 2008(b). Transhumance in Sardinia: An Ethno-archaeological View on Transhumant Pastoral Economies. *Revista Valenciana d'Etnologia* 4, 109-126.

Murru Corriga, G. 1990. *Dalla Montagna ai Campidani. Famiglia e Mutamento in una Comunità di Pastori*. Edes, Cagliari.

Olla, D. 1969. *Il Vecchio e il Nuovo dell'Economia Agro-Pastorale in Sardegna*. Feltrinelli, Milano.

Pasquinucci, M. 1979. La Transumanza nell'Italia Romana. In E. Gabba & M. Pasquinucci (eds.), *Strutture Agrarie e Allevamento Transumante nell'Italia Romana (III-I Sec. a.C.)*, 75-184, Giardini, Pisa.

Payne, S. 1973. Kill-off Patterns in Sheep and Goats: the Manibles from Asvan Kale. *Anatolian Studies* 23, 281-303.

Potter, T.W. 1979. *The Changing Landscape of South Etruria*. P. Elek, London.

Reid, A. 1996. Cattle Herds and the Redistribution of Cattle Resources. In D.T. Kenneth (ed.), *Zooarchaeology: New Approaches and Theory*, World Archaeology 28 (1), 43-57.

Skydsgaard, J.E. 1974. *Transhumance in Ancient Italy*. Analecta Romana Instituti Danici VII, Copenhagen, 7-36.

Solmi, A. 1967. Ademprivia: Studi sulla Proprietà Fondaria in Sardegna. In A. Boscolo (ed.), *Il Feudalismo in Sardegna*, 47-144, Editrice Sarda Fossataro, Cagliari.

Vroom, J. 1998. Early Modern Archaeology in Central Greece: The Contrast of Artefact-rich and Sherdless Sites. *Journal of Mediterranean Archaeology* 11 (2), 131-64.

Willis, K.J. & K.D. Bennett. 1994. The Neolithic Transition – Fact or Fiction? Palaeoecological Evidence from the Balkans. *The Holocene* 4, 326-330.

3.3 The prehistoric peopling process in the Holocene landscape of the Grosseto area: How to manage uncertainty and the quest for ancient shorelines

Author

Giovanna Pizziolo

Dipartimento di Archeologia e Storia delle Arti, University of Siena, Siena, Italy
Contact: pizziolo@unisi.it

ABSTRACT

In this paper, we intend to discuss the evolution of the Holocene prehistoric landscape in the Grosseto area (Southern Tuscany, Italy) and the interaction with the pre- and proto-historic peopling process. The study area consists of an alluvial plain pertaining to the Ombrone and Bruna rivers, demarcated by the Hills of Castiglione della Pescaia and Grosseto towards the north and east and by the Uccellina Mountains towards the South. The area was characterised by a marine and lagoon environment until 2800 BP. Then a progressive transformation led to the formation of the Prile Lake, which afterwards gave place to salt marshes. The present alluvial plain seems to be the result of several reclamation activities, which occurred during the last four centuries.

Our ongoing research is focused on the reconstruction of the Holocene prehistoric landscape, highlighting what we consider to be those features of the past that are still observable in the present, and most likely responsible for the formation of the present-day landscape. From an archaeological point of view, the evidence found in the area consists mainly of prehistoric funerary remains found in caves in the hills which surround the alluvial plain. The lack of information relating to settlement and production activities introduces some uncertainty. Which features of the landscape still preserve evidence from the Neolithic to the Bronze Age? How can we make use of the biased archaeological dataset?

To answer to these questions we have taken into consideration the relevant impact of sea level changes on the landscape as well as the land use. To the same end, we have devoted particular attention to identifying and defining the micro-topography of the Holocene prehistoric landscape. This research has been developed using GIS and drawing on a multi-scale dataset which includes archaeological excavation and survey data, historical cartography and aerial photographs.

KEYWORDS

prehistoric landscape, prehistoric settlement strategies, GIS analysis, BIAS factors

INTRODUCTION

This paper represents the initial step of our ongoing research into the interactive processes between people and landscape in the Holocene landscape of the Grosseto Area (Southern Tuscany, Italy).

The study area is an alluvial coastal plain that has been characterised by substantial changes during the last 20,000 years. From a Landscape Archaeology perspective, the understanding of these dynamic conditions is fundamental to the investigation of prehistoric settlement strategies as part of the general Man-Environment relationship. Moreover, analysis of the landscape evolution can contribute essential clues to understanding the settlement strategies that occurred from the Neolithic to the Bronze Age (6th-2nd millennium BC). Along with a description of this case study, we will present the methodological approach and the research strategies that we have developed within the Grosseto context, striving to elaborate upon our management of the uncertainties which characterise this archaeological data.

We have chosen, from the various meanings of uncertainty, the one that refers to doubts arising due to incomplete information (Foody & Atkinson 2002) and we have mapped out our variables of interest (e.g. prehistoric evidence of the Grosseto Plain) assessing these uncertainties in our maps.

GEOGRAPHIC CONTEXT

The study area consists of the alluvial plain of the Ombrone and Bruna rivers, an area demarcated by the Hills of Castiglione della Pescaia and Grosseto towards the north and east, and by the Uccellina Mountains towards the south (fig. 1). Several studies have been conducted by geologists in the Grosseto Plain (Innocenti & Pranzini 1993; Stea & Tenerini 1996; Bellotti et al. 1999; Bellotti et al. 2001; Bellotti et al. 2004; Lambeck et al. 2004). They were mainly interested in understanding the general dynamics of sea level and coastal changes on a large scale. Detailed information on soil characteristics, land unit interpretation (Arnoldus-Huyzendveld 2007) and palaeonvironmental reconstruction (Biserni & van Geel 2005; Arnoldus-Huyzendveld 2007) is also available on a more convenient scale for a landscape archaeology approach, and has been widely employed in our research. To summarise the landscape changes occurring in the plain, we should indicate that since the Wurm III (40,000-20,000 BP), when sea level was about 120 metres below the present-day level, a large area extending beyond the actual coastline was covered by fluvial and aeolian deposits, divided into two valleys by the main rivers, the Ombrone and the Bruna. Evidence of these eroded deposits is still visible in the outcropping of low terraces along the borders and in the central part of the plain. Subsequent to this occurrence, i.e. during the Versilian transgression (17,000-6,000 BP), an irregular rising of the sea level caused dramatic changes in the landscape, and as a consequence, the valleys, previously eroded by the above-mentioned rivers, became inlets. Quoting Bellotti (2004, 85), around 10,000 BP 'the palaeogeographic setting of the area was characterised by a lagoon extending perpendicularly to the shoreline which was 2.5 km further inland than

Figure 1. Location of the study area (rectangle) and of the city of Grosseto (star).

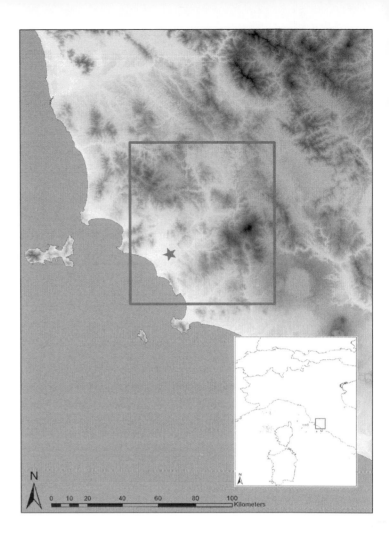

today. The Ombrone river flowed into the lagoon, building up gravelly–sandy fingerprint deltaic bodies (fig. 2).

It may be said that around 6,000 BP, when the sea level reached present-day heights, the role of the fluvial sediments was very determinant in shaping the landscape (fig. 3). The Ombrone River mouth prograded inside the lagoon, causing large accumulations, whilst the Bruna River, less rich in clastic sediment accumulations than the Ombrone, created a wider lagoon at the northern end of the Grosseto Plain. The Pleistocene deposits, accumulated in the central part of the plain, were still surrounded by water – as they had been since Etruscan times (8th-3rd century BC). It is after the 3rd-2nd century BC that the southern part of the plain was filled with Ombrone deposits, whilst the northern part was still a lagoon, which then converted to the lake *Lacus Prilius* (Curri 1978). Afterwards the coastal lake gave way to salt marshes, which underwent several reclamation efforts during the last four centuries for conversion into arable land. Medicean levees (17th century) and several channels regulated the flow of water into the plain, whilst the 'colmata' reclamation activities (18th-20th century) carried out sediment distribution in the area of the plain previously occupied by water.

When we take the above into consideration, it is clear that we are dealing with a context in which the incursion of environmental changes and human activity on Holocene pre- and protohistoric settings

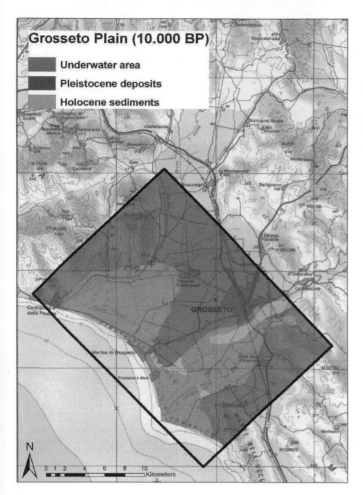

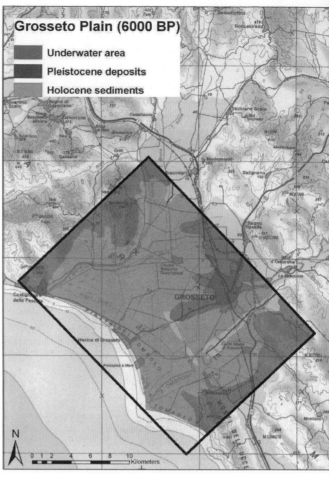

Figure 2. A GIS reconstruction of palaeogeographical settings of Grosseto Plain during 10,000 BP (after Bellotti 2004). *See also full colour section in this book*

Figure 3. A GIS reconstruction of palaeogeographical settings of Grosseto Plain during 6,000 BP (after Bellotti 2004). *See also full colour section in this book*

is quite significant. The present-day surface in the Grosseto Plain has little resemblance to the Holocene prehistoric surface. In fact, some areas of the plain were submerged at that time by the inlet basins and gradually transformed into lagoons. In addition, the prehistoric shorelines may have been disturbed by erosion and accumulation processes related to sea level changes and palaeo-hydrographical activity. Furthermore, reclamation activities and continued agricultural exploitation may also produce bias factors which affect the analysis of archaeological data obtained by field survey activities.

PREHISTORIC CONTEXT

Southern Tuscany, and in particular the Grosseto district, is an area that has been part of the diffusion process of Holocene prehistoric cultures in Central Italy. Since the Neolithic period, this area may have played the role of 'coastal bridge' between the Tuscan archipelago islands, which lay in front of the Grosseto Plain, and the interior zones of Central Italy composed of Monte Amiata, the Siena district and the

Appennine area. In fact, during the first part of the Neolithic, the raw materials and many kinds of artefacts arriving from the Mediterranean Sea (e.g. obsidian from Lipari Island) and by north-western regions (green stone and incised ware) have been found in the Archipelago and along the coastal zone from Pisa to Grosseto (Tozzi & Weiss 2000). As concerned the Neolithic cultures, the Grosseto district is included in the zone of Mediterranean Neolithisation of 'Cardiale-ware' from the Tyrrhenian area and incised pottery from the Northern Italy.

Prehistoric evidence in the Grosseto district is very scarce, and very few obsidian blades dating to the Early Neolithic (middle of 6th millennium BC) have been found in the inland site of Manciano. Little evidence related to the generic Neolithic period has been discovered in the interior part of the district (Monte Amiata and Massa Marittima) and along the coastal area at Monte Argentario, Castiglion della Pescaia, Follonica, and in the urban centre of Grosseto and Roselle (Fugazzola Delpino et al. 2004; Grifoni Cremonesi 1970; Mazzolai 1960). Cardial pottery in caves has been found at Grotta dello Scoglietto (Cavanna 2007) and at Grotta del Fontino where two burials (dated 6420 ±40 BP) are associated with incised pottery of the *facies* Sarteano-Sasso (Vigliardi 2002). Ongoing rescue archaeology research reveals an Ancient Neolithic site located on the watershed hill on the north-eastern side of the Grosseto district (preliminary data, unpublished). Also, the findings of Cardial Ware in the site of Pienza – Siena district – (Calvi Rezia 1972, 1973; Calvi Rezia et al. 2000, Calvi Rezia et al. 2007), in the interior part of Tuscany, testify to exchanges between inland and coastal areas. The connection between the Siena and Grosseto districts – the interior and the coastal zone – is very important. We can assume that the communication system could have been developed through the hydrographic network formed by the Ombrone and Orcia Rivers.

Figure 4. The Copper Age sites in the Grosseto District.

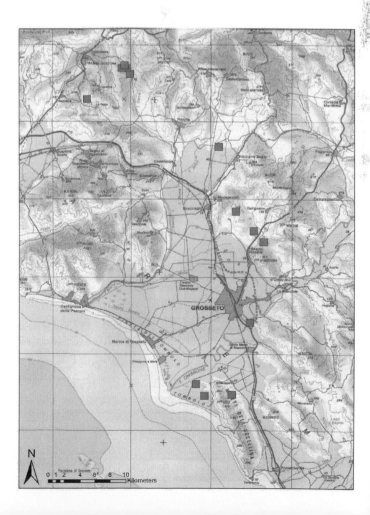

In order to better understand the peopling dynamics during the Neolithic period (Volante 2007) in the Grosseto Plain, it is necessary to improve and augment our information on prehistoric evidence, which has up till now been very scarce in our study area. One of the challenges of our research is to identify, in the Landscape of today, the shorelines and landing places that could have been exploited and crossed during the Neolithic period in the framework of maritime – inland communication. As concerns the Copper Age (3rd millennium BC), we can observe other types of problems caused by the lack of prehistoric information. In fact, settlement areas and dwelling activities have not been uncovered or recorded. Instead, funerary complexes, as collective burials, have been discovered. These have been mainly found inside natural caves located on the calcareous hills which surround the alluvial plain (fig. 4).

The improbable lack of settlement and production activities occurring in the Grosseto Plain gives rise to several questions and forces us to deal with uncertainty. Such partial results can also be explained by archaeological preservation problems and by a general trend of the 20th-century archaeological research to investigate mainly cave contexts. Taking into account these biases, we must also investigate which features of the landscape may preserve the pre-protohistoric evidence.

PREHISTORIC LANDSCAPE: PROBLEMS OF UNCERTAINTY

The absence of prehistoric data recorded in the plain is probably caused by several factors. Some are surely related to the history of the researches, often performed in an unsystematic way and focusing exclusively on the caves. But the more effective ones are the dynamic changes of the geographical settings over the last 10,000 years ago.

Because of these factors, it is crucial to follow a landscape archaeology approach and try to reconstruct the prehistoric settings of the Landscape. From our perspective, it is very important to start with those specific features of the past, still observable in the present, that might have characterised the prehistoric scenarios during the Holocene Period. The palaeogeographic changes indicate that some part of the Pleistocene terraces were not underwater during the Holocene Period and are still visible today. The evidence of these deposits can be found in the outcroppings of low terraces along the borders and in the central part of the Grosseto Plain. The shift in sea level during the Neolithic and Copper Ages played a fundamental role in shaping the shorelines which have been delineated along the Pleistocene terraces. Their study has become one of the focuses of our research. To investigate these settings, we have constructed a GIS framework which includes a range of information on geological maps, topographical maps, historical cartography and colour orthophotos (2004).

Geological maps spanning a scale from 1:100,000 (Servizio Geologico d'Italia) to 1:10,000 (Carta Geologica Regionale – Regione Toscana) have been input into the system in order to acquire data on Pleistocene terraces and Holocene features which, together with recent alluvial deposits, characterise the plain area.

A detailed Digital Elevation Model (DEM) has been elaborated from the topographical maps available at a scale of 1:10,000 (Carta Tecnica Regionale – Regione Toscana) with the purpose of highlighting the surface and the edges related to the Pleistocene terraces. In order to best detect these edges, we have built up an 'historical' DEM, selecting only the contours and high points that are free from modern disturbances, i.e. that are not related to modern activities such as drainage channels, elevated infrastructures or channel banks, earthworks etc.

Nevertheless, we have to consider that within the last four centuries the area has been involved in several reclamation processes which, using artificial flooding, deposited a large quantity of sediment in the plain. The results of these activities may have hidden the edges of the terraces. Consequently, it is very important to map the areas that have undergone the reclamation process in order to better understand the sequence of alluvial sedimentation. However, it is worth remembering that our geological maps have not been very accurate in providing detailed information on the temporal sequence of the Holocene deposits. From this perspective the mapping of the reclamation initiatives has become an important goal. Several historical documents and maps have been used to define the reclamation areas (De Silva & Pizziolo 2004; De Silva 2006, 2007) in particular, the analysis of the 19th-century cadastral maps. An interesting comparison is then possible when we overlay this information on the artificial alluvial deposits over the colour orthophotographs. This comparison may detect crop-mark or damp-mark anomalies which confirm the presence of different types of soils. In the GIS environment we can perform spatial analyses between these various data within a multisource approach. In this composite framework, we try to focus on features of the landscape that may still retain the pre-protohistoric evidence, and to determine why some information has been invisible in the plain area. Our interpretation of the Prehistoric Landscape suggests that Pleistocene deposits, free from reclamation activities and with a very low slope gradient (according to our 'historical' DEM), can be defined as areas which could be 'walkable' surfaces during the Neolithic and Copper Age period. Moreover, these areas should not have been affected by massive depositional actions after the Holocene Prehistory. In other words, if we select areas that satisfy all our three criteria based on geological, historical and morphological variables we can identify portions of landscape that have a high archaeological potential, i.e. areas with a strong possibility for preserving prehistoric evidence or a 'high potential for prehistoric preservation'. The selection of these variables has been performed through GIS tools, and the overlay of these areas highlights the portion of landscape where we can further focus our attention.

PREHISTORIC EVIDENCE: PROBLEMS OF UNCERTAINTY

We have already stressed that the prehistoric evidence in the Grosseto Plain has scarcely been assessed and very little data has been collected during previous research. However, the recent field activities carried out in a rescue archaeology framework, by the Department of Archaeology and History of Art of the University of Siena, suggested new interpretations of the peopling process of the area. According to the requests of the City Council, the archaeological surveys were carried out in the outskirts of Grosseto, in areas selected by the urban planning authorities and not according to our research priorities. Despite this procedure, the field survey, begun in October 2009, confirmed the presence of prehistoric people in the plain and, regardless of the sporadic nature of the archaeological evidence, it offered an encouraging perspective.

In the first stage, we tested the reliability of data collected during the survey. We analysed the lithic artefacts, examining their characteristics, such as the presence of gloss and the state of the surface preservation of each item. These types of analyses helped us to assess that the majority of artefacts were found in a primary deposition. We also dedicated particular attention to the identification of raw materials that may substantiate connections with the archipelago (e.g. obsidian) or with the inland area.

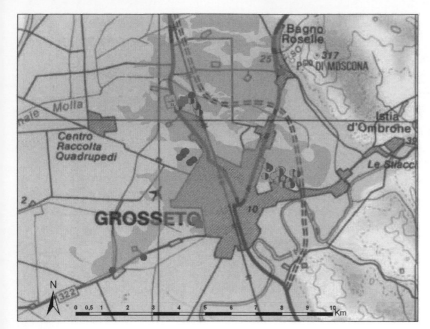

Figure 5. Distribution of off-site evidence in the Grosseto area. *See also full colour section in this book*

In general terms, we can confirm the presence of prehistoric evidence in the Grosseto Plain. However, the typology and quantity of these artefacts still force us to deal with uncertainty. Indeed, the analysis of the findings indicates that often we are not dealing with diagnostic artefacts which could help us to figure out a well defined settlement pattern structured into functional areas or defined in clear chronological phases. In other words, this prehistoric evidence is not related to large sites but probably is the result of off-site activities.

The typological analysis allows us to subdivide these sites into simple chronological classes organised as follows: Holocene (often represented by Neolithic/Copper Age evidence), Holocene/Pleistocene (when we can identify both periods in the lithic assemblage), Pleistocene (only a few cases relate to Middle Palaeolithic), and generic ones (when no significant chronological attribution can be made). In order to exploit all the information gathered, we produced thematic maps (fig. 5), which show the reliability of data and their chronological assessment.

DATA INTEGRATION: AN ANSWER TO UNCERTAINTY

The observation of uncertain archaeological data in a GIS environment helps us to change our perspective. Off-site evidence, within acceptable parameters of accuracy and according to a landscape archaeology approach, may provide very valuable information (e.g. Bintliff et al. 2000) when trying to understand the use of territory. In our case study, the data collected during the field survey offers us the possibility of verifying whether or not our interpretation of palaeo-geographical settings is feasible (fig. 6). All prehistoric evidence has been discovered in areas with 'high potential for prehistoric preservation' (see section 4). Thus we shift to a more accurate scale to explore in detail the prehistoric settings, and specifically, landscape morphology. Today the field surface seems clearly uniform and very flat due to the modern use of powerful mechanical ploughing tools, so that ancient bumps and shallow areas have been levelled out. Thus in order to create the morphological context for these archaeological finds we need to reconstruct the terrain surface by referring to previous settings.

Precious information has been obtained from topographical maps published during the 1930s at a scale of 1:10,000, to ascertain the reclamation activities of the Grosseto plain (fig. 7). With the input of these historical maps into the GIS, we acquired contour lines and high points from which we were able

Figure 6. A 3D view of the Pleistocene terrace (grey line) overlayed to a DEM and to off-site Neolithic evidence in the Central part of Grosseto Plain (grey blocks). The image shows the edge location of the off-site in respect of the micro-morphologies and of the Pleistocene deposit.

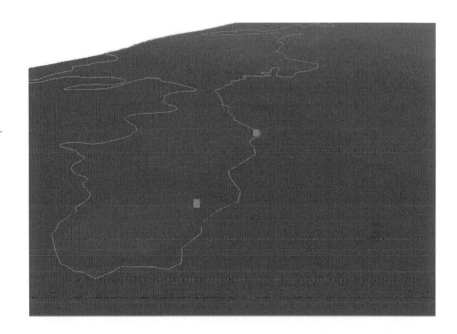

Figure 7. An example of data integration: off-site distribution and Pleistocene deposits overlaid on the historical topographical maps (1930) which show a variety of morphologies. *See also full colour section in this book*

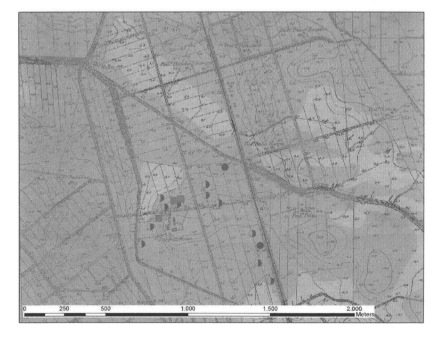

Legend
▨ Pleist. terraces

**Prehistoric
Topographical Unit**
● Holocene
◑ Pleist/Holocene
● Pleistocene
▣ Generic

Figure 8. An example of data integration: off-site distribution and Pleistocene deposits overlaid on the historical aerial photograph. The contour lines derived by the historical topographical maps (1930) allow us to appreciate the little hillock and to imagine the inlet when the sea level was 6 metres above the present-day level. We can also appreciate damp-marks and soil marks. *See also full colour section in this book*

to visualise the previous morphological settings. These appear to be much more complex compared with the present-day morphology. Actually, according to the 1930s maps, all the prehistoric finds related to Holocene periods are located on micro reliefs forming small/low hills (fig. 8). Moreover, it is worth high-lighting that the sea level during the Neolithic period was supposedly 6 metres above the present-day level. In this respect, these little hillocks also show noteworthy inlets fronting onto the inferred Neolithic shorelines.

This morphological interpretation matches up with the results of the analysis of historical aerial photographs (1943, 1954) which show distinct damp-mark or soil-mark anomalies which we can assign to wetland or dryland features. Evidentially, in this case, the combination of historical sources provided excellent information, essential to the understanding of palaeo-geographical settings. Moreover the combination of these interpretations with prehistoric evidence provided new clues for our study.

CONCLUSION

This research, still in its initial stages, shows that, through the integration of data derived from field activities, archaeological analysis and GIS elaboration, it is possible to produce and sustain an effective cycle of understanding. This process implies a continuous shift in the data input from general to local scale and moreover a shift in data interpretation from local to general scale and *vice versa*. In our case study, for example, the off-site finds helped in the assessment of prehistoric occupation of the area, while the morphological analysis provides a possible insight into settlement strategies and location choices.

In conclusion, even though we are dealing with uncertainty, the landscape archaeology approach has ultimately eliminated many implausible interpretive strategies and helped define Neolithic shorelines while creating a context for off-site evidence. The correlation between the data furnished by historical sources (namely 1930s maps and historical aerial photographs) and the data furnished by archaeological means has been crucial to the research. In our endeavour to define the nature and shape of the ancient shorelines, the elaboration of these data allow us to imagine, despite the present-day flatness of the Grosseto Plain, a complex and indented coastline, which could have been an attractive feature during the peopling process of the area.

ACKNOWLEDGEMENTS

The investigation on prehistoric contexts of Grosseto area is developed by the Prehistoric Section of the Department of Archeologia e Storia delle Arti, University of Siena and directed by Dr Nicoletta Volante. I am very grateful to Dr Volante for the analysis of lithic artefacts collected during the surveys and for all the interesting discussions on the prehistoric peopling process in the Grosseto area, especially the ones during our field activities in Maremma. Her comments have been a precious help in the formulation of this work. I also give my heartfelt thanks to Dr Sabina Viti for reviewing this text, providing excellent suggestions.

REFERENCES

Arnoldus-Huyzendveld, A. 2005. Il paleoambiente storico di Grosseto. In Citter, C. (ed.) *Lo scavo della Chiesa di S. Pietro a Grosseto, Nuovi dati sull'origine e lo sviluppo di una città medievale*. All'Insegna del Giglio, Firenze, 59-68.

Arnoldus-Huyzendveld, A. 2007. Le trasformazioni dell'ambiente naturale della pianura grossetana. In Citter, C. & Arnoldus-Huyzendveld. A. (eds.), *Archeologia urbana a Grosseto. Origine e sviluppo di una città medievale nella 'Toscana delle città deboli'. Le ricerche 1997-2005. Vol. I: la città di Grosseto nel contesto geografico della bassa valle dell'Ombrone*. All'Insegna del Giglio, Firenze, 41-62.

Bellotti, P., C. Caputo, L. Davoli, S. Evangelista & P. Valeri. 1999. Lineamenti morfologici e sedimentologici della piana deltizia del Fiume Ombrone (Toscana Meridionale). *Boll. Soc. Geol. It.* 118, 141-148.

Bellotti, P., C. Caputo, L. Davoli, S. Evangelista, E. Garzanti, F. Pugliese & P. Valeri. 2004. Morpho-sedimentary characteristics and Holocene evolution of the emergent part of the Ombrone River Delta (southern Tuscany). *Geomorphology* 61, 71-90.

Bellotti, P., G. Belluomini, L. Bergamin, M.G. Carboni, L. Di Bella, S. Improta, P.P. Letunova, L. Manfra, T.G. Potemkina, P. Valeri & P. Vesica. 2001. Nuovi dati cronostratigrafici sul sottosuolo della piana deltizia del Fiume Ombrone (Toscana meridionale). *Studi Costieri* 4, 31-40.

Bintliff, J., M. Kuna & N. Venclovà. 2000. *The Future of Surface Artefact Survey in Europe*. Sheffield Academic Press, Sheffield.

Biserni, G. & B. van Geel. 2005. Reconstruction of Holocene palaeoenvironment and sedimentation history of the Ombrone alluvial plain (South Tuscany, Italy). *Review of Palaeobotany and Palynology* 136, 16-28.

Bravetti, L. & G. Pranzini. 1987. L'evoluzione quaternaria della pianura di Grosseto (Toscana): prima interpretazione dei dati del sottosuolo. *Geografia Fisica e Dinamica Quaternaria* 10, 85-92.

Calvi Rezia, G. 1972. *I resti dell'insediamento neolitico di Pienza*, Atti XIV Riun. Sc. I.I.P.P., Puglia 13-16 October 1970, 285-300.

Cavanna, C. (ed.) 2007. La preistoria nelle grotte del Parco Naturale della Maremma. *Supplemento agli Atti del Museo di Storia Naturale della Maremma* 22, Grosseto.

Calvi Rezia, G. 1973. *I resti dell'insediamento neolitico di Pienza*, Atti XV Riun. Sc. I.I.P.P., Verona-Trento 27-29 October 1972, 169-180.

Calvi Rezia G., G. Grandinetti, S. Vilucchi. 2000. L'insediamento preistorico della Cava Barbieri di Pienza (Siena): il sito e le ricerche. In Tozzi, C., R. Grifoni, F. Fedeli (eds.), *I rapporti tra l'Italia centrale tirrenica e la Corsica in età antica: il neolitico a ceramica impressa cardiale*. Piombino, 26-27.

Calvi Rezia, G., L. Sarti, M. Rosini, L. Agostini. 2007. *Nouvelles données sur Pienza dans le cadre du Néolithique de l'aire haute tyrrenienne*. Actes 128°Congrès National des Sociétés Historiques et Scientifiques, Bastia 14-21 April 2003.

Curri, G.B. 1978. F.I.: Regio VII, volumen IV: Vetulonia, Olschki, Firenze.

De Silva, M. & G. Pizziolo. 2004. *GIS Analysis of Historical Cadastral Maps as a Contribution in Landscape Archaeology*, in Magistrat der Stadt Wien – Referat Kulturrelles Erbe – Stadtarchäologie Wien (eds.), *[Enter the Past] – The E-way into the Four Dimensions of Cultural Heritage CAA 2003 – Computer Applications and Quantitative Methods in Archaeology – Proceedings of the 31st Conference, Vienna, Austria, April 2003*, BAR International Series 1227, Archaeopress, Oxford, 294-298 and CD-ROM.

De Silva, M. 2006. The Fourth Dimension of Places: Landscape as an Environmental and Cultural Dynamic Process in the Maremma Regional Park. In S. Campana & M. Forte (eds.), *From Space to Place. 2nd International Conference on Remote Sensing in Archaeology, CNR*. Rome, Italy, 4-7 December 2006. BAR- International Series 1568, Archaeopress, Oxford, 285-290.

De Silva, M. 2007. *La cartografia storica per l'archeologia del paesaggio in ambiente GIS. Il caso dell'area grossetana*. Phd Thesis, Dottorato di Ricerca in Archeologia Medievale - XVII ciclo, Università degli Studi di Siena, unpublished.

Foody, G.M., P.M. Atkinson (eds.). 2002. *Uncertainty in Remote Sensing and GIS*. Wiley, Chichester.

Fugazzola Delpino, M.A., V. Lattanti, A. Pessina, V. Tinè. 2004. *Il Neolitico in Italia – Ricognizione, catalogazione e pubblicazione dei dati bibliografici, archivistici, materiali e monumentali*. Origines, Roma.

Grifoni Cremonesi, R. 1970. *I materiali preistorici della Toscana esistenti al Museo Civico di Grosseto*. Atti della Società Toscana di Scienze Naturali, Memorie, Serie A LXXVII: 78-91.

Innocenti, L. & Pranzini, E. 1993. Geomorphological Evolution and Sedimentology of the Ombrone River Delta, Italy. *Journal of Coastal Research* 9 (2), 481-493.

Lambeck, K., F. Antonioli, A. Purcell & S. Silenzi. 2004. Sea-level change along the Italian coast for the past 10,000 yr. *Quaternary Science Reviews* 23, 1567-1598.

Mazzolai, A. 1960. *Roselle e il suo territorio*, Grosseto, [s.n.]

Stea, B. & I. Tenerini. 1996. L'ambiente naturale della pianura grossetana e la sua evoluzione dalla preistoria alla cartografia rinascimentale. In C. Citter (ed.), *Grosseto, Roselle e il Prile. Note per la storia di una città e del territorio circostante*, Documenti di Archeologia, 8, Mantova: Società Archeologica Padana, 13-24.

Tozzi, C. & M.C. Weiss (eds.) 2000. *Il primo popolamento olocenico dell'area corso-toscana*. Edizioni ETS, Pisa.

Vigliardi, A. 2002. La grotta del Fontino.Un cavità funeraria eneolitica del grossetano. *Millenni, Studi di Archeologia Preistorica* 4, Firenze.

Volante, N. 2007. Neolitico ed Età del Rame. In Citter, C. & Arnoldus-Huyzendveld, A. (eds.) *Archeologia urbana a Grosseto. Origine e sviluppo di una città medievale nella 'Toscana delle città deboli'. Le ricerche 1997-2005. Vol. I: la città di Grosseto nel contesto geografico della bassa valle dell'Ombrone*. All'Insegna del Giglio, Firenze: 122-127.

Applying concepts of scales

4.1 Landscape scale and human mobility: Geoarchaeological evidence from Rutherfords Creek, New South Wales, Australia

Authors
Simon Holdaway[1], Matthew Douglass[2] and Patricia Fanning[3]

1. Anthropology Department, The University of Auckland, Auckland, New Zealand
2. Anthropology Department, University of Nebraska-Lincoln, Lincoln, NE, United States of America
3. Graduate School of the Environment, Macquarie University, Sydney, Australia
Contact: sj.holdaway@auckland.ac.nz

ABSTRACT

The surface archaeological record is abundant in some parts of arid Australia and, if analysed with attention to the history of deposition, it provides an accessible resource with which to assess past landscape use. Here, we report results of studies of the mid-late Holocene Aboriginal occupants of one part of the Australian arid zone, based on analyses of the archaeological record in the 62 km² catchment of Rutherfords Creek in western New South Wales (NSW). We consider the types of behavioural information that can be derived from this record and how interpretation varies when considering different spatial and temporal scales. Those hunter-gatherers, who were highly mobile, traversed areas that were orders of magnitude larger than the areas that can be studied in detail by archaeologists. This requires the development of techniques for inferring the extent of landscape use from isolated spatial and temporal samples. We describe some of these techniques and consider the implications of the results obtained from their application.

INTRODUCTION

The arid regions of western NSW Australia have an abundant archaeological record that is highly visible due to the significant loss of topsoil caused by overgrazing since the late 19th century. Erosion has exposed stone artefacts and the remains of heat retainer hearths that are the legacy of intermittent occupation by Aboriginal people stretching back into the Late Pleistocene. An extensive, visible archaeological record, where stone artefacts and hearths reflect people's presence in the landscape, offers the opportunity for understanding mobility within a settlement system. However, the temptation to study mobility via a 'dots on maps' approach, i.e. where a map of recorded archaeological remains is interpreted directly as the outcome of the spatial scale of past human behaviour, runs into obvious logistic problems. Australia, for example, has almost the same landmass as the whole of Europe, meaning that geographic regions span areas that incorporate several of the smaller European nation states. It is not feasible to study the archaeological record left by people who, through time, might have traversed thousands of square kilometres. The first aim of this study is, therefore, to consider archaeological approaches that do not require the a priori definition of the geographic boundaries of the region used by hunter-gatherers when considering settlement systems. We illustrate that a great deal can be learnt about how people used space by studying archaeological remains left at individual locations. In so doing, we describe a landscape archaeology that involves making inferences about what went on at a number of locations from what is found at one location.

People undertook different activities in different parts of the landscape at different times and for a variety of reasons. Variability in the archaeological record therefore reflects the variety of these activities. This variability is time-dependent, in the sense that it will tend to increase the longer the period of time considered. It is the outcome of different activities that leads to pattern in the archaeological record, therefore time has to pass for archaeologists to have something to interpret (Holdaway & Wandsnider 2006, 2008).

Previous approaches to landscape definition sometimes sought to identify functional settlement pattern models typified in hunter-gatherer archaeology by reconstructions of a 'seasonal' round of activities and the identification of functional site types (e.g. Thomas 1973). These reconstructions were one-dimensional in the sense that they were a-temporal, with spatially defined sets of localities thought to be behaviourally linked in a single system. People were thought to undertake the same activities repeatedly in the same locations each season. In many environments, however, including arid Australia, shifts in the environment do not conform to annual seasonal changes and are better characterised as temporally non-linear (Stafford-Smith & Morton 1990). The second aim of this paper is, therefore, to illustrate landscape approaches that are multidimensional and, at the same time, do not rely on patterns generated by regularly recurring seasonal environmental shifts. Human-environment interaction is reciprocal such that humans both affected, and were affected by, the environment, meaning that through time the nature of the interaction may change. These interactions with the environment are in turn closely tied to the formation of the archaeological record. The analysis of regions by studying single locations provides the means to understand a range of human-environment interactions without proscribing fixed boundaries within which these interactions took place, or requiring that the only mechanism for change is the shift from one stable system to another.

In the research reported here, the understanding of human-environment interactions are geomorphologically and chronologically based since it is from the sediments and other datable materials that we have the best evidence for how humans interacted with the landscape. Processes of erosion and deposition help explain why the abundant artefact record is exposed in the way that it is in Australia, and provide the basis for deriving inferences about how this abundance is to be interpreted. Time and space are closely connected to inferences about mobility, the investigation of which helps to explain how and when people utilised places within the landscape.

THE STUDY AREA: WESTERN NSW, AUSTRALIA

Deposits of stone artefacts are currently exposed on eroded surfaces along the edge of watercourses in the western NSW rangelands of Australia (fig. 1), with individual exposures varying in extent from a few hundred to more than 10,000 square metres (Holdaway & Fanning 2008). Erosion is largely attributable to overgrazing by introduced domesticated animals (mostly sheep) beginning in the late 19th century (Fanning 1999). Some of the stone artefact exposures are associated with the eroded remains of heat retainer hearths that provide charcoal for radiocarbon age determinations (fig. 2). When combined with sediment and hearthstone ages obtained using optically stimulated luminescence (OSL), these age determinations permit a chronology of occupation to be established (Holdaway et al. 2005, 2008; Fanning et al. 2008; Rhodes et al. 2009, 2010). OSL age estimates and sediment analyses indicate that watercourses were periodically eroded removing artefacts into sediment deposits in lakes downstream. OSL age estimates of the surfaces upon which the artefacts are now resting provide an indication of the maximum time period over which these artefacts accumulated (Fanning et al. 2007, 2009a) (fig. 3). Radiocarbon age esti-

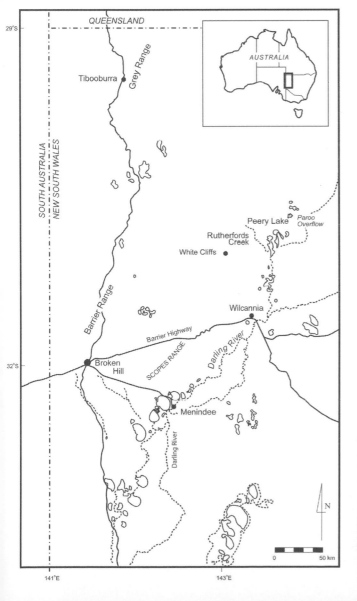

Figure 1. Western NSW, Australia showing places mentioned in the text.

Figure 2. Eroded remains of heat retainer hearths. Hearths are indicated by flags.

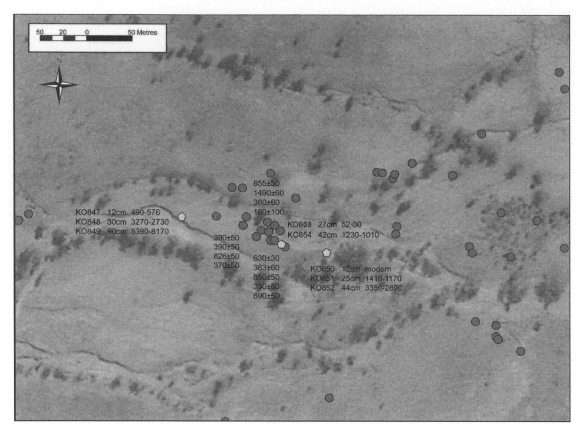

Figure 3. OSL and hearth radiocarbon determinations, respectively indicated by light grey polygons and dark grey circles, from one section of Rutherfords Creek.

mates from the hearths post-date the OSL ages obtained from sediments that lie immediately beneath the hearths. Hearth age estimates tend to cluster temporally as well as spatially, suggesting discrete episodes of repeated hearth construction at the same location, with those episodes separated by periods measured in decades to centuries (Holdaway et al. 2002, 2005).

Rutherfords Creek is an ephemeral stream network draining a catchment area of 62.5 km² in the semi-arid north-west of the state of New South Wales (fig. 1). Over the four year period from 2005 to 2008, we conducted archaeological and geomorphological surveys along a 15 km length of the main channel, from near the headwaters to the mouth at its junction with Peery Lake, an ephemeral lake basin (fig. 4). The eroding valley floor margin, with an area of 37.8 km², is comprised of multiple overlapping units of fluvial sediments deposited in episodes spanning the Holocene (Rhodes et al. 2010). Artefact deposits are visible on eroded patches (locally referred to as 'scalds' – fig. 5) comprising approximately 2 km² of the valley floor. The archaeological surveys were confined to a randomly selected sample of approximately 2,364 mapped scalds, amounting to approximately 4.5% of the eroded valley floor by area (fig. 4). A total of 27,108 artefacts were analysed, with an overall density of approximately 0.3 artefacts per square metre. Figure 4 also shows the distribution of the 1,054 hearths that were recorded during the archaeological surveys. Each hearth was initially identified on morphological grounds and confirmed by comparison

Figure 4. Rutherfords Creek, western NSW, showing the location of scalds (i), hearths (ii) and analysed stone artefact assemblages from randomly selected scalds (iii). *See also full colour section in this book*

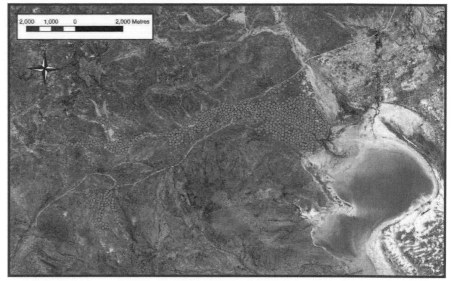

(i) Scalds

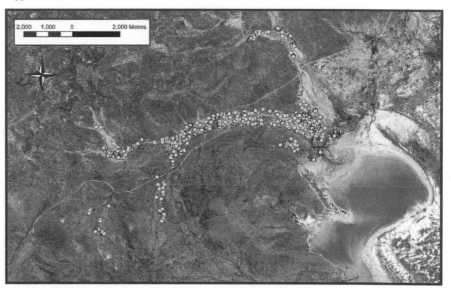

(ii) Hearths

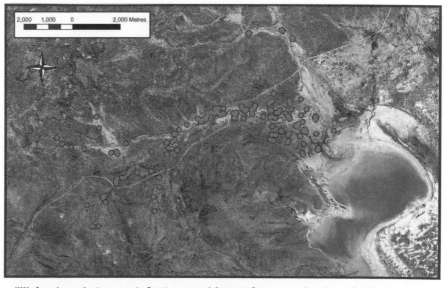

(iii) Analysed stone artefact assemblages from randomly selected scalds

Figure 5. Eroded sediment patch locally described as a 'scald'.

of the magnetic field over the hearth, obtained using a fluxgate gradiometer (Fanning et al. 2009b), with that of the surrounding land surface. A sub-sample of 256 hearths was excavated to obtain charcoal samples, with about one third of these yielding sufficient charcoal for radiocarbon age determinations.

Results indicate that hearths date to the last two and a half thousand years, a pattern that reflects the age of the sediments into which the hearths were originally dug (Holdaway et al. 2008a; Rhodes et al. 2009, 2010) (fig. 3). Unconformities in OSL ages from sediments beneath the surface suggest that erosion and deposition of sediments has probably resulted from periodic floods in the past (Fanning et al. 2008). This also explains the lack of deeply stratified archaeological deposits in the valley: artefacts were periodically removed by erosion and transported into sediment-choked stream channels, ultimately becoming buried in the sediment deposits in Peery Lake (Fanning et al. 2009a), rather than in stream terraces that are common, for example, in parts of Europe. The end result of similar processes occurring over large parts of the interior of Australia is an archaeological record where preservation of ancient deposits is relatively rare. As a consequence, datable archaeological features appear to show a dramatic increase in abundance during the last 1000-2000 years, thus implying a late Holocene increase in population, whereas in fact this may be a consequence of erosion (Holdaway et al. 2008; Surovell et al. 2009 report a similar pattern in radiocarbon age determination accumulation for North America).

HUMAN RESPONSES TO ENVIRONMENTAL CHANGE

The environment of the Australian arid zone is marginal for human subsistence, with a distribution of resources that is patchy, of low density, and subject to large fluctuations in abundance (Gould 1991). This partly reflects the episodic nature of rainfall, but also low fertility derived from poor soil nutrients. Places where water and fertile soils co-occur are rare, leading to a patchy distribution of resources. As well, much of the available carbohydrate occurs as wood making it unavailable to mammals (Stafford- Smith & Morton 1990). Western NSW, for example, has an average annual rainfall of less than 250mm and pan evaporation commonly exceeding 2,000mm, meaning that there is a significant net loss of moisture. These averages are further subject to major fluctuations largely connected to El Niño-Southern Oscillation (ENSO) episodes which bring alternating periods of severe drought and higher than average rainfall to large parts of eastern Australia. Most palaeoenvironmental evidence suggests that, during the mid-to-late Holocene, a climatic optimum occurred around 4000 BP, followed by desiccation and aridity from about 3000 BP to about 1500 BP and amelioration up to the present (Holdaway et al. 2002, 2010a). However, the region has always been subjected to marked local climatic variability.

The majority of available water is found in ephemeral creeks, rock holes, and other features that hold water for varying lengths of time following rain. Drought conditions prevail, but are broken by the punctuated and highly localised rainfall events discussed above (e.g. Dunkerley 1999). Areas receiving rain consequently experience brief increases in water availability and relative, though limited, increases in land productivity, while adjacent areas remain dry (Fanning et al. 2007). Aridity placed great limitations on the use of large portions of the landscape. In the Western Desert of Western Australia, with the extended absence of rain, residential groups contracted to more reliably watered portions of the landscape (Gould 1991). If severe drought conditions persisted, populations employed a strategy of territorial abandonment.

A similar pattern is replicated in the NSW archaeological record, albeit at a different temporal scale to that discussed by Gould (1991). Radiocarbon age determinations, obtained from the previously described hearths in Rutherfords Creek, show statistically significant correlations with peaks in the record of Sea Surface Temperatures (SSTs) derived from deep sea cores that provide an indication of Eastern Asian Monsoon activity as well as shifts in the position of the Inter-Tropical Convergence Zone (ITCZ) (Holdaway et al. 2010a). Changes in these systems have a direct impact on Australian continental rainfall, where the southward positioning of the ITCZ and periods of higher SSTs across northern Australia is associated with higher rainfall in northern and central Australia (Sturman & Tapper 2006). In contrast, the same hearth radiocarbon ages show negative correlations with the record of dust movement from the Australian continent retrieved from an ombrotrophic peat bog in the Old Man Range, Central Otago, New Zealand (Marx et al. 2009). The greatest flux in dust deposition occurs during dry periods that are punctuated by flood events, when the sediment deposits that produce the dust are replenished. A high dust flux therefore occurs when La Niña (high rainfall) periods, resulting in fluvial sediment transport to the Australian continental interior, precede El Niño (low rainfall) periods when dry conditions result in aeolian sediment transport offshore. We interpret the results from the SST and dust correlations to mean that hearths were constructed more frequently during periods with higher moisture levels but were constructed less frequently when dust transport was higher and conditions in the interior were likely to have been more arid. A marked variation in occupation intensity is suggested, from little or no occupation during dry periods to more frequent activity during more moist times (Holdaway et al. 2010a). However, the timing and degree of these environmental changes would have made the prediction of resource abundance particularly difficult. This must be kept in mind when interpreting the pattern presented by the stone artefact assemblages.

MOBILITY

From the Rutherfords Creek research described above, it appears that short-term increases in the availability of water and food resources within the valley systems of western NSW, triggered by unpredictable and localised rainfall, produced, over the long term, a patchy archaeological record. In the desert regions of interior Australia, dense artefact concentrations near water courses are separated by diffuse artefact scatters and isolated occurrences. The sparse and unpredictable character of resource distributions in the short-term meant that it was neither easy to spatially target foraging expeditions, nor possible to know precisely where stone suitable for the production of artefacts might be found. While raw material is locally abundant (cobbles of silcrete and quartz are readily available in dry creek channels, and as extensive stone pavements on adjacent slopes, and silcrete cobbles also occur in the form of boulder mantled outcroppings (Douglass & Holdaway 2010), stopping to find and produce tools as each foraging opportunity was encountered had a time and energy cost. The response was a technological strategy based on a generalised tool-kit (Kuhn 1992), or personal gear (Binford 1977, 1979), carried at all time as a 'hedge' against unforeseen needs, but one that did not involve large quantities of retouched tools.

Retouch is rare on stone artefacts in NSW. Most of the retouched tools in the NSW assemblages are categorised as lightly retouched pieces, notches and denticulates, with little indication of substantial resharpening. Even the retouch on typologically defined scrapers is not very invasive. The exceptions are tula adze slugs (bits for flake adzes – Holdaway & Stern 2004), but these account for only a small proportion of the total number of tools. However, while retouched tools are rare, flakes are abundant. Flakes were struck from cores, but there is only limited evidence for substantial predetermination of form through the shaping of core surfaces. Most flakes were struck in a single direction, although the dorsal scars on flakes indicate some flaking from opposed platforms and some indication of core rotation (Holdaway et al. 2004). The exceptions are flakes struck from flake cores (tranchet cores) where core form acted to limit flake morphology. But these forms are rare. Despite the lack of complexity in technology, assemblages show evidence for the selective removal of larger flake blanks and the transportation of these flakes over a variety of distances.

Movement of flakes in the Rutherfords Creek study area was assessed using a method for determining mobility based on the proportion of cortex present on artefacts, the details of which have been widely published and discussed (Dibble et al. 2005; Douglass et al. 2008; Douglass & Holdaway 2010; Holdaway et al. 2010b; Lin et al. 2010). Three steps are involved:

1. The total mass (or weight) of an assemblage is divided by an estimate of the total mass of the individual nodules from which cores were flaked, thus giving an estimate of the number of nodules reduced.

2. Estimated nodule frequency (or number) is multiplied by the average surface area of the nodules. This provides an estimate of the expected cortical surface area that should be present in an assemblage.
3. This value is compared to the actual quantity of cortex observed in the assemblage and is expressed as a ratio.

A variety of test studies have been conducted to demonstrate the accuracy of the Cortex Ratio method for determining cortex proportions in assemblages. Experimental testing by Dibble et al. (2005) and Douglass et al. (2008) provided an initial verification that the method could accurately gauge archaeological cortex proportions. Further testing by Lin et al. (2010) demonstrated the suitability of the measurement of flake surface area and volume used in the calculation of the Cortex Ratio through comparison with results obtained with laser scanning. Douglass & Holdaway (2010) further investigated the method by considering how variation in raw material size estimates might affect Cortex Ratio calculations. The results of these studies have demonstrated the overall robustness of the method and its suitability for investigating archaeological cortex proportions

Application of the methodology to assemblages from western NSW returns values of the cortex ratio consistently below one, indicating that cortex is underrepresented (Douglass et al. 2008, their table 2). The addition of many non-cortical flakes and cores to assemblages would explain the observed values of the cortex ratio. However, the local abundance of raw materials, as well as surveys of the size of stone cobbles, makes the scenario of extensive raw material importation seem unlikely (Douglass & Holdaway 2010). A more probable explanation for the disparity in cortex proportions is the removal of material from assemblages for use elsewhere. Cortex Ratios less than one reflect a tendency towards the removal of artefacts with a greater cortical surface area to volume ratio than the nodules from which they were produced. This would result from the selective removal of large blanks, blanks that would tend to have cortex on their dorsal surface as a consequence of their size (Douglass et al. 2008). Artefacts with a high proportion of cortex and large surface area, but which are also thin and therefore have a low volume, most affect the Cortex Ratio. The ratio therefore also informs on the shape of the flakes that were removed, i.e. those having a high edge to mass ratio.

The results of the Cortex Ratio calculated for the Rutherfords Creek assemblages indicate some variation in the degree to which flakes were removed (fig. 6). For some scalds, substantial artefact removal is indicated, while for others many fewer flakes were taken away. A few scalds with low artefact densities have Cortex Ratios that show an over-abundance of cortex. Interpreting the spatial pattern indicated by the different Cortex Ratio values requires that the nature of movement be considered, something that is made easier if movement is considered in a number of different ways. Movement frequency (both on an individual and group level) affects the potential for artefacts to be moved in that less movement will result in more artefacts discarded where they were manufactured and greater movement will increase the probability that any one artefact is moved away from its manufacturing location. Therefore, all things being equal, fewer moves will tend to produce increased Cortex Ratio values and more moves the opposite. Movement linearity will affect the linear distance that an artefact will be moved. Movements that are not linear may cover a great total distance but not move the artefact far from the place where it was produced. Linearity can be expressed using the concept of tortuosity. The degree of movement tortuosity ranges from a non-tortuous straight line movement between points to a movement path so tortuous as to

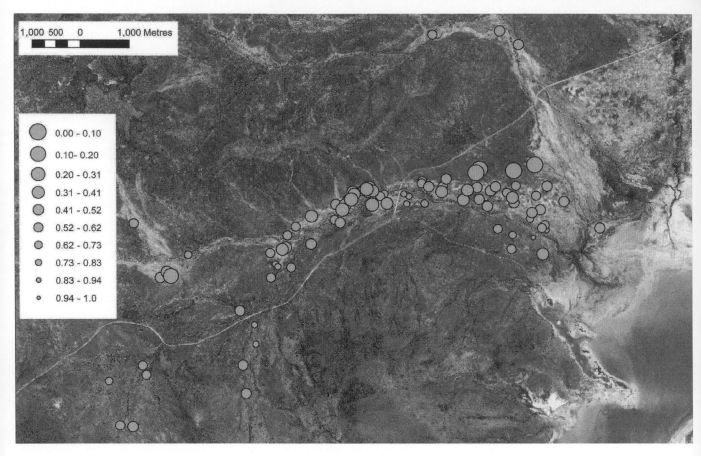

Figure 6. Scalds selected randomly for artefact analysis in Rutherfords Creek with values calculated for the Cortex Ratio.

cover a plane without crossing itself. Differences in movement tortuosity therefore represent differences in the thoroughness of landscape usage (Roshier et al. 2008). Low tortuoisty (i.e. more linear movement paths) is associated with a higher velocity of movement across the landscape, which will lead to lower values for the Cortex Ratio at any one location. In contrast, high tortuosity, while still potentially involving the transport of flakes, will lead to local deposition and therefore an increased chance that an artefact assemblage at any one location will have a higher Cortex Ratio value. Finally, artefact longevity will have an effect on artefact movement since those artefacts that exist for longer periods of time before being discarded will have a greater chance of being transported over greater distances.

These concepts of movement may be related to environmental resource availability. Concentrated resource patches will encourage higher tortuosity and lower velocity while sparse resource distributions will lead to less tortuous movement and a higher velocity. Between-patch movements across less desirable portions of the landscape will have low tortuosity, while within-patch movements are apt to be more tortuous as more intensive searching for resources occurs.

Applying this reasoning to the Rutherfords Creek study area, we find that there is a lack of pattern in relation to the Peery Lake 'resource patch'. Rutherfords Creek flows into Peery Lake, a large ephemeral lake basin that when flooded supports a diversity of resources such as fish, birds and grazing animals. What is striking about the distribution of Cortex Ratio values calculated for individual scalds in Rutherfords Creek is that there is no uniform gradation of values towards or away from the lake (fig. 6). In addition, the majority of the scalds, including those that contain the vast majority of the artefacts, have cortex values well below one, suggesting considerable artefact movement. Despite the existence of the lake 'resource patch', the Cortex Ratio values suggest that tortuosity of movement was relatively low and velocity high. This suggests that the lake and its immediate surroundings was not sufficiently desirable to encourage place to place movement around its shores in what might be described as a 'daisy chain' of occupations. If it had been, transport of flakes from one location to the next would inevitably have led to Cortex Ratios closer to one through repeated deposition of artefacts as people moved from one place to the next in what Binford (1980) has described as a 'leapfrog' pattern. Instead, the daisy chain was broken and artefacts were moved a sufficient distance away so as not to be returned.

Both the low Cortex Ratio values displayed by the artefact assemblages, and the results of the hearth radiocarbon age estimates previously described, suggest a series of episodic occupations. Thus, despite the chronology of occupation for Rutherfords Creek spanning around 2000 years, the total amount of time that it was occupied was probably much less. However, the outcomes of a series of short-term manufacturing and movement events are still apparent in a palimpsest deposit that accumulated over more than two millennia.

The long-term consequences of mobility are reflected in variations in the hearth radiocarbon chronologies. As discussed previously, gaps in the distribution of radiocarbon age determinations correlate with multi-decadal shifts in the late Holocene Australian climate (Holdaway et al. 2010a). While not explainable at the temporal scale by the movement of individuals, this pattern does indicate the long-term consequences of a mobility strategy practiced by groups of individuals: populations abandoned locations right across western NSW for prolonged periods of time in keeping with multidecade-long shifts in rainfall.

The transport of large, thin flakes demonstrates an emphasis on artefact selection and transport despite the relative lack of retouched tools, and reflects an organisation of technology geared not to the conditions that persisted in the vicinity of the assemblages where the artefacts were manufactured, but instead to the conditions that existed elsewhere where the artefacts were intended to be used. It suggests that use often occurred at the 'in-between' places in the wider Australian landscape. At these places resources might be encountered but these were never concentrated enough or predictable enough to allow prolonged occupation. Interpreted at the scale of a single site or, indeed, a single geographic feature like a drainage basin, the archaeological record shows little understandable pattern in the sense that artefact concentrations do not pattern neatly with geographic features. There is for instance, no simple relationship between the size or complexity of assemblages with increasing stream order (Shiner 2008). However, this lack of pattern does make more sense when used to support an inference of mobility as a response to an unpredictable climate and therefore unpredictable resource base. The same general location might be repeatedly occupied but not necessarily used to exploit the same resource suite or abundance (Holdaway et al. 2008b), and individual targeted areas never represented more than just a part of the greater expanses where stone artefacts might be needed.

LANDSCAPE

Demonstrating that the archaeological record from western NSW shows high levels of mobility should not come as a surprise. It has been known for many years that Aboriginal people in western NSW moved extensively (e.g. Allen 1974) as they did in other parts of the Australian arid zone (e.g. Veth 1993). However, demonstrating mobility poses significant problems for archaeologists. As Close (2000) points out, it is one thing to propose that people should have moved extensively, but quite another to indicate how in fact they did move. The sheer scale of movement provides significant methodological problems. How big should the research area be to encompass the likely size of the settlement system? Some writers in Australia, for instance, have proposed that the spatial organisation of society is best considered at the scale of entire drainage systems (Peterson 1976; Sutton 1990). However, one such drainage system, the Murray-Darling Basin (MDB), covers an area of more than a million square kilometres or 14% of continental Australia. While one individual is unlikely to have traversed such an area, many individuals contributed to the formation of archaeological records found in sub-catchments within the MDB, like that at Rutherfords Creek. The people who formed this record over more than two millennia might be expected to have traversed very large regions at one time or another. If archaeologists cannot possibly survey regions with land areas the size of the MDB, how should they go about placing limits on their study areas and still be able to meaningfully assess the settlement pattern?

When considering this problem, it is useful to distinguish between landscape as a place where people lived, and landscape as a means of understanding variability in the archaeological record. People certainly occupied space; they acted within it, and 'organised' it in their own ways. Archaeologists can make inferences about how past peoples conceived of landscape, as we have done based on our western NSW study area (see, for example, Holdaway & Allen in press). However, such inferences come from the results of analysis, and not simply from record description. How people interacted with a landscape is not inscribed directly by the distribution of archaeological materials identified across it. Using landscape as a means to assess variability is a form of comparative analysis. The value that comes from incorporating space into analysis comes from the ability this provides to compare more than one unit of observation. Increasing the number of units will either increase the variability apparent in the record or not. Both results are potentially interesting. Because the incorporation of space increases variability it also enhances the ability to integrate a diversity of data sets and hence processes that produce variability by their operation at different spatial scales. This in turn enhances the ability to determine the outcome of interactions between multiple processes.

Deriving inferences about how people used and perceived a landscape involves investigating a set of places where people lived, among other things. But it is not necessary for the analytical space to equate with the actual space utilised by either multiple groups through time or a single group at one time. As indicated above, the areas involved in either case may be impossibly large. Instead, it is necessary to analyse a sample of places that were used with techniques that allow inferences to be drawn about what the sample means in its wider spatial context. In the Rutherfords Creek study, for example, the archaeological survey ranged across a 15 km long drainage line measurable in a GIS as an area of 62 km^2. But the inferences about the use of space by Aboriginal people derived from this study area extended over regions and orders of magnitude larger than this.

Considering a sampling unit with a larger area increases the potential that a more variable archaeological record will be encountered. However, there is no rule on how large or small a sample area should be. At one extreme, one could imagine a group of artefacts representing the activities of one or a group of individuals. Studies of the Meer site published a number of years ago provide one example (Cahen et al. 1979). Precise behavioural inference is possible from such examples but the explanatory power of such inference is limited because it represents the behaviour of only a handful of people at one point in time. There is no indication of the degree to which this behaviour can be generalised to larger populations. At the other extreme, surveying the archaeological record within a geographically defined space like Rutherfords Creek provides information on the intensity of use, but it cannot provide information on the operation of specific groups of people at any one point of time. Stated in this way, it is easy to see that any one analysis is no more correct than the other: both the examples relate to different types of research questions. What should be concluded is that analyses need to be situated at a scale (or set of scales) where the results are not so specific that they are trivial and not so generalised that they appear to apply to nearly every type of archaeological record (Holdaway & Wandsnider 2006).

Much the same point can be made when considering time. More time passing may increase variability, but variability requires increases in the amount of behaviour accumulated at one place, not simply a greater period of elapsed time. The prolonged accumulation of the western NSW archaeological record, all the more apparent because it is both exposed and is a palimpsest lagged onto a single surface, represents relatively little accumulated behavioural time even though it is spread across more than two millennia of elapsed time. The type of behaviour evidenced by the Cortex Ratio suggests movements of artefacts by individuals 'gearing up' for long trips that also could be measured in days, although, as argued above, the movements involved sometimes took people away from Rutherfords Creek and they did not return for periods measured in decades or centuries. In any event, analysis reveals examples of a palimpsest of short-term events that occurred over long periods of elapsed time. In contrast, climatic correlations are only apparent when many behavioural episodes are combined together to investigate elapsed time periods measured in centuries. Similarly to the analysis of space, neither one of these analyses is more correct than the other since both address different types of questions. Neither is the behaviour indicated by spatial analysis to be preferred over analyses based on time. Rather, it is the variability introduced by comparing all the types of analyses, aimed at determining behaviour operating at both different spatial and temporal scales, which produces the long-term record of human behaviour.

CONCLUSION

In western NSW, Australia, analysis of archaeological remains in a 15 km long drainage system has provided information on the nature of land use and movement over areas that are orders of magnitude larger than those actually studied. The movement of flakes, interpreted as examples of 'gearing up behaviour', occurred across distances that moved people well beyond the limits of Peery Lake. On the longer term, the accumulation of hearths indicates how the accumulations of behavioural events relate to long-term shifts in the climate and therefore resource availability.

Hunter-gatherers moved over substantial areas in arid parts of Australia, complicating the task of archaeologists interested in understanding behavioural changes related to human-environment interactions. It is rarely possible to define spatial units that equate with behavioural units in such situations. In addition, the archaeological record accumulates through time, meaning that multiple behavioural events accumulate within one area. Landscape approaches in archaeology provide the means to deal with the palimpsest of behaviour represented in the archaeological record both spatially and temporally. No one spatial or temporal scale is more suitable than the other but, as both scales are increased, the potential for a more variable record is increased. Archaeologists are able to investigate the variability at different scales as a means for deriving different types of inferences about past behaviour.

ACKNOWLEDGEMENTS

We are indebted to the Indigenous traditional owners of country in western NSW for their permission to undertake research, and their help and support with fieldwork. The initial impetus for the research came from Dan Witter, formerly with the NSW National Parks and Wildlife Service, and initial funding came from the Australian Institute for Aboriginal and Torres Strait Islander Studies. Students from Auckland and Macquarie Universities provided invaluable labour in the field. We thank the NSW National Parks and Wildlife Service for permission to conduct the research at Paroo-Darling National Parks. Funding was provided by the University of Auckland, Macquarie University, the NSW National Parks and Wildlife Service and the Australian Research Council. All radiocarbon determinations were performed by the University of Waikato Radiocarbon Dating laboratory. We thank Alan Hogg and his colleagues for their help with the samples. OSL determinations were carried out by Ed Rhodes. Briar Sefton drew the figures. Two anonymous reviewers provided very helpful comments on an earlier version of this paper.

REFERENCES

Allen, H. 1974. The Bagundji of the Darling Basin: cereal gathers in an uncertain environment. *World Archaeology* 5, 309-22.

Binford, L. 1977. Forty-seven trips: a case study in the character of archaeological Formation Processes. In Wright, R.V.S. (ed.), *Stone Tools as Cultural Markers*, 24-36. Australian Institute of Aboriginal Studies, Canberra.

Binford, L. 1979. Organization and formation processes: looking at curated technologies. *Journal of Anthropological Research* 35, 255-73.

Binford, L. 1980. Willow smoke and dogs' tails: hunter-gatherer settlement systems and archaeological site formation. *American Antiquity* 45, 4-20.

Cahen, D., L. Keeley. & F.L. Van Noten. 1979. Stone tools, toolkits and human behavior in prehistory. *Current Anthropology* 20, 661-683.

Close, A.E. 2000. Reconstructing movement in prehistory. *Journal of Archaeological Method and Theory* 7, 49-77.

Dibble, H. L., U.A. Schurmans, R.P. Iovita & M.V. McLaughlin. 2005. The measurement and interpretation of cortex in lithic assemblages. *American Antiquity* 70, 545-560.

Douglass, M.J. & S.J. Holdaway. 2010. *Quantifying stone raw material size distributions: an archaeological application for investigating cortex proportions and stone artefact curation in lithic assemblages*. Papers in Honour of Val Attenborough. Australian Museum Technical Series, Sydney.

Douglass, M.J., S.J. Holdaway, P.C. Fanning & J.I. Shiner. 2008. An assessment and archaeological application of cortex measurement in lithic assemblages. *American Antiquity* 73, 513-26.

Dunkerley, D.L. 1999. Banded chenopod shrublands of arid Australia: modelling responses to interannual rainfall variability with cellular automata. *Ecological Modelling* 121, 127-138.

Fanning, P.C. 1999. Recent landscape history in arid western New South Wales, Australia: A model for regional change. *Geomorphology* 29, 191-209.

Fanning, P.C., Holdaway, S.J. & R. Philipps. 2009a. Heat retainer hearth identification as a component of archaeological survey in western NSW, Australia. In Fairbairn, A. & O'Connor, S. (eds.), *New Directions in Archaeological Science*, 13-23. ANU E Press, Terra Australis 28, Canberra.

Fanning, P.C., S.J. Holdaway, & E. Rhodes. 2007. A geomorphic framework for understanding the surface archaeological record in arid environments. *Geodinamica Acta* 20, 275-86.

Fanning, P.C., S.J. Holdaway, & E. Rhodes. 2008. A new geoarchaeology of Aboriginal artefact deposits in western NSW, Australia: Establishing spatial and temporal geomorphic controls on the surface archaeological record. *Geomorphology* 101, 526-532.

Fanning, P.C., S.J. Holdaway, E.J. Rhodes & T.G. Bryant. 2009b. The surface archaeological record in arid Australia: geomorphic controls on preservation, exposure and visibility. *Geoarchaeology* 24, 121-46.

Gould, R.A. 1991. Arid-land foraging as seen from Australia: adaptive models and behavioral realities. *Oceania* 62,.12-33.

Holdaway, S.J. & H. Allen. (in press). Placing ideas on the land: Practical and ritual training amongst the Australian Aborigines. In Wendrich, W. (ed.) *Apprentices*. University of Arizona Press, Tuscon.

Holdaway, S.J. & P.C. Fanning. 2008. Developing a landscape history as part of a survey strategy: A critique of current settlement system approaches based on case studies from Western New South Wales, Australia. *Journal of Archaeological Method and Theory* 15, 167-189.

Holdaway, S.J., P.C. Fanning, M. Jones, J. Shiner, D. Witter & G. Nicholls. 2002. Variability in the chronology of late Holocene Aboriginal occupation on the arid margin of southeastern Australia. *Journal of Archaeological Science* 29, 351-363.

Holdaway, S.J., P.C. Fanning & E. Rhodes. 2008a. Challenging intensification: human – environment interactions in the Holocene geoarchaeological record from western New South Wales, Australia. *The Holocene* 18, 411-420.

Holdaway, S.J., P.C. Fanning & E. Rhodes. 2008b. Assemblage Accumulation as a Time Dependent Process in the Arid Zone of Western New South Wales, Australia. In Holdaway, R.V.S. & Wandsnider, L. (eds.) *Time in Archaeology*, 110-133, University of Utah Press, Salt Lake City.

Holdaway, S.J., Fanning, P.C., Rhodes, E. Marx, S. Floyd, B. & M. Douglass. 2010a. Human response to palaeoenvironmental change and the question of temporal scale. *Palaeogeography, Palaeoclimatology, Palaeoecology* 292, 190-210.

Holdaway, S.J., Wendrich, W. & Philipps, R. 2010b. Variability in low level food production societies as a response to resource uncertainty. *Antiquity* 84, 185-194.

Holdaway, S.J., Fanning, P.C. & J. Shiner. 2005. Absence of evidence or evidence of absence? Understanding the chronology of indigenous occupation of western New South Wales, Australia. *Archaeology in Oceania* 40, 33-49.

Holdaway, S.J., Shiner, J., & P.C. Fanning. 2004. Hunter-gatherers and the archaeology of the long term: An analysis of surface, stone artefact scatters from Sturt National Park, New South Wales, Australia. *Asian Perspectives* 43, 34-72.

Holdaway, S.J. & N. Stern. 2004. *A Record in Stone: The Study of Australia's Flaked Stone Artefacts*. Museum Victoria and Aboriginal Studies Press, Melbourne and Canberra.

Holdaway, S.J. & L. Wandsnider. 2006. Temporal scales and archaeological landscapes from the Eastern Desert of Australia and Intermontane North America. In Lock, G. & Molyneaux, B. L. (eds.), *Confronting Scale in Archaeology*, 183-202, Springer Science and Business Media, New York.

Holdaway, S.J. & L. Wandsnider. 2008. Time in archaeology: an introduction. In Holdaway, S.J. & L. Wandsnider (eds.), *Time in Archaeology*, 1-12, University of Utah Press: Salt Lake City. Kuhn, S.L. 1992. On planning and curated technologies in the Middle Paleolithic. *Journal of Anthropological Research* 48, 185-214.

Lin, S., M. Douglass, S.J. Holdaway & B. Floyd. 2010. The application of 3D laser scanning technology to the assessment of ordinal and mechanical cortex quantification in lithic analysis. *Journal of Archaeological Science* 37, 694-702.

Marx, S.K., H.A. McGown & B.S. Kamber. 2009. Long-range dust transport from eastern Australia: A proxy for Holocene aridity and ENSO-type climate variability. *Earth and Planetary Science Letters* 282, 167-177.

Morrow, T. 1996. Lithic refitting and archaeological site formation processes. A case study from the Twin Ditch Site, Greene County, Illinois. In Odell, G. (ed.), *Stone Tools Theoretical Insights into Human Prehistory*, 345-376, Plenum, New York.

Peterson, N. 1976. The natural and cultural areas of Australia: a preliminary analysis of population groupings with adaptive significance. In Peterson, N. (ed.), *Tribes and Boundaries in Australia*, 50-71, Australian Institute of Aboriginal Studies, Canberra.

Rhodes, E.J., Fanning, P.C. & S.J. Holdaway. 2010. Developments in optically stimulated luminescence age control for geoarchaeological sediments and hearths in western New South Wales, Australia. *Quaternary Geochronology* 5(2-3), 348-352.

Rhodes E.J., Fanning P.C., Holdaway S.J. & C. Bolton. 2009. Ancient surfaces? Dating archaeological surfaces in western NSW using OSL. In Fairbairn, A. & O'Connor, S. (eds.), *New Directions in Archaeological Science, Terra Australis 28*, 189-200, ANU E-Press, Canberra.

Roshier, D.A., V.A.J. Doerr & E.D. Doerr. 2008. Animal movement in dynamic landscapes: interaction between behavioural strategies and resource distributions. *Oecologia* 156, 465-477.

Shiner, J.I. 2008. *Place as Occupational Histories an Investigation of the Deflated Surface Archaeological Record of Pine Point and Langwell Stations, New South Wales, Australia*. BAR International Series 1763, Oxford.

Surovell T.A., J.B. Finley, G.M. Smith, P.J. Brantingham & R. Kelly. 2009. Correcting temporal frequency distributions for taphonomic bias. *Journal of Archaeological Science* 36, 1715-1724.

Stafford Smith D.M. & Morton S.R. 1990. A framework for the ecology of arid Australia. *Journal of Arid Environments* 18, 255-278.

Sturman, A.P. & N.J. Tapper. 2006. *The Weather and Climate of Australia and New Zealand*. Oxford University Press, Oxford.

Sutton, P. 1990. The pulsating heart: large scale cultural and demographic processes in Aboriginal Australia. In Meehan, B. & White, W. (eds.), *Hunter-Gatherer Demography Past and Present*, 71-80, University of Sydney, Sydney.

Thomas, D.H. 1973. An empirical test of Steward's model of Great Basin settlement patterns. *American Antiquity* 38, 155-176.

Veth, P.M. 1993. *Islands in the Interior: The Dynamics of Prehistoric Adaptations Within the Arid Zone of Australia*. International Monographs in Prehistory 3, Ann Arbor.

4.2 Surface contra subsurface assemblages: Two archaeological case studies from Thesprotia, Greece

Authors
Björn Forsén[1] and Jeannette Forsén[2]

1. Department of Philosophy, History, Culture and Art Studies, University of Helsinki, Helsinki, Finland
2. Department of Historical Studies, Göteborg University, Göteborg, Sweden
Contact: bjorn.forsen@helsinki.fi

ABSTRACT

This paper focuses on two archaeological sites in the Kokytos valley in north-western Greece, discovered by the Thesprotia Expedition in 2004. We stress the discrepancies between surface and subsurface assemblages and try to explain these differences. Initially the sites were found and recorded in a surface survey. After this several multiscale datasets were obtained through different methods, e.g. phosphorus sampling, trial excavations and different geophysical techniques. In combining the different datasets it became quite clear that the surface assemblages are biased mainly through erosion and modern landscaping, but also due to so-called 'walker effects'. Both sites proved to be more extensive in size and have a longer occupational time span when adding multiscale datasets to the surface collections made in the surface survey.

KEYWORDS

intensive survey, surface contra subsurface, post-deposition, hidden landscape, windows

INTRODUCTION

Anyone conducting an intensive archaeological field survey has been forced to consider to what degree an adequate picture of the past can be built on the basis of the finds detected on the surface. Or to put it another way: to which extent do surface assemblages concur with subsurface assemblages, and how can

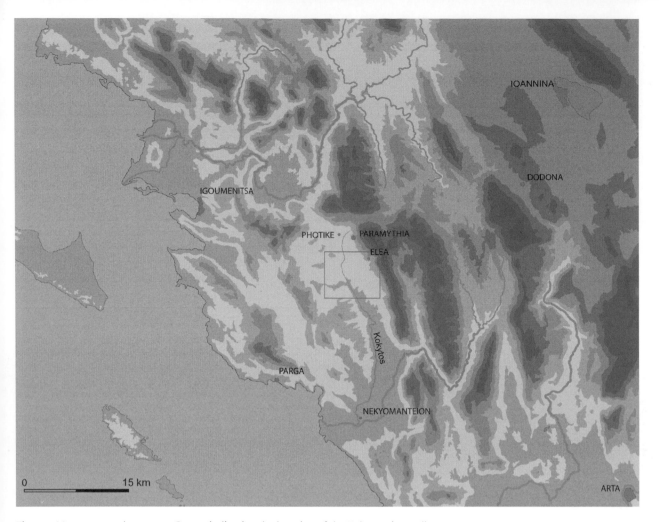

Figure 1. Map over north-western Greece indicating the location of the Kokytos river valley.

possible differences be explained and taken into account? Scholars working in the Aegean have in general believed that developments in the regional history could be constructed on the basis of surface assemblages, although at the same time being aware of the fact that such assemblages to some degree are shaped by post-depositional factors, the most common being erosion.

One of the first scholars to express doubt concerning the reliability of surface assemblages was Jeremy Rutter (1983, 138-139), who coined the expression 'low-visibility ceramic phases' for certain prehistoric periods that produced pottery less likely to be preserved and/or recognised. Bintliff, Howard & Snodgrass (1999, 139-168) brought the discussion further by suggesting that part of the prehistoric sites in Greece are not noted at all in intensive field surveys, thus creating a 'hidden landscape' that can be visualised only by increased attention. However, this explanation may be applicable only to certain regions of Greece, such as Boiotia (Davis 2004, 22-34). Nevertheless intensive field surveys are today often accompanied by other techniques, such as phosphorus sampling, different geophysical methods and occasionally even trial excavations.

We will in this paper focus on two sites discovered by the Thesprotia Expedition in the Kokytos valley in Thesprotia, north-western Greece (fig. 1), with the aim of highlighting discrepancies between surface and subsurface assemblages as well as discussing possible explanations of these discrepancies. At the onset of the Thesprotia Expedition the strategy was to collect surface finds as intensively as possible and to return and focus on sites that seemed to produce finds, however few, from chronological periods previously absent in the archaeological record of the region. This emphasis has proven to be a productive way of opening up windows to the otherwise hidden parts of the past landscape.

SITE PS 12

Site PS 12 was detected in 2004 by our survey team. In a field with a scatter of Middle Palaeolithic chipped stone a small concentration of poorly preserved prehistoric sherds were sampled that were given a preliminary date to the later part of the Neolithic period or the Early Bronze Age. Several revisits to the site produced more abraded sherds and finally also a flint blade of similar date. All these finds were concentrated in a small area along the upper edge of the field, indicating that the site might continue in that direction. The field in question is located at the lowermost eastern slope of the Liminari Hill, separated from the hill by a dirt road. On the other side of the road there is a small sheltered nook which could have been a perfect setting for a prehistoric settlement (fig. 2). Unfortunately this area, abandoned today but at one stage under cultivation, is badly overgrown and used as the setting for some 50 beehives! In order to confirm the existence of a site at this location we decided to open up some trial trenches.

The Liminari Hill consists of limestone, denuded of all soil. Only some prickly oak bushes and mountain tea grow on the slopes today. The situation may however have been different some thousand years ago before the topsoil eroded away, creating an alluvial fan in the nook, where the site is located. It is unknown at which time the erosion took place, but it may have consisted of several different phases, thus covering Neolithic or Bronze Age layers with more recent sterile soil.

Figure 2. The setting of PS 12 at the foot of the Liminari hill with initial findspot of prehistoric artefacts across the dirt road indicated with an arrow.

First surface finds

Figure 3. Examples of local and non-local pottery from the Early Bronze Age Epirus.

The initial trial trenches revealed the existence of a clear cultural layer with dark soil, charcoal, burnt mud-brick, pottery, chipped stone and spindle-whorls stretching some 50m higher up than the edge of the field where the first sherds were detected. This cultural layer is covered by a sterile top soil layer ranging in thickness from 30-50cm. In one of the trenches we found a rudimentary wall, which was exposed for a length of nine metres. Most of the pottery from these trial trenches dates to the Early Bronze Age (EBA) (fig. 3), although some pottery of the Middle (MBA) and Late Bronze Age (LBA) also was found next to the wall. Three C-14 samples taken from trenches A and D date to the EBA, whereas two samples taken next to the wall, date to the MBA and LBA. Useful references regarding the material culture of the Bronze Age in Epirus are for instance Wardle (1997), Tartaron (2004), and finally Dousougli & Zachos (2002).

Encouraged by these results we decided to proceed by taking phosphorus samples as well as conducting a magnetometer survey. So far we have the results of fifteen phosphorus samples. According to these results the site stretches at least 100m in north-south direction and has at least two clear concentrations of phosphorus anomalies, one close to trench D and the second one further to the south (fig. 4).

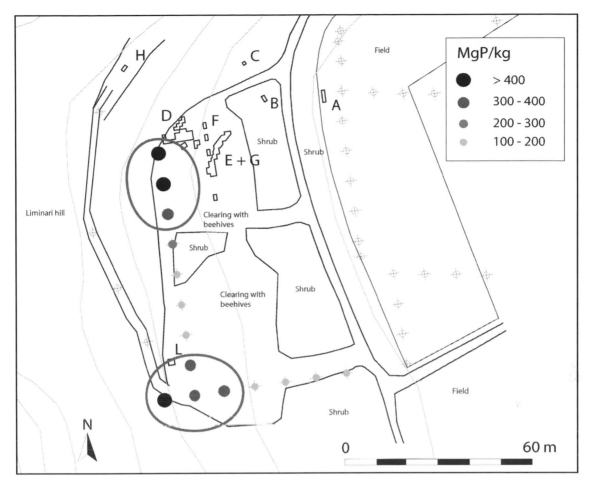

Figure 4. Phosphorous samples from PS 12 obtained and analysed by M. Lavento, Helsinki, indicating two clear anomalies in concentrations.

Due to magnetic disturbances caused by the beehives, the magnetometer could be used only in the northern part of the site. The resulting magnetometer map indicates that the rudimentary wall continues for at least another 15m. As it follows a contour line it is most likely a terrace wall. Indications of other walls were visible close to trench D, accompanied by one of our concentrations of phosphorus anomalies.

Guided by these results we decided in 2009 to open trial trenches near the terrace wall and at the two concentrations of phosphorus anomalies in the hope of finding remains of prehistoric houses. To our great surprise the walls close to trench D, faintly visible on the magnetometer map, turned out to be part of a grave tumulus with a diameter of ca. nine metres and with a central cist grave dating to the very end of the Middle Bronze Age (fig. 5). Below the tumulus a thick cultural layer of Early Bronze Age was found, with large quantities of pottery and chipped-stone, as well as numerous spindle-whorls (fig. 6) and pieces of daub. For similar tumuli in the general area of north-western Greece and Albania as well as discussions concerning their origin see for instance, Prendi (1993, 17-28), Papadopoulos (1999,141) and most recently Kilian-Dirlmeier (2005, 5-46, 82-89).

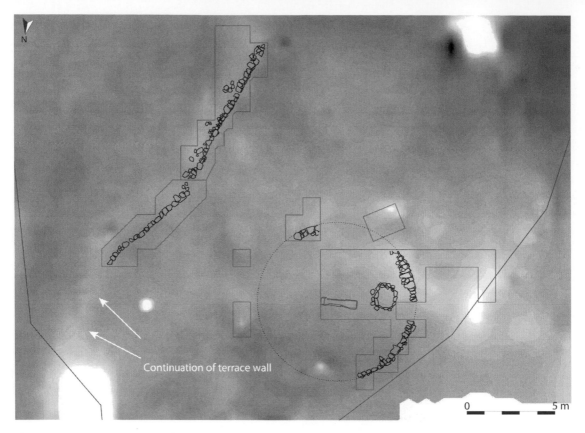

Figure 5. Magnetometer map indicating the terrace wall of Middle to Late Bronze Age and the early Late Bronze Age tumulus at PS 12.

Figure 6. Spindle whorls found at PS 12 of Early Bronze Age date.

The small test pit opened up in the southern concentration of phosphorus anomalies hit straight on a previously robbed cist grave, which is dated by a C-14 sample to the transition from MBA to LBA (this cist grave could also be part of a tumulus, but time restrictions prevented further work at this locus). Finally, a charcoal sample obtained in the trench near the terrace wall produced a radiocarbon determination associated with the wall at the beginning of the LBA, making it only slightly later than the graves to which it probably is connected.

Thus, the first trial trenches at PS 12 confirmed the impression received by the scant survey finds, meaning that we were dealing with a prehistoric site. However, the trial trenches taken together with the phosphorus- and geophysical survey results showed that the site was much larger and had been settled for a much longer period than the surface finds indicated. Furthermore, it is noteworthy that only last year's trial trenches revealed that this EBA site, after a long hiatus, had been reused as the setting for a communal burial place towards the end of the Middle Bronze Age

SITE PS 36

Site PS 36 is located in Mavromandilia (fig. 7). While surveying some fields just to the east of the river Kokytos we found a scatter of Archaic to Classical pottery which we regarded as a site, PS 31. When we had finished walking the tract we stumbled just by chance on another very small, but clear concentration of pottery, assigned PS 36. The small size of this concentration, only some 10x10m, explains why we missed it initially while surveying. On the basis of the pottery collected on the surface it seemed to date to the Early Iron Age.

Figure 7. Mavromandilia, just east of the Kokytos river, the setting for PS 31 and PS 36.

Figure 8. Two vessels where the smaller was found inside the larger. Both are of ninth or eight century BC date.

Interestingly enough the Greek Archaeological Service had in 2003, in connection with drainage work, at a depth of ca one metre below surface, found a rich dump of pottery of Early Iron Age date at the edge of the field of PS 36 (Tzortzatou & Fatsiou 2009, 39-43). No finds were visible between PS 36, PS 31 and the spot where our Greek colleagues had excavated. As hardly any sites dating to the Early Iron Age or the Archaic period were known previously in all of Thesprotia we decided to put more stress on this area by carrying out a trial excavation at PS 36 and by taking phosphorus samples around it.

The soil at Mavromandilia consists mostly of loam and silty clay. The depth of the fine top soil varies considerably, sometimes being only some 30-40cm, whereas at the spot of the excavation of the Greek Archaeologicl Service it is 1.5m. Below the top soil coarser sediment layers are found, including limestone particles and gravel, as well as cultural layers.

In the excavations of PS 36 we found remains of fire places and/or shallow bothroi consisting of dark soil mixed with charcoal, ash, animal bones and large amounts of broken pottery. The different features found at PS 36 can on the basis of C-14 analyses be dated to between 1100 and 700 BC (Forsén 2009, 57-59). Part of the pottery dating to the ninth and eighth centuries BC was well preserved, such as a large fragment of a Thapsos ware stirrup-handled crater (Forsén 2009, 64 no. 9), a small trefoil mouthed pitcher (Forsén 2009, 62 no. 5) and two vessels, where one was found inside the other (fig. 8; Forsén 2009, 64-66 nos. 13-14). We also found daub/clay lining, indicating huts nearby (Forsén 2009, 60 fig. 5). On the basis of the pottery sequence, the site seems to have been used in one capacity or another at least until the fourth century BC, as evidenced by a stamped Corinthian amphora handle (Forsén 2009, 66 no. 21) and a small black glazed bowl (Forsén 2009, 66 no. 20).

In order to study the surroundings of PS 36 in more detail a total of 89 augering holes were made to a depth of three metres maximum below the surface. Phosphorus samples were taken from all holes, either from cultural layers if such were found or otherwise at a depth of ca. 50-60cm below the surface. The spread of phosphorus anomalies give us a better idea of the size of the three find concentrations at Mavromandilia. Thus, PS 36 appears to be ca. 40x20m large, whereas the spot for the Greek Archaeological Ser-

Figure 9. Map showing the location of the augering holes where the phosphorous content indicates the size of the three sites PS 31, PS 36 and the so-called Greek Archaeological Service site.

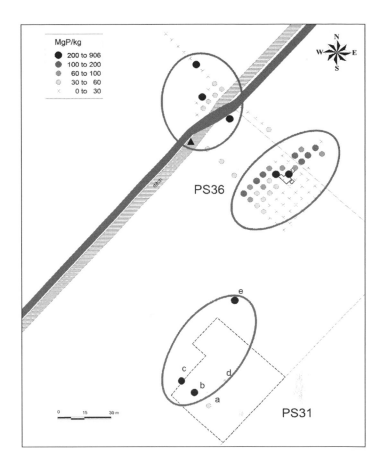

vices excavation continues on both sides of the modern ditch, covering at most ca. 40x40m. The size of PS 31 is more difficult to estimate, but it is at a maximum 40x60m (fig. 9; Lavento & Lahtinen 2009, 77-83).

The setting of Mavromandilia is and has always been affected by the group of small springs that exist nearby (fig. 10). At springs 4 and 7, also today water rises up to the surface, developing small seasonal ponds, from where it runs towards the river Kokytos. Most water has apparently run from spring 7, from where at some stage it has formed a small stream leading to the Kokytos, between PS 36 and the Greek Archaeological Service site. This stream was recently covered by soil, but on the basis of the drillings it may originally have been 2-3m deep.

Springs such as 4 and 7 may through time be covered and can also find new outlets. Thus the farmers have covered a similar natural spring, 3, and channelled the water to two artificial springs, 1 and 2. A close study of the satellite photograph reveals two other possible springs, 5 and 6, visible as shallow depressions. They are located along the 'old' stream running from spring 7 to the Kokytos, one of them being located between PS 36 and the Greek Archaeological Service site (fig. 10).

In the case of Mavromandilia it appears that the small stream located between PS 36 and the spot excavated by the Greek Archaeological Service existed already during the Early Iron Age, and that the existence of running water made the place attractive for human occupation. The deposition of sterile fine soil on top of the settlement took place during a time period of some 2,500 to 3,000 years. The sedimentation

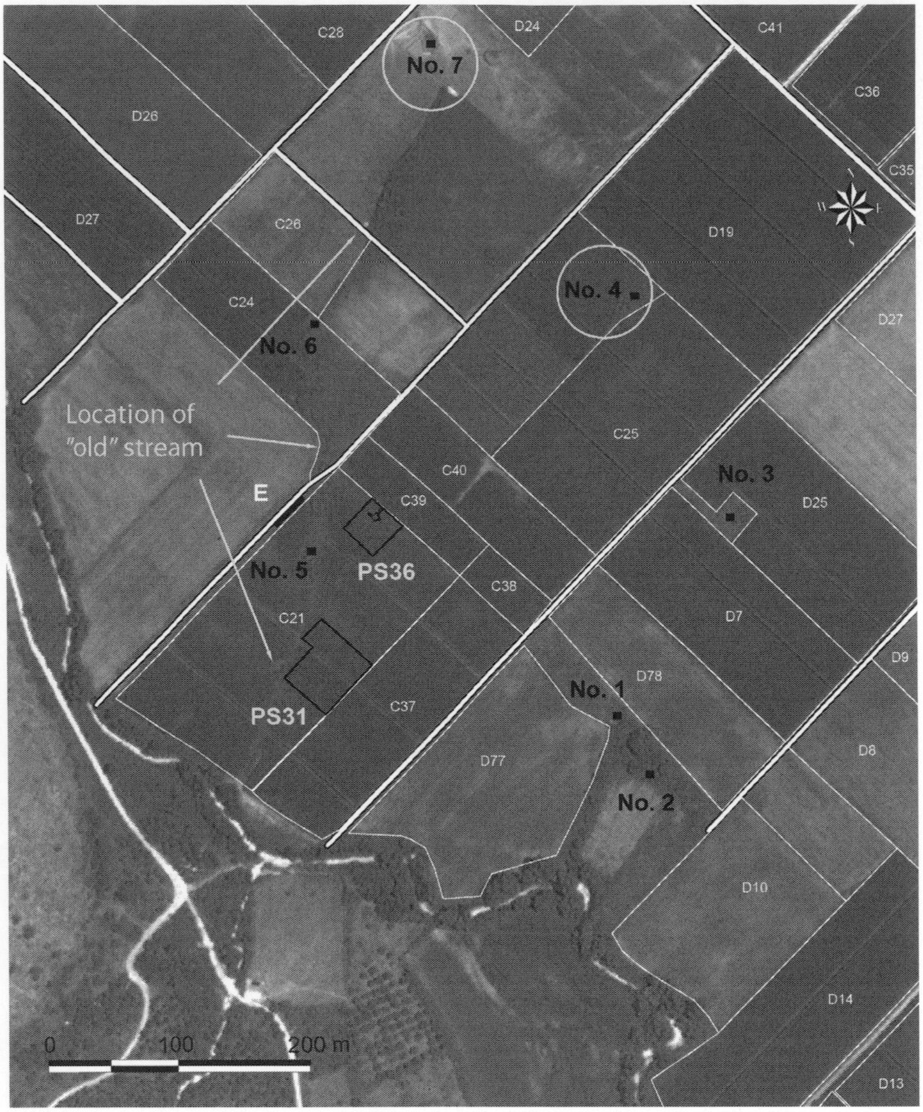

Figure 10. Map showing the springs and 'old' stream near the three sites at Mavromandilia.

may have been caused by the small stream while it meandered and changed course through time. Another factor influencing the sedimentation is the fact that the region in general slopes towards the Kokytos River, thus facilitating the movement of finer particles into that direction through the centuries.

As the pottery found at PS 36 and in the Greek excavation nearby is roughly contemporaneous it seems likely that they represent different parts of one and the same site. The people here probably lived on both sides of the stream and above spring 5. The relationship between this site and PS 31 on the other hand is more problematic and can only be clarified through further work. However, what speaks for PS 31 being a separate site is the fact that the pottery here seems to be Archaic to Classical in date, thereby slightly later than the main period of activity at PS 36 and the Greek excavation site.

The case of PS 36 at Mavromandilia reveals a similar pattern as at PS 12. Thus the trial excavations and the phosphorus sampling also here showed that the site was much larger and had been settled for a much longer period than the surface finds indicated. However, this time the discrepancy was not due only to lush vegetation and poorly preserved artefacts as in PS 12, but rather to strong erosion and remodelling of the landscape in order to create new and larger fields for the farmers, something that had covered part of the site with soil void of artefacts. Our two cases from Thesprotia confirm that the surface assemblages of prehistoric sites often reveal only part of the reality. A more complete picture can only be obtained by adding multiscale datasets to the surface finds. This article was written in 2010. For more information concerning PS 12 that has appeared since then, see Forsén and Tikkala 2011 or http://www.thesprotiaexpedition.com.

REFERENCES

Bintliff, J.L., P. Howard & A.M. Snodgrass. 1999. The Hidden Landscape of Prehistoric Greece. *Journal of Mediterranean Archaeology* 12, 139-168.

Davis, J.L. 2004. Are the Landscapes of Greek Prehistory Hidden? A Comparative Approach. In Alcock, S.E. & J.F. Cherry (eds.), *Side-by-Side Survey. Comparative Regional Studies in the Mediterranean World*, 22-35. Oxbow Books, Oxford.

Dousougli, A. & K. Zachos. 2002. L'Archéologie det Zones Montagneuses: Modèles et Interconnexions dans le Néolithique de l'Épire et de l'Albanie Méridionale. In Touchais, G. & J. Renard (eds.), *L'Albanie dans l'Europe préhistorique*, 111-143. Athens: École française d'Athènes.

Forsén, J. 2009. The 'Dark Age' in the Kokytos Valley-Not So Dark After All. In Forsén, B. (ed.), *Thesprotia Expedition I. Towards a Regional History*, 55-72. Foundation of the Finnish Institute at Athens, Helsinki.

Forsén, B. & Tikkala, E. (eds.). 2011. *Thesprotia Expedition II. Environment and Settlement Patterns*. Foundation of the Finnish Institute at Athens, Helsinki.

Kilian-Dirlmeier, I. 2005. *Die Bronzezetlichen Gräber bei Nidri auf Leukas: Ausgrabungen von W. Dörpfeld 1903-1913*. Verlag des Römisch-Germanischen Zentralmuseums, In Kommission bei Habelt in Mainz, Bonn.

Lavento, M. & M. Lahtinen. 2009. Geo-archaeological Investigations at Mavromandilia of Prodromi In Forsén, B. (ed.), *Thesprotia Expedition I. Towards a Regional History*, 73-87. Foundation of the Finnish Institute at Athens, Helsinki.

Papadopoulos, T.J. 1999. Tombs and Burial Customs in Late Bronze Age Epirus. In Betancourt, P.P., V. Karageorghis, R. Laffineur & W.-D. Niemeier (eds.), *Meletemata. Studies in Aegean Archaeology Presented to M.H. Wiener as he enters his 65th Year*, 137-143. Université de Liège, University of Texas at Austin, Liège/Austin.

Prendi, F. 1993. La Chaonie préhistorique et ses rapports avec les regions de l'Illyrie du sud In Cabanes, P. (ed.), *L'Illyrie méridionale et l'Epire dans Antiquité-II*, 17-28. De Boccard, Paris.

Rutter, J.B. 1983. Some Thoughts on the Analysis of Ceramic Data Generated by Site Surveys. In Keller, D.R. & D.W. Rupp (eds.), *Archaeological Survey in the Mediterranean Area*, 137-142. British Archaeological Reports, Oxford.

Tartaron, T. 2004. *Bronze Age Landscape Society in Southern Epirus, Greece*. British Archaeological Reports, Oxford.

Tzortzatou, A. & L. Fatsiou. 2009. New Early Iron Age and Archaic Sites in Thesprotia. In Forsén, B. (ed.), *Thesprotia Expedition I. Towards a Regional History*, 39-53. Foundation of the Finnish Institute at Athens, Helsinki.

Wardle, K.A. 1997. The Prehistory of Northern Greece: A Geographical perspective from the Ionian Sea to the Drama Plain. In ΑΦΙΕΡΩΜΑ ΣΤΟΝ N.G.L. Hammond, 509-540. Etaireia Makedonikon Spoudon, Thessalonike.

New directions in digital prospection and modelling techniques

5.1 Biting off more than we can chew? The current and future role of digital techniques in landscape archaeology

Author

Philip Verhagen

Research Institute for the Cultural Landscape and Urban Environment, VU University Amsterdam, Amsterdam, The Netherlands
Contact: j.w.h.p.verhagen@vu.nl

ABSTRACT

In this paper, a broad overview is given of the recent development of digital techniques in landscape archaeology, and of the way in which these have effectively revolutionised the way in which we do landscape archaeology nowadays. Within this development, a number of fields can be identified where computer techniques are highly successful in producing better scientific results more efficiently. The main contribution of computer techniques to landscape archaeology is found in their application to the prediction and detection of archaeological remains, to exploratory data analysis and to the visualisation of research results. A number of examples are shown illustrating this. The paper also tries to address the more fundamental issue if the application of digital methods and techniques is actually helpful for developing new interpretations and theory. It is concluded that we are still facing some stiff challenges there that are closely related to the attitude of archaeology as a science to theory, but also to the difficulties of developing software tools that actually do what we need them to do.

KEY WORDS

digital techniques, landscape archaeology, archaeological theory

INTRODUCTION: THE DIGITAL REVOLUTION IN LANDSCAPE ARCHAEOLOGY

For those not so closely involved in the development of digital techniques within and outside (landscape) archaeology, it may not always be appreciated how quickly the digital world is changing. A short look at the amount and variety of software tools available for landscape archaeologists nowadays shows the speed at which these processes move. Freeware and open source packages like Python, Meshlab, Land-Serf, Depthmap, gvSIG, Google Sketchup, Whitebox GAT, NetLogo, R and GeoDa may all be used for specific tasks that are of interest to (landscape) archaeologists. This list is far from exhaustive; and yet, almost all of these packages were either not available ten years ago, or have gone through significant modification and development. And more significantly: there will be very few landscape archaeologists who have used all of them, or even know what they can be used for. We are now far removed from the days when an archaeological 'computer specialist' could be relied on to be proficient in all the computer skills necessary for archaeological research.

Looking back however at how computers have slowly invaded the archaeological work process, one could be excused to say that this has not been a revolution at all, but a gradual integration of digital techniques in research, just like they have more and more influenced our daily lives without us really noticing. In practice, however, I think that for most archaeologists there has been a moment when certain digital techniques were really 'discovered' and fundamentally changed the way of doing research. And while the process of technological innovation may have been relatively slow, there can be no doubt that it has really revolutionised landscape archaeology. As an example, we can look at the way databases entered archaeology somewhere in the early 1970s: they were the domain of experts with arcane knowledge who needed days of effort to code field forms in a format that mainframe computers could read, and then again spend days trying to explore what went wrong. While the latter aspect certainly has not disappeared from daily research practice, the casual use that is now made of a package like MSAccess by archaeologists is far removed from those early days. And so it went with a number of techniques: at first slowly creeping into the research process, and now being fully accepted as indispensable to do landscape archaeological research. One of the developments that have truly revolutionised academic research over the past few years for example is the ongoing digitisation of paper sources, to the point where new and old academic journals are now mostly digital publications, and searching for information on virtually any subject has become much easier.

In this paper I want to take a closer look at what the digital revolution has achieved for landscape archaeology over the past five to ten years. In doing so, I will inevitably only discuss the broadest trends, and will gloss over certain aspects of computing that are of interest to landscape archaeologists. I will largely try to confine my conclusions to the areas where I feel that computer techniques have made the greatest contribution to landscape archaeology: prediction, detection and visualisation. And finally, I will try to address the issue if the application of digital techniques, and in particular quantitative modelling, is actually helpful for developing new interpretations and theory.

A REVOLUTION IN PREDICTION

One of the fields in landscape archaeology where computer techniques, in particular GIS and statistical software, have always been of primary importance is archaeological predictive modelling (see e.g. Judge & Sebastian 1988; Van Leusen & Kamermans 2005). It became firmly established in the time of what could be called the first wave of the digital revolution, where the availability of affordable software on personal computers suddenly opened up computing power to a much wider public. In its long association with digital techniques predictive modelling has witnessed significant changes as a consequence of developments in digital technology. A major factor has been the increasing availability of digital data sources. Only twenty years ago, there were almost no geographical and archaeological data available that could be used as input for predictive models, and almost everything that was needed to create predictive models had to be digitised by the archaeologists themselves. Nowadays, an enormous amount of digital data sources including a wealth of historical maps, digital elevation models and aerial photographs is available on-line, and increasingly it can be downloaded free of charge, or for highly reduced rates compared to the 1990s. Obviously, this makes it much easier to combine data sources and search for patterns and correlations between the archaeological data and the various environmental and historical sources available.

A second factor is the increasing availability of complex analysis methods in software. Geomorphometric indices for example, like the calculation of slope and curvature from digital elevation models, have always been part of GIS software, but it is only in the last few years that landscape archaeologists have become aware of the further analysis capabilities that are offered by packages like LandSerf (see also Hengl & Reuter 2009). These include the possibility to calculate scale-dependent, 'fuzzy' measures of landform (Fisher et al. 2004), like the degree of 'peakness' of the higher elevations in a landscape. And statistical analysis methods that are able to integrate expert judgment and 'hard' data have become available as well, like Bayesian statistical methods and Dempster-Shafer modelling (Ejstrud 2005; Van Leusen et al. 2009). These 'new' predictive modelling techniques are especially well suited to deal with the aspect of uncertainty of predictions, which could potentially provide a much better understanding of the value of the models made (Verhagen et al. 2010).

Powerful 3D 'solid modelling' software is now gradually becoming better affordable and easier to use, and will open up a whole new range of predictive modelling techniques based on both stratigraphic and geographic relationships (de Beer et al. 2011). An impressive example of what 3D modelling techniques can do for the prediction of archaeological resources is found in the North Sea Palaeolandscape Project (Gaffney et al. 2007). This project originally started as an experiment to see whether seismic profiles taken by oil companies in the North Sea could be used to map the palaeogeographical landscape features hidden under the sea floor. The seismic profiles in fact turned out to be very useful for this. However, the effort that was eventually undertaken to map much of the British North Sea was only possible through the use of high bandwidth computer networks and large data storage facilities.

But prediction is not all about formal statistical techniques and using big computers. Before going into the field for any type of fieldwork, no landscape archaeologist today will miss the opportunity to consult the available digital map resources. Of these, LiDAR-based elevation models are currently the most valued. But we have to realise that only ten years ago these were hardly available. In fact, the quality of LiDAR images has dramatically improved over the last few years as well. We can now get LiDAR data with a horizontal resolution of less than 1m, allowing us to detect microtopographical features that are com-

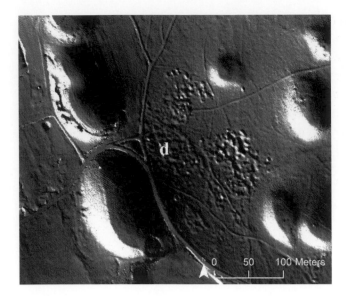

Figure 1. Example of the use of LiDAR images for the detection of archaeological features. Taken from: Opitz et al. (this volume).

pletely invisible from the ground and from aerial photographs (see fig. 1). Especially in forested areas, the contribution of LiDAR to the detection of archaeological features can be enormous (see e.g. Devereux et al. 2005; Doneus et al. 2008). Furthermore, image filtering techniques are still developing, and will hopefully lead to an increased application in areas where LiDAR is usually less successful, like ploughed fields. In an increased appreciation for remote sensing, in the 1990s and early 2000s, satellite images were considered to be perhaps useful for the detection of large archaeological features in arid countries, but not a true match for good aerial photographs (see a.o. Scollar et al. 1990; Fowler 2002). Because of the increased resolution of the available images, and the capture of multispectral images from airplanes, this has clearly changed (see e.g. Winterbottom & Dawson, 2005; Aqdus et al. 2007). With these multispectral and hyperspectral airborne images, we can now look at the landscape at the level of detail that is needed for archaeological prospection, and at the same time get all the added value of using image interpretation techniques, like band and image combination and (semi-)automated classification. And obviously LiDAR images and geophysical measurements can be combined successfully with multispectral images as well (e.g. Kvamme 2008).

A REVOLUTION IN FIELDWORK

In (geo-)archaeological fieldwork, the influence of digital technology is now more and more visible as well. Until the late 1990s, there were no practical systems for setting up wireless communication, and the available equipment was fragile and lacked easy-to-use software applications for use in the field. This has now completely changed, largely because the availability of affordable weather-resistant equipment has increased enormously. Especially in commercial archaeology, the use of field computers has by now become standard procedure, as it not only increases the speed at which measurements are taken and finds and features are recorded, but also diminishes the amount of error in registration, as it is possible to compare the results of recording with the situation in the field at virtually the same moment (Wagtendonk et al. 2009).

The accuracy of GPS measurements used to be an obstacle as well to the use of mobile field computing in archaeology; but the sub-metre precision that is in most cases required has now become affordable as well. And there will be more in the future: miniature helicopters equipped with GPS and stabilising platforms can now be bought at competitive prices and used to take photographs and make photogrammetric measurements of the earth's surface at a high level of detail (Eisenbeiss 2009).

One of the fieldwork activities that has always been highly dependent on computing techniques is geophysical prospection. In this field, the developments sketched on mobile computing have also increased its applicability. Especially the availability of high-precision GPS measurements has now made it possible to take geophysical measurements much quicker than before. It is now common practice to see archaeologists doing geophysical measurements with what basically looks like a high-tech lawnmower, or on a quad bike equipped with a GPS antenna. The level of detail at which these measurements are taken is increasing as well, although there still is a considerable amount of difference in the resolutions and speeds that can be obtained with the various methods available. Resistivity measurements for example always need physical contact of the probes with the earth's surface, which severely reduces the speed of measurement. Low-resolution methods using electromagnetic and self-potential techniques however can easily cover up to 5 hectares per day. Furthermore, signal detection and filtering methods have improved to the point where geophysics (and especially ground-penetrating radar) are now close to being able to detect individual features and larger artefacts, and can also be applied in soil types that until recently were not considered to be very well suited, like clay and peat.

Geophysical measurements are now routinely combined with other GIS-based and remote sensing data to aid interpretation. Furthermore, the increased processing power of computers, coupled to 3D processing software allows us to create 3D images of the subsoil from GPR measurements at a level of detail that was unthinkable 10 years ago. However, in spite of the use of clever image enhancement techniques, the delineation of subsoil features and their interpretation still seems to require a lot of expert judgment. One challenge for the future in geophysics would therefore be to investigate whether automated classification tools can be used to the same effect as in (airborne) remote sensing.

A REVOLUTION IN VISUALISATION

Landscape archaeologists are all familiar with the powerful mapping capabilities of GIS and there is an increasing awareness that visualisation of complex numerical data sets is becoming easier through the availability of powerful data mining software like Weka. However, 3D modelling is probably the field where developments are processing at the highest speed when it comes to more powerful ways of visualising research results. 3D modelling of especially ancient buildings and artefacts has been around for almost as long as GIS has been in place (see e.g. Reilly 1990; Forte & Siliotti 1997; Barceló et al. 2000). Up to now it has failed to make a real breakthrough to landscape archaeology. Partly this has been because of the complexity of creating 3D reconstructions, and the difficulty of using 3D reconstruction software in combination with GIS and/or solid modelling packages. It is however a commonly used and appreciated technique in urban archaeology and architecture. In recent years this field has experienced a massive growth in application and it is now becoming more and more accessible to users that are not skilled 3D modellers. The major reason for this has been the development of free and relatively easy to use software to make architectural reconstructions. Google Sketchup is currently the most widely used package to quickly create 3D models, that may not be up to the standards of real architects, but nevertheless can do a decent job in visualising the main features of buildings. And even game development software can be used to this effect, as is shown by the reconstruction of the Maya city of Chunchucmil by David Hixson using the Unreal Engine software (see en.wikipedia.org/wiki/File:Chunchucmil-reconstruction-2.jpg).

A relatively new development is the way in which photographs can now be combined into 3D images. The technique is quite simple, as it only needs overlapping photos. The information on focal distances contained in digital photographs is then used to compare overlapping photos and decide how they should fit together. This information can be used further to create 3D models from the pictures through photogrammetric techniques. This development will undoubtedly become very important in the near future, too, as it is a highly affordable alternative to traditional photogrammetry that can be used in all kinds of settings, including excavation pits. We can therefore expect an increasing synergy between the approaches used in architectural reconstruction, landform modelling and subsurface modelling, as they all can strengthen each other.

A REVOLUTION IN THEORY?

There is no doubt that powerful image capturing, manipulation and visualisation techniques will continue to attract the attention of landscape archaeologists in the near future. And in fact, we need time and experience with these new techniques to fully judge their applicability to the questions of prediction, detection, analysis and visualisation. But the question is: do all these developments really help us in interpreting the landscape and the archaeological record, and thus lead to new insights? Obviously, the detection of new, unsuspected features through, e.g. geophysics or LiDAR can lead to new insights, and a number of new descriptors of the landscape can easily be obtained from especially digital terrain models, including the now almost common calculations of viewsheds and cost surfaces in GIS.

One of the most interesting lines of research in landscape archaeological GIS in this respect has been the development of spatial descriptions of visibility of the landscape, in particular the work done by Llobera (2001, 2003, 2007). The use of viewshed techniques has now become standard issue in many GIS-based studies (see fig. 2), but it has also shown us how complex it is to use visibility as a measure of perception of the landscape. First of all, it is only one of the senses we use, be it probably the most important one. Modelling soundscapes (Mlekuž 2004) has not really caught on in landscape archaeology. The lack of reliable data on palaeo-vegetation is also an important issue to take into account here, especially since a human takes up a very small portion of space; views can be blocked very easily by a strategically placed obstacle, a fact that has been used to effect for many military and political purposes. Nevertheless, the analysis of the visual characteristics of the space that people lived in is often thought to contribute to a better understanding of how the landscape is structured from a human point of view. The use of the well-known Higuchi visibility ranges (see e.g. Wheatley & Gillings 2000) has therefore become quite popular in archaeological GIS-studies. And as there are very few researchers interested in both phenomenology and quantitative modelling, perhaps it would be good to look at a combination of both approaches, especially since virtual reality has now also become a major business in computing in and outside archaeology.

What is true for viewshed analysis probably becomes even more serious when we are looking at

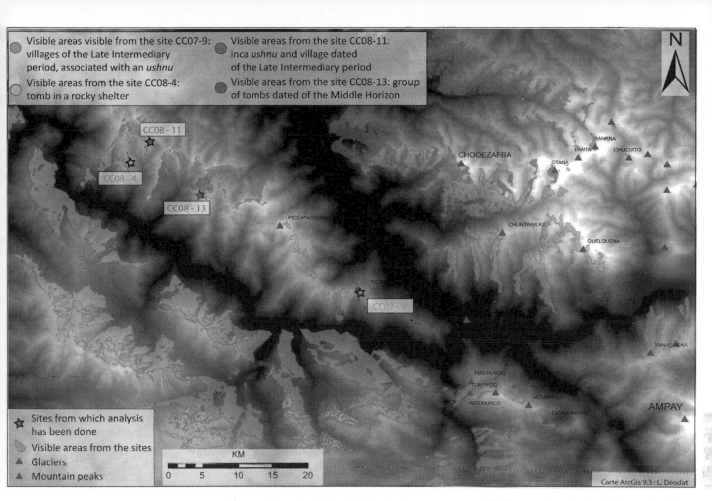

Figure 2. Example of viewshed analysis to map areas visible from archaeological sites. Taken from: Déodat and LeCocq (this volume). *See also 5.2 Figure 8 in the full colour section in this book*

modelling of prehistoric travel routes (see fig. 3). While it has in fact been one of the earliest archaeological uses of GIS to model terrain accessibility through cost surfaces (Gaffney & Stančič 1991), and finding the least cost paths to try to find out where people might actually have travelled, it is a field of research that has been almost completely neglected outside the community of archaeologists working with GIS, as is for example demonstrated in the papers in Snead et al. (2009). We know that there is much more to path finding than just a straightforward least cost path calculation, but no one seems to know what factors are actually influencing travel patterns. Furthermore, a whole field of research in space syntax (Hillier & Hanson 1984) and network analysis studies (Lock & Pouncett 2007) has mostly bypassed landscape archaeology and would need to be integrated with standard GIS analyses, and probably also with agent-based modelling, in order to make more progress in this respect.

New modelling techniques also try to capture the dynamics of complex socio-natural systems. As with all the other techniques shown, the development in this realm has gone fast over the past ten years. Tools that are currently available, for example for agent-based modelling, are very powerful and can easily be used with spatial data sets. Partial success can be observed, especially in cases where we can try to link human activities to natural processes like erosion. A much more difficult and controversial issue however is found when we consider the effects of the environment on human behaviour itself.

The debate on the value of dynamic simulation models however is far from new: in the 1970s a number of computer-based studies appeared (Doran 1970; Hodder 1977; Sabloff 1981) that claimed to model socio-natural systems into some detail, and these raised a wave of criticism. This is partly because all things quantitative fell out of grace with a large part of the archaeological community in the late 1980s with the rise of post-processual archaeology; but perhaps also because we do not really know whether these models offer realistic descriptions of how socio-natural processes operated in the past. There is an uneasy marriage between modern, post-processual and holistic archaeology and quantified modelling. On the one hand we have our (sometimes vague) ideas on how people lived and behaved within their cultural background, and how this related to their socio-cultural and natural surroundings. On the other hand we have our limited data sets of the environmental settings in prehistory, and in many cases an even more limited data set about the socio-cultural settings. Getting from relatively simple land use models based on relatively simple theories about subsistence economy to conclusions about the identity of prehistoric people may then seem a bit of a tall order, regardless whether we use quantitative modelling or not. However, I want to argue that this kind of modelling can certainly be used to effect, provided we do not use it as the ultimate means to reconstruct the past.

In other disciplines quantitative modelling is primarily seen as one of the potential tools to develop theories and interpretation, whereas archaeologists are usually much more concerned with the formulation of theory from logical argument. The scientific research process can in essence be seen as a slow movement along the lines shown in figure 4, sometimes going back and forth between the different stag-

Figure 3. Example of least cost path calculations using different specifications of movement costs, in a study area in Cappadocia, Turkey. Adapted from Verhagen & Polla (2010). *See also full colour section in this book*

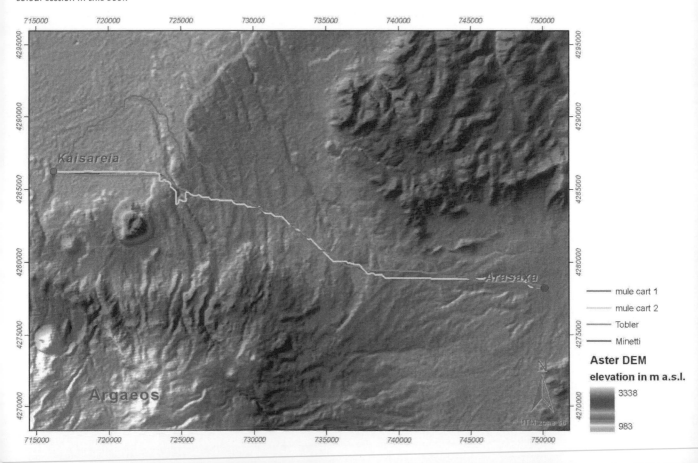

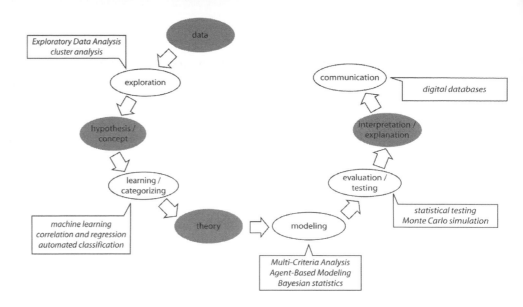

Figure 4. The position of modelling in the scientific research process. Quantitative methods and models can play a role in almost any stage of scientific research, but are best applied in the stages where theory is developed and made suitable for testing. The separation between hypothesis and theory however may not always be as clear as suggested. Adapted from O'Sullivan & Gahegan (2007).

es of theory formation and interpretation. In this particular scheme, the role of statistics is deliberately included to show what can be done with formal, computer-based methods, much more than is normally understood by archaeologists. However, using these techniques to effect takes a lot of effort since many of them are quite complex and need training to be properly executed and interpreted.

Most important in the scheme however is the position of modelling: it is situated after the formulation of theory. Modelling is a way of making a theory explicit and to open it up to testing. The true power of modelling techniques nowadays is that they allow us to quickly generate a number of scenarios and compare these to actual empirical data. Obviously, this also means that the role of modelling is limited to cases where we can specify scenarios quantitatively, and where we at least have the potential to test the models with data. If this is not the case quantitative models will be no better than speculation.

The need for quantitative modelling was eloquently phrased by Joshua Epstein (2008), one of the pioneers of agent-based modelling:

> The choice, then, is not whether to build models; it's whether to build *explicit* ones. In *explicit* models, assumptions are laid out in detail, so we can study exactly what they entail. On these assumptions, *this* sort of thing happens. When you alter the assumptions *that* is what happens. By writing explicit models, you let others replicate your results.
> (...)
> It is important to note that (...) models do *not* obviate the need for judgment. However, by revealing tradeoffs, uncertainties, and sensitivities, models can *discipline the dialogue* about options and make unavoidable judgments more considered (...).
> (...)
> Models can surprise us, make us curious, and lead to new questions (...).

He basically sees an important role for models as exploratory tools to judge the value of theories and concepts, and makes a plea for looking at these models as heuristic tools. However, this way of thinking about modelling has not made a real breakthrough in landscape archaeological research yet. The best-known example of this approach in landscape archaeology is probably the application of agent-based modelling techniques by Kohler et al. (2005) to simulate and better understand the population dynamics of the American south-west. In their case however, they had excellent data at their disposal to test the simulated population development.

So it seems that landscape archaeology is caught somewhere between the shining perspective of using quantitative models as exploratory, heuristic devices that can be used for a number of different research questions, and the technical and theoretical limitations of what we are trying to do with them.

A REVOLUTION IN ARCHAEOLOGICAL COMPUTING?

We also have to be aware that we are dealing with issues that are not only complex from a theoretical point of view, but from a computing point of view as well. There is an enormous amount of software tools available, but as a general rule, they will not work together seamlessly. In practice, an experienced 'computer archaeologist' will always have more than one option at hand to do the required task. But this comes at the price of having to transfer data between different packages, and to be able to judge what all these packages can do for you. The lack of good data standards both inside and outside archaeology is something that we have to live and deal with as best as we can.

More serious is the fact that the archaeological community has very few people who can do a decent job in programming. This is really something that should be at the top of our priority list if archaeologists do not want to be stuck with tools that were originally designed for completely different questions. The least cost path issue is an example of this: it has only dawned on archaeologists fairly recently that the standard tools for calculating these paths can produce different results, depending on the software package chosen (Herzog & Posluschny 2008). A tool that allows us to use all different techniques available and compare these still has to be released.

It seems that many computing solutions developed by archaeologists have been made on an ad hoc basis, and lack a firm integration in existing software platforms. Some quite useful software tools are therefore no longer available, like the agent-based modelling tools for GRASS made by Lake (2000). This is a real problem, since developing software is quite expensive, and maintaining it will always be cheaper than having to reinvent the wheel, even when in most cases it will mean redesigning the software every 5 to 10 years. We therefore need to prepare ourselves for a revolution in archaeological computing. Ignoring how to make software is not going to get us anywhere, and our best bet currently is to use the power of open source software development. While there is still a lot to be improved with regard to open source software, especially where it concerns user friendliness, it is the only way in which archaeologists can actively participate in the way in which software develops. Open source software plays a key role in making innovative methods and tools available to the scientific community. A main reason why a number of useful tools are not more widespread is because of software licensing costs and the fact that closed source software cannot be modified, enhanced and then freely shared by its users, thereby severely limiting its usefulness for science. A first step in this direction has been the establishment of the Open Archaeology website, sponsored by Oxford Archaeology (www.openarchaeology.net). However, this approach will only work if archaeologists start to contribute actively to its maintenance and provide ideas for further development.

ACKNOWLEDGMENTS

The author would like to thank the organisers for inviting him to present this paper at the LAC2010 conference. Figures 1 and 2 were made available by kind permission from Rachel Opitz and Laure Déodat. Furthermore, the author would like to thank two anonymous reviewers for reading and commenting on the paper.

REFERENCES

Aqdus, S.A., W.S. Hanson & J. Drummond. 2007. Finding archaeological cropmarks: a hyperspectral approach. *Remote sensing for environmental monitoring, GIS applications, and geology VII* 674908(1-11).

Barceló, J., M. Forte & D.H. Sanders (eds.) 2000. Virtual reality in archaeology. BAR International Series 843, Oxford.

Déodat, L. & P. LeCocq. 2011. Using Google Earth and GIS to survey in the Peruvian Andes. In Kluiving, S.J. & Guttmann, E.G.B. (eds.) 2012 *LAC2010 Proceedings, Landscape & Heritage Series*. Amsterdam University Press.

Devereux, B.J., G.S. Amable, P. Crow & A.D. Cliff. 2005. The potential of airborne LiDAR for detection of archaeological features onder woodland canopies. *Antiquity* 79, 648-60.

Doneus, M., C. Briese, M. Fera & M. Janner. 2008. Archaeological prospection of forested areas using full-waveform airborne laser scanning. *Journal of Archaeological Science* 35, 882-893.

Eisenbeiss, H. 2009. A Model Helicopter over Pichango Alto – Comparison of Terrestrial Laser Scanning and Aerial Photogrammetry. In Reindel, M. & Wagner, G.A. (eds.), *New Technologies for Archaeology: Multidisciplinary Investigations in Palpa and Nasca, Peru*, 339-358. Springer, Berlin.

Ejstrud, B. 2005. Taphonomic Models: Using Dempster-Shafer theory to assess the quality of archaeological data and indicative models. In Leusen, M. van & Kamermans, H. (eds.), *Predictive Modelling for Archaeological Heritage Management: a research agenda*, 183-194. Rijksdienst voor het Oudheidkundig Bodemonderzoek, Amersfoort.

Epstein, J. 2008. *Why Model?* The Journal of Artificial Societies and Social Simulation 11: http://jasss.soc.surrey.ac.uk/11/4/12.html

Fisher, P., J. Wood & T. Cheng. 2004. Where is Helvellyn? Fuzziness of multi-scale landscape morphometry. *Transactions of the Institute of British Geographers* 29, 106-128.

Forte, M. & A. Siliotti. 1997. *Virtual archaeology: Re-creating Ancient Worlds*. Harry N. Abrams, New York.

Fowler, M.J.F. 2002. Satellite remote sensing and archaeology: a comparative study of satellite imagery of the environs of Figsbury Ring, Wiltshire. *Archaeological Prospection* 9, 55-69.

Gaffney, V. & Z. Stančič. 1991. *GIS approaches to regional analysis: A case study of the island of Hvar*. Znanstveni inštitut Filozofske fakultete, Ljubljana.

Gaffney, V., K. Thomson & S. Fitch (eds.) 2007. *Mapping Doggerland: The Mesolithic Landscapes of the Southern North Sea*. Archaeopress, Oxford.

Hillier, B. & J. Hanson. 1984. *The Social Logic of Space*. Cambridge University Press, Cambridge.

Hengl, T. & H.I. Reuter (eds.) 2009. *Geomorphometry. Concepts, Software, Applications*. Elsevier, Amsterdam.

Herzog, I. & A. Posluschny. 2008. *Tilt – Slope-Dependent Least Cost Path Calculations Revisited*. Paper presented at the 36th International Conference on Computer Applications and Quantitative Methods in Archaeology, Budapest.

Hodder, I. (ed.) 1977. *Simulation Studies in Archaeology*. Cambridge University Press, Cambridge.

Judge, W.J. & L. Sebastian (eds.) 1988. *Quantifying the Present and Predicting the Past. Theory, Method and Application of Archaeological Predictive Modeling*. U.S. Department of the Interior, Bureau of Land Management, Denver.

Kohler, T.A., G.J. Gumerman & R.G. Reynolds. 2005. Simulating ancient societies. *Scientific American* 293, 76-82.

Kvamme, K.L. 2008. Archaeological prospecting at the Double Ditch State Historic Site, North Dakota, USA. *Archaeological Prospection* 15, 62-79.

Lake, M.W. 2000. MAGICAL Computer Simulation of Mesolithic Foraging In Kohler, T.A. & Gumerman, G.J. (eds.), *Dynamics in Human and Primate Societies: Agent-Based Modeling of Social and Spatial Processes*, 107-143. Oxford University Press, New York.

Llobera, M. 2001. Building Past Landscape Perception with GIS: Understanding Topographic Prominence. *Journal of Archaeological Science* 28, 1005-1014.

Leusen, M. van & H. Kamermans (eds.) 2005. *Predictive Modelling for Archaeological Heritage Management: a research agenda*. Rijksdienst voor het Oudheidkundig Bodemonderzoek, Amersfoort.

Leusen, M. van, A.R. Millard & B. Ducke. 2009. Dealing With Uncertainty in Archaeological Prediction. In Kamermans, H., Leusen, M. van & Verhagen, P. (eds.), *Archaeological Prediction and Risk Management*, 123-160. Leiden University Press, Leiden.

Llobera, M. 2003. Extending GIS-based visual analysis: the concept of visualscapes. *International Journal of Geographical Information Science* 17, 25-48.

Llobera, M. 2007. Reconstructing Visual Landscapes. *World Archaeology* 39, 51-69.

Lock, G. & J. Pouncett. 2007. Network analysis in archaeology. In Clark, J.T. & Hagemeister, E.M. (eds.), *Digital Discovery. Exploring New Frontiers in Human Heritage. CAA 2006: Computer Applications and Quantitative Methods in Archaeology*, 123-154. Archaeolingua, Budapest.

Mlekuž, D. 2004. Listening to Landscapes: Modelling Past Soundscapes in GIS. *Internet Archaeology* 19, http://intarch.ac.uk/journal/issue19/mlekuz_index.html

Opitz, R., Nuninger, L. & C. Fruchart. 2011. Thinking topographically about the landscape around Besançon (Doubs, France). In Kluiving, S.J. & Guttmann (eds.) *LAC2010 Proceedings, Landscape & Heritage Series*. Amsterdam University Press.

O'Sullivan, D. & M.N. Gahegan. 2007. *Issues around verification, validation, calibration, and confirmation of agent-based models of complex spatial systems*. Paper presented at the NSF-ESRC Workshop on Agent-based Models of Complex Spatial Systems, Santa Barbara. http://www.m8p.com.

Reilly, P. 1990. Towards a virtual archaeology. In Lockyear, K. & Rahtz, S. (eds.), *Computer Applications in Archaeology*, 133-139. BAR International Series 565, Oxford.

Sabloff, J.A. (ed.) 1981. *Simulations in Archaeology*. University of New Mexico Press, Albuquerque.

Scollar, I. A. Tabbagh, A. Hesse & I. Herzog. 1990. *Archaeological prospecting and remote sensing*. Cambridge University Press, Cambridge.

Snead, J.E., C.L. Erickson & J.A. Darling (eds.) 2009. *Landscapes of Movement. Trails, Paths, and Roads in Anthropological Perspective*. University of Pennsylvania Museum of Archaeology and Anthropology, Philadelphia.

Verhagen, P., H. Kamermans, M. van Leusen & B. Ducke. 2010. New developments in archaeological predictive modelling. In Bloemers, T., Kars, H., Valk, A. van der & Wijnen, M. (eds.), *The Cultural Landscape & Heritage Paradox. Protection and Development of the Dutch Archaeological-Historical Landscape and its European Dimension*, 431-444. Amsterdam University Press, Amsterdam.

Verhagen, P. & S. Polla. 2010. *Climb every mountain, ford every stream. Combining least cost path modelling and theories on (pre-)historic travel*. Paper presented at the PECSRL 2010 conference, Riga/Liepaja 23-27 August.

Wagtendonk, A., P. Verhagen, S. Soetens, K. Jeneson & M. de Kleijn. 2009. Past in Place: The Role of Geo-ICT in Present-day Archaeology. In Scholten, H.J., Velde, R. van de & Manen, N. van (eds.), *Geospatial Technology and the Role of Location in Science*, 59-86. Springer, Dordrecht.

Wheatley, D. & M. Gillings. 2000. Vision, perception and GIS: developing enriched approaches to the study of archaeological visibility. In Lock, G. (ed.), *Beyond the Map*, 1-27. IOS Press/Ohmsha, Amsterdam.

Winterbottom, S.J. & T. Dawson. 2005. Airborne multi-spectral prospection for buried archaeology in mobile sand dominated systems. *Archaeological Prospection* 12, 205-219.

1.1 Figure 5. The Cist with the Engraved Circled Cross

1.2 Figure 6. Simplified canal system

1.2 Figure 7. Relative certainty of water delivery

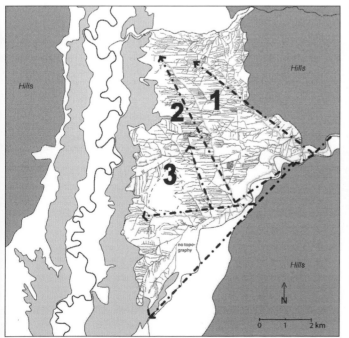

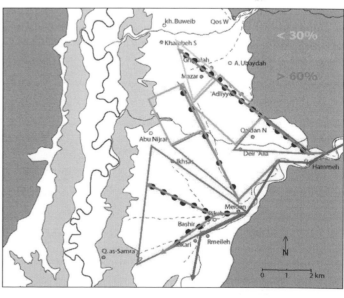

the Bryony　Milešovka Hill　Radovesice tailings pile　Panorama of the Czech Central Mountains　Bořeň Hill　Village of Liběšice　Kaňkov Hill　Želenice Hill　Zlatník Hill　Red Hill

Town of Bílina　Bílina Mine　Mine pit lake with artefacts of the swamped area (power pole)

1.3 Figure 5. Analysis of visual and perceptive characteristics – panorama: South-eastern view from the reference point no. 17. (Hájek et al. 2009). Corel Draw X4: Matáková 2009.

1.3 Figure 6. Register of historical and cultural values, register of road system based on analysis of old maps and archive materials. Part of the Map of the historical communication network and minor landmarks (Hájek et al. 2009), ArcGIS map: Matáková 2009.

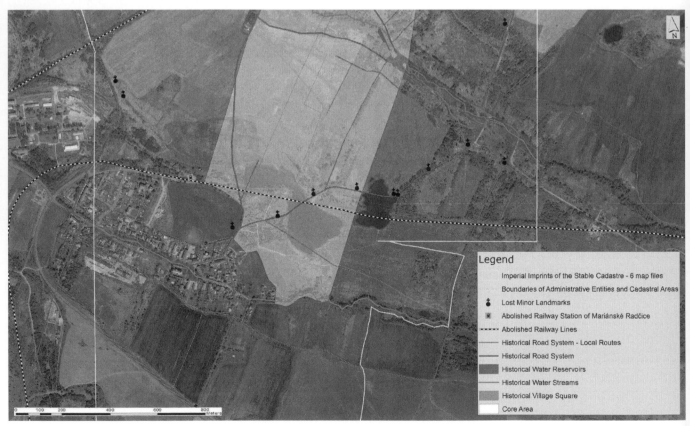

Legend

Imperial Imprints of the Stable Cadastre - 6 map files

Boundaries of Administrative Entities and Cadastral Areas

Lost Minor Landmarks

Abolished Railway Station of Mariánské Radčice

Abolished Railway Lines

Historical Road System - Local Routes

Historical Road System

Historical Water Reservoirs

Historical Water Streams

Historical Village Square

Core Area

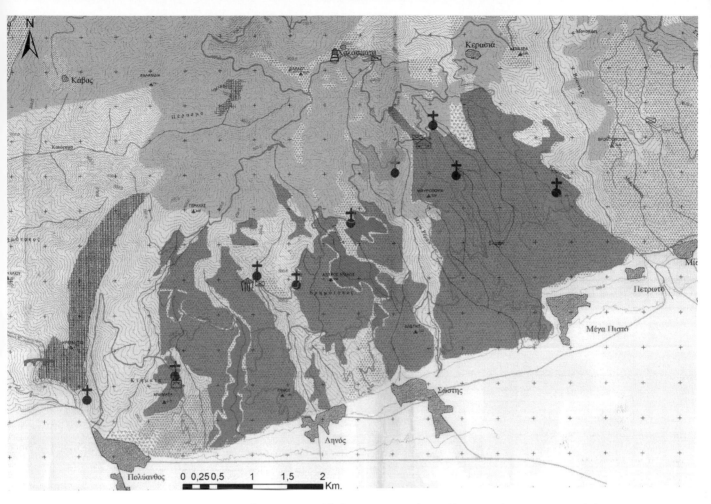

1.4 Figure 2. Map with the monuments located in areas where remnants of vine and wheat cultures and pastures were spotted (red= cultivated areas, purple= pastures). North is directed towards the top of the map.

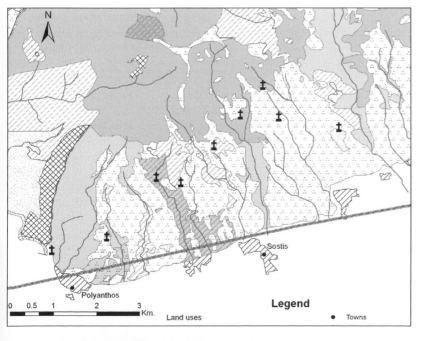

1.4 Figure 3. Vegetation map showing the location of the Byzantine monuments, the monastic centres located near water springs and streams and the afforestated areas. Source: The Research Program Pythagoras II – Environment.

Legend

Land uses

▨	Partly forested oak woodland	● Towns
▨	Partly forested beech woodland	✝ Monasteries
▨	Barren	— Streams
▨	Agricultural areas	═ Via Egnatia
	Afforestations	▲ Springs
	Evergreen broadleaved woodland	
	Oak woodland	
	Bare lands	
	River bank	
	Mixed woodland of evergreen broadleaved-coniferous plantations	
▨	Mixed oak-evergr. broadleaved woodland	
	Mixed woodland of oak-coniferous plantations	
▨	Mixed oak-beech woodland	
▨	Settlements	

1.5 Figure 2. Map of Ekrem Hakkı Ayverdi.
Produced by using Ekrem Hakkı Ayverdi maps.

LEGEND

- Garden
- Building
- Empty space
- Vegetable garden
- Mosque gardens
- Pasture
- Home gardens
- Public gardens
- Square
- Graveyard
- School gardens
- Private gardens
- Palace gardens

0 150 300   600   900   1.200
Meters

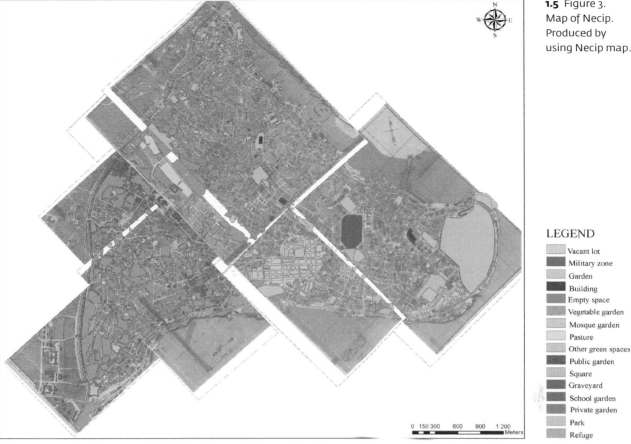

1.5 Figure 3. Map of Necip. Produced by using Necip map.

LEGEND

- Vacant lot
- Military zone
- Garden
- Building
- Empty space
- Vegetable garden
- Mosque garden
- Pasture
- Other green spaces
- Public garden
- Square
- Graveyard
- School garden
- Private garden
- Park
- Refuge

0 150 300   600   900   1.200
Meters

1.5 Figure 4. Produced by using present maps (1965).

Eyüp

Beyoğlu

Haliç

Zeytinburnu

Marmara Denizi

LEGEND

- Court
- Building
- Mosque garden
- Other green spaces
- Graveyard
- Refuge
- Sporting area

0 155 310   620   930   1.240
Meters

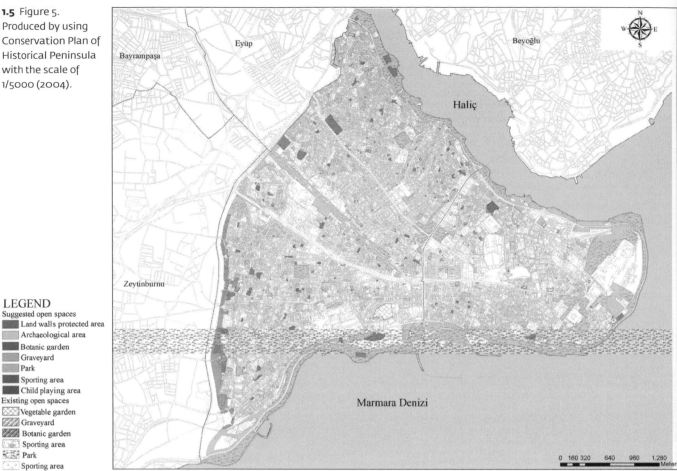

1.5 Figure 5.
Produced by using
Conservation Plan of
Historical Peninsula
with the scale of
1/5000 (2004).

LEGEND

Suggested open spaces
- Land walls protected area
- Archaeological area
- Botanic garden
- Graveyard
- Park
- Sporting area
- Child playing area

Existing open spaces
- Vegetable garden
- Graveyard
- Botanic garden
- Sporting area
- Park
- Sporting area

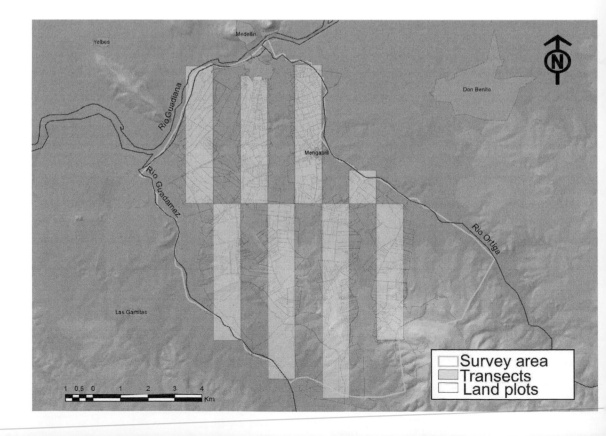

1.6 Figure 5.
Definition of the
survey area.

1.6 Figure 7. Overall picture of survey areas of 2009 campaign. Gray lines represent GPS tracks. White polygons are areas of interest.

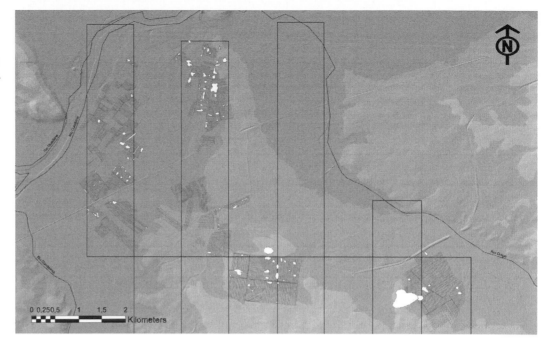

1.7 Figure 1. Plan of Palmse manor centre (1753). Source: Pahlen, G.F. 1753. Plan der Hoflage von dem Guthe Palms.

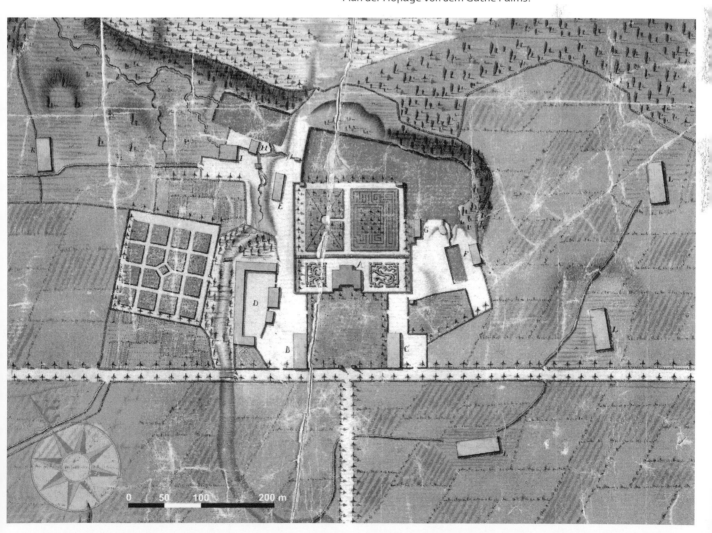

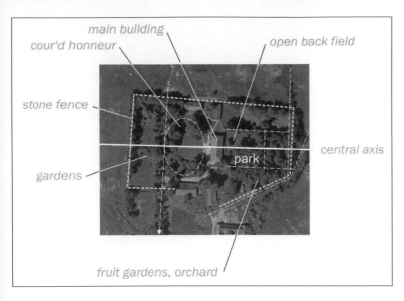

1.7 Figure 3. Schematic map of spatial composition of Vasta manor centre.
Source: Nurme 2007.

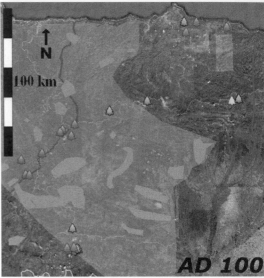

1.8 Figure 1: Location of the *Conventus Asturum* and its gold deposits.

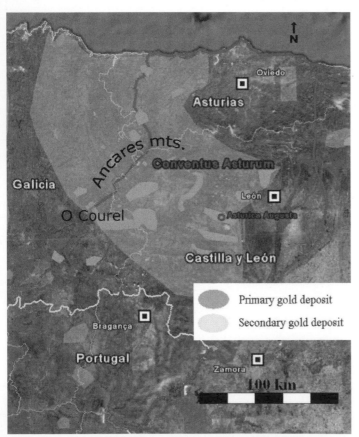

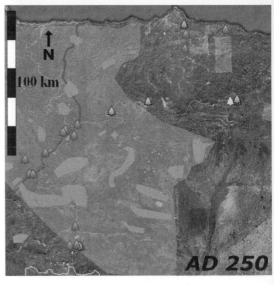

1.8 Figure 3: Forest evolution as attested by pollen records.

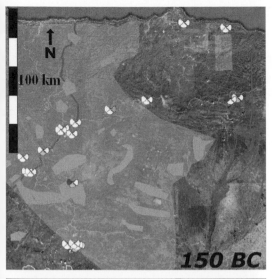

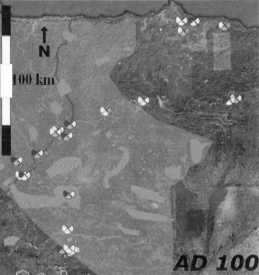

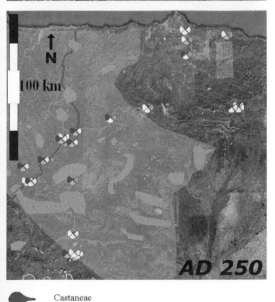

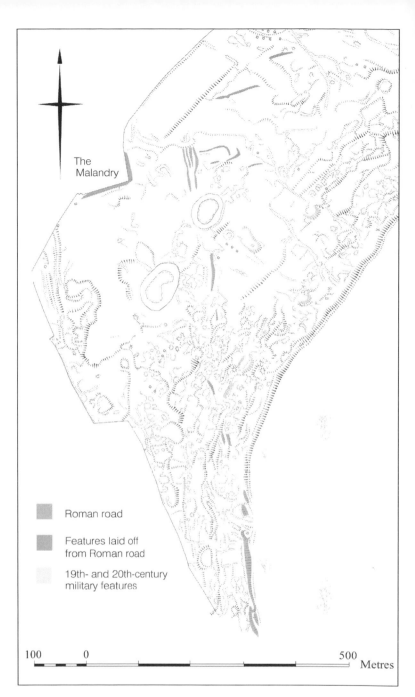

Castaneae

Cerealia

Juglans

1.8 Figure 4: Absence/presence of cultivated species as attested by pollen records.

1.9 Figure 4. On the South Common, Lincoln, the route of the Roman Ermine Street is delimited by earthworks. At right angles to this are fragments of an extensive field system and the boundary of a medieval hospital, The Malandry (© English Heritage).

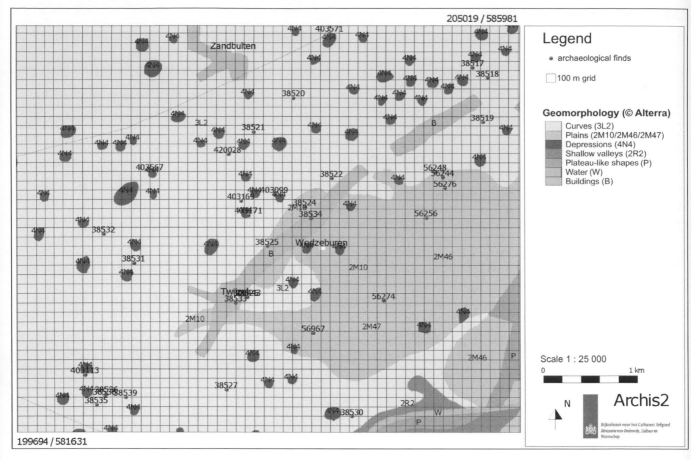

1.10 Figure 5. Example of one of the maps made in Archis, showing pingo scars as water-filled depressions on the geomorphological map. Archaeological finds present in the Archis database are shown on the map as red dots. One square on the map represents one hectare (100 x 100m). Created in Archis, Rijksdienst voor Cultureel Erfgoed.

2.1 Figure 3. Aerial photograph of the apron of the opencast pit Jänschwalde with archaeological longitudinal sections and ground plans of charcoal piles (photo: H. Rösler).

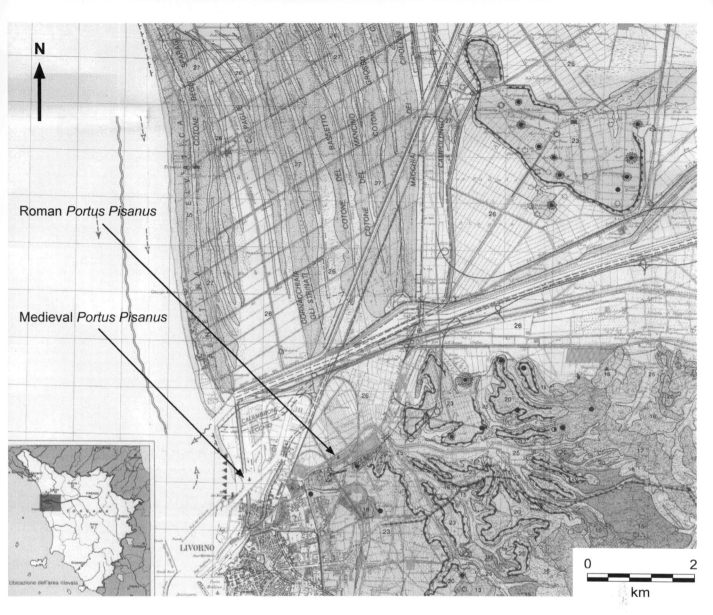

N

Roman *Portus Pisanus*

Medieval *Portus Pisanus*

LIVORNO

0   2
km

2.2 Figure 4. Coltano and *Portus Pisanus* area shown in the lower Arno Valley in a detail of the geomorphological map of Mazzanti (1994). The numbers identify geomorphological units (Mazzanti, 1994) and the colored symbols identify archaeological sites (Pasquinucci, 1994) dated to prehistory (black), archaic and classical period (red), middle ages (violet), and modern times (green).

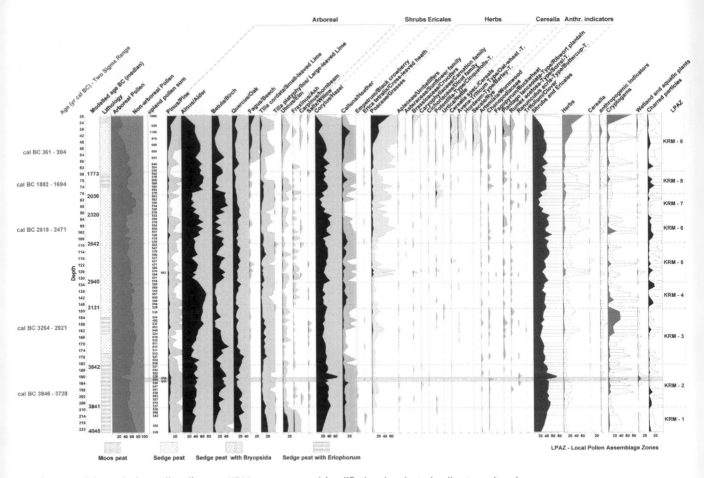

2.3 Figure 5. High resolution pollen diagram KRM 0.30 – 2.24 m (simplified, only selected pollen types/taxa).

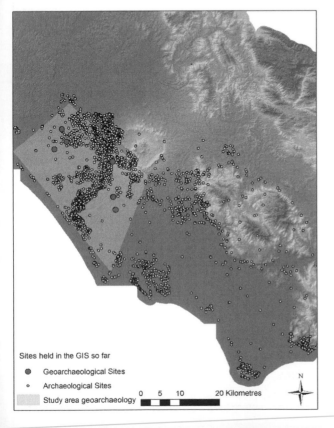

2.4 Figure 1. Overview study area with geoarchaeological and archaeological sites entered so far.

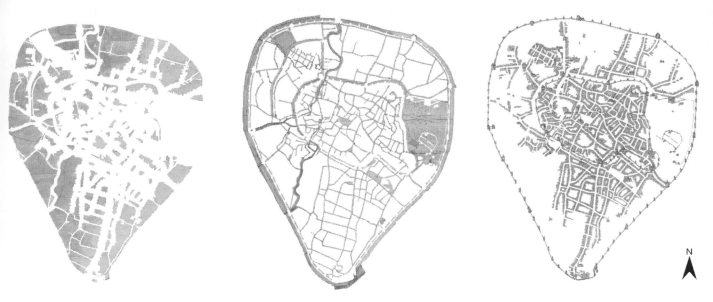

2.5 Figure 2. Cartographic analysis of the 16th-century town plan of Brussels by Jacob of Deventer, realized in a scale of approximately 1:8600 (© Royal Library of Belgium, Brussels).

From left to right are shown respectively the open spaces (36%), the road network, watercourses and green spaces (34%) and buildings (30%).

2.5 Figure 3: a) Soil profile of Dark Earth on the site of *Hôtel de Lalaing-Hoogstraeten*; b) Graph showing enhanced phosphorus levels for the Dark Earth units (US 7338 and US 7321); c) Granulometric data showing the high similarity between the units US 7338 and US 7321 and the natural soil (US 7340), suggesting they share the same matrix; d) Thin section

micrograph showing phosphorus-rich excrement proving the addition of manure (plain polarised light); e) Thin section micrograph showing a textural pedofeature enabling its identification as former at least temporary unprotected topsoil (plain polarised light); f) Thin section micrograph showing dendritic phytoliths (plain polarised light).

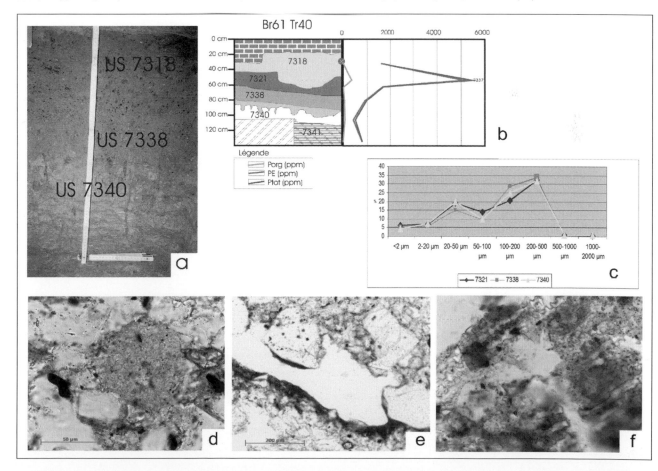

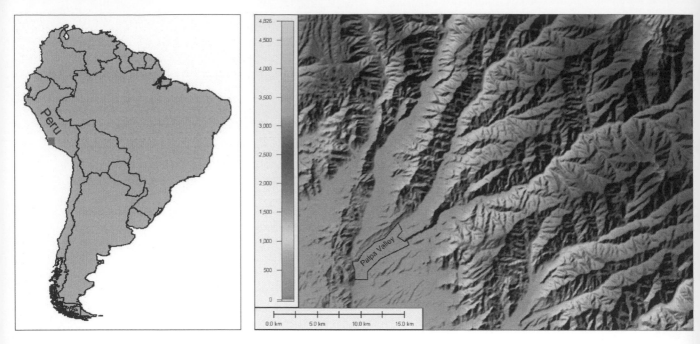

3.1 Figure 1. Map showing the location of the Palpa Valley. The two valleys reaching the Palpa Valley from the east are the valleys of Rio Palpa and Rio Vizcas.

3.2 Figure 5. View of the mountainous zone of Monte Novu in the south-eastern part of the village territory of Fonni with *Cuile su Seragu* visible in the centre of the photograph.

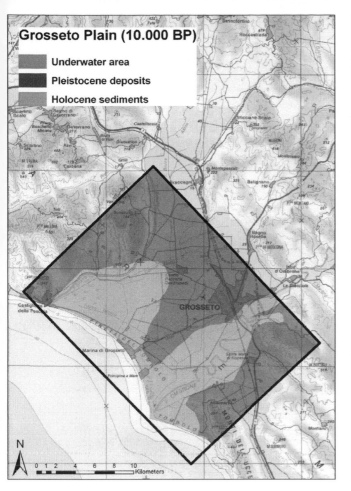

Grosseto Plain (10.000 BP)

Underwater area
Pleistocene deposits
Holocene sediments

GROSSETO

N

0 1 2   4   6   8   10
Kilometers

Grosseto Plain (6000 BP)

Underwater area
Pleistocene deposits
Holocene sediments

GROSSETO

N

0 1 2   4   6   8   10
Kilometers

3.3 Figure 2. A GIS reconstruction of palaeogeographical settings of Grosseto Plain during 10,000 BP (after Bellotti 2004).

3.3 Figure 3. A GIS reconstruction of palaeogeographical settings of Grosseto Plain during 6,000 BP (after Bellotti 2004).

3.3 Figure 5. Distribution of off-site evidence in the Grosseto area.

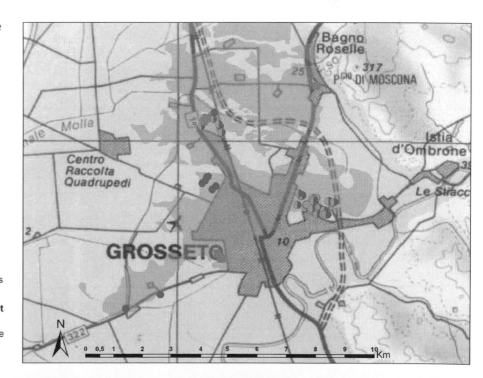

Legend
Pleist. terraces

Prehistoric Topographical Unit
● Holocene
◑ Pleist/Holocene
○ Pleistocene
■ Generic

GROSSETO

N

0 0,5 1   2   3   4   5   6   7   8   9   10 Km

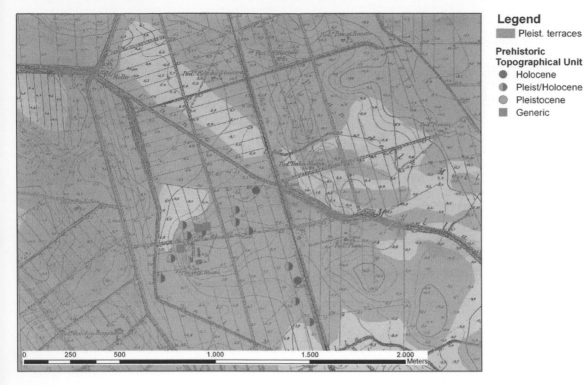

3.3 Figure 7. An example of data integration: off-site distribution and Pleistocene deposits overlaid on the historical topographical maps (1930) which show a variety of morphologies.

3.3 Figure 8. An example of data integration: off-site distribution and Pleistocene deposits overlaid on the historical aerial photograph. The contour lines derived by the historical topographical maps (1930) allow us to appreciate the little hillock and to imagine the inlet when the sea level was 6 metres above the present-day level. We can also appreciate damp-marks and soil marks.

Legend

Prehistoric Topographical Unit
● Holocene
◑ Pleist/Holocene
○ Pleistocene
■ Generic

Contour lines 1930
— 9,000
— 8,500
— 8,000
— 7,500
— 7,000
— 6,500
— 6,000
— 5,500
— 5,000
— 4,500

N

4.1 Figure 4. Rutherfords Creek, western NSW, showing the location of scalds (i), hearths (ii) and analysed stone artefact assemblages from randomly selected scalds (iii).

(i) Scalds

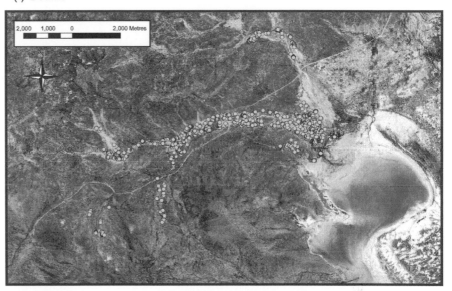

(ii) Hearths

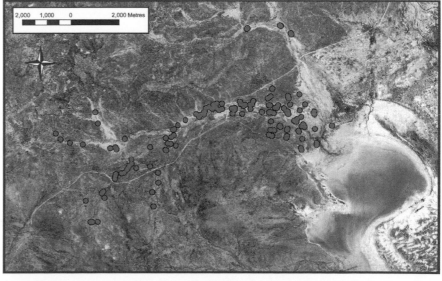

(iii) Analysed stone artefact assemblages from randomly selected scalds

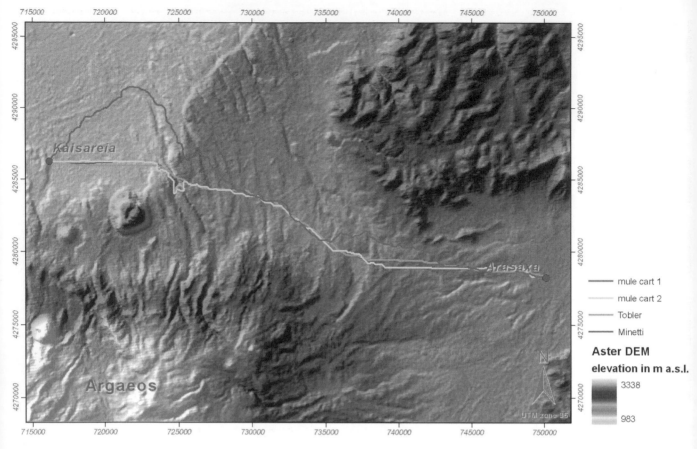

5.1 Figure 3. Example of least cost path calculations using different specifications of movement costs, in a study area in Cappadocia, Turkey. Adapted from Verhagen & Polla (2010).

Right page >

5.2 Figure 6. Localisation of sites in relation with ecological floors, extracted from GIS.

5.2 Figure 7. Example of visibility analyses between two or more sites, extracted from GIS.

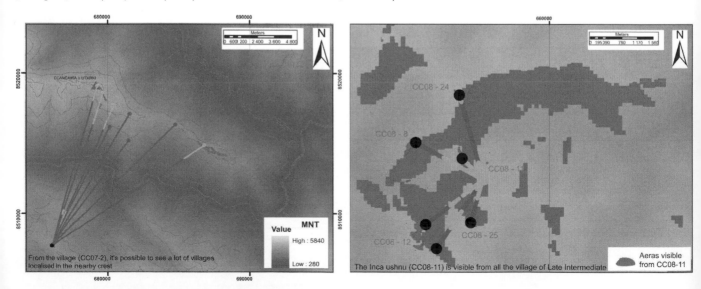

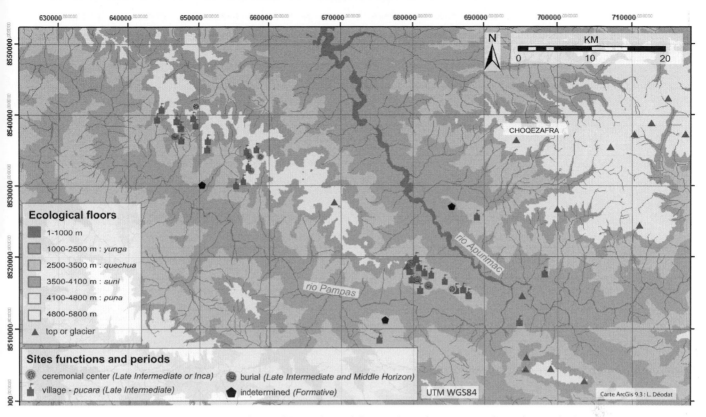

The schematic drawing above shows the different floors of the Andes and their ecological resources. It also indicates the barter system in which llamas or mule caravans are used to transport products between zones. Archaeological sites of the Formative period are located in the lowlands and the fertile ecological floor (Yunga). Villages of the Late Intermediate are located in the highlands (Suni, Puna and top of the Quechua zone).

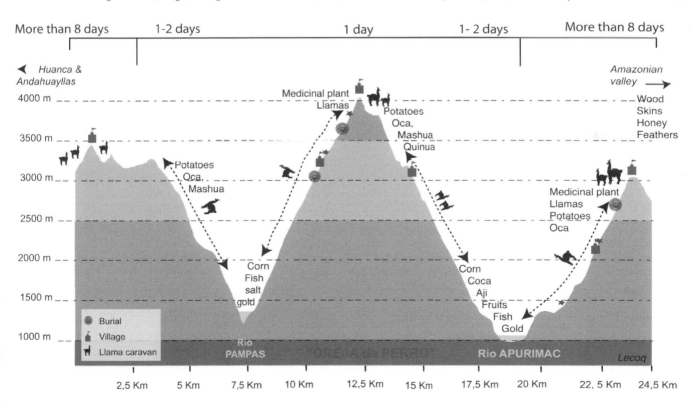

5.2 Figure 8. Visibility map based on a DTM, extracted from GIS.

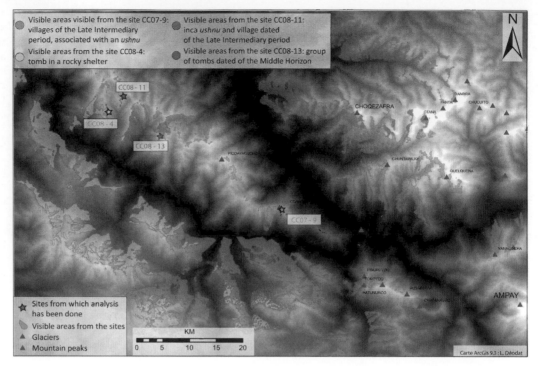

Visible areas visible from the site CC07-9: villages of the Late Intermediary period, associated with an *ushnu*

Visible areas from the site CC08-4: tomb in a rocky shelter

Visible areas from the site CC08-11: inca *ushnu* and village dated of the Late Intermediary period

Visible areas from the site CC08-13: group of tombs dated of the Middle Horizon

★ Sites from which analysis has been done

Visible areas from the sites

▲ Glaciers

▲ Mountain peaks

KM
0   5   10   15   20

Carte ArcGis 9.3 : L. Déodat

5.3 Figure 2. Example of viewshed: shade indicates the non-visible areas within the fifteen km radius around the analysed site, in this case *Singilia Barba*. Similar visibility analyses were carried out for each of the sites included in the project, as well as for a random distribution of sites, in order to compare them statistically.

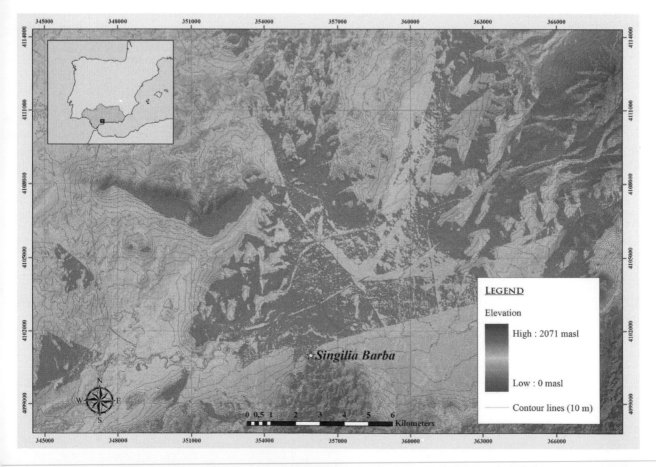

LEGEND

Elevation

High : 2071 masl

Low : 0 masl

—— Contour lines (10 m)

★Singilia Barba

0 0,5 1   2   3   4   5   6
Kilometers

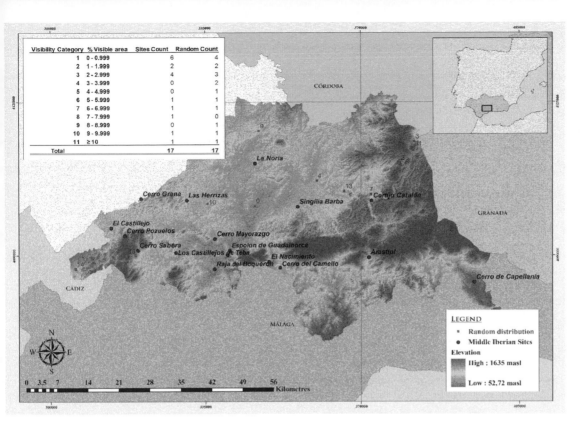

Visibility Category	% Visible area	Sites Count	Random Count
1	0 - 0.999	6	4
2	1 - 1.999	2	2
3	2 - 2.999	4	3
4	3 - 3.999	0	2
5	4 - 4.999	0	1
6	5 - 5.999	1	1
7	6 - 6.999	1	1
8	7 - 7.999	1	0
9	8 - 8.999	0	1
10	9 - 9.999	1	1
11	≥10	1	1
Total		17	17

5.3 Figure 4. Map showing both the random and archaeological distributions of sites during the Middle Iberian period, combined with a table showing the different categories of visible areas, and the number of sites counted in each classification. Both distributions were employed for investigating the randomness/relationship of the Iberian settlement pattern with regard to visibility and relative height.

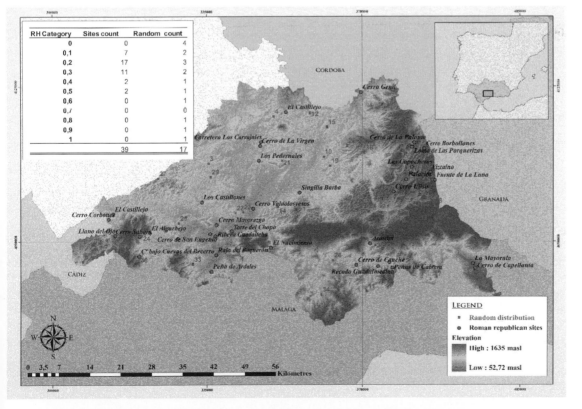

RH Category	Sites count	Random count
0	0	4
0,1	7	2
0,2	17	3
0,3	11	2
0,4	2	1
0,5	2	1
0,6	0	1
0,7	0	0
0,8	0	1
0,9	0	1
1	0	1
	39	17

5.3 Figure 5. Map showing both the random and archaeological distributions of sites during the Roman Republican period in the Antequera Depression, combined with a table showing the different categories of relative height, and the number of sites counted in each category. Both distributions were employed for investigating the randomness/relationship of the Republican settlement pattern with regard to visibility and relative height.

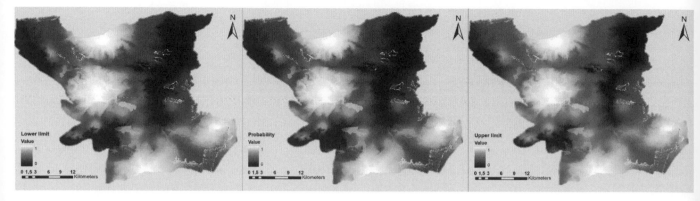

5.4 Figure 3. Probability map for the study area, with upper and lower values (at different scales) corresponding to the 95% confidence interval.

5.5 Figure 1. Data processing steps for the generation of the Local Relief Model (LRM). See text for the description of data processing. The image series shows a group of burial mounds in the Schönbuch area.

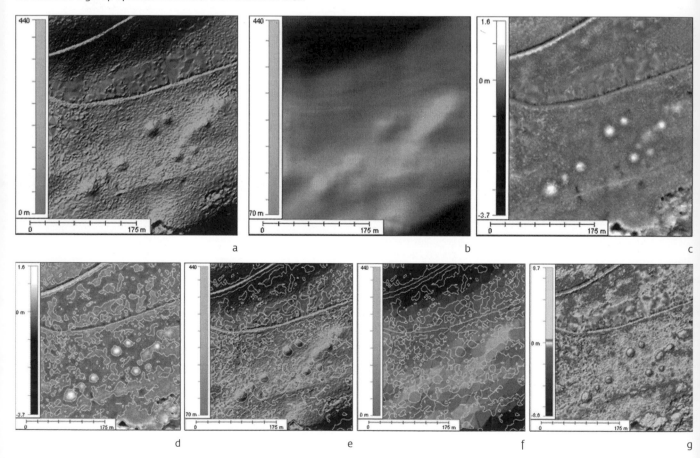

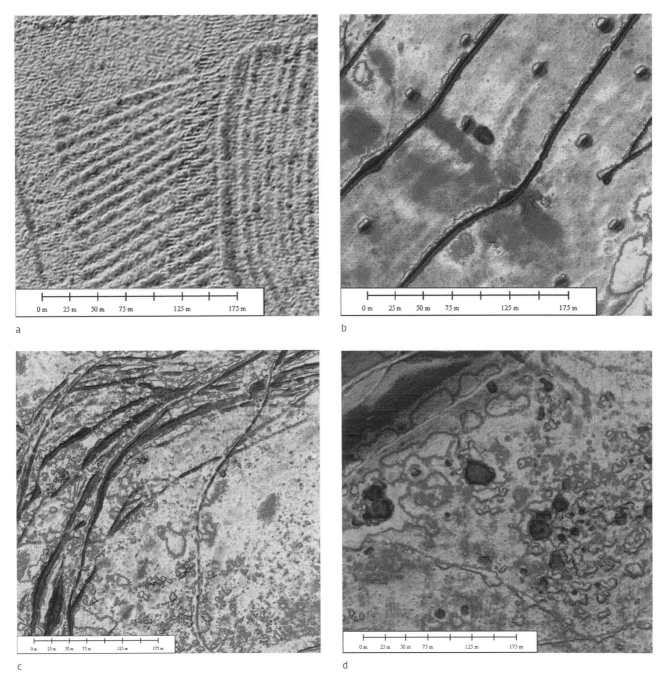

5.5 Figure 3. LRM colour maps showing (a) ridge and furrow, (b) kiln podia, (c) sunken roads and (d) mining traces.

5.6 Figure 4. Znióváralja, location of the old layout and uses of land. The photos display remaining elements: walls, fences, ruins of an edifice, a meadow and some old trees.

5.6 Figure 5. Bratislava, the digitalised map of the original layout of the former Jesuit garden, mid-18th century (*Hungarian National Archives*, T2 No. 1495.), stretched on the aerial photograph: 1. previous garden 2. formal buildings. The area is partly a green space today as well. The terrain levels and the location of the buttress are the same.

5.6 Figure 7. Szécsény, Franciscan cloister. Digitalised drawing of the garden of 1777 (*Historia Domus*, *Szécsény*), compared to the present state seen with red lines. 1. vegetable garden 2. fruit garden 3. fish ponds 4. canals. The cross section shows the presumed height difference of the embankment after 1950.

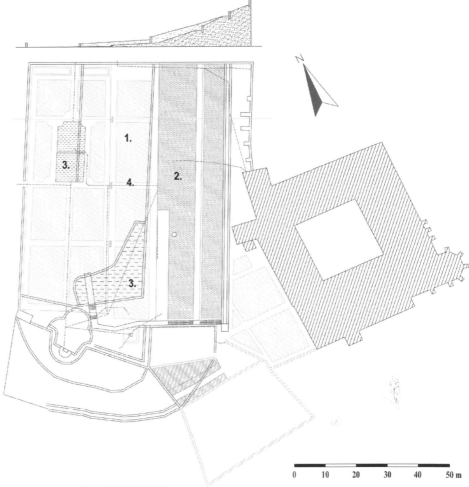

0   10   20   30   40   50 m

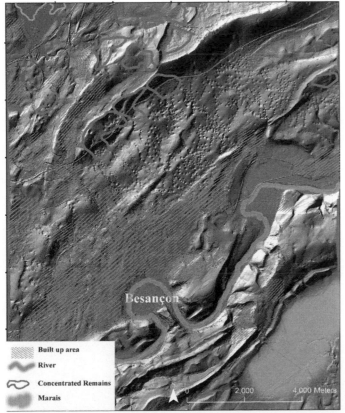

5.7 Figure 4. Areas with concentrations of stone piles and the remains of stone, rock-cut, or ditch and mound field boundaries in the forest are outlined in red. Almost a quarter of the modern forest contains evidence of organised field systems predating the current, and long established, system of parcels. *Image: R. Opitz, C. Fruchart, Lieppec / MSHE C.N. Ledoux*

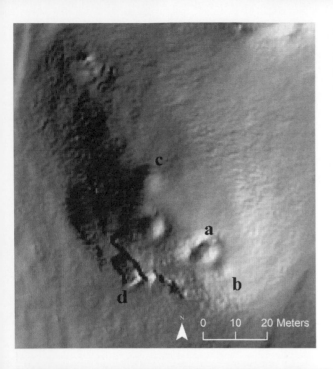

5.7 Figure 8. The 'feature set' of limekiln, quarry, claypit and charcoal burning platform, grouped together inside a doline, occurs frequently in the Forêt de Chailluz. a: limekiln; b: claypit; c: charcoal burning platform; d: quarry. *Image: R. Opitz, C. Fruchart, Lieppec / MSHE C.N. Ledoux*

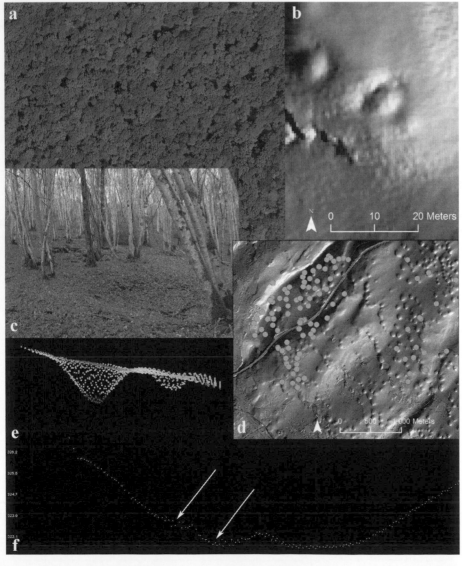

5.7 Figure 11. Multiple visualisations of the same data are used to explore relationships between feature, site, landscape and distribution. (a) Air photo of the forested location of the limekiln. (b) Hillshaded DTM of the limekiln. (c) Photo of the remains of the limekiln taken from about 4M away. (d) Yellow dots show the distribution of lime kilns in the local area. (e) The limekiln's appearance in the point cloud, with points coloured by elevation. (f) A profile section through a limekiln situated in the bottom of a doline. *Image: R. Opitz, C. Fruchart, Lieppec / MSHE C.N. Ledoux*

5.8 Figure 7. Soil classification in the area around the 'Fürstensitz' Heuneburg (Baden-Württemberg). – Fischer et al. 2010.

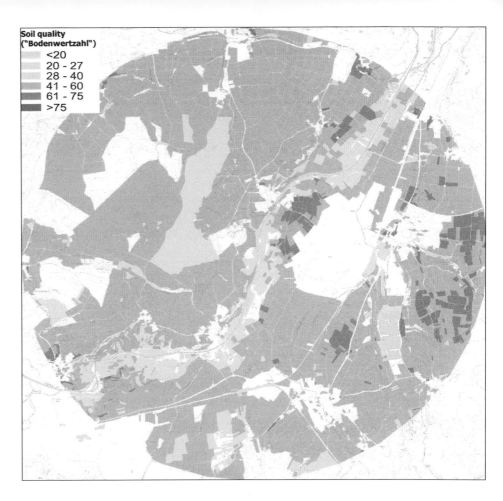

Soil quality ("Bodenwertzahl")
<20
20 - 27
28 - 40
41 - 60
61 - 75
>75

5.8 Figure 8. Slope classification in the area around the 'Fürstensitz' Heuneburg (Baden-Württemberg). The dark brown slopes indicate areas with more than 10 degrees of slope which are not suitable for ploughing. – DEM D-25 (25 m grid), © German Federal Office for Cartography and Geodesy 2004.

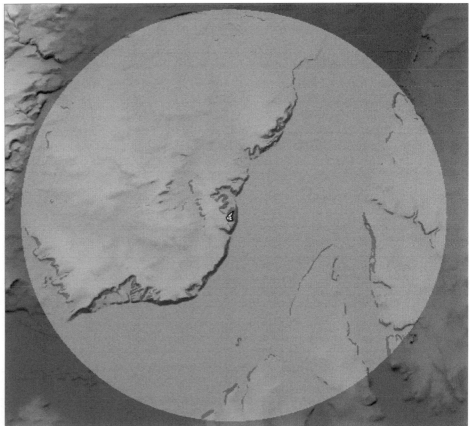

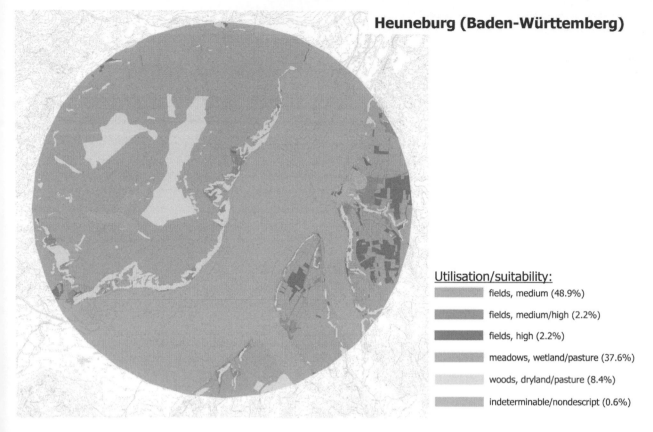

Heuneburg (Baden-Württemberg)

Utilisation/suitability:

- fields, medium (48.9%)
- fields, medium/high (2.2%)
- fields, high (2.2%)
- meadows, wetland/pasture (37.6%)
- woods, dryland/pasture (8.4%)
- indeterminable/nondescript (0.6%)

5.8 Figure 9. Combined classification of soil and slope values in the area around the 'Fürstensitz' Heuneburg (Baden-Württemberg). – Fischer et al. 2010.

5.8 Figure 10. Bronze Age and Iron Age settlement sites with archaeobotanical investigations in Baden-Württemberg. The steadiness of types of carbonised grain is represented in the diagrams for different periods: BZ = Bronze Age, BZ3 = Late Bronze Age/Urnfield Culture, HA = Early Iron Age/Hallstatt Period, HaLa = Early Iron Age/Hallstatt-Latène Period, La1 = Early Iron Age/Early Latène Period, La2/3 = Late Iron Age/Middle & Late Latène Period. – Fischer et al. 2010.

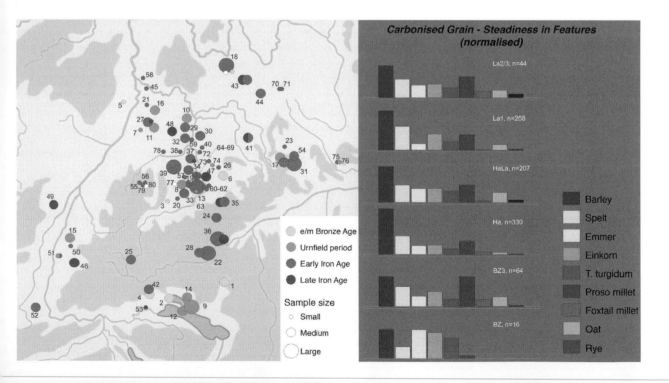

5.9 Figure 4. *Ammaia*. 3D reconstruction of the 'buried' structures of the forum detected with the GPR survey (elaboration by L. Verdonck).

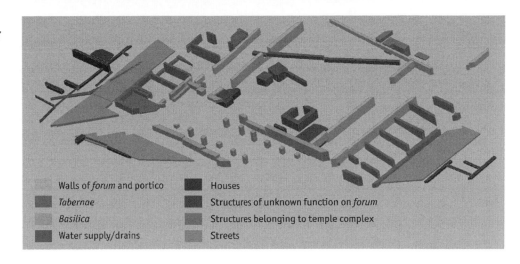

Walls of *forum* and portico	Houses
Tabernae	Structures of unknown function on *forum*
Basilica	Structures belonging to temple complex
Water supply/drains	Streets

5.9 Figure 5. *Ammaia*. Interpretation of geophysics survey and excavated areas with reconstruction of the street network: 1. forum, 2. baths, 4. area of the southern gate.

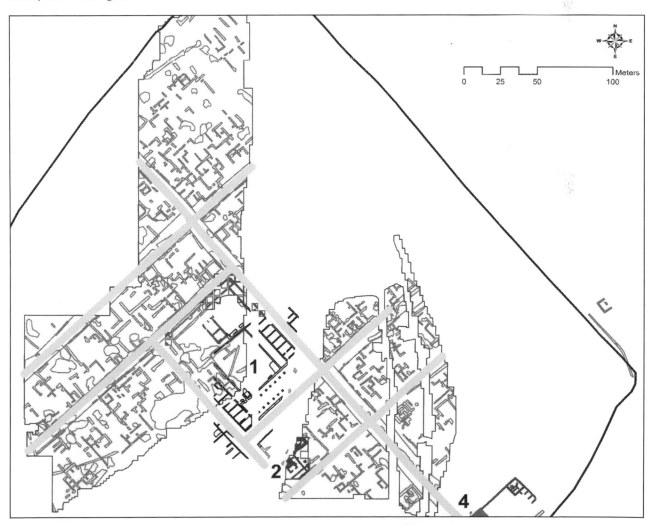

5.10 Figure 5. A DEM of a dunefield in the valley of the river IJssel. Note that the relief is exaggerated by a factor 5.4 and illuminated from the north-west.

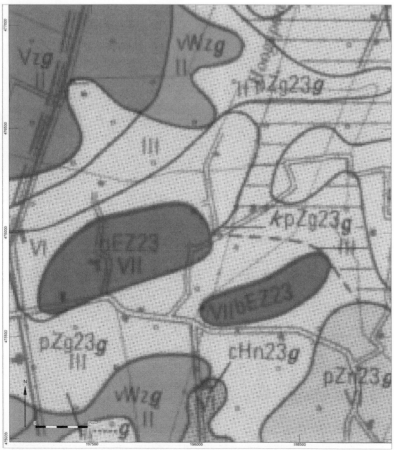

5.10 Figure 6. Comparing the soil map of the same area with a DEM, a strong relationship between topographical height and soil type can be recognised. The brown areas represent soils heightened by sods, which are characteristic for cultivated cover sand dunes.

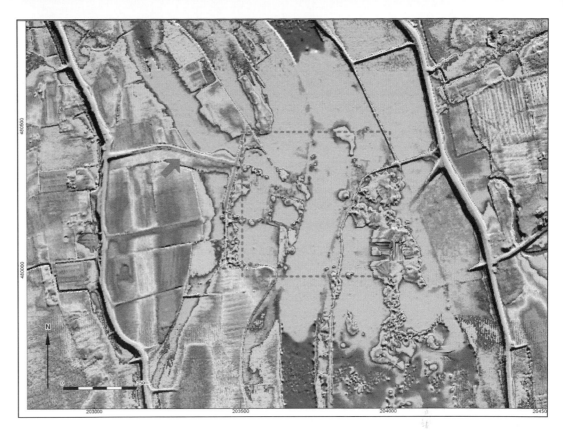

5.10 Figure 13. A DEM of the IJssellinie. Note that the relief is exaggerated by a factor 5.4 and illuminated from the north-west.

5.10 Figure 14. The IJssellinie on a topographical map from 1976.

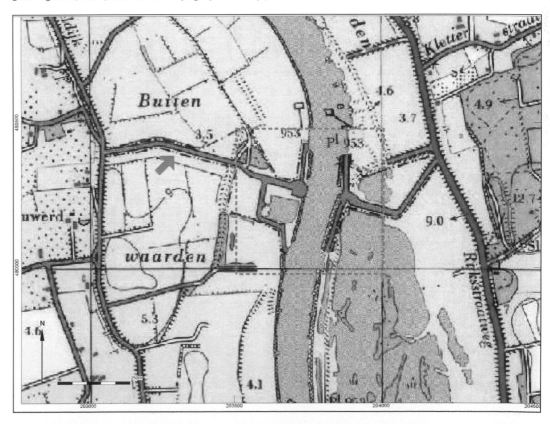

6.3 Figure 2. One of several late medieval carvings on the wooden rood screen at Sancreed church, west Cornwall, that depict individuals facing, like the Roman god Janus, and like increasing numbers of landscape archaeologists, both forward and back, looking into the past and the future. In this case the figure, a triciput, is also in the present looking out and thus, like us, responsible for bridging the two.

6.3 Figure 4. The Cornwall and Devon HLCs combined and simplified to create a regionalised characterisation. Patterns in and relationships between the several phases of enclosed land, rough ground and settlement suggest numerous regional and more local landscape archaeology research issues. Closer examination of the detail of each parent HLC would identify many more. (Derived from material that is the copyright of Cornwall Council and Devon County Council.)

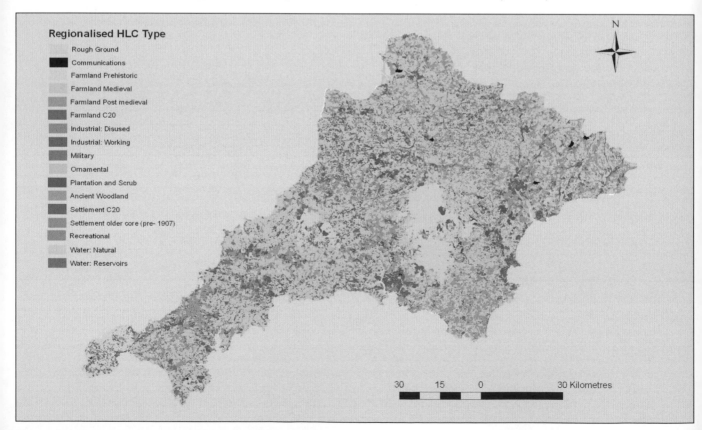

Regionalised HLC Type
- Rough Ground
- Communications
- Farmland Prehistoric
- Farmland Medieval
- Farmland Post medieval
- Farmland C20
- Industrial: Disused
- Industrial: Working
- Military
- Ornamental
- Plantation and Scrub
- Ancient Woodland
- Settlement C20
- Settlement older core (pre- 1907)
- Recreational
- Water: Natural
- Water: Reservoirs

30 15 0 30 Kilometres

5.2 Using Google Earth and GIS to survey in the Peruvian Andes

Authors

Laure Déodat and Patrice Lecoq

Archéologie des Amériques, Université de Paris, Paris, France
Contact: laure.deodat@voila.fr

ABSTRACT

In the Central Cordillera of the Andes in Peru, the Ayacucho region offers a landscape of mountains and deep, warm valleys, whose vast ecological diversity has encouraged human settlement from the Archaic period, 10,000 years ago, up to the present day. The Choquek'iraw-Chanca project, initiated in 2007, aims at understanding the process of occupation in one part of this region, the 'La Mar' province (so-called *Oreja de Perro*, 'dog's ear'), which is located between 1,000 and 4,500m altitude, in the north-east of Ayacucho (the regional capital).

Two survey campaigns were carried out in this area. These indicated a significant regional occupation that began soon after the Formative Period (± 500 BC) and continued through the Inca era (1532 AD), with a stronger presence of sites from the Late Intermediate period (1000 to 1400 AD), probably belonging to the Chanka. Today, there are still a few hamlets of people speaking the Quechua dialect and their rural lifestyle seems to be similar to that of pre-Columbian times.

Preliminary work on high definition satellite images from Google Earth enabled us to identify a few archaeological sites (villages and ceremonial centres). By surveying on foot, we discovered other sites that were not visible on Google Earth (burials in rock shelters, agriculture terraces and ancient roads). Each site has been properly recorded and referenced by GPS points (UTM WGS 84). This helps us integrate all the information recorded on the ground into a computer database and a Geographic Information System (GIS). The distribution of archaeological sites – most of which are Chanka villages – can then be shown on different base maps. The GIS also provides the opportunity to make thematic maps carry out spatial analysis via digital terrain model (DTM), e.g. slopes, site inter-visibility and visibility between sites and their environments, which allowed us to understand the different patterns of landscape occupation.

KEYWORDS

prospection, Google Earth, GIS, spatial analysis, Chanka, settlement

In the heart of the central Cordillera of the Andes in Peru, the region of Ayacucho displays a landscape of mountains and enclosed hot valleys with a great ecological diversity, which have been favourable to human settlement since the Archaic period 10,000 years ago (Mac Neish et al. 1983). After the formative period (500 BC to 200 AD), which is still little known, it is during the middle Horizon, between 500 and 1000 AD, that the region became the centre for the emergence and development of the Wari culture which spread across a large part of the central Andes (Isbell 1978, 2000). It was eventually occupied by warrior groups related to the Chanka in the 12th century, and these groups have left the most important number of traces in these areas. It was subsequently occupied by the Incas in the 14th century (Lumbreras 1975; Gonzalez Carré et al. 1987; Bauer 2010).

The Choquek'iraw Chanka project, begun in 2007, aims to understand the process that populated a province in this region – La Mar, locally called 'la *Oreja de Perro*', lying between 1,000 and 4,500m above sea level, at the western end of the department of Ayacucho and 130 km west of Cuzco and the famous Ma-

Figure 1. 3 D satellite image showing the study zone, taken from Google Earth.

chu Picchu (fig. 1). It follows the excavations done from 2003 to 2006 on the site of Choqek'iraw in order to determine the nature and age of its occupation (Lecoq 2007, 2008).

The study zone's extremely steep relief and limited – if not non-existent – road network make access to this region very difficult. In addition, up till a few years ago the presence of armed groups, successors of the Maoist guerrilla group, *Sendero Luminoso* (*Shining Path*), and associated with drug traffickers, made it a dangerous region where it was difficult to go far from the towns to study or excavate the existing archaeological sites. What is more, most of the IGN (*Instituto Geográfico Nacional*) maps are still incomplete, not having been verified in the field by cartographers, and there are only a few aerial photographs of the area available.

These difficulties are at the origin of the methodology adopted for surveying 'la Oreja de Perro' – consisting of using high resolution satellite imagery from Google Earth to prepare the field work, and developing a DataBase Management System (DBMS) and Geographic Information System (GIS); the first time such an approach has been envisaged for Peru.

The limited time available for field surveys – only possible during three weeks in the summer university holidays and having to overcome formidable logistics problems – is another reason for the methods used.

THE STUDY CONTEXT

The province of La Mar is a mountainous, particularly steep region extending between the beds of the *ríos* Pampas in the south and Apurimac in the north. The average altitude is 3,000m, but there is a steep gradient. The elevation at the valley bottom is 1,200m, which rises within a few kilometres to more than 4,000m at the top of the mountains. The existence of the two valleys, but also their situation on the line dividing the Andean highlands from the Amazonian lowlands, make this sector ideal as a place for understanding how it was populated. The group(s) who lived there must have tried to control several ecological levels – especially the Amazonian piedmont – as well as the communication axes such as the valleys of the *ríos* Pampas and Apurimac. But these valleys were probably more than mere thoroughfares from one region to another. There are good reasons for thinking that during the Late Intermediate period, in the 12th century and at the moment of Inca expansion from the 14th, these regions were also the main occupation centres for various ethnic groups, Chanka for example (Rostworowski 2001[2008]; Bauer 2010) – a hypothesis to be checked in the field. The surveying work, effected by M. Saintenoy, former VSI (International-al Solidarity Volunteer) at the French Institute of Andean Studies, while working on his doctoral thesis, on the northern bank of the *río* Apurimac should very shortly cast new light on the occupation of this valley.

ANTECEDENTS AND METHODOLOGY: FROM SATELLITE IMAGES TO ON-THE-GROUND SURVEY

The archaeological data relating to the region are rare, mainly dating from the 1970s, before the emergence of the *Sendero Luminoso* prevented field research. They are due in particular to researchers such as Grossmann (1967), Scott (1972), Lumbreras (1975) and Isbell (1978), who identified formative tradition occupations, Wari and Chanka, in various sectors of the valley. In 1987, González Carré, Pozzi-Scot and Vivanco (Gonzalez Carré et al. 1987) outlined a map showing the distribution of Chanka sites at Ayacucho, sites characterised by the presence of domestic structures with a circular plan.

So, in 2007 and 2008, the project Choquek'iraw Chanka began, with two prospecting campaigns in this region, in close collaboration with the University of San Cristobal de Huamanga of Ayacucho and financial support from the French Ministry of Foreign Affairs.

The first campaign focused on the confluence of the *ríos* Pampas and Apurimac, the second on the vicinity of Chungui, the regional centre, some 60 km farther to the west. The aim in both these sectors was to bring to light archaeological sites of all types and periods. We decided to define a 'site' either as a major concentration of archaeological material, generally ceramic, or as an isolated structure or group of structures of varied character. Consideration was also given to all the paths, probably ancient, of which certain sections were constructed (retaining wall and paving; Hyslop 1984).

To succeed, the chosen prospecting method used all the available resources, in particular exploiting high resolution satellite images collected by Google Earth, topographic mapping, and collecting information orally from the local population. These three types of data then oriented exploration on foot in this craggy region, where systematic prospecting was impossible – the very rough relief and the dangers associated with the presence of groups of armed rebels limited walking expeditions dramatically.

The exceptional quality of the satellite images Google Earth (fig. 2) supplied for a large part of the surveyed region – giving very high quality close-up images up to a scale of 1/1000th – has enabled us to locate eleven sites and a set of paths actually before even arriving on the spot; the lack of strong vegetation in our zones of study facilitated the surveying. There are two sets of images for our study area, dated 17 and 28 June 2005 respectively, both belonging to the GeoEye company; the pictures were probably taken by the Ikonos satellite, at a spatial resolution of 0.82 x 3.2 m. Interpretative sketches of sites were then obtained from these very images. The structures clearly identified with recognisable forms have been indicated with unbroken lines; dots were used to represent the uncertain forms. This preliminary photo-interpretation work supplied relatively precise sketches of sites whose functions could now be suggested.

Figure 2. The site of Corral Corral, a village. Google Earth satellite image (left) and its interpretation (right), with sketches of the refined image made on the ground.

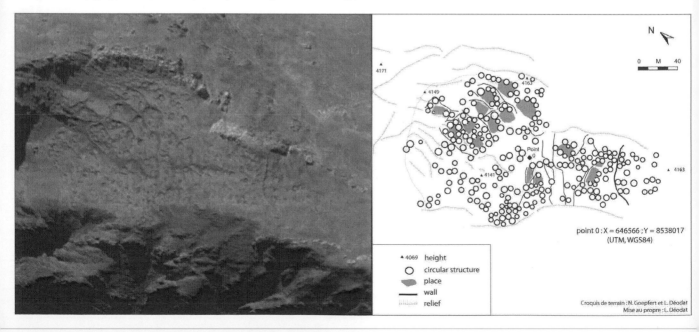

The forms that were delineated – essentially circles – belonged to structures already known from previous research: we had in front of us several villages made up of houses with circular plans, associated with plot boundaries and sometimes surrounded by outer walls. On each sketch, strategic points were targeted to obtain their geographic co-ordinates on Google Earth before being entered on the GPS, so as to facilitate their localisation in the field surveying. The GPS used, a Garmin 60CX, has an accuracy of 2m in the best of cases. All that was needed in the field was to follow the GPS to find the sites and carry out a classic survey on foot.

For each site discovered, several procedures were carried out in parallel:

- A series of GPS points were logged: in the centre of some structures, at each wall-end, in the centre of certain public spaces and on all the boundary of the site; the measurements were done using UTM projection coordinates and WGS 84 geodetic system. This *tracking*, systematically done, made it possible to define the sites and to calculate their area and perimeter. *Tracking* is a GPS procedure: when moving, the GPS automatically traces your route by logging a multitude of points.
- Site information sheets with six categories of data – general characteristics, location, geographic situation, site description, archaeological interpretation and equipment – were completed. A few structures – depending on their interest, their state of preservation, and their comparative originality – were also given a detailed description on a specially designed sheet. Sketches (plan and cross-section) and photos accompany each structure and each site sheet. These information sheets were done using the DBMS FileMaker Pro 8.5.
- The site sketches were either drawings made on site, or plans developed through photo-interpretation from satellite images retaken and completed in the field. In this way all the visible structures were represented, while recording the position of doors and other details. When it seemed of interest a cross-section complemented these sketched plans, which turned out to be very useful as these outlines assemble a great deal of information: the natural lie of the land, the organisation of the structures in relation to the dip with visibility of the artificial terraces built to level the spaces, and dimensions of the structures themselves. Let it be noted, the sketches made from photo-interpretation have proved to be relatively accurate in the light of on-site observation, and the errors in interpreting the satellite images were minimal.
- Ceramic and lithic material were taken from each of the sites we examined, and were then studied in the laboratory to improve the precision of the relative dating and to establish a chronological framework for the whole region.

It can be considered today that using Google Earth to survey such an isolated and dangerous region has saved a lot of time for two essential reasons. First of all, it has made it possible to detect sites and develop sketches before surveying on the ground, and next, to speed up the long work normally necessary with the local population to collect information on sites that may be located near the communities they come from.

Nevertheless, these satellite images cannot be used as a substitute for traditional surveying methods for the simple reason they do not enable all types of sites to be detected. On the photos only villages or sufficiently important ceremonial centres can be made out by a reasonably trained eye. On the other hand, funerary sites remain invisible on satellite images, which is also the case for all the other smaller sites or those reduced to a simple concentration of archaeological material. We have only been able to discover this kind of site thanks to the local population.

Last but not least, even though this tool has allowed relatively precise plans to be drawn up, it is still unable to go into the finest details. Thus, reconnaissance on the ground has proved essential, not just for checking the data gathered and improving the plans, but also for taking away the archaeological material (ceramic, lithic etc.) needed to identify each site's period or function.

Google Earth should therefore be considered as a tool with a triple function:

- Detection of sites thanks to high resolution satellite images (Madry 2006; Garrison et al. 2008; Goossens et al. 2008), of the same type as aerial photographs (Deletang 1999);
- Data storage – since it makes data collecting possible;
- Sharing, since it facilitates the instant transmission of all available data to other users, via the Internet (Conroy et al. 2008).

It is an extra tool for archaeologists to use in developing surveying strategies, to be used parallel with other more traditional methods (Ferdière & Zadore-Rio 1986; Jung 1998; De Laet et al. 2007).

RESULTS

Regional occupation and identification of ethnic groups
The data obtained through these surveys show an important occupation of the region that began in the Andean formative period, between 500 BC and 200 AD, and continued up to the Inca period. In all, 46 sites were discovered in two years.

The Formative period
Two sites from this period (CC08-14 and CC07-1) have been located; they correspond to concentrations of ceramics scattered across the fields on the upper and middle slopes of the río Pampas at the limit of the middle valleys between 1,800 and 3,000m. They also include vestiges of agricultural terraces and/or other structures difficult to characterise, possibly connected to the culture of maize and tubers of potato type (*oca, mashua* etc.).

The Middle Horizon
The regional occupation continued during the Middle Horizon (600 to 1000 AD), characterised by a Wari tradition site located in the same biotope. The only site attributable to this period (CC08-13) is positioned at the confluence of two watercourses on the lower slopes of a hill. The site comprises the remains of a structure with a circular plan (5m in diameter) to which a dozen circular cists (± 1-1.20m in diameter) are associated, constructed with randomly selected stones, and spaced at 1 to 1.50m. These types of varying depth cist burials are typical of the Wari and Tiwanaku cultures that flourished during the Middle Horizon, when they were generally placed near inhabited sectors – most often next to water courses, as is the case here, for ideological and ritual reasons.

The Late Intermediate period

The following period, called Late Intermediate, extending in the Andes from 1000 to 1400 AD, was heavily settled with 34 listed sites. These were generally fortified villages (29 sites), called *pucara*, characteristic of the settlements of the Late Intermediate period in the central and southern Andes, which left a deep mark on the landscape and are quite visible on Google Earth images or the available aerial photos (fig. 3). Veritable eagle's nests at the tops of cliffs and crest lines in the main regional massifs, they are evidence of a large population together with the strategic role that this out-of-the-way region appears to have had (Lumbreras 1975; Gonzalez Carre et al. 1987; Vivanco 2005). From these beetling citadels, apparently multifunctional – both defensive and ritual – it was easy for the inhabitants to overlook all their territory and most especially the ríos Pampas and Apurimac. Certain villages seem to have had, in addition, a defensive system consisting of walls and ditches. They include circular planned structures, 4 to 6m in diameter, often placed one against the other. When the villages were built on mountain sides or strongly sloping lines of crests, small terraces supported by walls had to be constructed to build the houses on level ground. These circular structures were organised around more or less cramped squares or esplanades with, in this case, their doors systematically opening into the square.

Frequently, these constructions had only one room, with a single entrance and a beaten-earth floor. The walls, 45 to 60cm thick, were made of ungraded large limestone stones often very roughly hewn, or of

Figure 3. The site of Corral Corral, a village with characteristic circular structures. *See for the full colour version also the front cover of this book*

Photo N. Goepfert

blocks of basalt, piled up more or less regularly on the ground or on a bed of rubble. They were faced on both sides, and the central interval was filled with chippings, sometimes with earth mortar.

On certain sites (CC08-1 and 6) walls – probably folds for animals or boundaries for fields – are noted. Nearby remains of agricultural terraces, old paths – difficult to date owing to their constant use by the local population – and simple or multiple burials under rock shelters (8 tombs listed) are also found. Small stone funerary constructions, of *chullpa* type, typical of the Late Intermediate period, could also be included (CC07-3). These sites are generally attributed to very warlike regional groups associated with the Chanka and seemingly originating from the Amazonian piedmont.

The Inca period

The Inca occupation – marking the beginning of the Late Horizon and extending in the Andes from 1438 to 1532 AD – does not seem to have had much effect on the regional landscape. Five sites date from this period and Inca material is found in a few villages of the previous period. The Incas seem to have been content to subjugate and control most of the Chanka villages, occupy some strategic positions, and develop the road network. Only the construction of ceremonial complexes, of *ushnu* type, seems to have transformed the landscape in any way (fig. 4).

These *ushnu* – still not clearly dated – are platforms, more or less complex, located on the summits of massifs, probably connected to mountain cults and star-gazing (Martinez 1976; Reinhard 2002). Two of these sites – one of which can be seen on the satellite images – are clearly attributable to the Incas (Zuidema 1980).

The first (CC08-5), located at 4,200m altitude, consists of a set of three platforms, about 20.10m x 19.80m and 1.20m high, surrounded by a large wall open to the east formed of well-squared stones perfectly fitting together. The monument is oriented to the north-east and towards the peak of a great massif – which attests to the decisive role of the surrounding mountains (fig. 5). It is dominated to the south by a rocky spur and to the north-east by another important massif with a singularly tapering profile, both nowadays considered locally sacred as receptacles for the spirits of the ancestors – the *Apu*. The rest of the site includes diverse circular structures partly protected to the north and north-west by a large wall; these structures pre-date the *ushnu*, as they were destroyed by it.

The second *ushnu* (CC08-11) shows much the same configuration. Located at 4,280m, on a rocky

Figure 4. Ceremonial centre or *ushnu* (CC08-11).

Figure 5. Choquezafra glacier and surrounding area, visible from the *ushnu* (CC08-11).

spur dominating the whole region, it is based on an ancient circular structure and comprises two plat-forms, one on top of the other. Here again the main façade is oriented to the north-east and the snowy peaks of the Choqesafra massif, 36 km away beyond the Apurimac valley. Yet again, this ceremonial edi-fice is located on a crest-line right between the highlands and lowlands of the Amazonian piedmont and close to an ancient path – a particularity of most of the listed *ushnu* in other regions of Ayacucho and the Andes (Vivanco 2004; Meddens et al. 2008).

In both cases it appears that circular structures had been built by Chanka settlers before the Inca *ushnu*, since they were destroyed by it, as if the Incas had wanted to mark the landscape durably and thereby estab-lish themselves in the region at several strategic points. Where no Inca *ushnu* was built, nevertheless, Inca tradition villages with rectangular structures are found beside the villages with circular structures (3 sites).

DATA INTERPRETATION

The tools

An issue inherent to a study of this kind, carried out on a micro-regional scale, is the concept of choice of settlement. How, and according to what criteria, do a group of people decide at a given moment to settle in a particular place? These are the criteria that archaeologists working on spatial analysis seek to iden-tify. The use of GIS made it easier to envisage and analyse factors as varied as the nature of the terrain, the slopes, their orientations, the various ecological levels and the capacities the various sites had to observe one another. All of these factors had to be taken into account, and were integrated into a global synthesis. This is why – right from the start – integrating the data into a GIS appeared essential in order to effect this analysis of how the space was occupied. Accordingly, topographic, geological and ecological maps were integrated into the GIS so as to superimpose all our data on these maps. Furthermore, a digital terrain model (DTM) – a representation of the topography of a given zone of the Earth's surface accurate to 90m – was downloaded from the NASA site: http://srtm.csi.cgiar.org. This DTM enables very interesting spatial analyses to be generated within the GIS.

The different patterns of land occupation (fig. 6)

Out of the 35 villages identified, 28 were built at an altitude of more than 3,000m; 17 at more than 4,000m. Altitude, then, was the essential criterion governing the choice of site. This systematic predilection for commanding heights – which is not limited to our micro-region, as numerous previous studies have already shown, such as those on the Asto settlements to the north of the region studied (Lavallée & Julien 1973) or those concerning the region Intersalar, in Bolivia (Lecoq 1999) – must not obscure the need for the minimal vital necessities: water, food, wood for fuel, building materials, etc.

The presence of rocky outcrops, abundant in the zones surveyed (essentially sedimentary rocks such as limestone and some volcanic stones) allowed the populations to use local materials to build their houses. They did not have to bring in stones from elsewhere, which made it easier for them to occupy these isolated sites. Only the materials used for making a few objects, such as hand mills, may have been transported over a few kilometres.

Concerning water needs, the Chungui zone does not lack lagoons, nor small streams; the Mollebamba zone, on the other hand, is more arid. Some springs however, well known to the present inhabitants, exist close to the sites discovered. These springs probably existed in the Late Intermediate, although this is not certain.

If today the vegetation of the zones in which the sites are situated are almost deserts – the present inhabitants travel several kilometres to get the wood they need for cooking – without any precise research, it is still hard to imagine what the vegetation was like in the early periods. Data collected by A.J. Chepston (2009) from cores taken from a dry lagoon in the Cuzco region at 3,350m, all show the same: from circa 1100 AD (Late Intermediate) – following a long, cold and relatively dry period – temperatures rose considerably and precipitation was heavier, making it easier for vegetation to grow. These factors seem to have allowed populations to settle at high altitudes, on territory which until then was uninhabitable, to develop these new plots by growing potatoes or quinoa. The villages' circular structure fits into this scheme perfectly.

Moreover, ArcGIS software enables us to create maps of classes of slopes based on the DTM. When the sites listed in the two surveyed regions are superimposed on this map, it quickly becomes apparent that all the village sites – with or without ceremonial centre – are placed on slopes of less than 10°.

All these data are very helpful for understanding how the sites were occupied. Thus it appears that the inhabitants of these villages favoured the crests, in altitude, and also preferred fairly flat zones, which was not so simple given the rugged nature of the terrain throughout the region. The occupation zones were therefore quite restricted in the last analysis – if all the necessary, or even indispensable, occupation

Page 331 >
Figure 6. Localisation of sites in relation with ecological floors, extracted from GIS. *See also the full colour section in this book*

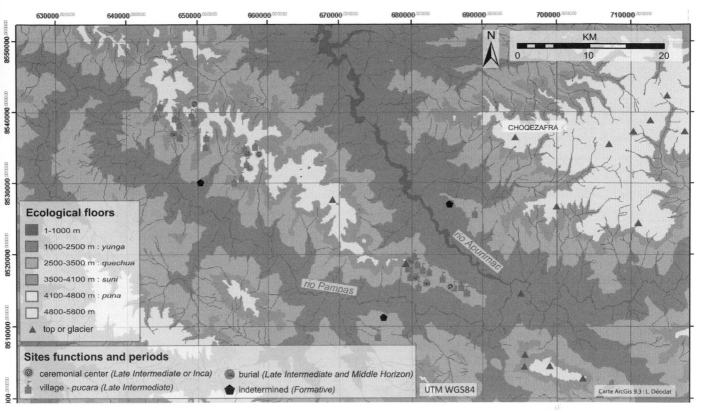

The schematic drawing above shows the different floors of the Andes and their ecological resources. It also indicates the barter system in which llamas or mule caravans are used to transport products between zones. Archaeological sites of the Formative period are located in the lowlands and the fertile ecological floor (Yunga). Villages of the Late Intermediate are located in the highlands (Suni, Puna and top of the Quechua zone).

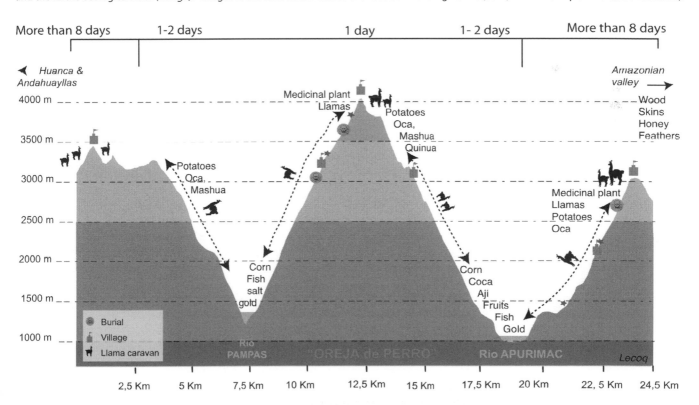

criteria are summed up: elevated zone for keeping watch and defence, flat space for ease of construction, rocky outcrop for construction materials, and proximity to springs or a stream for vital needs. As for food resources, another parameter must be taken into account – the ecological levels.

The ecological levels: an anthropological approach (fig. 6)

In the Andes, the ecological levels play a very important role, which had to be taken into account. To take full advantage of varied resources each community sought to exploit various ecological levels. This particular exploitation of space, dating from the Pre-Hispanic period, has been defined as 'verticality' by the American anthropologist Murra in 1975.

Accordingly, we created a map showing the various levels of the zone studied, using as a reference the work that Pulgar Vidal proposed in 1941 (Pulgar Vidal 1996) on the eight natural regions of Peru, taken partly from those developed by Morlon (1992, 122-202). Each ecological level has its own characteristics – climate, fauna, flora, soil types – which affect how the land is exploited – agriculture or pastoralism – and the vegetable and tree varieties cultivated, which may be summed up as follows:

- 500 to 1,000m, generally the domain of the forest, where coca, cocoa and coffee are exploited. This level has been defined as coastal (or *challa*) on the Pacific littoral or low and high *selva* (*omagua* and *ruparupa*) on the eastern foothills of the Andes.
- 1,000 to 2,500m, on the *Yunga* fluvial level valleys, with a hot, dry climate are found, suitable for fruit trees such as orange, lemon and avocado. The natural vegetation is shrub and forest.
- 2,500 to 3,500m, or *Quechua* level, the slopes bordering the valleys are highly suitable for agriculture, which is why most of the population today lives on this ecological level, which has given its name to the language (Quechua) of the populations living there. Maize, wheat, quinoa, tomatoes, squash, etc. are cultivated on terraces. Fruits such as the papaya and pomegranate are also grown. Some trees grow naturally in this region: the *molle*, and the eucalyptus (though the latter was introduced after the conquest).
- 3,500 to 4,100m, *Suni* level, a zone with a temperate to cold climate is found with a very rough relief and many ravines and small enclosed valleys. This is the realm of the tubers with their countless varieties (*oca, masua, papa*, etc.), which are cultivated traditionally with the digging stick or *chaquitaclla* (Morlon et al. 1992). This is also the region, near streams and springs, where *tunta* and *chuño* – dehydrated potatoes – are prepared. Vegetation here is essentially shrub with various prickly plants, as well as dwarf trees.
- 4,100 to 4,800m, the *Puna* stage, corresponds to a mountain desert region, with a shallow arable layer and a very cold climate. Not very suitable for agriculture, the vast natural pastures covered with grass like the *ichu* (*Poaceae, stipa Ichu*) found here are therefore used for livestock.
- Above 4,800m, the pastures give way to the high mountains and glaciers – often considered sacred. Human occupation is generally limited to a few high altitude sanctuaries.

It may be observed that all Late Intermediate sites, attributable to the Chanka and Inca groups, are located in the *Quechua* (8 sites) *Suni* (15) and *Puna* (11) regions, between 2,500 and 4,500m altitude – with a large majority above 3,500m.

The Andean verticality model fits the region especially well, where today each community still possesses a parcelled out space, adapted to the various ecological levels it controls. This space extends along the *Ríos* Pampas and Apurimac until the Amazonian piedmont, and probably the same was the case in the Pre-Hispanic era, even if this is difficult to determine. It should be made clear that the world of the *Puna* may only have started at 4,300m – according to D. Lavallée (Lavallée &Julien 1973, 86) – which increases the arable land surface area considerably. Now, 8 sites are located precisely between 4,100 and 4,300m, which is hardly significant; for depending on their level, they would have been turned towards either pastoralism or tuber growing. Unfortunately, we still lack precise local data on this subject. We can, nonetheless, sum up the situation as follows: the 8 villages in the *Quechua* zone could have a wide and varied diet of vegetables and cereal; the 15 villages in the *Suni* region seem to have oriented their activities mainly towards tuber growing, and the 11 in the *Puna* region probably devoted themselves to stock-rearing. It is tempting to imagine systems of exchanges between all and sundry, and even to suppose two villages distinct in space may have formed a single community.

Numerous questions are still to be answered. We have no example of a village located in the lower levels, especially favourable for fruit groves. That said, we suppose the villages discovered corresponded to their 'urban centre' and that in the other levels, as is the case today, the inhabitants only had a few sheds of perishable materials next to their fields and/or groves, types of remains difficult to detect when prospecting on foot. Cores perhaps should be considered, in places suitable for farming for instance, in the hope of finding vestiges of these temporary occupations.

Another aspect raises some issues concerning the sites located above 3,500m altitude: why establish the 'administrative' centre of a region, or of an ethnic group (if such is the case) at such an elevated altitude, far away from the resources in the valley, instead of 'in the middle' so as to be at an equal distance from the resources of the *Yunga* and *Puna* levels? The modern village of Mollebamba offers a good example of the efficient exploitation of the levels. Its inhabitants live in the main hamlet at 3,000m, where they can cultivate maize and all sorts of vegetables; they also have a house at 4,000m to produce many varieties of tubers and keep animals; lastly, they possess a few fruit-trees (e.g. orange and lemon) at 1,000m in the valley.

Now most of the villagers in the Late Intermediate lived in very out-of-the-way places, which seems to suggest the principal factor behind the choice of sites was not the availability of varied resources – the inhabitants being quite content with a meat-tuber diet – but rather a need for defence and control. One argument tends to strengthen these hypotheses, and it concerns fields of view.

Fields of view, inter-visibility (fig. 7)

DTM allows work on notions of fields of view, and thus fosters more profound spatial and three-dimensional thinking. The software can be asked to draw all the zones visible from a precise point, so visibility maps can be created for each site so as to define several parameters:

- Inter-site visibilities, which allows an approach to notions of inter-population relations or even relations of domination and submission between different sites that may have belonged to two ethnic groups, for example.
- The zones visible from a site, such as glaciers and other sacred mountains or *Apu* and the water courses – which opens a door into the ritual world of these cultures.

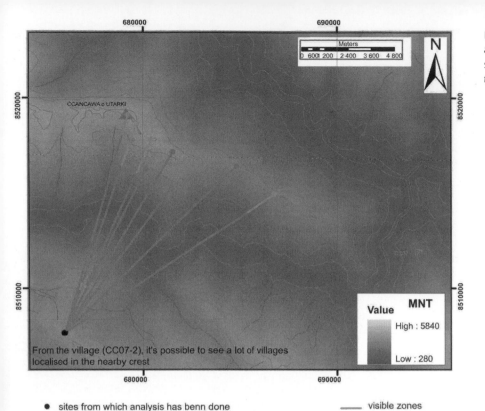

Figure 7. Example of visibility
analyses between two or more
sites, extracted from GIS. *See also
the full colour section in this book*

- sites from which analysis has benn done
- observed and visible sites from observation points (black)
- observed but no visible sites from observation points (black)

—— visible zones
—— no visible zones

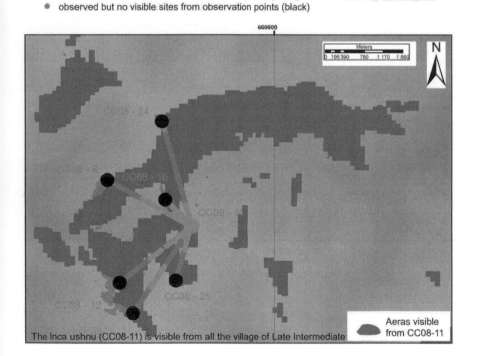

This type of analysis was carried out for all the Late Intermediate villages, and the map presented here (fig. 8) shows the results for four quite different sites. It should be noted that the analyses were done within a range of 300 km around each site, and that a certain number of tests were first carried out and considered

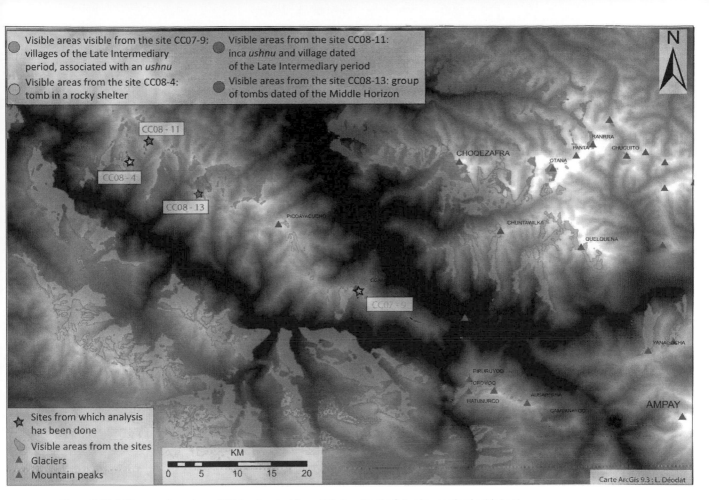

Figure 8. Visibility map based on a DTM, extracted from GIS. *See also the full colour section in this book*

together with our field notes and the topographic maps before defining a rigorous method. The examples chosen allow the method to be validated, while considering a sample of sites with different functions and chronologies: villages, ceremonial centres, tombs belonging to the Middle Horizon and the Late Intermediate period.

The first site to attract attention, because it offers a perspective of 360°, is a ceremonial platform or *ushnu* (site CC08-11) from which it is possible to observe one of the most important glaciers of the region: the Choquesafra. Site CC07-9, which is an elevated village with a small ceremonial platform at its highest point, offers a good perspective too, with (yet again) a wide view of the Choquesafra but also of a large part of the Pampas valley. From site CC08-4, which is a tomb arranged in a rocky crevice, probably associated with the village CC08-6 some tens of metres above, a good view is obviously to be had over the whole south-west zone (since the site is built against a cliff on the north-east side) and therefore over the *Rio* Pampas. Lastly, from site CC08-13, which is a group of Middle Horizon tombs, the perspective is evidently non-existent – except for a few summits in the immediate vicinity – which goes to show that visibility was not necessarily looked for in this site's position. Those who chose it probably did so because of the confluence of two small streams about 50m below, which was a decisive criterion for locating necropolises and other tombs of this period throughout the central and southern Andes (Lecoq 1999, chapter VI).

The negative result, as far as this site is concerned, reinforces the importance of visibility for the Late Intermediate villages – whether fortified or not, with or without ceremonial centre: during this period people settled in places where they could keep an eye on their neighbours and the communication routes and see the principal *Apu*.

CONCLUSIONS

All these data help us to understand better what could push people to prefer settling at one site rather than another. Strictly based on archaeological evidence, the populations living at between 3,500 and 4,500m either had extensive pastures and consequently abundant meat, or a large variety of tubers; exchanges may have existed between the ones and the others. At the same time, living this way in the high places made it possible to overlook and monitor a large part of the surrounding territory, as well as coming under the protection of the *Apu* (the sacred mountains) that could be seen from most of the sites. As for the villages between 2,500 and 3,500m, their inhabitants had more varied agricultural resources but less visibility. Possessing lands at other ecological levels, they may well have had a greater variety of resources, but this remains to be proven, archaeologically speaking, since no trace of Late Intermediate occupation below 2,500m has been found (except for site CC07-19).

Moreover, if these results concerning modes of settlement were proven, they could be used in the future as new criteria for detecting the presence of sites. Numerous parameters would then be available, when reading a map, for suggesting zones to survey that are more likely to contain undiscovered sites. For the Late Intermediate period this would mean, for instance, crest zones above 2,500m with slopes of less than 10°. That said, it is also indispensable to survey the low zones more systematically, in order to have more chances of discovering other types of sites of the same period at these altitudes.

Be that as it may, the Google Earth satellite images have proved to be prospecting tools of prime importance, and spatial analysis has made it possible to define how sites were settled and so better understand the populations we are interested in. These methods – still little used in the Peruvian Andes – offer very promising prospects for the future.

ACKNOWLEDGMENTS

The authors wish to thank Timothy Seller for the English translation from the original version in French and Jean-Luc Daunac for his revision of the English text. The co-direction of the project Choquek'iraw Chanka was done by Cirilo Vivanco Pomancanchari, Professor at the University S.C. of Huamanga, whom we thank for the help he gave us. We wish to thank Thibault Saintenoy for having supplied us with a map of glaciers.

REFERENCES

Bauer S., L.C. Kellet & M. Aráoz Silva. 2010. *The Chanka. Archaeological Research in Andahuaylas (Apurimac), Perú*. Monograph 68, Cotsen Institute of archaeological Press, UCLA.

Conroy, G.C, Anemone, R.L., Vanregemorter, J. & Addison, A. 2008. Google Earth, GIS, and the Great Divide: A new and simple method for sharing paleontological data. *Journal of Human Evolution* 55, 751-755.

De Laet, V., Paulissen, E. & Waelkens, M. 2007. Methods for the extraction of archaeological features from very high-resolution Ikonos-2 remote sensing imagery, Hisar (southwest Turkey). *Journal of Archaeological Science* 34, 830-841.

Deletang, H. (ed.) 1999. *L'archéologie aérienne en France. Le passé vu du ciel*. Paris, Errance.

Ferdiere, A. & Zadore-Rio, E. 1986. *La prospection archéologique*. Paris, DAF 3, Éditions de la maison des sciences de l'homme.

Garrison, T.G., Houston, S.D., Golden, C. & Inomota, T. 2008. Evaluating the use of IKONOS satellite imagery in lowland Maya settlement archaeology. *Journal of Archaeological Science* 35, 2770-2777.

Gonzales Carré, E., Pozzi-Escot, M., Pozzi-Escot, D. & Vivanco, P.C. 1987. *Los Chankas: Cultura material*. Ayacucho, Laboratorio de Arqueología, Escuela de Arqueología e Historia, Facultad de Ciencias Sociales, Universidad Nacional de San Cristóbal de Huamanga, 224.

Goossens, R., De Wulf A., Bourgeois J. Gheyle W. & Willems T. 2008. Satellite imagery and archaeology: the example of CORONA in the Altai Mountains. *Journal of Archaeological Science* 33, 745-755.

Grossmann, G. 1967. *Early ceramic cultures of Andahuaylas, Apurimac*. Berkeley, Ph.D., University of California.

Hyslop, J. 1984. *The Inka road system*. New York, Institute of Andean Research.

Isbell, W.H. 1978. El imperio Huari ¿estado o ciudad?. *Revista del Museo Nacional* 43, Lima, 227-241.

Isbell, W.H. 2000. Repensando el Horizonte Medio: el caso de Conchopata, Ayacucho, Perú. *Boletín de Arqueología, Pontificia Universidad Católica del Perú* 4, 9-68.

Jung, C. 1998. La photo et carto-interpréation. In Ferdière, A., *La prospection archéologique*. Paris, Errance, Collection Archéologiques, 129-160.

Lavallee, D. & Julien, M. 1973. *Les établissements Asto à l'époque préhispanique*. Tome X, Travaux de l'IFEA, Lima, 143.

Lecocq, P. 1999. *Uyuni préhispanique, Archéologie de la Cordillère Intercalar (Sud-Ouest Bolivien)*. Oxford, British Archaeological Report, International Series 798.

Lecocq, P. 2007. Choqek'iraw, la merveille inca des Andes. *Archéologia* 444, 20-35.

Lecocq, P. 2008. Le site inca de Choqek'iraw (Pérou), Nouvelles données sur l'histoire précolombienne. *Les nouvelles de l'Archéologie* 111-112, 122-128.

Lecocq, P. 2010. Terrasses aux mosaïques de Choqek'iraw, Pérou. Description générale et premières interpretations. *Journal de la Société des Américanistes*, 96-2, pp. 7-73.

Lumbreras, L.G. 1975. *Las fundaciones de Huamanga, Hacia una prehistoria de Ayacucho*. Lima, Editorial Nueva Educación.

Macneish, R.S., Vierra, R.K., Nelken-Turner, A. & Phagan, C.J. (eds.) 1983. *Prehistory of the Ayacucho Basin, Peru, Vol IV: The Preceramic Way of Life*. Place, University of Michigan Press.

Martinez, G. 1976. El sistema de los Uywiris en Isluga *Anales de la Universidad del Norte* 10, 255-327.

Madry, S. 2006. An Evaluation of Google Earth for Archaeological Exploration and Survey. *Digital Discovery. Exploring New Frontiers in Human Heritage*. CAA 2006, Computer Applications and Quantitative Methods in Archaeology, Edited by Jeffrey, T. Clark, & Emily M. Hagemeister, Archaeolingua, Budapest.

Meddens, F.M., Branch, N. P.; Vivanco, C., Riddiford, N. & Kemp. R. 2008. High Altitude Ushnu Plarforms in the Department of Ayacucho Peru, Structure, Ancestors and Animating Essence. In Staller, J.E. (ed.), *Pre-Columbian Landscapes of Creation and Origin*. London, Springer Sciences + Business Media, LLC, 315-355.

Morlon, P. (ed.) 1992. *Comprendre l'agriculture paysanne dans les Andes Centrales. Pérou-Bolivie*. Paris, INRA Editions.

Murra, J.V. (ed.) 1975. El control vertical de un máximo de pisos ecológicos en la economía de las sociedades andinas. In *Formaciones Económicas del Estado Inca*. Lima, Instituto de Estudios Peruanos, 59-115.

Pulgar Vidal, J. 1996. *Geografía del Perú*. 10th ed. Lima, Peisa.

Reinhard, J. 2002 [1991]. *Machu Picchu, el centro sagrado*. Cuzco, Instituto Machu Picchu.

Rostworowski, M. 2008 [2001]. *Le grand Inca Pachacútec Inca Yupanqui*. Paris, Tallandier.

Saintenoy T., 2011. *Choqek'iraw et la vallée de l'Apurimac. Paysages et Sociétés préhispaniques tardives*. Thèse de l'Université Paris 1.

Scott Raymond, J. 1972. *The cultural remains from the Granja de Sivia, Peru: an archaeological study of tropical forest culture in the Montaña*. Illinois, Ph.D., University of Illinois at Urbana-Champaign.

Vivanco P.C. 2004. Ushnu o lugares sagrados del imperio Inka en territorio Chanka, Ayacucho (Perú). *Investigación* 12 (12), Universidad Nacional de San Cristóbal de Huamanga, Vicerrectorado Académico, Oficina de investigación, Ayacucho, 149-159.

Vivanco P.C. 2005. El tiempo de los purun runas o chankas en la cuenca de Qaracha, Ayacucho (Perú). In Tomoeda, H. & Millones, L. (eds.), *Pasiones y desencuentros en la Cultura andina*. Lima, Fondo editorial del Congreso del Perú, 13-29.

Zuidema, T. 1980. El Ushnu. *Revista de la Universidad Complutense* 28 (117), 317-362.

5.3 The occupation of the Antequera Depression (Malaga, Spain) through the 1st millennium BC: A geographical and archaeological perspective into Romanisation

Author
Maria del Carmen Moreno Escobar

Department of Geography, History and Philosophy, Pablo de Olavide University, Seville, Spain
Contact: mcmoresc@upo.es

ABSTRACT

This article presents a GIS based approach to investigations of Romanisation in the Antequera Depression, (Málaga, Spain). Two geographic variables, visibility and relative height, were appraised to investigate territoriality and social dynamics during the Iberian and Roman periods. The results indicate both the continuity of Pre-Roman tendencies and the appearance of new trends in settlement, interpreted as the re-orientation of competitive behaviour amongst local communities from warfare and conflict to more symbolic concerns.

KEYWORDS

Antequera Depression, settlement patterns, Geographical Information System, statistics, Iberian period, Roman Republican period

INTRODUCTION

Over the last 20 years, Geographical Information Systems (GIS) have been widely but unevenly applied to archaeological data across Europe (Wagtendonk et al. 2009, 75-78), mostly in Northern Europe. Spanish, and more precisely Andalusian, archaeology has been slower to take advantage of GIS methods. This does not to imply a total lack of GIS application; a number of researchers in the Guadalquivir valley and the *campiña* of Seville have been using GIS to positive effect (e.g. González Acuña 2001; Keay et al. 2001).

An insight into the development of human occupation in the Pre-Roman and Roman periods in Central Andalusia is considered invaluable due to the central position of the Antequera Depression, which has served as a natural crossroad since prehistoric times. These factors are crucial for gaining a better understanding of the process of Romanisation in Andalusia, as well as an insight into the changing territorial organisation of the area during a historically dynamic period in the Mediterranean basin. In working towards this improved understanding, well-established computing and quantitative methodologies were applied to the Antequera Depression, an area in which spatial analysis has not been previously carried out.

GEOGRAPHY AND ARCHAEOLOGY OF THE ANTEQUERA DEPRESSION

The significance of the Antequera Depression is hard to interpret without a clear image of its spatial context, i.e. its location in relation to the Andalusian region. Andalusia occupies the southern part of the Iberian Peninsula, having an area of 87,268 km². Its topography is diverse, resulting in varied ecological niches, from fertile lowlands close to the Guadalquivir River, to the mountainous chain of the *Sierras Béti-*

Figure 1. Map showing the location of the study area, the Antequera Depression (Malaga, Spain). The location of the Roman town of *Singilia Barba* is shown in the detailed map (see Fig. 2 for its calculated viewshed).

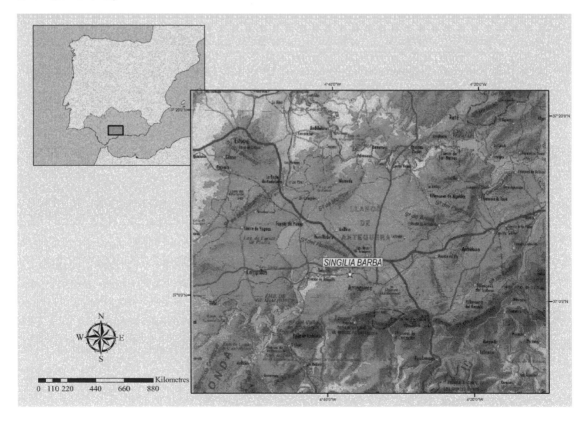

cas, which hosts the highest peak in the Iberian Peninsula, and the desert of Tabernas in *Almería* province. In such a geographical and ecological mosaic, ecotones have played an important role in the historical development of the entire region. One of these areas of interest is the Antequera Depression, which links the fertile plains of the Guadalquivir valley through Fuente Piedra, the Mediterranean coast through the Guadalhorce valley, and the mountains of Eastern Andalusia through the Intrabetic Hollow (it is noteworthy that several modern communication links, including the newly built high speed railway, cross the Antequera Depression). The Antequera Depression itself has a marked agricultural character due to the richness of soils and the relative abundance of water, with an emphasis on crops and olive trees farming (Guarnido Olmero 1977; Mata Olmo et al. 2003, 578-579).

The geographical and ecological setting has favoured human occupation since the Palaeolithic (SIPHA). However, several periods stand out in the historical development of the area, specifically, Late Prehistory (Neolithic, Copper Age and Bronze Age) and the Roman period. In the first case, outstanding evidence of the density of human occupation across the landscape is evidenced by the megalithic monuments of *El Romeral*, *Viera* and *Menga*, which are amongst the most important expressions of the megalithic phenomenon in Southern Europe (refer to García Sanjuán & Wheatley 2009 for a recent update of research in this field). The prominence of the area during the Roman period is shown by its high density of urban centres, such as *Arastipi*, *Antikaria*, *Singilia Barba*, *Nescania*, *Ulisis* and *Osqua*, amongst others.

In addition to these characteristics, it is important to note the abundance of archaeological studies in the Antequera Depression, thanks both to academically led research and rescue archaeology. The creation of local inventories of archaeological sites in the municipalities of this territory has provided a useful mechanism to investigate settlement dynamics of the study area, notwithstanding their limitations and problems as products of field survey (See Barker 1991, and Terrenato 2004 for a detailed account). Rescue archaeology performed at different points of the Antequera Depression in the last three decades has provided high-quality information about the nature and organisation of settlement dynamics across the area.

Although research has focused on the Antequera Depression, few projects have performed spatial analysis, with the exception of the ongoing *Societies, Landscapes and Territories in Late Prehistory of the Antequera Depression* project (García Sanjuán & Wheatley 2009). However, this has not prevented the development of theories regarding settlement patterns and occupation dynamics in this area. Corrales (2002) postulates a progressive transformation of the countryside from the Roman conquest to Imperial times. In her hypothesis, alterations primarily occur in the most fertile areas, and secondly in regions where agricultural exploitation would be less suitable. An alternative hypothesis considers continuity of the Pre-Roman territorial layout during Roman Republican times (Prieto et al. 2001). Two major periods of change are identified: first, the governments of Caesar and *Octavianus*, when important transformations would affect *Baetica* after the civil war between Pompeius and Caesar; and second, the Flavian period (second half of the 1st century AD), after the extension of *Ius Latii* to the provinces of Hispania. Neither of these studies made use of spatial analysis as a means of testing these potential transformations.

Given that the archaeology of the Antequera Depression is relatively well known and in light of the suggested transformations as a consequence of the Roman conquest, spatial analysis was applied to investigate the changing territorial layout of the area.

RESEARCH QUESTIONS AND METHODOLOGIES

This study serves as an initial attempt to investigate transitions and continuities in the settlement patterns of the Antequera Depression after the Roman conquest. It offers an overview of their nature and organisation in both Pre-Roman (Iberian period) and Roman Republican times (5th-3rd century BC, and end of 3rd-1st century BC, respectively). The spatial distribution of archaeological sites across the study area were analysed in relation to different geographical variables, such as visibility and relative height. This had the aim of investigating whether the location of settlements in both periods was chosen in relation to these variables or not, therefore investigating the logic behind settlement patterns and possible changes that the introduction of Roman territorial models might have brought to the Antequera Depression.

These two variables were chosen to integrate into a specific pre-existing framework of research on Pre-Roman societies in Spain. Studies by Parcero Oubiña (2002) on Pre-Roman communities in the north-western Iberian Peninsula (*sociedades castreñas*) emphasise the relevance of defensibility/accessibility, relative height and visibility to site location during the Iron Age. Research carried out in the former *Contestania* (Valencia province, Spain) by Grau Mira (2006) and in the former *Layetania* (Barcelona province, Spain) by Ruestes i Bitrià (2006) have demonstrated the same tendency for Iberian settlements, with a significant emphasis on visibility, as demonstrated by the existence of 'visibility networks' between the main *contestanos* sites.

Based on this past research, the need to examine visibility and relative height for the Antequera Depression is clear. Visibility is important because of the relationship between visual dominance and territoriality, i.e. what can be seen, can be more efficiently controlled. Relative height, on the other hand, is important as a way to show the prominence of site locations in relation to their surroundings, as well as being partially associated with concepts like defensibility and accessibility. However, these latter factors should be investigated in more detail through the application of GIS techniques like Cost Surface Analysis.

Concerning the first of the tools employed in this study, GIS are 'collections of interrelated computer programs designed for the handling and processing of spatially referenced information.' (Kvamme 1999, 154). Since its introduction in archaeology, it has been employed for managing historical and archaeological heritage, for organising and recording the evidence produced in surveys and excavations and for analytical purposes (Conolly & Lake 2006, 33-50). In terms of its application across Europe, GIS has not been widely applied to Roman Archaeology, although some specific examples are worthy of being highlighted: in the *Arroux* valley (France) (Madry & Rakos 1996), on the island of Brač (Croatia) (Stančič & Veljanovski 2000), and, more recently, the Western *Baetica* by Keay and Earl (2007). However, it should be noted that, despite the advantages of GIS in managing and analysing archaeological data, notes of caution have been expressed (Llobera 1996; Tschan et al. 2000; Wheatley & Gillings 2000). GIS have been used in this project in three different ways:

- As a management tool: given the multiplicity of data sources for archaeology in the study region and the importance of its spatial dimension, the use of GIS was employed primarily as a means for organising this information.
- As an extraction tool for deriving information related to the organisation and nature of the archaeological data, such as the prominence of sites in their surroundings, calculated through an index of relative height.

– As an analytical tool to discover relationships in the data, like the percentage of visible area from each site location in order to relate this to hypothetical territorial control measures from individual settlements.

Regarding the use of statistics, no studies focused on the Pre-Roman and Roman archaeology of the Antequera Depression have made use of this as a research tool. However, combining statistics and GIS is of considerable value to landscape archaeology since it allows the validation of hypotheses relating settlement locations to other factors (e.g. environmental, symbolic) (Kvamme 1999; Keay et al. 2001). In general, statistics have not been widely applied in archaeological research, in part due to their association to Processual archaeology by post-processual archaeologists (Shennan 1998, 2-3) and the increasing complexity of the analyses applied to social research (Fotheringham et al. 2000, 8-9) amongst other arguments exposed. By offering a rigorous means of hypothesis testing, simple statistical analyses can offer insights into issues of continuity/discontinuity in territorial layouts, encouraging their use in landscape studies such as that developed for the Antequera Depression. Therefore, statistics were employed to identify patterns in the archaeological data and the comparisons between the characteristics of the archaeological samples. Given the ordinal nature of the data (both the relative height index and high visibility areas were expressed in numerical format), the Kolmogorov-Smirnov (K-S) test was considered to be the most suitable significance test. This is employed for determining the existence or lack of association between the variables and the site distributions in both periods, as a means to interpret site locations as related to the singular characteristics of their surroundings in terms of visual control and prominence in the landscape, or as a matter of chance.

THE ANALYSES AND THEIR PARAMETERS

An indirect selection was introduced into this study based on the chronological classification of sites, due to the poor quality of surface remains in some of the sites. This resulted in a reduction in sites studied from 108 to 47, since these were occupied during either Middle Iberian, Roman Republican or both periods. From these, 17 were inhabited in the period before the Roman conquest (Middle Iberian) and 39 during Roman Republican times, with nine occupied throughout both periods.

ESRI ArcGIS version 9.3 was employed to derive information from the archaeological data, such as the visible area from any site as well as their absolute and relative heights. These data were used in subsequent statistical tests to investigate the existence or absence of correlation between the geographical variables studied and site locations.

In these analyses a Digital Elevation Model with a resolution of 20m was employed. This model is based on data derived from the Andalusian Cartographic Institute. Regarding visibility analyses, individual viewsheds were created for sites in both periods (fig. 2). Two main parameters were established: the radium of visibility and observer height. For the former, a maximum of 15 km was set, which contrasts the two or three km calculated as a maximum of reliable human visibility (García Sanjuán 1999, 133), since the aim of these experiments was to establish the maximum visible area for each site. The environmental and climatic characteristics of the Antequera Depression (more than 300 days of sunshine, dry climate, somewhat hilly landscape) make the *radius* of visibility considerably greater than in other areas of the Iberian Peninsula and Europe, as acknowledged for the Guadalquivir Valley (Keay et al. 2001).

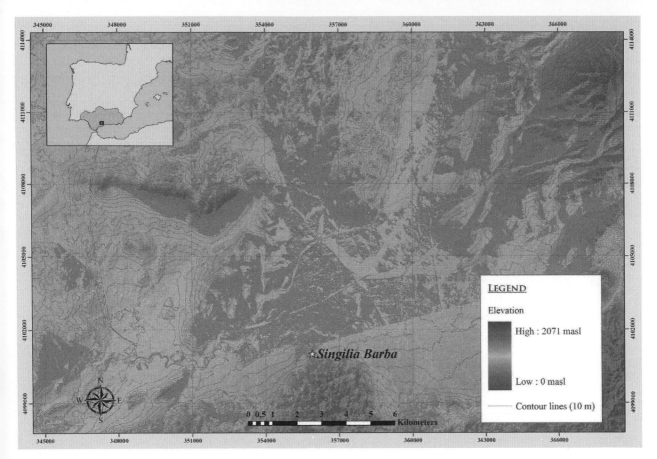

Figure 2. Example of viewshed: shade indicates the non-visible areas within the
fifteen km radius around the analysed site, in this case *Singilia Barba*. Similar
visibility analyses were carried out for each of the sites included in the project, as
well as for a random distribution of sites, in order to compare them statistically.
See also the full colour section in this book

Regarding observer height, due to the lack of information about structure heights in Pre-Roman and Roman times, various estimates were calculated based on the partially-preserved Iberian wall located in *Cortijo Catalán* (Archidona) (Recio Ruiz 1984), to which the height of an adult was added, resulting in 6.20m (which falls between the theoretical heights of Pre-Roman walls considered within research on Iberian architecture (Moret 1996, 95; Zamora Merchán 2006, 35)). This theoretical height was applied to analyse visibility from sites classified as *Fortress*. In the case of sites classified as *Settlement* or *Not determined* the observer height was set at 1.7m. Additionally, the curvature of the Earth was also considered, since this influences visibility by approximately 7.86m for every 10 km from the viewpoint (Conolly & Lake 2006, 229).

Once all the viewsheds were calculated, the visible areas from each site were translated into percentages, tabulated and analysed using the K-S test. It is a significance test that allows the comparison of two datasets with ordinal format in order to acknowledge these as randomly generated or as related (a detailed account of this test can be found in Shennan 1997, 56-61). In this case, two random distributions were created (one for each period) through the Monte-Carlo simulation in order to tabulate the characteristics of the study area (Wheatley & Gillings 2002, 136-137). Viewsheds were calculated for each point of these distributions and the visible areas from each location translated into percentages and tabulated to provide a second entity to compare to the archaeological site distributions. Once the percentages were tabulated and categorised, the K-S test was carried out.

In the case of the sites' relative height, an index was created through the equation:

$$RH = (h_{site} - h_{min}) / (h_{max} - h_{min})$$

Where

h_{site} is the site elevation

h_{max} is the maximum height in the surroundings of the site in a radius of 15 km

h_{min} is the minimum height in the surroundings of the site in a radius of 15 km

The application of this equation generated a result between 0 and 1 as an index of relative height for each site in regards to its surroundings, which was then grouped in different categories.

As with visibility analyses, two random distributions were created for checking the characteristic relative height of the study area, again using the Monte-Carlo simulation and applying the same equation to the heights of each point randomly generated. These were tabulated against the indexes of the archaeological distributions in each period to perform the K-S test.

RESULTS AND THEIR INTERPRETATION IN THE REGIONAL CONTEXT

A number of interesting outcomes were achieved from this study and these will be discussed in relation to research carried out in the municipalities of: Marchena (Seville) (García Vargas et al. 2002); El Coronil (Seville) (González Acuña 2001; Keay et al. 2001), located in the Guadalquivir valley, and in the *campiña de Jaén* (Jaén) (Castro López & Gutiérrez Soler 2001), as well as in the Iberian areas of Eastern Spain *Contestania* and *Layetania* (Ruestes i Bitrià 2006) (fig. 3).

Overall, the results demonstrate a change in patterns of territorial organisation in the Antequera Depression after the Roman conquest.

Figure 3. Map showing the location of Marchena and El Coronil (Seville) (A), campiña de Jaén (Jaén) (B), Contestania (Valencia province) (C) and Layetania (Barcelona province) (D). The territorial layouts of these areas have been studied for the Iberian and/or Roman times, and therefore, their territorial dynamics have been compared to the ones developed in the Antequera Depression.

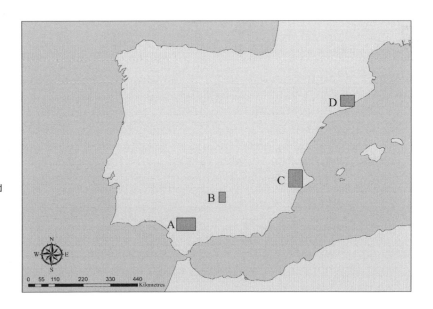

Pre-roman period: the Iberian settlement pattern

The viewshed and statistical analyses demonstrate that siting of settlements took into account visual dominance of the surrounding areas. This would have allowed for greater control of the territory of each community (van Leusen 2002, 16-?1). Since each had the capability for self-organisation and governance, and a tendency towards increased complexity, it is likely that this would have been reflected in the territorial layouts (Recio Ruiz 1994; Grau Mira 2006; Ruestes Bitrià 2006) (fig. 4).

Regarding the relative height of site locations, this factor had an obvious influence for Iberian settlements, and would likely have been key to increasing social complexity. By choosing highly visible areas for settlement locations, it could be speculated that this served as symbolic competition amongst the Iberian communities for social enhancement, as well as providing greater defensibility.

Roman period: the republican settlement pattern.

In contrast with the previous phase, during the Roman Republican period settlements do not appear to be located based on visibility as a parameter. This, together with the abandonment of eight sites (repre-

Figure 4. Map showing both the random and archaeological distributions of sites during the Middle Iberian period, combined with a table showing the different categories of visible areas, and the number of sites counted in each classification. Both distributions were employed for investigating the randomness/relationship of the Iberian settlement pattern with regard to visibility and relative height. *See also the full colour section in this book*

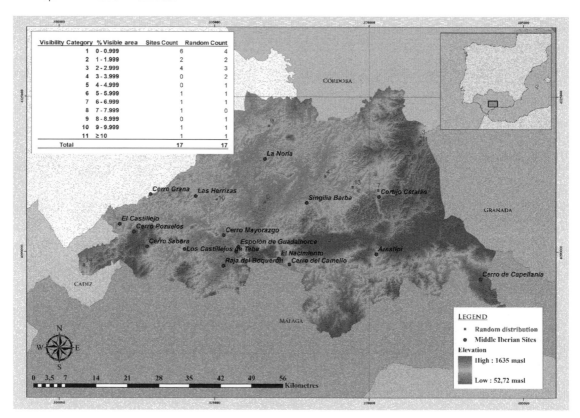

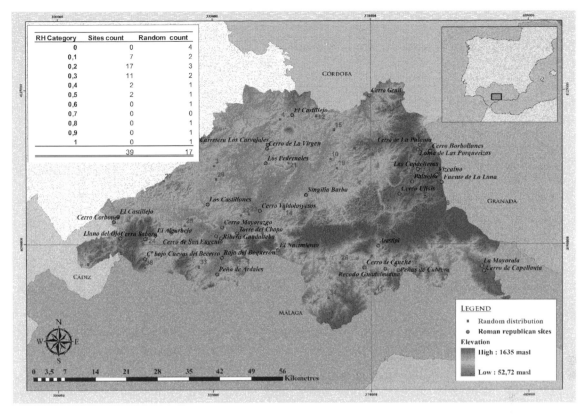

Figure 5. Map showing both the random and archaeological distributions of sites during the Roman Republican period in the Antequera Depression, combined with a table showing the different categories of relative height, and the number of sites counted in each category. Both distributions were employed for investigating the randomness/relationship of the Republican settlement pattern with regard to visibility and relative height. *See also the full colour section in this book*

senting 47% of the Iberian sites in the area) and the appearance of 30 new settlements (almost 77% of the Republican sites), demonstrates the potential impact that incorporation into the Roman world could have had for the communities established in the Antequera Depression, since they were the ones who perceived, organised and modified their landscape (Tort 2006; Delgado Bujalance & Ojeda Rivera 2009).

However, in spite of the transformations in the territorial layout, some elements of continuity remain, for example, the influence of relative height in relation to settlement location (fig. 5). The results demonstrate that this was still taken into account, probably in relation to symbolic competitions amongst communities as they were in prominent places across the landscape. The dynamics involved in these processes are more clearly understood when considering the wider contexts of Southern Iberia and the Iberian Peninsula.

Assuming visibility from settlements towards their surroundings as a locational factor, its importance in the Pre-Roman period and its irrelevance in Republican times may be interpreted as a consequence of the imposition of a higher level of political power (the Roman state) over the region. The indigenous communities may have lost a certain degree of autonomy and the capacity of controlling their own territories. As such, the interest of locating the settlements on places with high visibility decreased, explaining the lack of association between the site distribution in the Roman Republican period and the factor 'high visibility' shown by the statistical analysis. Notwithstanding the previous, it is not necessary to suppose the passive submissiveness of local communities to the Roman state. On the contrary, several interpretative models of Romanisation developed in the last decades highlight the dynamism and prominence of local communities in the transformation of the provinces after the Roman conquest, as well as in the process of constructing Roman Imperial culture (Millett 1990; Woolf 1995; Keay & Terrenato 2001; Bendala 2005; van Dommelen & Terrenato 2007; Revell 2009).

Research undertaken on visibility in Andalusia and other areas of the Iberian Peninsula may illustrate the dynamics shown in the Antequera Depression, although further work is necessary to validate the hypotheses proposed herein. In a number of these investigations, the changes in the territorial organisation has emphasised the decomposition of the Pre-Roman territorial layouts as a consequence of the Roman conquest, as in the case of coastal *Layetania* (Barcelona) (Ruestes i Bitrià 2006). However, the changes in settlement patterns and the influence of visibility upon them after the Roman conquest may be understood also as the development of a new perception, comprehension and organisation of the landscape where the settlements are located (Ojeda Rivera 2003). Visibility and its relation to site location were also investigated within the former province of *Baetica* for Pre-Roman (Turdetanian) and Roman periods by Keay et al. (2001). The authors demonstrated a visual relationship between rural settlements and the urban centres of *Salpensa* (El Casar), *Siarum* (Torre del Águila) and *Callenses* (El Molino Pintado), all located in the municipality of El Coronil (province of Seville). They argued that the visibility of urban areas might have been a means of enhancing Roman authority, instead of the more traditional economic or strategic concerns underlying the development of the Roman territorial layout.

In the *Campiña* of Seville, visibility has been suggested as a factor in selection of settlement location both the in Late Bronze Age and Early Iron Age, as shown by González Acuña (2001). It may have been a means of territorial control of areas of higher agricultural potential and communication routes in the area, mostly between the coast and the interior of the Andalusian region. However, in this case it was not demonstrated statistically that settlement locations were chosen due to their visibility properties, although sound GIS-viewshed analyses were carried out for settlements in the area, relating site location to agricultural potential and communication routes. Still within the provincial context, although visibility analyses have not been made in *Marchena* (García Vargas et al. 2002) or the *Campiña de Jaén* (Castro López & Gutiérrez Soler 2001) the authors suggest in both cases the potential role of visibility in choosing sites dated in Pre-Roman times, such as some Turdetanian sites located on hilltops and the set of towers surrounding the site of *Atalayuelas*, respectively. In the case of *Marchena*, the sites were progressively abandoned or their size and status decreased throughout the Republican period, whereas in *Atalayuelas* the towers were in use during Roman times.

Apart from visibility as a means of control of the surrounding area, additional arguments concerning the location on hilltops emphasise not only the visual control of the surrounding areas, but also the desire of avoiding the occupation of areas suitable for agricultural exploitation by the communities (Zamora Merchán 2006, 37). This will be the subject of future research on the Romanisation of the Antequera Depression.

Regarding the analyses focused on sites' relative height, the results show the importance of this topographical factor in relation to settlement locations for both Middle Iberian and Roman Republican periods. This may be related to different concerns throughout these periods, such as social prominence of the community and defensibility. The location of sites in areas of on high ground could be interpreted in Pre-Roman times as a means of competition between communities, through the symbolic lens, in addition to other concerns such as defensibility, within a wider context of conflict (as shown also by the high relative height index of some sites, such as *Cerro Pozuelos* (Cañete la Real) and *Cerro Sabora* (Cañete la Real). In contrast, during Roman Republican times it could have represented the continuation of the social competition as an alternative for actual warfare between communities. In any case, computational simulation models should be used to investigate these hypotheses relative to other factors, such as the relationship with communication routes (both terrestrial and river) and the location and accessibility of other settlements.

Furthermore, the combined association of Middle Iberian site distribution with both high visibility and relative height factors offers support to the issue of territorial control held by the local communities that inhabited the Antequera Depression between the 5th and 3rd centuries BC.

CONCLUSIONS AND FURTHER WORK

The analyses demonstrate the changing situation of the Antequera Depression between the 5th and 1st centuries BC. Firstly, the Iberian communities would have experienced increased social complexity, reflected in the territorial layout as the appearance of settlements in highly specific locations. Iberian communities chose elevated positions above the surroundings, as a means of visual control, and as a display of their significance within the wider context of the Antequera Depression. Therefore, the territorial layout would show not only political and strategic issues underpinning the historical development of the area, but also some of the mechanisms of competition between the Iberian communities from a symbolic perspective. These social developments would be altered to a certain degree after the Roman conquest, leading to the transformation of the territorial structures within the Antequera Depression. These changes were represented by the disappearance of many of the Iberian sites and the development of many new sites, following a different spatial logic. In Roman Republican times, communities abandoned their previous interest in visual control of the surroundings (and hence, warfare) for an emphasis on the symbolic meaning of settlement locations as a means of social competition. As such, the incorporation of this area within the Roman Empire would have meant not only the transformation of the territorial layout, but also the change of the social dynamics within and between the local communities. However, further research is required, given that this was a pilot study on the Romanisation process in the Roman province of *Baetica*. Some aspects to be studied in the future will be, amongst others:

- The study of other geographical variables, such as the proximity of settlements to springs and rivers, accessibility, proximity to routes of trade and communication, and cultural variables such as proximity to sanctuaries and other meaningful places.

- Explore in greater detail the implications of visibility: smaller *radii* of visual control and the existence of visibility networks.
- Explore the settlement pattern for the Imperial period, as a means of detecting the continuation of trends and new developments.
- The analysis of these aspects in other parts of former *Baetica*, to investigate similarities and differences in the Romanisation process over areas of contrasting characteristics.

REFERENCES

Barker, G. 1991. Approaches to archaeological survey. In G. Barker & J. Lloyd (eds.) *Roman landscapes. Archaeological survey in the Mediterranean region*, 1-9. British School at Rome, London.

Bendala, M. 2005. Urbanismo y romanizacion en el territorio andaluz: aportaciones a un debate en curso. *Mainake* XXVII, 9-32.

Castro López, M. & L. Gutiérrez Soler. 2001. Conquest and Romanization of the Upper Guadalquivir valley. In S. Keay & N. Terrenato (eds.) *Italy and the West: Comparative issues in Romanization*, 145-160. Oxbow, Oxford.

Conolly, J. & M. Lake. 2006. *Geographical Information Systems in Archaeology*. Cambridge University Press, Cambridge.

Corrales Aguilar, P. 2002. La articulación del espacio en el sur de Hispania (de mediados del siglo II a.C. a mediados del siglo II d.C.). *Mainake* XXIV, 443-456.

Delgado Bujalance, B. & J.F. Ojeda Rivera. 2009. La comprensión de los paisajes agrarios españoles. Aproximación a través de sus representaciones. *Boletín de la Asociación de Geógrafos Españoles* 51, 93-126.

García Sanjuán, L. 1999. *Los orígenes de la estratificación social: Patrones de desigualdad en la Edad del Bronce del suroeste de la Península Ibérica (Sierra Morena Occidental c. 1700-1100 A.N.E./2100-1300 A.N.E.)*. Oxford, Archaeopress.

García Sanjuán, L. & D. Wheatley. 2009. El marco territorial de los Dólmenes de Antequera: valoración preliminar de las primeras investigaciones. In B. Ruiz González (ed.), *Dólmenes de Antequera. Tutela y valorización hoy*, 128-143. Sevilla, Junta de Andalucía.

García Vargas, E., M. Oria Segura & M. Camacho Moreno. 2002. El poblamiento romano en la Campiña sevillana: el término municipal de Marchena. *Spal* 11, 311-340.

González Acuña, D. 2001. Análisis de visibilidad y patrones de asentamiento protohistóricos. Los yacimientos del Bronce Final y periodo orientalizante en el sureste de la campiña sevillana. *Archaeologia e calcolatori* 12, 123-142.

Grau Mira, I. 2006. Transformaciones culturales y modelos especiales: aproximación SIG a los paisajes de la romanización. In I. Grau Mira (ed.), *La Aplicación de los SIG en la arqueología del paisaje*, 211-226. Universidad de Alicante, San Vicente del Raspeig.

Guarnido Olmero, V. 1977. La depresión de Antequera. *Cuadernos geográficos de la Universidad de Granada* 7, 39-70.

Keay, S. & G. Earl. 2007. Structuring of the provincial landscape: the towns in central and western Baetica in their geographical context. In G. Cruz Andreotti, P. Le Roux & P. Moret (eds.), *La invención de una geografía de la Península Ibérica*, 305-358. Universidad de Málaga/Casa de Velázquez, Málaga,/Madrid.

Keay, S. & N. Terrenato. 2001. Preface. In S. Keay & N. Terrenato (eds.), *Italy and the West: Comparative issues in Romanization*, ix-xii. Oxbow, Oxford.

Keay, S., D. Wheatley & S. Poppy. 2001. The territory of Carmona during the Turdetanian and Roman periods: some preliminary notes about visibility and urban location. In A. Caballos Rufino (ed.), *Carmona romana*, 397-412. Ayuntamiento de Carmona – Universidad de Sevilla, Carmona.

Kvamme, K. 1999. Recent directions and developments in Geographical Information Systems. *Journal of Archaeological Research* 7 (2), 153-201.

Llobera, M. 1996. Exploring the topography of mind: GIS, social space and archaeology. *Antiquity* 70, 612-622.

Madry, S.H.L. & L. Rakos. 1996. Line-Of-Sight and Cost Surface techniques for regional research in the Arroux River Valley. In H.D.G. Maschner (ed.), *New Methods, old problems. Geographic Information Systems in Modern Archaeological Research*. Southern Illinois University at Carbondale.

Mata Olmo, R., C. Sanz Herráiz, J. Gómez Mendoza & F. Allende Álvarez. 2003. *Atlas de Paisajes Españoles*. Ministerio de Medio Ambiente, Madrid.

Millett, M. 1990. *The Romanization of Britain: an essay in archaeological interpretation*. Cambridge University Press, Cambridge.

Moret, P. 1996. *Les fortifications ibériques de la fin de l'âge du Bronze à la conquête romaine*. Collection de la Casa de Velázquez, 56. Madrid.

Ojeda Rivera, J.F. 2003. Epistemología de las miradas al paisaje. Hacia una mirada humanista y compleja. In J. Fernández Lacomba, F. Roldán, F. Zoido (eds.), *Cuadernos del Instituto Andaluz de Patrimonio Histórico: Territorio y patrimonio. Los paisajes andaluces*, 192-199. Junta de Andalucía, Granada.

Parcero Oubiña, C. 2002. *La construcción del paisaje social en la Edad del Hierro del Noroeste Ibérico*. CSIC - Xunta de Galicia, Ortigueira.

Prieto, A., J. Cortadella & O. Olesti. 2001. Aproximación a la organización territorial de la Depresión de Antequera en época romana. In F. Wulff Alonso, G. Cruz Andreotti & C. Martínez Maza (eds.) *Comercio y comerciantes en la historia antigua de Málaga (siglo VIII a.C. - año 711 d.C.). II Congreso de Historia Antigua de Málaga*, 627-638. Diputación de Málaga, Málaga.

Recio Ruiz, Á. 1984. Aportación a la carta arqueológica del T.M. de Archidona: Estudio de un nuevo yacimiento ibérico. *Mainake* 6-7, 91-103.

Recio Ruiz, Á. 1994. Prospecciones arqueológicas: un modo de aproximación al conocimiento de los procesos de interacción indígenas/fenicios en el valle del Guadalhorce (Málaga). *Mainake* 15, 85-107.

Revell, L. 2009. *Roman imperialism and local identities*. Cambridge University Press, New York – Cambridge.

Ruestes i Bitrià, C. 2006. El poblamiento ibérico y romano en la Layetania litoral (del río Besòs a la Riera de Teià): aplicación arqueológica de un SIG. In I. Grau Mira (ed.), *La Aplicación de los SIG en la arqueología del paisaje*, 227-245. Universidad de Alicante, San Vicente del Raspeig.

SIPHA: Base de datos del Patrimonio Inmueble de Andalucía. Available at http://www.juntadeandalucia.es/cultura/iaph/bdi/frmSimple.do. (accessed on 20 October 2010).

Stančič, Z. & T. Veljanovski. 2000. Understanding Roman settlement patterns through multivariate statistics and predictive modelling. In G. Lock (ed.), *Beyond the map. Archaeology and spatial technologies*, 147-156. Ravello, IOS Press.

Terrenato, N. 2004. Sample size matters! The paradox of global trends and local surveys. In S.E. Alcock & J.F. Cherry (eds.) *Side-by-Side Survey. Comparative regional studies in the Mediterranean World*, 36-48. Oxbow, Oxford.

Tort, J. 2006. Del *pagus* al paisaje: cinco apuntes y una reflexión. In R. Mata & A. Tarroja (eds.), *El paisaje y la gestión del territorio. Criterios paisajísticos en la ordenación del territorio y el urbanismo*, 699-712. Diputació de Barcelona, Barcelona.

Tschan, A., W. Rackzowski & M. Latalowa. 2000. Perception and viewsheds: are they mutually inclusive? In G. Lock (ed.), *Beyond the map: Archaeology and spatial technologies*, 28-48. IOS Press, Amsterdam.

Van Dommelen, P. & N. Terrenato. 2007. Introduction: Local cultures and the expanding Roman Republic. In P. Van Dommelen & N. Terrenato (eds.), *Articulating local cultures. Power and identity under the Roman Republic*, 7-12. *Journal of Roman Archaeology* (Supplementary series), Portsmouth.

Van Leusen, M. 2002. *Pattern to process: Methodological investigations into the formation and interpretation of spatial patterns in archaeological landscapes*, Doctoral thesis, Rijksuniversiteit Groningen, URL: http://dissertations.ub.rug.nl/faculties/arts/2002/p.m.van.leusen/ (accessed 22 October 2010).

Wagtendonk, A.J., P. Verhagen, S. Soetens, K. Jeneson & M. de Kleijn. 2009. Past in Place: The Role of Geo-ICT in Present-day Archaeology. In H.J. Scholten, R. Velde & N. van Manen (eds.), *Geospatial Technology and the Role of Location in Science*, 59-86. Springer, London.

Wheatley, D. & M. Gillings. 2000. Vision, perception and GIS: developing enriched approaches to the study of archaeological visibility. In G. Lock (ed.), *Beyond the map: archaeology and spatial technologies*, 1-27. IOS Press, Amsterdam.

Wheatley, D. & M. Gillings. 2002. *Spatial technology and Archaeology: The archaeological applications of GIS*. Taylor & Francis, London/New York.

Woolf, G. 1995. *Beyond Romans and natives. World Archaeology*, 339-350. London, Routledge.

Zamora Merchán, M.M. 2006. *Territorio y espacio en la Protohistoria de la Península Ibérica. Estudios de visibilidad: El caso de la cuenca del Genil*. Doctoral thesis. Universidad Autónoma de Madrid. Madrid.

5.4 Mapping the probability of settlement location for the Malia-Lasithi region (Crete, Greece) during the Minoan Protopalatial period

Authors

Ricardo Fernandes[1, 2], Geert Geeven[3], Steven Soetens[4,5] and Vera Klontza-Jaklova[6, 7]

1. Leibniz-Laboratory for Radiometric Dating and Isotope Research, Christian-Albrechts-Universität, Kiel, Germany
2. Graduate School "Human development in landscapes", Christian-Albrechts-Universität, Kiel, Germany
3. Department of Mathematics, Faculty of Sciences, VU University Amsterdam, Amsterdam, The Netherlands
4. Institute for Geo- and Bioarchaeology, Faculty of Earth and Life Sciences, VU University Amsterdam, The Netherlands
5. Research Institute for the Cultural Landscape and Urban Environment, VU University Amsterdam, Amsterdam, The Netherlands
6. INSTAP-EC, Pacheia Ammos, Crete, Greece
7. IIHSSA Priniatikos Pyrgos Project, Crete, Greece
Contact: rfernandes@gshdl.uni-kiel.de

ABSTRACT

The current study considers a mixed environmental/historical statistical model to establish a probability map for settlement locations in Crete's Malia-Lasithi region during the Minoan Protopalatial period. The work represents the continuation of previous research that focused on site location choices during the Protopalatial and whereby a comparison was made between the performances of a purely environmental over a mixed environmental/historical model. Statistical modelling consisted of fitting a logistic regression model using a Deletion/Substitution/Addition (DSA) algorithm for model selection. Model uncertainty was assessed through calculation of confidence intervals at the 95% confidence level and the results are presented as probability maps that show upper and lower interval endpoints for the study area. Assessment of the model's predictive performance, on both the study area and on an independent validation area, indicates that the model is able to capture some underlying structure that determines preferences for site locations. Moreover there is a general agreement between the generated settlement probability map and many of the existing published survey results. The results obtained demonstrate the usefulness of the modeling approach and we expect that the existing model can be further improved in the future by incorporating more survey data.

KEYWORDS

Minoan, Protopalatial, Predictive Modelling, Logistic Regression, DSA algorithm

INTRODUCTION

Archaeological predictive modelling (APM) is often used with the underlying meaning of predictive modelling applied to heritage management. For this reason predictive modelling is the focus of much controversy and debate. The main issue of this debate is whether predictive modelling should be used as a tool in establishing a heritage managing policy. The purpose of this paper is not to enter such a debate but rather to emphasise that the role of predictive modelling is not limited to heritage management.

Foremost predictive modelling provides a systematic approach to understand settlement location choices of past populations. With the use of a good theoretical framework based upon the existing archaeological data it then becomes possible to pursue appropriate statistical modelling. A major benefit of a statistical model is that it can be used to both systematically establish the major factors determining site location and to define probability maps that can subsequently be used to guide survey strategies which may further validate the statistical model.

The results presented in this paper represent an extension of previous work (Fernandes et al. 2011) that was strictly focused on the understanding of past location choices for Minoan populations during the Protopalatial period (1900-1650 BC) in central-eastern Crete. That research was able to establish a model with an excellent statistical fit and a sound archaeological interpretation. One of the major results was the establishment of a historical/environmental model in which the proximity to major centres (urban and agricultural) was introduced as a historical variable.

For the current paper the research area was extended as to include the region of Malia-Lasithi as defined by Knappett (1999). This region presents a certain degree of cultural uniformity when compared to the neighbouring regions (fig. 1). An independent validation dataset was reserved in order to assess the model's predictive performance. The fitted model was used to create a probability map for the research area indicating the settlement location probability for the Protopalatial period. The quantitative assessment of model uncertainty based on confidence intervals, which is part of the approach that we present here, is usually lacking in settlement probability mapping.

The settlement probability mapping and model uncertainty assessment provided in the current paper should provide an opportunity to test the statistical model and thus provide a more profound insight into the settlement location choices for the Malia-Lasithi region in Crete during the Protopalatial period. Furthermore, archaeological validation is also needed to establish that the introduction of historical parameters, namely the distance to major centres, is valid or that such parameters are actually the result of an existing surveying bias.

HISTORICAL BACKGROUND

The Protopalatial Minoan period in Crete (Middle Minoan IB–IIB, c. 1900-1700/1650 BC) followed the long Prepalatial period (Early Minoan I-Middle Minoan IA, c. 3500-1900 BC). The Protopalatial period is marked by a gradual social differentiation with the gathering of population in villages culminating in the rise of the first major Minoan urban centres. After a series of earthquakes before 1700 BC, these urban centres were destroyed and rebuilt during the beginning of the Neopalatial period (Middle Minoan IIIB-Late Minoan IB, c. 1650-1480/1425 BC, after Warren & Hankey 1989, 1750-1500 BC, after Rehak & Younger 2001). Further information on Minoan chronology and terminology can be found in Watrous (2001) or Shelmerdine (2008).

As a whole the Protopalatial period represents a period of increased prosperity and tremendous population expansion (Hayden et al. 2004; Watrous 2001). Settlement hierarchy is suggested with information retrieved from architectural elements, high status artefacts and documentary information (Driessen 2001). Settlement hierarchy is connected to the formation of more or less centralised states, linking a city to its hinterland and perhaps extending its influence on a more regional level (Haggis 1999).

STUDY, SAMPLE AND VALIDATION AREAS

An area in central eastern Crete (fig. 1), as assessed by Knappett (1999), was selected for the current study, guided by the premise that material culture reflects a certain degree of unity when considering political, economic and ideological aspects. Knappett (1999) established the area borders, based mainly on studies done on the regional distribution of artefact styles (most notably pottery, and fine tableware). Within the area assessed by Knappett the most problematic area is the Lasithi plateau, with some authors asserting

Figure 1. Study, sample and validation areas.

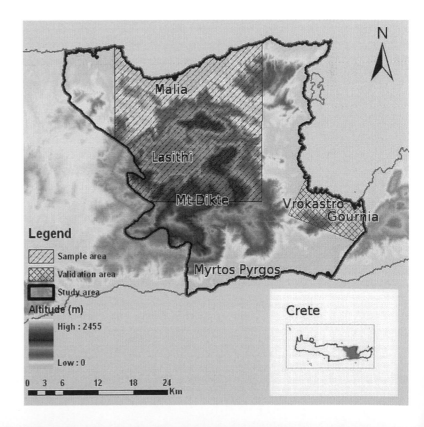

its independent status (Nowicki 1991, 1995) and others referring to the agricultural importance of Lasithi for the Malia state (van Effenterre 1980).

The study area has two interior areas, a sample area (also referred to as sampling area) used to train the statistical model, and a validation area used to verify the model's predictive performance (fig. 1).

GEOLOGICAL, GEOMORPHOLOGICAL AND CLIMATIC ATTRIBUTES OF THE STUDY AREA

At the core of the study area is the Dikti mountain range, one of the three large Cretan mountain complexes. Crystalline limestones make up the mountain range with phyllites-quartzites dominating to its north-east and north-west. The Dikti mountain range is extensively karstified, with common occurrence of caves, basins and several plateaus. The plateau basins consist of a red-brown Quaternary alluvium (Fassoulas 2000; Watrous 2001). The Lasithi plain, to the north of the Diktian massif, is the largest high altitude plateau on the island. The limestone mountain range rests upon a stratum of schist, which protrudes from the lower slopes around the edge of the plain. In the areas where the schist is exposed the water seeps out in the form of springs. The coastline is made up of alternating sandy beaches and steep rocky cliffs. The south-eastern part of the study area, roughly from the city of Agios Nikolaos to the Ierapetra basin, is formed mainly by a thick sequence of Neogene marine sediments, with great extents of Pleistocene deposits (Gaki-Papanastassiou et al. 2008). The extension directly south of the Dikti Mountain is formed by alternating areas of flysch, phyllite-quartzites, Neocene sediments and coastal alluvium. Crystalline limestones predominate in the north-eastern corner of the study area. The coastal area to the north of the Diktian mountain range is characterised by a gentle slope moving upwards to the south of the city of Malia. A phyllitic-quartzitic series, in the mountain slopes south of Malia, acts as a hydrological impermeable substrate of the region (Lambrakis 1998), where Upper Triassic-Upper Eocene karstified limestone and dolomite constitutes groundwater-bearing carbonate sequences that rest on impermeable layers (Lambrakis 1998).

Water is made plentiful available by the existence of a great number of springs and wells, collecting water from aquifers in the porous limestone.

Climatic conditions vary significantly from coastal to mountainous areas. The climate for coastal areas is typically mild Mediterranean, dry and warm. At higher altitudes, as for instance the Lasithi plain at 800m altitude, the climate is clearly of a non-Mediterranean type, with an average winter temperatures from 5°C to 10°C (Watrous 1982).

Data

The data used in this paper forms part of the Digital Crete Project database (Digital Crete 2008). For the current paper Protopalatial settlement locations were incorporated into a Geographical Information System (for the current study the software package ArcGIS® 9.2 was used). Sampling and validation areas were selected due to the higher quantity and quality of settlement data available, with respectively 132 and 118 selected sites.

Different GPS sensors were used to determine the site location with an accuracy of up to three metres. An ASTER-derived Digital Elevation Model (DEM) with a 30 metres resolution was employed. Supportive evidence for the sufficiency of the DEM resolution is provided by the Vrokastro survey (Hayden

2004) which included detailed information on site dimensions for the Protopalatial period and for which the calculated average for site diameter was of 70 metres.

Further details on the methods of data acquisition, satellite imagery, alongside other information on vector and raster datasets can be found in Sarris (2005) and Digital Crete (2008).

Variables

Seven independent variables (hereafter also referred to as predictors) were considered (table 1). The independent variables were selected from existing available digital data (Digital Crete 2008) as possible indicators of economic, environmental/climatic and cultural factors that would determine the site location. Including these variables resulted in a model with good predictive performance and we expect that further improvements can be realized by integrating more digital data that will become available in the:

Table 1. Variables considered in the current study.

Variable	Type	Categories
Altitude (X_1)	Continuous	
Slope (X_2)	Continuous	
Spring Density (X_3)	Continuous	
Cost distance to coastline (X_4)	Continuous	
Cost distance to major centres (X_5)	Continuous	
Surface Geology (X_6)	Categorical	Alluvium (1), tertiary deposits (2), limestone colluvium (3), hard limestones (4), mixed flysch (5), argillaceous flysch (6), schists (7), peridotites (8), granite (9), deposition cones (10), dolines (11), not identified (12), river beds (13), schist colluvium (14), gneiss (15)
Soil depth (X_7)	Categorical	Deep (1), shallow and deep (2), bare and shallow (3), deep and shallow (4), shallow and bare (5), shallow (6), deep and bare (7), bare and deep (8), bare (9), not identified (10)

Altitude: Climatic differences in the island of Crete are closely connected with altitude. Climate during the Middle Minoan period in the eastern Mediterranean was more humid and colder than at present (Moody & Rackham 1997; Moody 2000; Issar 1995; Issar & Makover-Levin 1996), most likely implying a decrease in the probability of finding high altitude settlements.

Slope: Minoan settlements are often located in a sloped area neighbouring an agriculturally valuable plateau, e.g. Lasithi plateau. This type of placement, when only considering data from a single cell location, would lead to wrong conclusions, which are overcome by considering cost surfaces or *de facto* site locations (see section *Least cost path analysis*).

Spring density: With rare examples of perennial rivers, modern Crete water supply relies heavily on pumped water from its several aquifers (Lambrakis 1998). With a colder and more humid climate during the Middle Minoan, the question arises whether past wells and springs represented also the main water source. In many cases there is a close proximity between a well or spring location and the site location. Hayden et al. (2004) reported for the Vrokastro area during the Protopalatial, that there was a close relationship between the site location and proximity to a well or spring. Archaeological evidence shows that water for the Malia cisterns was collected from springs and small mountain streams using a series of canals (Viollet 2003). Knossos was supplied both by a series of wells and by its aqueduct that initially connected to a spring (Angelakis et al. 2007). Seismic activity can affect aquifer conditions greatly and this in turn leads to variations in well levels, and on spring existence (Gorokhovich 2005), which might explain the lower percentage of springs neighbouring the settlements in the Malia region. The fluctuations in spring locations make it inadvisable to use distance to spring as a parameter. Therefore two assumptions were made. Firstly that there is a close relationship between spring and well locations, and secondly, that spring locations in current days still bear some relationship to the location of springs during the Protopalatial period. In order to parameterise the importance of spring locations, spring density was calculated for each map cell using Kernel density estimation (Silverman 1986) with a search radius of 1250 metres.

Cost distance to the coastline: The major urban Minoan centres are typically in the vicinity of the coastline, providing easy access to sea resources and maritime trade (Agouridis 1997; Cline 1994). Crete's coastline acquired approximately its present shape during the Lower Pleistocene (Moody & Rackham 1997; Lambeck 1996). Since the Bronze Age till present eustatic sea level changes, tectonic movements, and depositional/erosional processes have further induced changes in the coastline (Rapp & Kraft 1978). An isostatic model indicates that the sea level rose in Crete in the past 2000 years about 1.5 metres (Pirazzoli 2004). Seismo-tectonic movements have produced larger changes, with events between 4000 BP and 1500 years BP, consisting of both sudden small sea-level rises (around 25cm), and of sudden emergences like those registered in Western Crete (between 2.7m and 9m) (Moody & Rackham 1997). Further complexity is introduced when assessing effects of sedimentation/erosion at a local level. It should be expected that the Cretan coastline, since the Bronze Age, might have undergone changes up to several hundred metres, with variations along its extent. A probability map should be analysed at a regional scale that mitigates considerations into variations in ancient landscape. A cost distance analysis (see next section *Least cost path analysis*) was performed in order to define the least cumulative cost from each cell in the study area to the coastline.

Cost distance to major centres: Within a political/cultural unity it is assumed that communication routes would link settlements according to their hierarchical order. This is especially relevant within the Minoan context where Palace complexes probably played an important role in a form of centralised economy (Driessen 2001). Four locations were selected due to their expected historical importance and hierarchical level both within and in the vicinity of the study area, Malia, Lasithi plain, Myrtos Pyrgos, and Vrokastro/Gournia. A previous study (Fernandes 2009) considered the distance to least cost paths linking major centres as a parameter. However, given the limitations of the D8 spreading algorithm used in

ArcGIS evident especially at larger scales, an alternative was procured by considering the least cumulative cost distance from each cell in the study area to the previously defined major centre locations (see section *Least cost path analysis*). This approach is less sensitive to the determination of a specific route while providing an improved estimate over simple Euclidean distances of the costs of transposing the areas under consideration.

Surface geology: Surface geology serves as an indicator for available natural resources, namely agricultural potential (e.g. rich alluvial areas). With the overall stability of Cretan landscape during the Holocene (Moody & Rackham 1997; Moody 2000), and apart from coastal localities, it is not expected that since the Protopalatial important geological modifications have been operated at the regional scale that this study reflects.

Soil depth: Soil depth is an important indication of agricultural potential (Bulte & Soest 1999). This variable is however far more sensitive than surface geology to changes operated in the last four thousand years (Dietrich et al. 1995).

LEAST COST PATH ANALYSIS

A simple Euclidean distance does not reflect the real difficulties of traversing the area under study. Therefore we present here a more realistic approach that uses a least cost path analysis. This type of approach has already been done in the context of Minoan Crete (Siart et al. 2007; Soetens et al. 2003; Soetens et al. 2008).

The cost of traversing each map cell in the study area was defined by using as a cost function the value of the inverse horizontal speed ($1/v$) determined from the terrain's slope (m), with a, b, and c as parameters:

$$\frac{1}{v} = a + bm + cm^2 \ (1)$$

$a = 0.75ms^{-1}$
$b = 0.09ms^{-1}$
$c = 14.60ms^{-1}$

Equation (1) was derived by Rees (2004) and it was established empirically by traversing mountainous foot paths. This type of cost function is adequate for the Cretan mountainous terrain where slope plays a major role and rivers do not constitute significant obstacles. It was also assumed that surface vegetation and surface geology play comparatively a lesser role in the determination of the cost function.

A Geographical Information System (GIS) implements algorithms and provides tools to determine the least cumulative cost of traversing all cells from a chosen cell location to a destination. For the current study least cost calculations were done using the Cost Distance tool in the Spatial Analyst extension of ESRI's ArcGIS© 9.2 GIS software package.

Least cost path analysis was employed both as a distance measure and also to define cost surface areas. Site locations were defined as de facto site locations consisting of catchment areas or cost surfaces buffering each site location considering the equivalent in cost distance units of a Euclidean planar distance of 1250 metres. The distance of 1250 metres was selected after an informal assessment of the model's performance based on two main criteria: an accessible walking distance and limiting catchment overlap to nearest neighbours. Under this approach we are interested in determining settlement choices considering the site's immediate hinterland. Hence the model's results should be interpreted at an appropriate scale.

LOGISTIC REGRESSION

To analyse the data and build a model that can be used for prediction, we use a multivariate Logistic Regression (LR) model. LR is the classical tool for APM, representing Kvamme's (1983a, 1983b, 1988) integrated approach. LR is part of a statistical modelling framework known as Generalized Linear Models (GLMs). GLMs (McCullagh & Nelder 1989; Dobson 1990) are natural extensions to ordinary least squares regression in which the response variable Y is related to predictor variables through some, possibly nonlinear, link function. GLMs are among the most widely used models in applications of statistics and analysis routines are well established. LR is a generally applicable method that can be used for predictive modelling and classification. It can naturally handle both categorical and continuous predictor variables. Since we do not want our results to depend in any way on a particular, subjective chosen prior, which may be the case in Bayesian inference, we use classical frequentist methods for model fitting and inference (Venables & Ripley 2002). In this paper we are interested in modelling the probability that a given cell on the map in the study area is a de facto site, given a number or relevant archaeological and geological covariates, e.g. spring density or cost distance to coastline. Let Y denote this probability and let X be a p dimensional vector of relevant covariates. The logistic model assumes a linear model for the logarithm of the odds of site presence, i.e.

$$\log\left(\frac{P(Y=1\,|\,X=x)}{P(Y=0\,|\,X=x)}\right) = \beta_0 + \beta_1\beta x_1 + \ldots + \beta_p\beta x_p \quad (2)$$

For a given model, where the relevant predictor variables are known, statistical inference such as point estimates for relevant parameters and confidence intervals can be obtained using standard methods. Most relevant to our application are point estimates and confidence intervals for the probability of de facto site presence, rather than estimates for the individual β's. However, since the relationship between site presence and the seven candidate predictor variables we consider is unknown, we need to implement a model selection procedure to first select an appropriate model on which we can base our inference. Apart from having a good predictive performance, which is a requirement in order to be practically useful, we need our model to be archeologically plausible. This means that, if predictive performance can be only marginally improved with a model containing many terms including higher order interactions, we prefer a smaller model without complex interactions for reasons of interpretability. In the final stage of model selection, we keep this pragmatic two-fold aim in mind.

MODEL SELECTION

Given the observed binary response Y, indicating site presence, and a covariate matrix $X = [X_1...X_7]$ containing five continuous and two categorical predictor variables, our goal is to find a model for Y with a good predictive performance. Here, we use the Deletion/Substitution/Addition (DSA) (Sinisi & Van der Laan 2004) algorithm for model selection.

The use of DSA offers an important methodological improvement over traditional stepwise model selection procedures. Stepwise procedures for variable selection in linear regression are known to be suboptimal and should be avoided. The DSA procedure systematically and progressively builds more complex models that contain more variables and interactions between them in an attempt to find an optimal model that fits the data well and has good predictive performance.

The DSA algorithm performs data-adaptive estimation through estimator selection based on cross-validation and the L_2 ('squared error') loss function. DSA generates predictors as linear combinations of tensor product polynomial basis functions. We believe polynomial regression is appropriate since it is not clear that predictors such as cost distance to coastline and spring density are linearly related to the (*logit* of) the probabilities of site presence/absence. Allowing a predictor which is a linear combination of products of terms of different powers of the original predictor variables enables us to approximate the true relationship between predictors and outcome. The DSA algorithm generates a sequence of candidate models by minimising the empirical loss function over subspaces indexed by three user-defined parameters. These are:

1. γ_1: The maximum number of terms in the model.
2. γ_2: The maximum order of interactions allowed in the model.
3. γ_3: The maximum allowed sum of the powers of variables involved in a single term.

Starting from a model with only an intercept, the DSA algorithm searches the model space by making different moves. At each step, it chooses between an addition, deletion or substitution move, hence its name. After the optimal model in each subspace is identified, it selects the final model based on V-fold cross-validation.

The sampling area was randomly sampled using 10,000 points and the model was trained and statistical analysis performed using the programming language R (R Development Core Team 2007) and the 'DSA' package (Sinisi & Van der Laan 2004). The model was tested for its validity by considering 5,000 random points from the validation area.

RESULTS AND MODEL EVALUATION

The DSA algorithm was applied to the data from the study area, with default 5-fold cross-validation. We experimented with different settings of the γ's. In the final run, we chose to set $\gamma_2 = 2$. Although a higher setting did result in (marginally) lower cross-validated risks, we find third order interactions unsatisfactory since the practical interpretation of the selected higher order terms is unclear. From an archaeological point of view, we prefer to keep the model simple and since adding more complexity by introducing higher order interactions does not significantly improve the predictive performance of the model, we decided to include only pair-wise interactions between the predictor variables. Because it is unsure that, especially for the two continuous cost distance variables, the original variables are linearly related to the response, polynomial terms were included up to order three in order to be able to approximate the true relationship between predictors and response. Thus, the required parameters were finally set as follows: $(\gamma_1, \gamma_2, \gamma_3) = (8, 2, 3)$. Note that the DSA algorithm generates candidate models that include candidate predictors which are functions of the original input variables.

The final model returned by the DSA algorithm for the sampling area was:

$$\log\left(\frac{P(Y=1|X)}{P(Y=0|X)}\right) = 6.23 - 0.00335X_5 + 3.00 \times 10^{-7}X_5^2 + 0.946X_3 - 8.82 \times 10^{-5}X_4X_6 +$$

$$0.238X_2X_6 - 8.95 \times 10^{-12}X_5^3 + 4.16 \times 10^{-09}X_4^2X_7 \quad (3)$$

The LR model fitted to the data from the study area can be viewed as a binary classifier. Given values for the predictor variables in the model at any location, the LR model outputs the probability that an archaeological site is present at that location. Given a specific threshold, these probabilities can be converted in site predictions. Given the truth, i.e. information regarding whether a site is actually truly present at each location, predictions made by the model can be classified as either *TRUE* or *FALSE* respectively depending on whether the predictions *do* or *do not* resemble the truth. Hence, positive predictions (which are predictions of site presence at a location) are classified as either True Positive (TP) or False Positive (FP) predictions. Analogously, negative predictions (which are predictions of site absence at a location) are classified as either True Negative (TN) or False Negative (FN) predictions. The performance of binary classifiers is often evaluated using Receiving Operator Characteristic (ROC) curves (Peterson et al. 1954), also used within an archaeological context (Finke et al. 2008). In these curves, TP rate (TPr) is plotted versus FP rate (FPr),

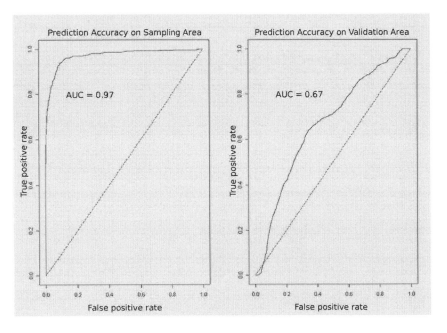

Figure 2. Model evaluation by ROCs for the sampling area (left) and the validation area (right).

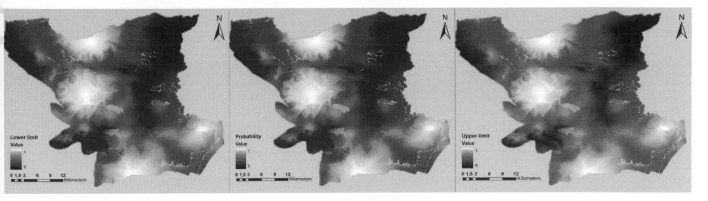

Figure 3. Probability map for the study area, with upper and lower values (at different scales) corresponding to the 95% confidence interval. *See also the full colour section in this book*

for all thresholds in the range (0, 1). The area under this curve (AUC) is a widely used performance measure. The AUC is a value between 0 and 1. An AUC of 1 corresponds to a 'perfect' predictor, whereas 0.5 is the expected AUC value of a 'random' predictor. Hence the predictive performance of our model was evaluated by plotting ROC curves and computing the AUC values (fig. 2).

PROBABILITY MAP

Through the establishment of a predictive model it becomes possible to create a map providing probability estimates for the location of archaeological sites. In this respect the approach here presented is no different from previously mentioned studies. However often lacking, within current research, is a probability error assessment which constitutes an important methodological deficiency. An analogy can be done, for instance, with the reporting of radiocarbon dates without indicating the type of distribution used or the associated standard deviation. Information on confidence intervals serves as a measure of uncertainty which can be very useful in the implementation of a survey strategy, namely by optimising the selection of preferential survey targets.

Given values for the five relevant geographical variables at any point of a map, the model can be used to estimate the probability of a site presence. Let, for any point k on the map, $x_k = (x_{1k},..., x_{7k})$ be the vector of local geographical data. Let $Y_k = 1$ denote the event that there is a site at point k. We use the model to predict

$$\hat{\eta}_k(x_k) = \log\left(\frac{P(Y_{k=1}|X_{k=k})}{P(Y_{k=0}|X_{k=k})}\right) = \hat{\beta}_0 + \hat{\beta}_1 x_{1k} + ... + \hat{\beta}_7 \beta x_{7k} \ (4)$$

Let $Z \sim N(0, 1)$ and $z_{\alpha/2}$, be the quantile of the standard normal distribution defined by $P(Z > z_{\alpha/2}) = 1 - \alpha/2$. Then, we use classical linear model theory which asserts that, under commonly made assumptions, that $[\hat{\eta}_k(x_k) - s_k z_{\alpha/2}, \hat{\eta}_k(x_k) + s_k z_{\alpha/2}]$ is a $(1 - \alpha)$ % confidence interval for $\hat{\eta}_k(x_k)$. Here s_k is the estimated standard error of $\hat{\eta}_k(x_k)$, which can be calculated from the obtained fit. We use the confidence interval, with significance level $\alpha = 0.05$, to quantify the uncertainty in the estimated probability at point k in the map (fig. 3). Figure 3 constitutes an illustration of the final results as these should be interpreted within a GIS environment, where for the entire region the probability and associated confidence interval values can easily be obtained. The upper and lower intervals are not symmetric towards the probability value (readily visible in some map areas), thus presenting both the upper and lower limits constitutes the most informative form of representation.

DISCUSSION AND CONCLUSION

The assessment of the model's performance using ROC curves (fig. 2) resulted in an AUC value of 0.97 for the sampling area which is indicative of an excellent fit (Hosmer & Lesmow 2004) and an AUC value of 0.67 for the validation area which represents a satisfactory result. This result indicates that the model is indeed able to capture some of the 'true' underlying structure that determines the preference for a site location. The lower performance of the model in the validation area is observed especially in the lower left corner of the graph of the ROC-curve (fig. 2, right), i.e. there are some locations with high probability values that are actually false positives. This may be due to differences in distributions of values for some of the predictor variables, i.e. overall differences in values for cost distance to major centres or surface geology between the sample and validation areas due to geological differences. Note that we did not re-estimate the coefficients of the trained model to fit the data from the validation area, which explains at least in part the reduction in the predictive performance of the model. We also point out that from a strictly archaeological point of view, the occurrence of some false positive predictions can be tolerated as long as the overall predictive performance is good and the model gives valuable insights. Under a theoretical perspective it can also be argued that a fully deterministic model is not be expected as several dynamic and cultural aspects are absent from model consideration, a critique often addressed within the post-processual movement (van Leusen et al. 2005).

In relation to the distribution of known archaeological sites the probability map corresponds well with settlement locations that were identified in the field, i.e. the Malia survey (Müller 1996, 1998, 2000), the Vrokastro Survey (Hayden et al. 2004), the surveyed areas of the Lasithi high plain (Watrous 1982) and the extensively surveyed surrounding mountains by Nowicki (1998). The probability map indicates to the north-west of the Lasithi plains areas of high probability for settlement location. As such the region west of Lasithi, the Pediada is known for a dense site distribution (Panagiotakis 2006), as well as the area towards Kavousi (Haggis 2002). High probability areas extended well beyond the natural limits of the Lasithi plain. This is particularly significant for the east and north-east of the Lasithi plain as Nowicki's surveys reported several archaeological sites along the valley of Amygdali to Mesa Potami (Nowicki 2000), which is the eastern natural passage that goes from Agios Nikolaos to the Lasithi high plain.

Protopalatial site locations of high probability outside sample and validation areas do also confirm a consistency with the archaeological reality, as there is an emphasis on the south coastal plain of Ierapetra, reaching as western edge of potential, the valley of Myrtos Pyrgos, from where higher potential is also predicted going north along the villages of Mythi and Males (Nowicki 1998) and even reaching Selakano. High archaeological potential is also defined further south in the area of the villages Meseleri, Prina, Kalamafka and Anatoli where archaeological sites have been reported by Watrous (1996) and Nowicki (1998, 2000).

The clear concentration of high probability areas surrounding the defined major centres is not surprising given the high importance of the variable cost distance to major centres in the model. The main issue to consider is to what extent the results reflect a surveying bias that may be caused by training the model in the sample area that includes the major centres of Malia and the Lasithi plain. It should be noticed however that the model results for the major centres outside the sample area (Myrtos Pyrgos & Vrokastro/Gournia) are, at least partially, in accordance with existing surveying data (Nowicki 1998; Hayden et al. 2004).

It is hoped that the results presented here will serve as a guideline for further field research, both to consider less surveyed areas but also to perhaps reassess previous survey results. In the case of Crete Bonnefant (1972) reported that only 12.5% of the island had been surveyed with only about 5% of the results published. These numbers have been improved in recent decades but unpublished results remain a major problem. The nature of the surveys also varies in terms of intensity, and there are many associated problems. The finds' visibility is often linked to the nature of the soil and whether ploughing has or has not disturbed the area under survey. Surveying is often made difficult by climate conditions or by terrains of hazardous access. It is often difficult to determine the site's extent, site's chronology, and to verify extent changes within a site's chronology. Even in the cases of well studied sites there is sometimes reluctance in destroying later layers in the pursuit of earlier ones.

Of special interest should be the areas outside the immediate proximity of major centres for which there is an indication of both high and low probability targets. In this respect the evaluation of the error maps is especially relevant as it provides a better insight into model uncertainty and thus allows for an optimisation of a survey strategy.

This work should be viewed as having a continuously ongoing nature, whereby the model here presented can be reassessed and improved by an evaluation done in the field and by the update of existing data. Particularly relevant is information obtained from paleo-environmental/climatic reconstructions as this would allow assessing settlement location choices as based on present-day data. A specific example of this type of reconstruction is the integration of data concerning Minoan spring locations which due to past seismic activity have subsequently often become inactive.

ACKNOWLEDGMENTS

The authors would like to thank the Institute for Mediterranean Studies – FORTH for having kindly provided most of the raw data used in this study. We would also like to thank the editors and reviewers of the LAC2010 proceedings for their kind comments which have greatly benefitted the current paper.

REFERENCES

Agouridis, C. 1997. Sea routes and navigation in the third millennium Aegean. *Oxford Journal of Archaeology* 16, 1-24.

Angelakis, A.N., Savvakis, Y.M. & Charalampakis. G. 2007. Aqueducts during the Minoan Era. *Water Science and Technology: Water Supply* 7, 103-111.

Bonnefant, J.C. 1972. *La Crète. Etude Morphologique*. Service de réproduction de thèses, Université de Lille III, Lille.

Bulte, E.H. & van Soest, D. 1999. A note on soil depth, failing markets and agricultural pricing. *Journal of Development Economics* 58, 245-254.

Cline, E.H. 1994. *Sailing the Wine-dark Sea: International Trade and the Late Bronze Age Aegean*. BAR International Series 591, Oxford.

Dietrich, W.E., Reiss, R., Hsu, M. & Montgomery, D.R. 1995. A process based model for colluvial soil depth and shallow landsliding using digital elevation data. *Hydrological Processes* 9, 383-400.

Digital Crete, http://digitalcrete.ims.forth.gr (accessed on 24 July 2008).

Dobson, A.J. 1990. *An Introduction to Generalized Linear Models*. Chapman & Hall, London.

Driessen, J. 2001. History and hierarchy. Preliminary observations on the settlement pattern in Minoan Crete. In Branigan, K. (ed.), *Urbanism in the Aegean Bronze Age*, 51-71. Sheffield Centre for Aegean Archaeology, University of Sheffield.

van Effenterre, H. 1980. *Le Palais de Mallia et la cité minoenne*. Edizione dell'Ateneo, Rome.

Fassoulas, C. 2000. *Field Guide to the Geology of Crete*. Natural History Museum of Crete Publications.

Fernandes, R. 2009. New ideas in predictive modelling: A Minoan case study. In Kalkers, A., Kiefte, D. & Termeer, M.K. (eds.), *SOJA Bundel 2009*, 92-103. Amsterdam.

Fernandes, R., Geeven, G., Soetens, S., Klontza-Jaklova, V. (2011), Deletion/Substitution/Addition (DSA) model selection algorithm applied to the study of archaeological settlement patterning. *Journal of Archaeological Science*.

Finke, P.A., Meylemans, E. & Wauw, J.V. 2008. Mapping the possible occurrence of archaeological sites by Bayesian inference. *Journal of Archaeological Science* 35, 2786-2796.

Gaki-Papanastassiou, K., Karymbalis, E., Papanastassiou, D. & Maroukian, H. 2008. Quaternary marine terraces as indicators of neotectonic activity of the Ierapetra normal fault SE Crete (Greece). *Coastal Geomorphology* 104, 38-46.

Gorokhovich, Y. 2005. Abandonment of Minoan palaces on Crete in relation to the earthquake induced changes in groundwater supply. *Journal of Archaeological Science* 32, 217-222.

Haggis, D.C. 1999. Staple Finance, Peak Sanctuaries, and Economic Complexity in Late Prepalatial Crete. In Chaniotis, A. (ed.), *From Minoan Farmers to Roman Traders: Sidelights on the Economy of Ancient Crete*, 53-85. Franz Steiner Verlag, Stuttgart.

Haggis, D.C. 2002. Integration and Complexity in the Late Prepalatial Period: A View from the Countryside in eastern Crete. In Hamilakis, Y. (ed.), *Labyrinth Revisited: Rethinking 'Minoan' Archaeology*, 120-142. Oxford.

Hayden, B., Dierckx, H., Harrison, G., Moody, J., Postma, G., Rackham, O. & Stallsmith, A. 2004. *Reports on the Vrokastro Area, Eastern Crete. Volume 2: The Settlement History of the Vrokastro Area and Related Studies*. University Museum Monograph 119, Philadelphia: University of Pennsylvania, 81-104.

Hosmer, D. & Lemeshow, S. 2004. *Applied Logistic Regression*, 2nd edition. John Wiley & Sons, New York.

Issar, A.S. 1995. Impacts of climate variations on water management and related socio-economic systems. In *Technical Documents in Hydrology*, International Hydrological Programme, UN Educational, Scientific and Cultural Organization, Paris, France.

Issar, A.S. & Makover-Levin, D. 1996. Climatic changes during the Holocene in the Mediterranean region. In Angelakis, A.N. & Issar, A.S. (ed.), *Diachronic Climatic Impacts on Water Resources with Emphasis on Mediterranean Region*, 55-75. Springer-Verlag, Heidelberg.

Knappett, C. 1999. Assessing a Polity in Protopalatial Crete: The Malia-Lasithi State. *American Journal of Archaeology* 103, 615-639.

Kvamme, K. 1983a. Computer processing techniques for regional modelling of archaeological site locations. *Advances in Computer Archaeology* 1, 26-52.

Kvamme, K. 1983b. *A manual for predictive site location models: examples from the Grand Junction District, Colorado*. Bureau of Land Management, Grand Junction District, Colorado, USA.

Kvamme, K. 1988. Development and Testing of Quantative Models. In Judge, W.J. & Sebastian, L. (eds.), *Quantifying the Present and Predicting the Past: Theory, Method, and Application of Archaeological Predictive Modelling*, 325-428. U.S. Department of the Interior, Bureau of Land Management, Denver.

Lambeck, K. 1996. Sea-level Change and Shoreline Evolution in Aegean Greece since Upper Palaeolithic Times. *Antiquity* 70, 588-611.

Lambrakis, N.J. 1998. The impact of human activities in the Malia coastal area (Crete) on groundwater quality. *Environmental geology* 36, 87-92.

van Leusen, M. & Kamermans, H. (eds.) 2005. Predictive Modelling for Archaeological Heritage Management: A research agenda. *Nederlandse Archeologische Rapporten* 29.

McCullagh, P. & Nelder, J.A. 1989. *Generalized Linear Models*. Chapman & Hall, London.

Moody, J. 2000. Holocene Climate Change in Crete: an Archaeologist's View. In Halstead, P. & Frederick, C. (eds.), *Landscape and Land Uses in Postglacial Greece*, 52-61. Sheffield Studies in Aegean Archaeology, Sheffield.

Moody, J. & Rackham, O. 1997. *The Making of the Cretan Landscape*. Manchester University Press, Manchester.

Müller, S. 1996. Malia. Prospection archéologique de la plaine de Malia. *BCH* 120 (2), 921-928.

Müller, S. 1998. Malia. Prospection archéologique de la plaine. *BCH* 122 (2), 548-552.

Müller, S. 2000. Malia. Prospection archéologique de la plaine. *BCH* 124(2), 501-505.

Nowicki, K. 1991. Report on Investigations in Greece VII. Studies in 1990. *Archaeologia* 42, 143-45.

Nowicki, K. 1995. Report on Investigations in Greece X. Studies in 1993 and 1994. *Archaeologia* 46, 63-70.

Nowicki, K. 1998. Lasithi (Crete): One Hundred Years of Archaeological Research. *AEA* 3, 27-47.

Nowicki, K. 2000. Defensible Sites in Crete c. 1200-800 B.C. *AEGAEUM* 21.

Panagiotakis, N. 2006. Oikistiki topografia stin eparchia Pediados apo ti Neolithiki periodo os tin Ysteri epochi tou Chalkou. In Kaloutsakis, E. & Kaloutsakis, A., *Pepragmena Th' Diethnous Kritologikou Synedriou, Elounta, 1-6 Oktovriou 2001. A2: Proïstoriki Periodos, Architektoniki, Tampakaki*, 167-182. Irakleio, Etairia Kritikon Istorikon Meleton.

Peterson, W.W., Birdsall, T.G. & Fox, W.C. 1954. The theory of signal detectibility. Transactions of the IRE Professional Group in Information Theory. *PGIT* 2 (4), 171-212.

Pirazzoli, P.A. 2005. A review of possible eustatic, isostatic and tectonic contributions in eight late-Holocene relative sea-level histories from the Mediterranean area. *Quaternary Science* 24, 1989-2001.

R Development Core Team. 2007. *R Development Core Team, R: a Language and Environment for Statistical Computing*. R Foundation for Statistical Computing, Vienna, Austria available from: http://www.R-project.org.

Rapp, G. & Kraft, J.C. 1978. Aegean Sea Level Changes in the Bronze Age. In Doumas, C. (ed.), *Thera and the Aegean World I*. International Congress Papers, London, 183-193.

Rees, W.G. 2004. Least-cost paths in mountainous terrain. *Computers & Geosciences* 30, 203-209.

Rehak, P. & Younger, G.J. 2001. Neopalatial, Final Palatial, and Postpalatial Crete. In Cullen, T. (ed.), *Aegean Prehistory. A Review. American Journal of Archaeology Supplement* 1. Archaeological Institute of America, Boston, 383-473.

Sarris, A., Karakoudis, S., Vidaki, C. & Soupios, P. 2005. Study of the Morphological Attributes of Crete through the Use of Remote Sensing Techniques. *IASME* 6 (2), 1043-1051.

Shelmerdine, C. (ed.) 2008. *The Cambridge Companion to the Aegean Bronze Age*. Cambridge, Cambridge University Press.

Siart, C., Eitel, B. & Panagiotopoulos, D. 2008. Investigation of past archaeological landscapes using remote sensing and GIS: a multi-method case study from Mount Ida, Crete. *Journal of Archaeological Science* 35, 2918-2926.

Silverman, B. 1986. *Density estimation for statistics and data analysis*. Chapman & Hall, London.

Sinisi, S.E. & van der Laan, M.J. 2004. Deletion/Substitution/Addition Algorithm in Learning with Applications in Genomics. *Statistical Applications in Genetics and Molecular Biology* 3 (1), 18.

Soetens, S., Sarris, A., Vansteenhuyse, K. &Topouzi, S. 2003. GIS Variations on a Cretan Theme: Minoan Peak Sanctuaries. In Foster, K.P. & Laffineur, R. (ed.), *AEGAEUM 24: Metron. Measuring the Aegean Bronze Age. Proceedings of the 9th International Aegean Conference / 9e Rencontre égéenne internationale, New Haven, Yale University, 18-21 April 2002*, 483-488. Université de Liège, Belgium.

Soetens, S., Sarris, A. & Vansteenhuyse, K. 2008. Between Peak and Palace. Reinterpretation of the Minoan Cultural Landscape in Space and Time. In Facorellis, Y., Zacharias, N. & Polikreti, K. (ed.), *Proceedings of the 4th Symposium of the Hellenic Society for Archaeometry, National Hellenic Research Foundation, Athens, 28-31 May, 2003, BAR-IS 1746*, 153-161. Archaeopress, Oxford.

Venables, W.N. & Ripley, B.D. 2002. *Modern Applied Statistics with S*. Springer, New York.

Viollet, P. 2003. The Predecessors of European Hydraulic Engineers: Minoans of Crete and Mycenaeans of Greece (2,100-1,200 B.C.) In Armanini, A. & Latinopoulos, P. (ed.), *XXX IAHR Congress, Theme E: Linkage Between Education Research and Professional Development in Water Engineering*, 337-344. Thessaloniki, Greece.

Warren, P. & Hankey, V. 1989. *Aegean Bronze Age Chronology*. Bristol Classical Press.

Watrous, L.V. 1982. Lasithi: A History of Settlement on a Highland Plain in Crete. *Hesperia Supplements* 18.

Watrous, L.V. 1996. The Cave Sanctuary of Zeus at Psychro: A Study of Extra-Urban Sanctuaries in Minoan and Early Iron Age Crete. *AEGAEUM* 15.

Watrous, L.V. 2001. Review of Aegean Prehistory III: Crete from Earliest Prehistory through the Protopalatial. In Cullen, T. (ed.), *Aegean Prehistory. A Review*. American Journal of Archaeology Supplement 1, 157 – 223. Archaeological Institute of America, Boston.

5.5 Using LIDAR-derived Local Relief Models (LRM) as a new tool for archaeological prospection

Author
Ralf Hesse

State Office for Cultural Heritage Baden-Württemberg, Esslingen am Neckar, Germany
Contact: ralf.hesse@rps.bwl.de

ABSTRACT

High-resolution Digital Elevation Models (DEM) based on airborne LiDAR have emerged as a valuable new data source in archaeology. While such data are becoming increasingly available on a regional to national scale, their potential is far from being fully utilised. One field in which improvements can be expected is the optimisation of data processing with the goal of extracting anthropogenic features for archaeological prospection. Until recent years, however, most archaeological applications of LiDAR have been limited to the visual interpretation of the DEM. In this case, the detection of potential archaeological features depends to a large degree on the chosen illumination angles. Here, a data processing approach is presented which produces Local Relief Models (LRM) from LiDAR-derived high-resolution DEM. The LRM represents local, small-scale elevation differences after removing the large-scale landscape forms from the data. The LRM greatly enhances the visibility of small-scale, shallow topographic features irrespective of the illumination angle and allows their relative elevations as well as their volumes to be directly measured. This makes the LRM an improved basis for spatially extensive archaeological prospection over a wide range of landscapes. The LRM raster map of local positive and negative relief variations can be used for the mapping and prospecting of archaeological features such as burial mounds, linear and circular earthworks, sunken roads, agricultural terraces, ridge and furrow fields, kiln podia and mining sites. This approach is currently being used in a project aimed at the spatially complete archaeological mapping and prospection of Baden-Württemberg, covering an area of 35,751 km². The goal is the verification and extension of the existing archaeological data base. An object-based local relief vector layer is produced as a by-product; however, due to the common agglutination of natural and anthropogenic features this cannot be efficiently used for archaeological prospection at present.

KEYWORDS

LiDAR, Local Relief Model, LRM, archaeological prospection, Baden-Württemberg, Germany

INTRODUCTION

The federal state Baden-Württemberg in south-western Germany is rich in archaeological heritage from the Palaeolithic onwards. Numerous Neolithic, Bronze and Iron Age, Roman, Merovingian, medieval and early modern sites make Baden-Württemberg a region of great archaeological importance (cf. LAD 2009, for an overview of recent archaeological research in Baden-Württemberg). Of particular importance are sites dating to the early Celtic period like the hill fort Heuneburg – perhaps the earliest urban settlement north of the Alps – and the Upper German and Rhætian segment of the Roman Limes, a part of the UNESCO World Heritage system.

However, it is unknown to what extent the current state of knowledge approximates the actual number of sites. This is particularly relevant given the high forest cover of 39% of the state which renders large areas as blank spaces for archaeological prospection by aerial photography. Residential and industrial sprawl, construction of roads, railway lines and pipelines, mechanised agriculture and forestry practices as well as looting pose serious threats to known and unknown archaeological sites. Before this backdrop of the urgent necessity for spatially extensive archaeological prospection, in 2009 the State Office for Cultural Heritage Management Baden-Württemberg launched a project aimed at the complete archaeological mapping of Baden-Württemberg using high-resolution airborne LiDAR (Light Detection And Ranging) data, covering an area of 35,751 km². The goal is the verification and extension of the existing archaeological data base. While it is recognised that aerial approaches to archaeological prospection have serious limitations for certain types of sites and particularly in areas of sediment accumulation, no other type of prospection allows a complete coverage of large areas with a single and consistent methodology.

METHODOLOGY

High-resolution Digital Elevation Models (DEM) based on airborne LiDAR (Light Detection And Ranging, also known as Airborne Laser Scanning, ALS) have in recent years become an important data source for the prospection, mapping and monitoring of archaeological sites. Such data are now becoming increasingly available on a regional or even national scale. In most archaeological applications, LiDAR is applied to relatively small areas (up to a few square kilometres). LiDAR DEM are mostly visualised as shaded relief images which allow viewing the land surface under different simulated lighting conditions (elevation and azimuth) as well as vertical exaggeration (e.g. Harmon et al. 2006; Bofinger et al. 2006; Boos et al. 2008; Risbøl et al. 2006). While the experimental manipulation of lighting conditions allows a visual optimisation of individual features, it is a time-consuming process as the visibility of potential archaeological features depends to a large degree on the chosen illumination angles (e.g. Devereux et al. 2005). For a spatially extensive prospection project covering thousands of square kilometres – like the present

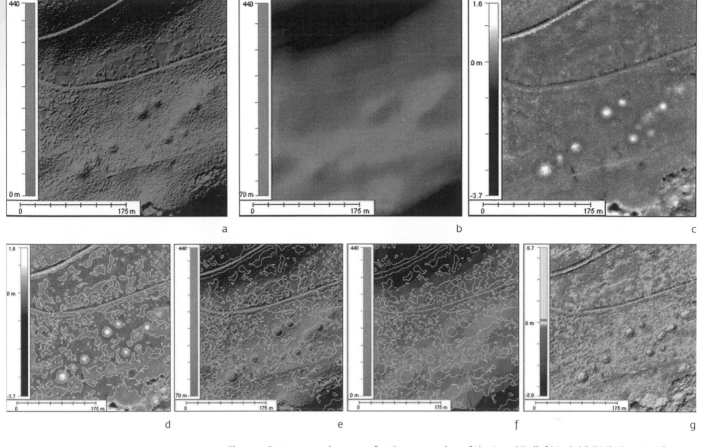

Figure 1. Data processing steps for the generation of the Local Relief Model (LRM). See text for the description of data processing. The image series shows a group of burial mounds in the Schönbuch area. *See also the full colour section in this book*

project in Baden-Württemberg – time is an important constraint. Therefore, and to enhance the reliability of the prospection results, improved visualisation techniques are required in archaeological applications of LiDAR. Experimentation with approaches for improved visualisation from other disciplines (Hiller & Smith 2008; Loisios et al. 2007; Rusinkiewicz et al. 2006) did not deliver satisfactory result for archaeological prospection.

In the first phase of the present project, a new approach was therefore developed and implemented (Hesse 2010). It is based on the observation that archaeologically relevant structures are usually characterised by very low relief relative to the elevation range of the surrounding landscape. Because they thus often only appear as subtle features in the conventional shaded relief visualisation, one goal of LiDAR data processing for archaeological prospection was identified as the problem of extracting local small-scale, low-relief features from the DEM and eliminate as far as possible the large-scale landscape forms from the data.

Several data processing steps (fig. 1) have to be applied to extract small-scale (detail) topographic features for archaeological interpretation:

a. A DEM is produced from the vegetation-filtered LiDAR point cloud data (in the present case with a pixel size of 1 x 1m).

b. A low pass filter is applied to the DEM. This smoothed elevation model represents a first approximation of the large-scale landscape forms. The kernel size of the low pass filter determines the spatial scale of features which will be captured in the LRM. In the present case, a kernel size of 25 metres is used for the low pass filter. This size was found experimentally to result in a good representation of many previously known archaeological features and is therefore assumed to work well for the detection of previously unknown features. Features much larger in diameter or cross-section are uncommon; furthermore, they would be conspicuous in conventional shaded relief images of the DEM. Degraded representation for much smaller features than the kernel size may become a serious issue if they are underlain by strongly convex or concave terrain (e.g. hilltops or ridges, valley bottoms), but is less pronounced on smooth slopes.

c. By subtracting this smoothed elevation model from the DEM, a first approximation of the local relief is achieved: only small-scale topographic features are preserved in the model while the large-scale landscape forms are eliminated. However, because small-scale features are smoothed rather than eliminated by the low pass filter, the model derived by this approach is biased towards small features, i.e. the local relief elevations are progressively underestimated as spatial extent of the features increases.

d. The zero-metre contour lines in the difference map represent the boundaries between positive and negative relief anomalies. Potentially, these lines can be used in vector data processing to derive shape parameters for individual anomalies.

e. The elevation values of the DEM are extracted along these lines. This results in a set of elevation values that do not belong to – positive or negative – relief anomalies.

f. A purged DEM is created from the extracted DEM point elevations by interpolation. This purged DEM represents the large-scale landscape forms after cutting out rather than smoothing small-scale features.

g. Subtraction of this purged DEM from the original DEM results in the final LRM which reflects less biased elevation information of small-scale features relative to the landscape at large. The visual rendering of positive and negative relief anomalies can be enhanced by colour-coding.

In comparison to using a simple difference map between the DEM and its low pass or median filtered derivate (Doneus & Briese 2006; Hiller & Smith 2008), the LRM derived using this approach results in a less biased representation of small-scale topographic features which reflects more truthfully the elevations of these features relative to the surrounding landscape and thus allows the direct measurement of feature volumes and heights. On the other hand, it is less computationally expensive and easier to implement than the kriging based filtering suggested by Humme et al. (2006) if an efficient workflow for data processing is applied.

DATA MANAGEMENT AND APPLICATION OF THE LRM WORKFLOW

The German state Baden-Württemberg has an area of 35,751 km². Vegetation-filtered LiDAR point cloud data were supplied by the State Surveying Office Baden-Württemberg. The enormous amount of data that had to be processed (more than one terabyte in approx. 160,000 separate files) required the acquisition of capable hardware (8-core Xeron with 16 GB RAM and 4.5 TB hard disc) and software (ENVI). Dedicated software for the efficient management of the data and the semi-automatic implementation of the LRM workflow was not available at the outset of the project.

Therefore, two graphical user interfaces were developed using VBA (Visual Basic for Applications). The first user interface allows the efficient interactive management of all raw, semi-processed and processed data. The current data processing and archaeological prospection status is documented for data segments of ten by ten kilometres. Several data processing steps implemented in VBA can be interactively executed for the selected data segments. For data processing steps that are executed in ENVI, the processing status is also documented in the user interface. Furthermore, relevant metadata like the number of known archaeological sites in each data segment or the average LiDAR raw data quality are displayed in the user interface, preview maps of the data segments can be displayed and short notes can be stored for each data segment.

Finally, data segments can be selected and all relevant data can be opened for visualisation and mapping. This includes:

- the LiDAR-derived DEM, DSM and LRM (DSM = Digital Surface Model);
- raster maps showing the LiDAR raw data quality in terms of point density and data gaps;
- raster maps showing surface depressions (e.g. dolines) derived from the LiDAR DEM;
- topographic and geological maps;
- aerial photographs;
- vector data of present settlements, roads etc.;
- vector data of known archaeological sites and find spots.

The mapping/prospection is based on the visual interpretation of a combination of all relevant data. Known archaeological sites and find points are used as reference for the qualitative and quantitative properties of archaeological features.

A second graphical user interface serves as a toolbox for actual prospection. It allows the orientation within the ten by ten kilometre segment and the documentation of the mapping status on the scale of one square kilometre as well as the interactive manipulation of the illumination in the shaded DEM visualisation. It also provides templates for the creation of new vector objects for the mapping of potential archaeological features.

FIRST RESULTS

After development and implementation of methodology, workflow and data management, the processing of the LiDAR point cloud data, the DEM and LRM for the entire state Baden-Württemberg were generated. Further LiDAR-derived raster maps showing raw data quality, data gaps as well as surface depressions

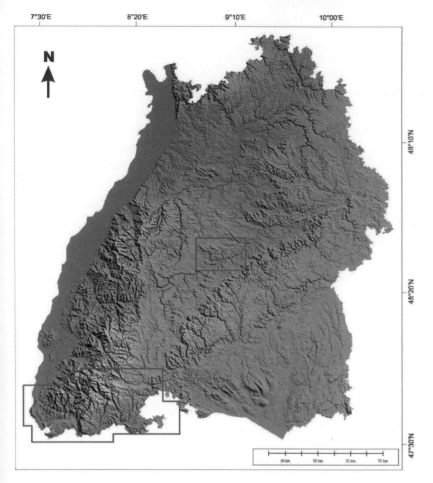

Figure 2. Map of Baden-Württemberg showing the current status of the project. Prospection has been carried out in the delimited areas Schönbuch (centre) and southern Black Forest/Upper Rhine (southwest).

were generated. Data processing was largely finished in January 2010. The subsequent archaeological prospection then concentrated on two large test regions, the forest Schönbuch region in the centre of Baden-Württemberg and the southern Black Forest and Upper Rhine region (fig. 2).

The Schönbuch region has an area of 600 km². Here, the 2,513 potential archaeological sites identified by LiDAR prospection compare with 1966 previously known sites and find spots. In the region southern Black Forest and Upper Rhine, prospection of an area of 2,700 km² resulted in 57,936 potential sites compared with 3,726 previously known sites and find spots.

Most features mapped as potential archaeological sites can be related to historic or prehistoric resource use. Terraced slopes as well as ridge-and-furrow document agricultural use; mining traces and slag heaps as well as thousands of kiln podia allow new insights into spatial patterns of mining, ore processing and related fuel supply. Furthermore, a large number of potential burial mounds as well as several previously unknown fortifications have been detected. Sunken roads as well as former field parcel patterns are also mapped as they allow inferences regarding the location of settlements.

EXAMPLES

The application of the LRM approach to the two extensive test regions Schönbuch and southern Black Forest/Upper Rhine indicates that in particular kiln podia, sunken roads and former agricultural structures (ridge and furrow, fig. 3a; terraced slopes) can in many cases be unambiguously identified because there are no natural morphological equivalents. Characteristics of kiln podia are a roughly circular outline, a steep upslope scar and a downslope lip (fig. 3b). Features created by the uprooting of large trees or mining

Figure 3. LRM colour maps showing (a) ridge and furrow, (b) kiln podia, (c) sunken roads and (d) mining traces. *See also the full colour section in this book*

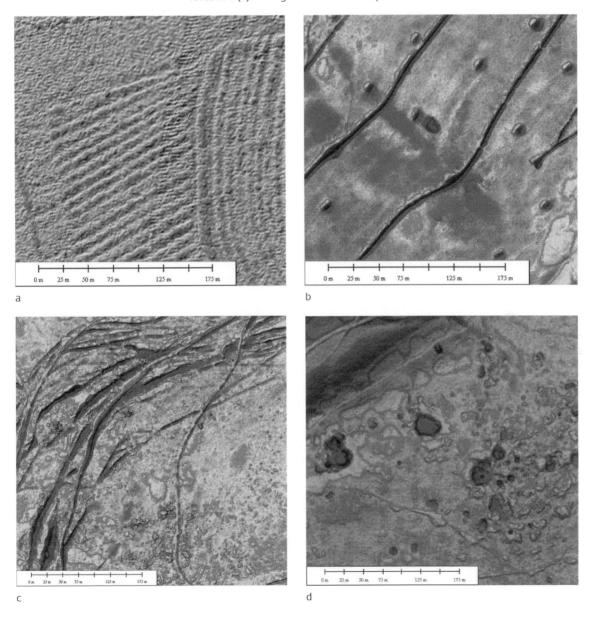

a

b

c

d

scars on slopes may occasionally be misinterpreted as small kiln podia, but in general the confidence of identification is high. The regular spacing typical of many kiln podia clusters can help to avoid misinterpretations. Sunken roads are often well-identified where they occur as swarms of incised linear features which climb slopes at oblique angles (fig. 3c).

The identification of burial mounds, however, is much less definite. There are many features which may appear as well-defined positive local relief in the LiDAR data, including small natural mounds, slag heaps related to ore processing, waste heaps related to mining or quarrying, wood piles or patches of low vegetation not filtered out by the vegetation removal algorithm. On the other hand, many burial mounds in agricultural areas have been deliberately removed to smooth the terrain or have been unintentionally levelled by ploughing. In these cases, they may survive as very low height anomalies of only a few decimetres. Definite identification of positive local relief features as burial mounds is therefore difficult and has to take into account their distribution in the landscape, with burial mounds often occurring in clusters. Traces of mining activities are often discernible as negative local relief features (fig. 3d); their size, shape and distribution may allow some age assignment based on the technology used. However, some morphological overlap exists between the superficial traces of collapsed mining galleries and geological features (sinkholes).

ADVANTAGES AND LIMITATIONS

Because large-scale landscape forms should not be neglected in archaeological prospection, the conventional method of shaded relief representation was used as an auxiliary tool. This allows direct comparison of shaded relief DEM and LRM visualisation and the assessment of advantages and limitations of each method.

While the shaded relief DEM appears visually more 'natural' for the observer, experimenting with variations of illumination elevation and azimuth angles is necessary. Illuminations using azimuth angles between 90° and 270° are often necessary to shed light on south-sloping terrain; however, this commonly leads to the optical illusion of relief inversion. While pronounced features are clearly visible under suitable illumination, the challenge is that a universally suitable illumination does not exist. Subtle features often become visible only under very constrained illumination conditions and therefore tend to be found accidentally rather than systematically.

By comparison, the LRM is well-suited to visualise very low topographic features independent of their topographic location and independent of their alignment. The LRM allows the visualisation of local relief variations in intuitively recognisable colours irrespective of the absolute elevation, making the LRM raster map almost as easily comprehensible as the DEM. Furthermore, by draping the LRM as a colour map over a 3D representation of the DEM, subtle local relief anomalies can be clearly visualised without the distortions caused by strong vertical exaggeration. Another advantage is that the elevations and volumes of small-scale topographic features relative to the surrounding landscape can be directly measured in the LRM. However, natural convex and concave landforms (e.g. hilltops, plateau edges, valley bottoms) also appear as positive or negative features in the LRM. Interpretation therefore requires some experience and the recognition of the effect on the natural landscape forms on the LRM. Furthermore, features on slopes suffer some distortion. Depending of the interpolation algorithm applied for creating the purged DEM, artefacts may be introduced. Another issue is the fact that the scale of features which are resolved in the LRM ultimately depends on the kernel size of the low pass filter applied in the workflow. Therefore, optimal adaptation of the LRM to features significantly larger or smaller than the standard kernel size used in a particular study may require much of the workflow to be applied repeatedly for different kernel sizes. This task can be partially or fully automated. At the present stage, a kernel standard size of 25 metres is used for the low pass filter. This kernel size was found experimentally to result in a good representation of many previously known archaeological features and is therefore assumed to work well for previously unknown features. Features much larger in diameter or cross-section are uncommon; degraded representation for much smaller features only becomes a serious issue if they are underlain by rather convex or concave terrain (e.g. sharp hilltops or ridges, valley bottoms), but is less pronounced on smooth slopes.

CONCLUSIONS AND OUTLOOK

Extracting Local Relief Models (LRM) from high resolution LiDAR data has the potential to greatly improve the potential of such data for the prospection, mapping and monitoring of archaeological sites. The value of the LRM approach lies in particular in the representation of local topographic detail detached from the large-scale landscape forms and in avoiding the necessity of experimenting with numerous combinations of illumination azimuth and elevation. Colour-coded maps of the LRM are found to be a valuable tool for the time-efficient prospection of extensive areas, particularly if used in combination (e.g., draped over) shaded relief. Another advantage is that heights and volumes of small-scale features can be directly measured. First results of the project in Baden-Württemberg confirm the feasibility of using LiDAR-derived Local Relief Models (LRM) for the archaeological mapping and prospection of very large areas.

Besides the ongoing archaeological prospection of Baden-Württemberg, further work is planned in the fields of data processing and interpretation. This future work will concentrate on (a) the statistical assessment of size and spatial distribution of sites, (b) the combined processing and interpretation of raster and vector data (i.e. taking into account shape and volume of relief anomalies): One goal of this work is (c) the (semi-)automated detection of selected features.

ACKNOWLEDGEMENTS

This paper presents the methodology and working progress of the project 'LiDAR-based archaeological prospection in Baden-Württemberg' at the State Office for Cultural Heritage in Baden-Württemberg. The manuscript benefitted from comments by Philip Verhagen and an anonymous reviewer.

REFERENCES

Bofinger, J., Kurz, S. & S. Schmidt. 2006. Ancient maps – modern data sets: different investigative techniques in the landscape of the Early Iron Age princely hill fort Heuneburg, Baden-Württemberg. In Campana, S. & Forte, M. (ed.), *From Space to Place. 2nd International Conference on Remote Sensing in Archaeology*. BAR International Series 1568, 87-92. Archaeopress, Oxford.

Boos, S., Müller, H., Hornung, S. & P. Jung. 2008. GIS based processing of multiple source prospection data in landscape archaeology. In *Proceedings of the 1st International Workshop on Advances in Remote Sensing for Archaeology and Cultural Heritage Management – Rome, 30 September – 4 October, 2008*, 113-117. Lasaponara, Rome

Devereux, B.J., Amable, G.S., Crow, P. & A.D. Cliff. 2005. The potential of airborne LiDAR for detection of archaeological features under woodland canopy. *Antiquity* 79, 648-660.

Doneus, M. & C. Briese. 2006. Full-waveform airborne laser scanning as a tool for archaeological reconnaissance. In Campana, S. & Forte, M. (ed.), *From Space to Place. 2nd International Conference on Remote Sensing in Archaeology*. BAR International Series 1568, 99-105. Archaeopress, Oxford.

Harmon, J.M., Leone, M.P., Prince, S.D. & M. Snyder. 2006. LiDAR for archaeological landscape analysis: a case study of two eighteenth-century Maryland plantation sites. *Antiquity* 71, 649-670.

Hesse, R. 2010. LiDAR-derived Local Relief Models – a new tool for archaeological prospection. *Archaeological Prospection* 17, 67-72.

Hiller, J.K. & M. Smith. 2008. Residual relief separation: digital elevation model enhancement for geomorphological mapping. *Earth Surface Processes and Landforms* 33, 2266-2276. DOI: 10.1002/esp.1659.

Humme, A., Lindenbergh, R. & C. Sueur. 2006. Revealing Celtic fields from LiDAR data using kriging based filtering. In *Proceedings of the ISPRS Commission V Symposium, Dresden, 25-27 September 2006*. Vol. XXXVI, part 5. ISPRS, Dresden.

LAD (Landesamt für Denkmalpflege Baden-Württemberg). 2009. *Archäologische Ausgrabungen in Baden-Württemberg 2008*. Thesis: Stuttgart.

Loisios, D., Tzelepis, N. & B. Nakos. 2007. A methodology for creating analytical hill-shading by combining different lighting directions. In *Proceedings of the International Cartographic Conference*, ICC, Moscow.

Risbøl, O., Gjertsen, A.K. & K. Skare. 2006. Airborne laser scanning of cultural remains in forests: some preliminary results from a Norwegian project. In Campana, S. & Forte, M. (ed.), *From Space to Place. 2nd International Conference on Remote Sensing in Archaeology*. BAR International Series 1568, 107-112. Archaeopress, Oxford.

Rusinkiewicz, S., Burns, M. & D. DeCarlo. 2006. Exaggerated shading for depicting shape and detail. ACM Transactions on Graphics. *Proceedings SIGGRAPH* 25 (3).

5.6 The use of digital devices in the research of Hungarian monastic gardens of the 18th Century

Author
Mária Klagyivik

Department of Garden Art, Corvinus University of Budapest, Budapest, Hungary
Contact: maria.klagyivik@uni-corvinus.hu

ABSTRACT

Hungarian monastic garden art of the 18th century included several distinct types of garden, which differed according to the specific religious orders and also to their geographical location, producing great diversity. The gardens were forced to change with history, and as a result, there are hardly any monasteries today that still preserve visible remains of the formal garden design. Therefore, the restoration of these gardens to a great extent has to rely on archival research to a great extent. However, owing to the inaccuracy and insufficient quantity of data, further methods are also needed to achieve authenticity. Though the use of digital devices is not new in garden research, it is still not used widely, especially not in Hungary. Yet, in the case of monastery gardens, its application can no longer be omitted, because this is the only way to get useful information on the gardens. This paper discusses how the most frequently used digital methods, GIS and geophysics can be applied in the case of monastic gardens. Generally, GIS software makes it possible to stretch period maps and layouts over today's more accurate, georeferenced maps and thus find the exact GPS coordinates of the garden elements, after which these can be found on site. Where nothing visible has remained, geophysical surveys can be applied. However, both methods face difficulties in the case of monastic gardens, as will be presented. Still, with the complex use of these digital devices, the models of the theoretical former gardens can be constructed, which is essential in understanding monastic garden art and also for the conservation of these gardens.

KEYWORDS

garden archaeology, monastic gardens, conservation, restoration, GIS, geophysics

INTRODUCTION

The Christian religious orders have bequeathed a great amount of spiritual and cultural heritage that embraces all the Christian parts of the European continent. Since their establishment, these orders were the determinants of the prevailing intellectual life. From the Middle Ages, all works that contributed to the spread of literacy were centred in the monasteries for centuries. The heritage of different monastic orders, however, manifests itself not only in intangible, but in material ways as well, seen directly in their environment. Depending on their monastic aims, religious orders settled either in towns or on the contrary, in peaceful, natural environments far from other human settlements. Therefore, partly by their effect on urban life and partly by their landscape-forming activity, nowadays their former estates form an integral part of research in both urban and landscape history.

UNESCO defines three main categories of cultural landscapes among which "the most easily identifiable is the clearly defined landscape designed and created intentionally by man. This embraces garden and parkland landscapes constructed for aesthetic reasons which are often (but not always) associated with religious or other monumental buildings and ensembles" (Mitchell et al. 2009, 20). Gardens, regardless of having a formal, Baroque design or a seemingly natural English style, are considered as cultural landscapes. The term can be applied to monastic gardens without hesitation, considering that most of these gardens have been primarily used for cultivation, although with aesthetic intentions, too.

Moreover, there is an associative value to these gardens, because the activities, symbols and objects of a religious order were reflected in the structure of their architecture and gardens. Nevertheless, as a result of the elapsed period, nowadays these historic gardens reflect how both ecclesiastic and secular culture have left their mark on them, and thus the task of conservation is compound: while preserving the remaining elements of the former Baroque cloister gardens, one has to keep in mind their later, secular use as well (which may last even up to the present day), in order to demonstrate the true history of these peculiar landscapes.

Comprehensive, interdisciplinary research on historical gardens, employing different tools and methods in the process, is common practice in Europe today; however, such research in Hungary is still in its infancy. There are hardly any examples of Hungarian garden restoration where all the necessary methods of historical research are applied together, although in recent years, some initiations have started to reach the European level of research and restoration works (Alföldy 2008; Fatsar 2009).

The conservation of gardens can never be a completely exact science, because there are many variable factors on which restoration depends (Goulty 1993), including their main building elements and the plants. And as a result of this, gardens necessarily fall into decay earlier than other works of art. Nevertheless, it is because of this unique feature that they can be renewed and regarded as a living historic monument (Florence Charter 1981, chapter 2, chapter 3). Thus, their restoration may be possible, even if apparently nothing has remained of them – but this requires comprehensive research, with complex on-site examinations, including garden archaeology.

Garden archaeology, as a special type of landscape archaeology applied to intentionally designed landscapes (Bowden 2006), introduces an interdisciplinary approach, the complex use of archival documents, diverse digital devices and excavations which helps to construct the theoretical models of the former historic gardens. Hence, this method can be used effectively in the case of the gardens of the religious orders, where it is essential for understanding monastic garden art and also for conservation. It is necessary to use these methods to achieve authentic results, which can then be used for the restoration and reconstruction works.

THE EVOLUTION OF MONASTIC GARDEN DESIGN

Monastic gardens are peculiar cultural landscapes, representing the designed and associative types of cultural landscapes all in one. These gardens were established for functional and contemplative aims since the Middle Ages. According to their monastic vows, the monks' environment reflected poverty, chastity and renunciation. The small, inner, enclosed places of the monasteries were used for herb gardens, orchards or kitchen gardens (Landsberg 1995), while in the outer land they cultivated cereals or had ponds for fish (Currie 2005). The surrounding walls of monasteries became a feature in the time of St Pachomius (292-348 AD), and have been a characteristic part of monastic structures ever since (Meyvaert 1986). The walls certainly limited the space and, therefore, it was usual to mix the different functions of places, for instance planting an orchard in the cemetery garden. Pleasure gardens also developed this way, through the beautification of vegetable or herb gardens. The important role of pleasure gardens created only for aesthetic reasons in monasteries was already implied by Albertus Magnus (cc. 1200-1280): "Nothing refreshes the sight so much as fine short grass. One must clear the space destined for a pleasure garden..." (English translation cited from Thacker 1979).

However, the simple design of the monastic environment changed over time. Since cultural and spiritual activity was centred around monasteries, they became quite wealthy, to which donations and legacies also contributed. As a result, by the time of the Renaissance their purity had vanished and hints of luxury appeared – although not to the same extent as in the case of rich aristocrats. The increased wealth could easily be traced in material ways, such as the architecture and decorations of their buildings and gardens. The proportion of places inside the enclosure, created especially for aesthetic reasons and for spending free time, became higher and higher. The cloister garth, once used for orchards or herb gardens, now gave way to ornate parterres and other highly decorative elements (Turner 2005), making these gardens equivalent to the general garden style of the era. The parterre de broderie, later the parterre à l'Angloise, the bosquets, alleys, trellises, clipped hedges, topiaries, fountains, pavilions and orangeries, shaped in an architectonic, formal way, became the dominant parts of the gardens, showing the pompous style of the era and the rule, the symbolic victory of humanity over nature.

This luxuriant behaviour resulted in a rapid decline of the religious orders. Since the opulent way of life was completely in contrast with their original aims of self-denial, it is not surprising that in the Era of Enlightenment, from the end of the 18th century, dissolution of monasteries spread Europe-wide (France, Germany, Hungary, Portugal, etc.). Secular dominance took over the power from the Church, and monastic communities could never again be as determinative as before.

Therefore, the 17th and 18th centuries can be regarded as the last peak of monastic culture and definitely the most valuable time concerning monastic garden art. Garden formations at that time were particularly diverse, but did not depend principally on the wealth and economic status of the order or the monastery. Instead, garden design was linked to the monastic regulations and their way of life, though with adaptation to the higher demands of aestheticism. The differences in monastic gardens were the result of the four main different groups of religious orders of the 18th century:

- *monastic orders* (living far from human settlements, devoting their time completely to spiritual activities)
- *mendicant orders* (living by physical work and begging, therefore settled in or close to towns)
- *canons regular* (clerks living in a particular community in a particular place)
- *clerks regular* (clerks dealing with teaching and pastoral care, hardly any fixed dwellers at a particular place).

These categories differed in their way of life, and thus gave rise to a wide variety of monastic gardens in the 18th century. Their characteristics are similar throughout Europe, though always with adaptations to the local historical, economic and political background as well as to the climate. Still, it can be generally ascertained that the monastic orders and the canons regular had more and larger decorative gardens, while the mendicant orders and the clerks regular generally restricted the adornments to the cloister garden, using the other parts of their estates for cultivation, though many times formed in an ornamental way.

WHY WE NEED DIGITAL DEVICES – HISTORICAL BACKGROUND

Hungarian monastic gardens have had to tide over hard times on several occasions, which have caused changes and destruction. As a result, many features of the earlier gardens have vanished, or appear to have vanished. To understand the necessity of digital devices in the research of Hungarian monastic gardens, a short historical review is required. This will help to explain the background of the present methods of research, and will show why a more comprehensive method of historical research is needed.

Destruction of the gardens was usually caused by cultural changes, primarily secularisation. Monastic gardens could become secular properties in two ways. On the one hand, they could be sold by the monks for financial purposes (a 'bottom-up' dissolution). This often happened, predominantly in the case of those orders that accumulated great wealth and could do business, such as the Jesuits, for instance.

The other way of secularisation was by force (a 'top-down' dissolution). The unfolding garden art of the Hungarian monasteries was already in crisis by the end of the 18th century, under the reign of Joseph II (1780-1790), who, in 1782, dissolved all those monastic orders which were not concerned with teaching or medicine. Taking into consideration the property of the Jesuits, who were abolished worldwide in 1773, and of the Pauline Fathers (the only order founded by Hungarians), who were extinguished in 1786, the number of the abolished monasteries was more than 150. The properties of the monasteries were distributed or auctioned, and fell into the hands of the state, the military forces or the municipalities. The building complexes gained a completely new function and were transformed into barracks, hospitals, warehouses or granaries (Velladics 2000), which, of course, launched irreversible processes in the gardens.

Though the 19th century was a peaceful, harmonious era for the remaining monasteries, the happenings of the 20th century brought about drastic changes. After World War I, the 1920 Treaty of Trianon reduced the territory of Hungary to about a third, which resulted in the loss of most of the remaining monasteries. The few still belonging to Hungary were further destroyed under the Communist regime from the 1950s onwards. The estates of the Church were taken into the ownership of the state again, and the new functions and uses devastated almost all value of the Hungarian cloisters. In better cases, the area of the gardens remained and was only transformed; the worst result was when the gardens were used for construction. The destructive processes at this time were so great that most of the gardens could not be saved even with their privatisation (return to the Church) after the change of regime in 1989.

As a consequence of this history, the number of surviving gardens is meagre, and many have been built up or detached from the monastic buildings. Those ones that still form one unit are very rare, and conservation is largely restricted to these cases; however, even for the more well-preserved, it is necessary to carry out comprehensive historic research.

DIGITAL DEVICES IN THE RESEARCH

Owing to the degradation and transformation of the monastic gardens, almost nothing has remained visible on the surface, and thus their recognition requires close examination. The primary sources of information on the design of these gardens are mainly the archival period documents, maps and images; however, these alone cannot give a picture precise enough for restoration. Written documents cannot show the exact geographical location of the features, moreover, the images are not wholly accurate, because even if they depict a state that was really attained, the proportions and scales can be wrong. To refine the data, other devices are needed.

Digital devices are used in both archival research and field work. An important aspect of their application in garden research is finding the best method for the job, since the monasteries are in differing states of decay and therefore cannot all be approached in the same way. Digital devices used in the research of landscape architecture include instruments for geophysical prospection, such as GPS and resistivity, and softwares, such as GIS, AutoCad and Photoshop, which allow the results to be plotted in a series of overlays.

A key aspect of the ongoing research on 18th-century monastic garden art in Hungary is trying to establish the most appropriate and informative methods of prospection and analysis. With the presentation of some case studies, the capability of different digital devices used in the context of monastic gardens will be demonstrated. AutoCad and Photoshop are mainly complementary devices and are used particularly in a later phase of the research. As both packages can be effective in design, they are mostly applied for creating visual models of the former gardens, which can form good bases for a restoration work. As they do not play such a great role in the actual phase of field research, they will not be detailed here.

GIS

Historical maps often hold information about a garden that cannot be found in any other written source. These may include boundaries, buildings or other physical features, or even land use or vegetation cover. The degree of accuracy of a map is, however, not always precise enough to be able to identify the different features today, because a few-metres precision is certainly not sufficient in gardens. It is almost impossible to align an old map to modern coordinate systems perfectly, as mapping methods of the earlier times often represented scale, distance, directions and angles very imprecisely. Therefore, not only the overall scale of the different maps has to be harmonised in order to compare them with each other, but also the proportions and inner scales of the old maps must be changed. The rubber sheeting process available in GIS software makes it possible to stretch period maps and layouts onto today's georeferenced aerial photographs or digital maps, by selecting control points on the original map that can still be recognised today. After a rough fitting of the map, further adjustments can be made to reach the best combination of the old and new (Rumsey & Williams 2002). The more maps we have of a place, the easier it will be to recreate the history of the garden.

Owing to the size of the monastic gardens, however, we may have problems finding many good maps or layouts. The plans of the design have rarely survived, and since the majority of these gardens were relatively small (compared to the estates of the aristocracy), we usually cannot gain information about their layout from town maps, nor from small-scale maps. These generally depict the gardens as a green area without any further details or with a schematic drawing, thus, we can draw conclusions concerning only the size and shape of the actual garden. Monasteries themselves did occasionally create maps, but usually only when they were required for some particular reason (e.g. fixing boundaries or settling litigious matters concerning the property, and so forth). These, however, still did not focus on the ornamental parts and layout of the garden. The images of monastic gardens found on engravings or planned layouts depict a state which may have never been implemented this way, hence, they should be handled sceptically and cautiously, and justified using other methods, too (e.g. the use of geophysics).

The best image sources for Hungarian monasteries of the 18th century are, therefore, the survey maps made by the Hungarian Treasury at the time of the orders' dissolution. These surveys are quite detailed, though mainly concentrate on the built elements of an actual property, and hence, other important minutiae of the gardens (such as the ornaments of the parterres or other decorative features) are not fully represented. The surveys were made with economic aims, and thus displayed only those features that were useful in this respect, while plants were depicted in a rather schematic way. Only the explicitly characteristic elements, such as alleys or solitary trees, were represented in detail.

The role of the rubber sheeting method is particularly useful in these cases, and can reveal hidden structures which are not otherwise perceptible. The identification process using the survey maps of the former Jesuit provostship in Znióváralja (today Kláštor pod Znievom, Slovakia) is a good example. The estate originated in the Middle Ages and was slowly transformed into a typical Baroque Jesuit estate, including three granges, several fish ponds, a brewery and a distillery. It also had meadows, arable land, hunting fields and gardens designed in a formal way, with elements such as an apiary, vegetable garden, ornamental garden, a bowling alley, a pool for tortoises and so forth, of which we can get information principally from archival documents (Klagyivik 2010). Today, however, most of the area has been overgrown by the village, destroying the formal unity of the estate, and seemingly leaving nothing of it but the main buildings.

In addition to the survey map, made around the time of the dissolution in 1773 (fig. 1), another survey from 1791 was at my disposal (fig. 2). The later one was much more precise, although it showed fewer features than the other, and only the main buildings and roads could be seen. In two steps – first adjusting the map of 1791 to the map of today, and then the survey map to the whole, all of the elements could be positioned correctly (fig. 3). The control points used during the process were angles of the cloister building, the main farm buildings which exist today as well, and some characteristic points of the road leading to the monastery.

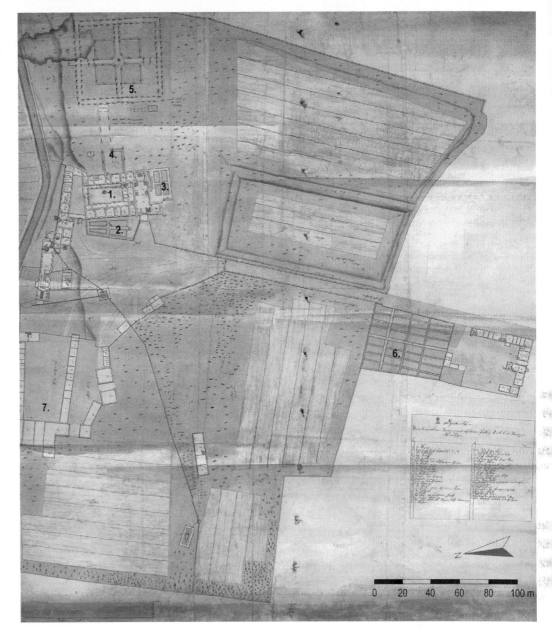

Figure 1. Survey map of the former Jesuit garden of Znióváralja, at the end of the 18th century. (*Hungarian National Archives*, S12 No. 277.) 1. monastery 2. ornamental garden 3. aviary 4. bowling alley 5. fruit and ornamental garden 6. vegetable garden 7. farm-buildings.

The use of the GIS software seemed to be effective and brought new results. It became evident that not only does the main road that leads to the cloister still exist today, but also the road running to the south is exactly the same as the old one. Furthermore, the majority of the fence has remained at the original location and still has the same formation as before, with stone pillars, shingled roofs and wooden lathes. At some points, boundaries have stayed in the form of vegetation lines, as can be seen on the southern part of the estate, next to the vegetable garden. However, only the shape and the manner of use of the formal ornamental garden can be detected today, as its layout has changed substantially. Though an old peculiar European hazel (Corylus avellana 'Pendula') is still standing in the middle of it, it cannot be identified with the solitary tree seen on the map; it must be the remnant of a later period, of which hardly any

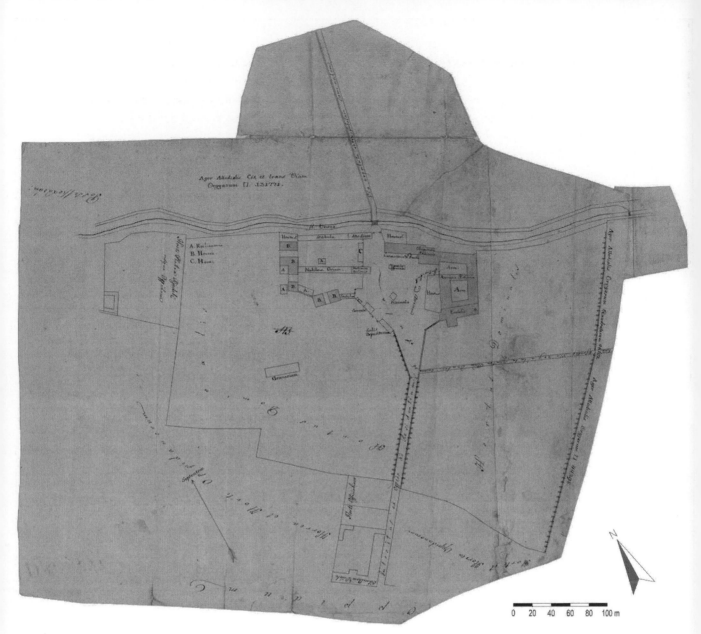

Figure 2. Map of the former Jesuit garden of Znióváralja, 1791. (*Hungarian National Archives*, S11 No. 91.)

sources could be found. Old oaks, beeches and limes are still living, some of which are probably survivors of a former alley. By georeferencing the old map and thus gaining the exact coordinates of the different features, one small edifice known only from the survey map but not from written sources could also be found on site. The little building which served originally as a place for keeping smaller animals is, however, today only a ruin overgrown by vegetation (fig. 4).

Another example, the former Jesuit garden in Pozsony (today Bratislava, Slovakia), has much less of a connection with the original layout. The garden lay in the suburbs of the town, right in the neighbour-

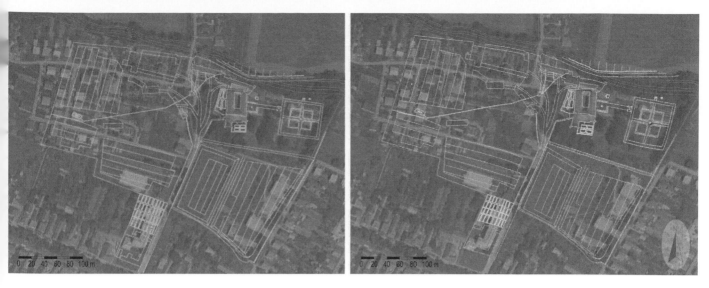

Figure 3. Znióváralja, digitalised period maps on aerial photograph, before (right) and after (left) the alignment with GIS rubber sheeting method.

Figure 4. Znióváralja, location of the old layout and uses of land. The photos display remaining elements: walls, fences, ruins of an edifice, a meadow and some old trees. *See also the full colour section in this book*

hood of the richest decorative gardens of the aristocracy, and it functioned first as a botanical garden and later as an ornamentally designed vegetable garden (Klagyivik 2007). Today, the area belongs to downtown Bratislava, and is partially built upon, seemingly having lost every mark of the earlier design.

Still, after layering the images of different periods over one another, some marks have turned out to be long-lasting and perceptible even today. The track of the streets, for instance, still follows the original boundaries of the garden. A surviving area of garden is sunken below street level, with a buttress which, though built in the 20th century, follows the line of the original one. The staircase leading to the street level is situated a bit farther on (figs. 5, 6). Nothing has remained, however, of the earlier buildings, which have been replaced by housing blocks built in the 1960-1970s. Neither did the formal botanical garden preserve its vegetation. Even the famous lime tree, which held an arbour built on its branches and existed until the 1960s, does not stand today.

As can be seen from these examples, the main benefit of using GIS in this kind of research is, first of all, to prove the authenticity of the image sources. If evidence is found – as in the above-mentioned case studies – that the drawing in question illustrates a state that really existed, further methods are worth applying in order to find non-visible remnants of the gardens as well. Since with the growth of the towns and villages, many of the monastic gardens have become more or less part of the built-up area, it is very difficult to apply useful methods for on-site examinations for those garden elements which are not visible on the surface. In these cases, we have to content ourselves with the results we can gain by using GIS and archival sources. Nevertheless, in areas which are still open spaces today, further methods like geophysics can also be used. These are presented in the following two case studies.

Figure 5. Bratislava, the digitalised map of the original layout of the former Jesuit garden, mid-18th century (*Hungarian National Archives*, T2 No. 1495.), stretched on the aerial photograph: 1. previous garden 2. formal buildings. The area is partly a green space today as well. The terrain levels and the location of the buttress are the same. *See also the full colour section in this book*

Figure 6. Bratislava, photo of today's layout with the buttress in the background.

GEOPHYSICS

In places which have not been built upon, on-site garden archaeology can play a significant role in the process of historical research. Historic gardens are a unique type of archaeological site, chiefly because they continue to evolve all the time. The living components of them are growing and altering constantly. As excavations are fairly expensive and time consuming, preparatory surveys should be carried out. The role of geophysics in garden archaeology is pretty new, having been applied throughout Europe only in the last two decades (Currie 2005), but it is becoming more and more essential, since it offers relatively good results without disturbing the remains under the ground.

The benefit of a geophysical survey relies on many factors. Apart from the climatic and soil conditions, the success depends on the materials of the garden features: the identification of walls and buildings can be quite good, culverts and drains can also bring nice results. Flower beds or gravel paths are fairly difficult to detect, while small-cut features like planting holes can hardly be revealed at all (Currie 2005). Furthermore, geophysics can be misleading on sites where features from later times can be found, because of the disturbance to the earlier layout. Owing to their historical background outlined above, this is a characteristic feature of many monastic gardens in Hungary. Where distortion of the soil is too high, geophysical survey cannot show good results, not even of the built elements of the garden.

This was the case, for instance, in the garden of the Franciscan cloister in Szécsény, Hungary, which still occupies roughly the same space as it did originally, but which was totally transformed during the era of Communism in the second half of the 20th century. In the beginning of the 18th century, a terraced garden was formed, of which some drawings and descriptions can be found in the Historia Domus of the cloister. According to this, more phases of the development of the garden can be determined. As being the cloister of a mendicant order, the gardens were made mainly for cultivation, comprising vegetable and fruit gardens, fish ponds and canals that led to the ponds (Szacsky 2002). The secularisation of the 1950s gave the ownership of the garden to an agricultural high school, which used it as a place for practice. The terraces were embanked and new terraces were formed at that time, which have remained until today (fig. 7).

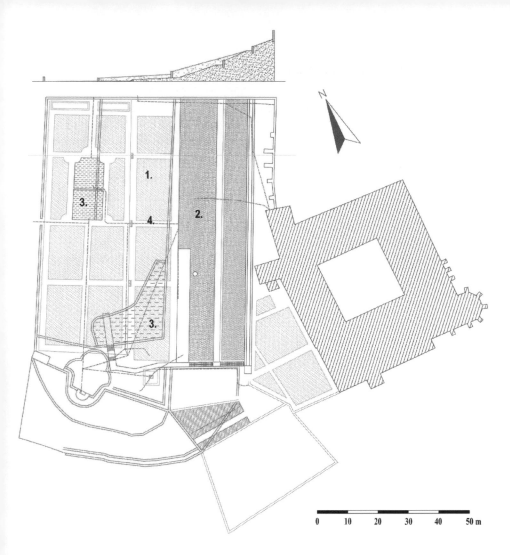

Figure 7. Szécsény, Franciscan cloister. Digitalised drawing of the garden of 1777 (*Historia Domus, Szécsény*), compared to the present state seen with red lines. 1. vegetable garden 2. fruit garden 3. fish ponds 4. canals. The cross section shows the presumed height difference of the embankment after 1950. *See also the full colour section in this book*

After aligning the rather inaccurate drawings of the 18th century to the present conditions with GIS methods, the exact locations of the garden elements could be read and on-site examinations could begin. The geophysical surveys of some parts of the garden have been executed by the Corvinus University of Budapest, Department of Garden Art in 2009. The extent of the 20th-century embankment had not been known beforehand, thus the success of the survey was doubtful. The aim of the survey was to find some traces of the formal buttresses and drains, features which could have been identified under good conditions. The results, however, were not promising: the elements that could have been found were not detectable at all, presumably because they were too deep under the ground, where detection could not penetrate. Moreover, the main material used for the embankment was construction debris, which also reduced the possibility of positive results, since the instrument displayed this spectacularly. Refuse on the surface was also a hindering factor for this method (fig. 8). The instrument used was an RM15 Multiplexer, which measures electrical resistance of the soil, and thus, it is particularly sensitive to debris that might cause anomalies. Hence, it seems that only excavations would be effective enough to show positive results in field research of this garden and to verify the drawings in the Historia Domus.

Nevertheless, on areas where disturbance has not been so comprehensive, geophysics can be really

Figure 8. Geophysical survey in the Franciscan garden of Szécsény, 2009. The inefficiency of the measurement: 1. unsuccessful survey due to the filled sediments of the embankment 2. ditch on the surface 3. debris on the surface. (*Corvinus University of Budapest, Department of Garden Art*. Leaders of the survey: Anett Firnigl, Mária Klagyivik).

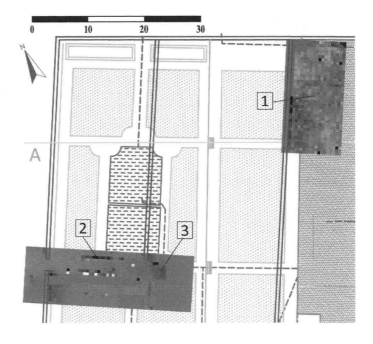

successful. The survey of the Camaldolese gardens in Majk, Hungary, proves this well. The hermitage was constructed in three phases between 1733 and 1770. Ornamental gardens lay next to the communal house and next to each small house. Orchards and vegetable gardens were also part of the estate. Changes made in the garden took place mainly because of the change in popular garden style. The owners followed the trends of the time, while in the meantime still preserving elements such as terraces from the Baroque period. As a result of this, the garden began to preserve the marks of more and more different eras (Baroque, English garden style, as well as the style of the late 19th and early 20th century), thus representing the continual history of itself. The greatest decay in the complex happened in the short period after the World War II. Its conservation and restoration began in 1979, and with some breaks it continues to this day.

The restoration project was based on an all-inclusive, multi-disciplinary research programme that comprised the archival research, on-site examinations and non-destructive methods of garden archaeology alike. The geophysical surveys have demonstrated the existence of many of the garden elements shown in the period maps. The resistivity survey was executed by the same RM15 Multiplexer instrument as the one in Szécsény, but seemed to be a much more effective method in this garden, because the disturbance was not so extensive. The terraced garden is covered with lawn and is out of use today. Depending on the depth of the measurements, different elements came to light. The survey in the garden next to the manor evinced Baroque elements such as canals connecting the building with a pool, as well as the axes of the parterres (fig. 9), while other parts showed the layout of the English style period (Fatsar 2004).

Geophysical survey, accordingly, can be especially effective if the disturbance of a garden has not been significant. This is, however, very rare in the case of monasteries, because after their secularisation, they have generally been managed in a rather unsuitable way. Instead of keeping the heritage the monks created, the principal aim has always been purely functional and economic. The number of monasteries

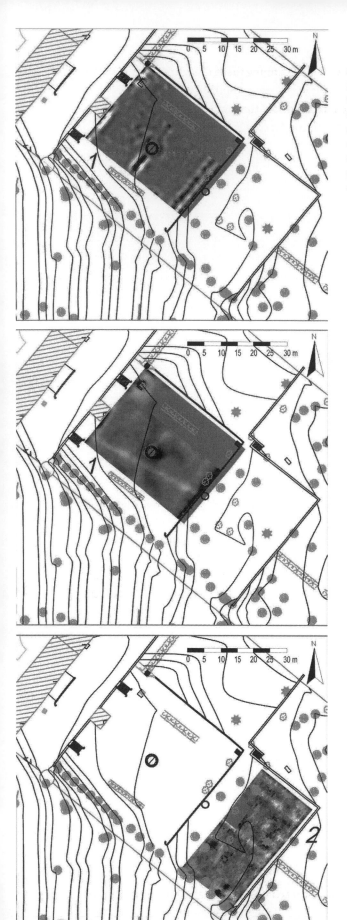

Figure 9. Geophysical survey in the Camaldolese garden of Majk, 2005. Measurements of 0,75m depth in the lower terrace showed the axes of the garden. The light lines can display marks of paths. The upper terrace shows differences according to the survey depth. At 0,5m depth, marks of a canal connecting the building with the well, while at 0,25m depth, the canals, perpendicular to the main axis of the garden can be seen. (*Corvinus University of Budapest, Department of Garden Art*. Leader of the survey: Dr Kristóf Fatsar).

where geophysics can be applied effectively are very few, but even among these there are some that can help us to reconstructing the theoretical models of the gardens of different religious orders. Nevertheless, the survey maps that are at our disposal provide so much information, that by correcting and analysing them spatially with GIS methods, we can get much closer to the understanding of monastic garden art.

CONCLUSION

Digital devices are an essential part of the research on Hungarian monastic gardens. As the gardens show hardly anything of their former layout, archival sources are needed – but the number of maps and layouts remaining is meagre, and even these do not present the gardens in much detail. The most useful image sources are survey maps, yet these are also not precise enough. In addition to image sources, one can rely on written documents, which rarely provide any exact locations of the garden elements. Therefore, a complementary method is needed that can combine the historical maps and today's maps and give the exact coordinates of the garden elements. This is especially important, because it is the only way to reveal such visible elements on the surface which, as a result of the great extent of changes, are not perceivable in any other way.

The use of GIS is not new in garden research, but it is still often omitted, especially in Hungary. However, due to the reasons mentioned above, it should be regarded as an essential tool in monastic garden research. As the presented case studies demonstrate, GIS methods can reveal many new, and sometimes surprising, data concerning the actual gardens. The difficulties in using GIS lie mainly in the lack of serviceable materials. A further step after analysing GIS data is the on site examination, which is extremely hard in the case of monastic gardens. Owing to their history, there are only a few gardens that are not built up nowadays, and even these few can cause problems for surveys because of the extent of the changes they have undergone. Non-destructive on-site examinations like geophysics are, hence, very difficult to implement effectively and thus the surveys are often unsuccessful.

The interdisciplinary methods discussed in this paper are powerful tools for exploring the principles and design of the different types of monastic gardens. If the theoretical models of several gardens can be created, these models can contribute to and facilitate the research of those other gardens about which nothing is known yet because of the lack of data. Moreover, as the comparison between monastic aims and their garden art was similar all over Europe, results may even be used successfully in other European monastic garden research, and thus aid in restorations as well.

REFERENCES

Alföldy, G. 2008. Történeti kertjeink integrált kezelése: elmélet és gyakorlat [An integrated methodology of the conservation and management of historic gardens: theory and practice]. 4D Tájépítészeti és Kertművészeti Folyóirat [4D Journal of Landscape Architecture and Garden Art] 11, 3-13.

Bowden, M. (ed.) 2006. Unravelling the Landscape. An Inquisitive Approach to Archaeology. Stroud, Tempus.

Currie, C. 2005. Garden Archaeology: A Handbook. York, Council for British Archaeology.

Fatsar, K. 2004. *Majk, egykori kamalduli remeteség, majd Esterházy-kastély kerttörténeti dokumentációja* [Documentation of the historic garden of the Camaldolese Hermitage in Majk]. [Manuscript] Budapest, Műemlékek Állami Gondnoksága.

Fatsar, K. 2009. A történeti kertekben végzett terepkutatások összetett eljárásai [A comprehensive approach to the restoration of historic gardens]. *4D Tájépítészeti és* Kertművészeti *Folyóirat* [4D Journal of Landscape Architecture and Garden Art] 13, 36-55.

Goulty, S. M. 1993. *Heritage Gardens: Care, conservation and management.* London, Routledge.

Klagyivik, M. 2007. A Heindl-örökség – egy XVII. századi botanikus kert története [The Heindl heritage – the story of a botanical garden in the 17th century]. *4D Tájépítészeti és* Kertművészeti *Folyóirat* [4D Journal of Landscape Architecture and Garden Art] 7, 52-58.

Klagyivik, M. 2010. A túróci jezsuita prépostság kertjei [The gardens of the Jesuit Provostry in Túróc]. In Sallay, Á. (ed.) 2010. *Ormos Imre Tudományos Ülésszak – Tájépítészeti Tanulmányok* [Imre Ormos scientific conference – studies in landscape architecture], 45-58. Budapest, Budapesti Corvinus Egyetem Tájépítészeti Kar.

Landsberg, S. 1995. *The Medieval Garden.* London, British Museum Press.

Meyvaert, P. 1986. The medieval monastic garden. In MacDougall, E.B. (ed.) 1986. *Medieval Gardens.* Dumbarton Oak Colloquium on the History of Landscape Architecture 9, 23-53. Washington D.C., Harvard University Press.

Mitchell, N., Rössler, M. & Tricaud, P.-M. 2009. *World Heritage Cultural Landscapes. A Handbook for Conservation and Management.* Paris, UNESCO.

Rumsey, D. & Williams, M. 2002. Historical maps in GIS. In Knowles, Anne K. (ed.) 2002. *Past time, past place: GIS for history*, 1-18. Redlands, CA, ESRI Press.

Szacsky, K. 2002. *Ferences kolostorkertek hagyománya és kialakítása a szécsényi rendház példáján* [Traditions and layouts of the Franciscan cloister gardens by the example of the cloister of Szécsény]. MSc thesis. Budapest, Corvinus University of Budapest.

Thacker, C. 1979. *The History of Gardens.* London, London Editions.

Turner, T. 2005. *Garden History. Philosophy and Design 2000 BC-2000 AD.* London/New York, Spon Press.

Velladics, M. 2000. Szerzetesrendi abolíció Magyarországon 1782-1790 [The abolition of religious orders in Hungary, 1782-1790]. *Levéltári Közlemények* 1(2), 33-52.

5.7 Thinking topographically about the landscape around Besançon (Doubs, France)

Authors
Rachel Opitz[1], Laure Nuninger[2] and Catherine Fruchart[1]

1. Centre national de la recherche scientifique, Université de Franche-Comté, Paris, France
2. Laboratoire de Chrono-Environnement, Centre national de la recherche scientifique, Université de Franche-Comté, Paris, France

Contact: rachel.opitz@mshe.univ-fcomte.fr

ABSTRACT

This paper focuses on the use of LiDAR (Light Detection and Ranging) data for the study of rural landscapes in the context of regional archaeological analyses. In particular, we concentrate on using LiDAR to highlight the importance of activities other than habitation, as well as the use of areas outside the modern ploughzone. It has frequently been said that one of the major challenges to archaeological landscape survey is the incorporation of uplands, marshes, forests and other areas we term 'outside the ploughzone'. Such areas are normally surveyed primarily through fieldwalking, but we suggest that LiDAR may make a significant contribution, although there are serious practical and methodological problems to overcome. Further, we argue that including these areas will alter the overall picture of rural landscapes in unexpected ways. The potential and challenges of integrating these areas and activities into landscape and regional scale research are sketched in this paper. We use a recent LiDAR survey as a case study to explore these issues. The project was funded by the Regional Council of the Franche-Comté for the LIEPPEC project, led by the USR 3124 and LEA ModeLTER, and is based in the hinterland of Besançon, Doubs, France.

The area surrounding Besançon is now largely forested, resulting in a dependence on the interpretation of the LiDAR model to guide field prospection. This paper provides some early results from the Forêt de Chailluz, north of Besançon; we use LiDAR to refocus the picture from one dominated by questions of settlement, settlement patterns and agriculture to one incorporating questions about complex networks of sites and activities, distributed across a wider range of landscape contexts. Using these initial results, we reflect on how LiDAR survey fits into the dynamic area of survey, landscape and regional archaeology.

KEYWORDS

LiDAR, survey, regional perspectives, remote sensing, rural landscapes

INTRODUCTION: LIDAR SURVEY IN REGIONAL AND LANDSCAPE RESEARCH

The archaeological study of local and regional long-term landscape change can be approached from many perspectives. Survey Archaeology, Regional Analysis and Landscape Archaeology are three major, inter-dependent approaches to this subject, employed to study how people exploited and experienced their sur-roundings, addressing questions including: How did natural and social resources and contexts influence the creation and development of settlement? Conversely, how did past societies manage and develop their surroundings to reshape the landscape? How are the cumulative results of these actions reflected in the modern landscape?

Based on a case study at Besançon (Doubs, France) (fig.1), for which we present some preliminary results, this paper attempts to illustrate some ways in which LiDAR survey can be used to address these questions. We open by situating our research in the context of fieldwalking and aerial survey archaeology,

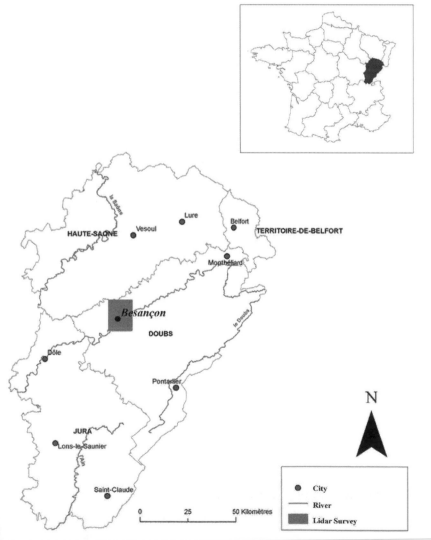

Figure 1. The location of the Besançon LiDAR survey. *Image: R. Opitz, Lieppec / MSHE C.N. Ledoux*

and Landscape and Regional Studies. We then focus on evidence, identified through the LiDAR survey, from the modern Forêt de Chailluz. This evidence provides new information on the development of the local rural economy, the organisation of the landscape, and the shifting boundaries of the forest. Based on these results, we discuss key methodological points on the use of LiDAR survey in micro-regional contexts, and note other areas where LiDAR can contribute.

RESEARCH CONTEXT

Many recent projects combine 'a landscape approach with traditional spatial and temporal systematics, to incorporate the dynamic scale of landscape analysis with the fine-scaled spatial and temporal analysis of patterns characteristic of traditional archaeological practice,' (Anschuetz et al. 2001, 191-192; Kantner 2008 for an overview) in producing regional studies. Within this framework, it is possible to ask a wide variety of questions. Regardless of the specific questions at hand, the biases of fieldwalking surveys and aerial photographic surveys will strongly influence the conclusions reached, because the vast majority of our systematically collected data on the settlement and, more broadly, on the organisation of the rural world comes from these surveys. The result is that many projects focus on patterns of habitation and agriculture, the two activities most often represented in survey data. In contrast, small features representing other, complementary aspects of the economy (e.g. surface mines, limekilns, charcoal platforms, and small quarries) are often under-represented. Far from a lack of interest in activities beyond settlement and the agriculture at the micro-regional and regional scale, the problem is the lack of data. Features and sites representing activities other than settlement and agriculture, although included in thematic studies and interpretations, are not identified in large enough quantities to be used in the type of formal spatial analysis widely employed in micro-regional and regional survey analyses. The bulk of the data describes settlements, while some local and larger studies include agriculture by incorporating information on field systems in addition to data on settlements (e.g. Favory 1988; Boyer et al. 2003) and off-site records such as ceramic scatters linked to agricultural practices, most commonly manuring (e.g. Wilkinson 1994, 2004; Nuninger 2003; Bintliff et al. 2007; Poirier et al. 2008; Bertoncello & Nuninger 2010). It is unusual to find a study which integrates all the structures which represent the many activities found in the rural landscape.

Secondly, systematic fieldwalking surveys usually try to sample across all landscape zones, but the majority of the data produced is concentrated in the modern agricultural landscape, where archaeological visibility is good (fig.2) but which represents only a small subset of past landscape zones. While aerial photography is effective in a broader range of landscape areas, its capabilities are significantly diminished in forests. This is particularly problematic for regions like the Franche-Comté in France where 45% of the surface is wooded, or Tuscany in Italy and Karst in Slovenia, where upwards of 60% of the area is under forest. Researchers have developed specific methodologies for uplands (e.g. Walsh & Richer 2006; Walsh et al. 2009; Riera et al. 2010 for examples), for wooded and scrub areas (e.g. Bommeljé & Doorn 1987; Doyen et al. 2004; Dupouey et al. 2007; Pautrat 2003), and at the other extreme for undersea areas (Gaffney et al. 2007). While the amount of research on 'marginal' parts of the landscape has greatly increased, projects focused on the ploughzone and different 'marginal' parts of the landscape are usually conducted separately and the two are not always well integrated.

Figure 2. Dense vegetation in both Mediterranean (left) and Continental (right) European landscapes poses serious problems for fieldwalkers wishing to survey outside the ploughzone.
Image: R. Opitz, P. Mosca Lieppec / MSHE C.N. Ledoux

The problematic implications of the lack of synthetic work incorporating both lowlands and uplands and other marginal zones have been noted many times (e.g. Cambi 2000; Bintliff & Kuna 2000; Bintliff 1994 and critically reiterated by R. Benton in his review of the Populus series; Benton 2001, 628-629). This disjunction calls for techniques that can be used to produce data on a large scale, analogous to those created through systematic fieldwalking, and so improve the integration of 'other' landscape areas – outside the ploughzone – into survey and subsequently regional analyses, while (ideally) opening up avenues for thinking more broadly about interactions between people and place and landscape beyond a collection of sites in an environmental background.

LiDAR (also known as light detection and ranging or airborne laser scanning) is a relatively new survey tool, increasingly used by archaeologists since 2004 (Sittler 2004). LiDAR is a technology that produces accurate and dense topographic data, similar to that recorded with a GPS or total station, by scanning the surface of any object with a laser and recording aspects of the returned waveform. LiDAR survey creates a very accurate and detailed model of the terrain in all unbuilt areas, including those obscured by forest or scrub, because some laser pulses will penetrate the vegetation canopy. It is particularly effective in areas outside the modern agricultural landscape, including uplands and other so-called marginal zones, because topographic remains are generally well preserved in these areas outside the ploughzone. LiDAR can be used for detailed planning of monumental remains or individual sites (e.g. Devereux et al. 2005; Corns & Shaw 2009), but it also can be used to systematically identify and characterise many small sites and features across the landscape. In this way it can vastly increase the amount of data we have on past human activities in the rural milieu (e.g. Sittler et al. 2007; Crutchley 2009).

We suggest that LiDAR can play a role, alongside fieldwalking data, in bringing these 'marginal' landscape areas more fully into the regional picture. At the same time, the continuous nature and detailed characterisation of the physical form of the terrain provide multiple avenues for getting off-site and exploring the physicality and experience of places. These applications are illustrated using examples from an ongoing study of the area around Besançon.

MODERN LAND USE AND RESEARCH HISTORY AT BESANÇON

The area surrounding the town of Besançon (Roman *Vesontio*) in eastern France (fig.3) is a heavily wooded karstic landscape, dominated by a large forest to the north of the town and a marsh and further woodland to the south. Besançon has been an urban centre since the first Iron Age and, after becoming *Vesontio* in the Roman period, continued as a regional centre, remaining an important town throughout the Late Antique, Medieval and Modern periods.

The city of Besançon itself is well studied, thanks to extensive rescue excavations. In contrast, little is known about the archaeology of the surrounding area. A LiDAR survey commissioned in April 2009 has revealed the remains of previously unknown buildings, almost 200 limekilns, more than 2,000 charcoal burning platforms, dozens of quarries of various types and networks of field boundaries in the

Figure 3. (left) The hillshaded bare earth DTM of the study area. (right) Air Photo mosiac of the study area. The LiDAR survey for Besançon covers an area of 140 km2 and includes the Marais Saone to the south and the Forêt de Chailluz to the north of the urban centre. The discrete return LiDAR data was collected at a nominal resolution of 0.5m (8pts/m2), with up to 4 returns per pulse. The data was initially processed by the providers (AeroData France), and reclassified by the authors using Terrascan and Terramodeler to generate improved bare earth models. Multiple hillshades were produced for visual inspection, in parallel with visualisation of the point clouds for individual areas. *Image: R. Opitz, L. Nuninger, Lieppec / MSHE C.N. Ledoux*

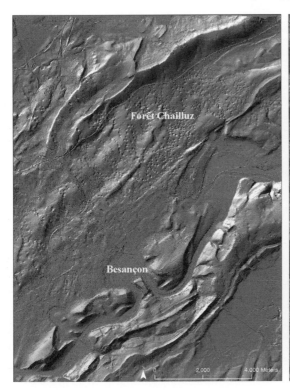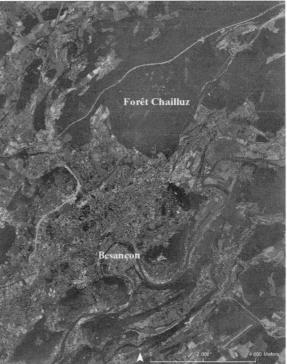

woods. These sites and features now represent the majority of archaeological information about the local rural area. The LiDAR survey is the basis for an ongoing campaign of fieldwork, concentrated on the Forêt de Chailluz and its adjacent pastures and meadows. The importance of the LiDAR dataset in this study area forces us to think seriously about the potential role of topographic data, the physical forms of past landscape elements encapsulated in the modern landscape, and the relationship between LiDAR and field survey.

It is widely agreed that one of the advantages of remote sensing survey, including LiDAR, is its rapidity compared to any other form of extensive survey. A further advantage is the significant increase in the number of features recorded in the area studied: often by as much as 50% outside areas heavily degraded by modern land use (Nuninger et al. 2008). However, systematic surface survey is needed to confirm and correct the interpretation of the LiDAR data and to provide chronological evidence (e.g. Risbol et al. 2006; Georges-Leroy et al. 2008a, b). The following sections present some initial results for part of the Forêt de Chailluz and outline two of the main challenges in the interpretation of combined LiDAR and surface surveys.

OUTSIDE THE PLOUGHZONE: EXAMPLES FROM THE BESANÇON SURVEY

Field clearance in the Forêt de Chailluz

The Forêt de Chailluz is 1,673 ha in its present-day extent. Systematic inspection of the LiDAR data shows that evidence of clearance and the establishment of field systems is concentrated in a 400 ha area in its northwest corner (fig.4). While this area is well within the modern and medieval forest and is now very much outside the ploughzone, it probably once formed part of the agricultural landscape. Through study-

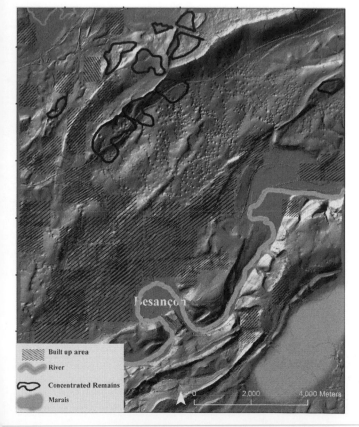

Figure 4. Areas with concentrations of stone piles and the remains of stone, rock-cut, or ditch and mound field boundaries in the forest are outlined in red. Almost a quarter of the modern forest contains evidence of organised field systems predating the current, and long established, system of parcels. *Image: R. Opitz, C. Fruchart, Lieppec / MSHE C.N. Ledoux. See also the full colour section in this book*

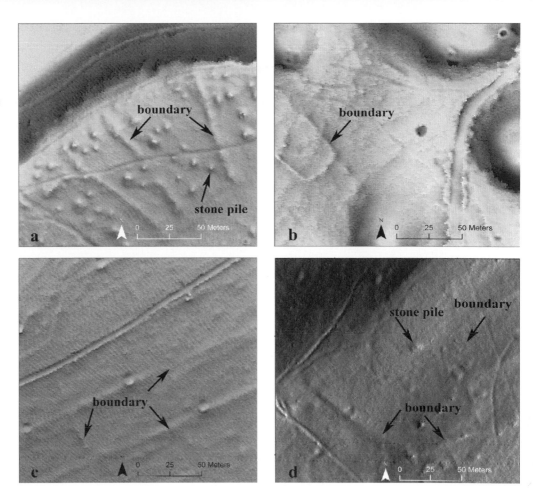

Figure 5. Traces of at least three field systems have been identified in the north-west part of the Forêt de Chailluz. (a) Field system characterised by broad linear boundaries and recut bedrock. (b) Field system characterised by narrow stone and earth mounds and small fields. (c) Field system characterised by long narrow fields. (d) Stone piles typical of those appearing throughout the field systems. *Image: R. Opitz, C. Fruchart, Lieppec / MSHE C.N. Ledouxa*

ing the evidence for clearance, we pursue a more detailed understanding of the changing character and shifting boundaries of the forest.

The LiDAR model reveals evidence for several phases of activity in the north-west area, including at least three separate field systems (fig.5). One extensive field system is characterised by a combination of broad, linear concentrations of regular width formed by seemingly unworked or heavily eroded stones and natural topographic features, possibly recut for emphasis, to define individual fields. A second distinct system has smaller, discontinuous fields bounded by narrow, linear earth and stone mounds. A third has a regular grid of long, narrow fields. Stone piles of various constructions, sizes and distributions appear throughout the field systems, and occasionally outside of them. This diversity suggests several distinct phases of activity, and clearance for different purposes. Field boundaries and stone piles are much less common in the rest of the forest, except for one small area close to the village of Braillans, which probably corresponds to activity surrounding that village. The absence of features representing clearance in the central part of the forest suggests that this area remained wooded.

Outside the forest within a mixed meadow and woodland area, known as L'Ermitage, we find a fossilised system of small fields, with further underlying linear field boundaries. The same pattern of small fields and occasional underlying linear boundaries can be found in the recently forested Les Vallieres, nearby. At this point any link between these areas and the remains in the Forêt de Chailluz is uncertain.

However, the possibility that the underlying linear boundaries are related to the broad linear field boundaries seen in the Forêt de Chailluz, and that they cross the border of the modern forest, is suggested based on their size and alignment. The marked difference in their appearances can be explained by the differences in later land use and preservation conditions inside and outside the forest. Studying these field systems through the LiDAR survey presents remains inside and outside the forest together in a single dataset, highlighting possible links and alignments across land use zones.

Habitation in the forêt de chailluz

Within the north-west area, the LiDAR survey has also revealed the remains of a collection of buildings, paths, associated enclosures and other features, shown in Figure 6. Concentrations of stone mounds in this area, if they belong to the same phase as the buildings, may represent either burial monuments or field clearance. These features, originally identified in the LiDAR model, were further characterised on the basis of fieldwalking evidence and are interpreted as delimiting buildings and enclosures. Identifications of the buildings were made on the basis of findings of numerous nails and metal objects, small quantities of ceramics retrieved through test pitting, and areas of rich dark earth, distinct from the typical clay soils in this area.

The chronology of these buildings and enclosures has not yet been established, and will likely require excavation and scientific dating. Based on the present evidence we can draw a few preliminary conclusions. This group of structures is cut by the pre-17th-century Chatillon-Besançon border trench (fig. 7). Further, these buildings and enclosures can be supposed to pre-date the 14th century as they are not mentioned in historical texts or maps, which provide fairly detailed records from the middle of the 14th century. The metal-detected and surface ceramic finds within these features suggest activity here in the late La Tene or Gallo-Roman period. Many of the lime kilns in this area cut the field boundaries and enclosures, and are probably later. These features are probably prior to the 17th century, as from that period the area was reserved for the supply of wood for buildings, as attested by archival evidence and a mid-18th-century map.

Figure 6. The buildings and enclosures shown here appear to belong to a separate phase than the extensive field system characterised by broad linear mounds. Surface and metal detecting finds suggest a late La Tene or Gallo-Roman date for these features. The main enclosures are indicated by arrows. *Image: R. Opitz, Lieppec / MSHE C.N. Ledoux*

Figure 7. A limekiln (a) cuts a field boundary (b) which in turn cuts the remains of a building (c) in Bois de la Lave, in the northwest part of the Forêt de Chailluz. This type of series of features can be used to establish relative dates. *Image: R. Opitz , C. Fruchart, Lieppec / MSHE C.N. Ledoux*

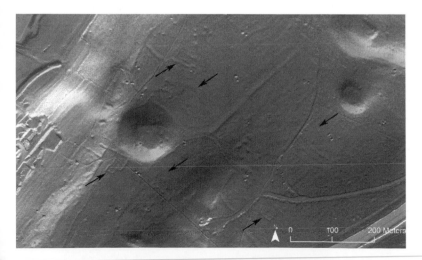

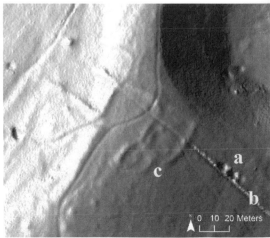

Figure 8. The 'feature set' of limekiln, quarry, claypit and charcoal burning platform, grouped together inside a doline, occurs frequently in the Forêt de Chailluz. a: limekiln; b: claypit; c: charcoal burning platform; d: quarry. *Image: R. Opitz, C. Fruchart, Lieppec / MSHE C.N. Ledoux. See also the full colour section in this book*

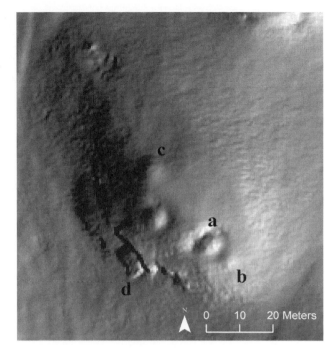

The possible presence of settlement in this area during the late La Tene or Gallo-Roman period, indicating the forest had been cleared (or not yet developed) at this time, would fit well with larger regional trends. This group of buildings can be compared with two other groups of small buildings from the local area, one excavated during the construction of the LGV (the new high-speed railway line) (Goy 2009) and dating to the Gallo-Roman period and the other with extensive Gallo-Roman surface finds, interpreted as a sanctuary or temple. The sites are located respectively 3.5 and 1.5 km further north along the road leaving Besançon for the north-west. Taken together, these sites begin to form a picture of suburban residences and religious sites along the main road outside an important town.

Focus on special purpose sites: limekilns, small quarries, clay pits and charcoal burning platforms
In addition to the buildings and field systems described above, that seem to indicate settlement and agricultural or pastoral activity within what is now forest, the LiDAR survey provides evidence for other economic activities which are likely to have been carried out in or near a forested environment. Notably, a large number of limekilns, quarries, charcoal burning platforms and claypits have been recorded. These features seem to broadly follow two organisational schemes. In one, a single feature of each type appears, creating a set of features that would allow the complete process of lime production to take place in a single locale. This 'feature set' arrangement is frequently found inside dolines, the broad natural basins typical of karstic terrain and found throughout the forest (fig. 8). In this process we can suppose that stone is cut from the side of the doline and broken into small pieces. The kiln is constructed, partly dug into the sloping ground. Clay to cover the layers of burning stone is dug just opposite the kiln, creating a depression which may also be used to store water. A space is cleared and levelled and wood is burned to create charcoal as a parallel activity, or to fuel the kiln.

While there is a remarkable concentration of 'feature sets' in the Bois de la Lave in the north-west part of the Chailluz forest and another concentration in its centre, there does not appear to be any internal organisation of the groups of features. Based on their shape and small size (4-5m), the kilns were probably built for one-time use, individual-scale production. This differs from what has been observed in Forêt de Haye (Lorraine, France) or in Karst (Slovenia), where bigger kilns with a different shape were more likely dedicated to production for the whole community (personal communication with M. Leroy, A. Marsetic). The dense concentration of kilns suggests the sporadic or continuous use of this area for lime production, possibly over a long period. The unorganised yet concentrated character of the distribution may point to a combination of collective rules or practices and individual actions.

The clearly organised pattern of many of the charcoal platforms, and their distinctive half-cut half-built construction, using semi-circular dry-stone walls used to create small terraces on the hillslope, indicates a strong collective framework which organised production, possibly linked to the management of the forest. The regularity of the structures and their distribution suggests a short period for this activity, and may point to regional production (see Beyrie et al. 2003 for a metallurgy example), rather than purely local industry. Further archival and field research is needed to provide confirmation for these interpretations.

The limekilns and charcoal platforms represent a substantial effort invested in non-agricultural production over an extended period or several discrete periods. The forest, which could easily remain largely blank on a survey-based map, is shown through the LiDAR survey to be an important production zone. This example illustrates that activities in the forest need to be understood to create a complete picture of the rural economy around Besançon.

CHALLENGES IN INTEGRATION AND INTERPRETATION

Systematic fieldwalking surveys have had a major impact on how we study the rural world (For an overview, see Ammerman 1981; Cherry 1983, 2003; Banning 2002; Athanassopoulos & Wandsnider 2004.) To analyse and interpret the findings from LiDAR surveys in the context of broader research on landscape change or regional development, they must be integrated with fieldwalking survey data. Perhaps a useful analogy for how this might be accomplished is the integration of off-site material within archaeological surveys and regional studies. Like records of manuring scatters, LiDAR survey provides data for areas that were previously empty and for non-habitation aspects of past landscapes. One way of integrating the two at the analytical level is the treatment of the different types of evidence as related distributions (e.g. Bertoncello & Nuninger 2010; Poirier et al. 2008), or the evidence could be combined during the final interpretation. However, there are two basic problems specific to the integration of LiDAR survey: first, the absence of good chronological information for many features detected through the LiDAR survey and second, the lack of overlap between LiDAR and fieldwalking survey distributions.

Chronologically-challenged features
While the data from the LiDAR survey permits the study of the overall intensity of the use of 'marginal' areas for various activities, a detailed chronology must be established for these features to truly integrate the results of LiDAR and fieldwalking surveys. While it is possible to study in some detail the purpose of

these features, based on close observation of their forms and comparison with excavated or otherwise well known examples, LiDAR models have a somewhat difficult relationship with chronology. As with any remote sensing or cartographic method, we can build relative chronologies based on the superposition of features, such as the lime kiln cut into a field boundary which in turn cuts the remains of a building, shown in Figure 7. The current fieldwork campaign combines surface prospection, GPS mapping, metal detection and test pitting, targeting locations where features have been identified in the LiDAR model. Experience shows that most features have few or no reliably datable surface ceramic or metal remains, so to obtain absolute dates further test trenching or coring and absolute dating are needed. An exploratory project has been established to date several lime kilns, using C14, in order to estimate a *terminus post quem* for the otherwise difficult to date structures cut by the kilns.

Despite difficulties in establishing a strong chronology for individual features in the Forêt de Chailluz, it has been possible to identify some broad trends based on spatial distributions, overlaps and relative positioning. For now, based on the relative chronology established by superimposed features and the limited dating evidence available, this local landscape can be studied within the overall picture, analysing relationships (proximity, accessibility, purpose) of features to the antique and medieval agglomerations, the existing road networks and religious landscapes. However, to integrate features found through LiDAR into phase-by-phase distributions widely used to interpret survey data at the micro-regional and regional scales, dates for individual features must be established.

Mutually exclusive distributions

We have noted that LiDAR data is almost inversely biased to fieldwalking data – it performs best outside the ploughzone (fig. 9). At the same time, LiDAR survey reveals a very different collection of features from those usually produced by fieldwalking survey; many lime kilns, quarries for stone and lime, charcoal burning platforms, field boundaries, drainage ditches and terraces (fig. 10) have been found.

Figure 9. Earthworks in the ploughzone (left), while still visible, have been substantially flattened and details of the feature are not clear. Contrast this with the details visible in the well preserved remains in the forest (right). *Image: R. Opitz, Lieppec / MSHE C.N. Ledoux*

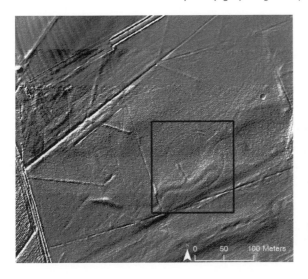
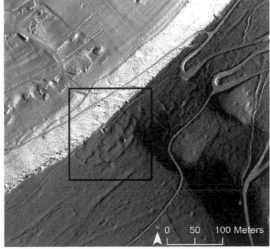

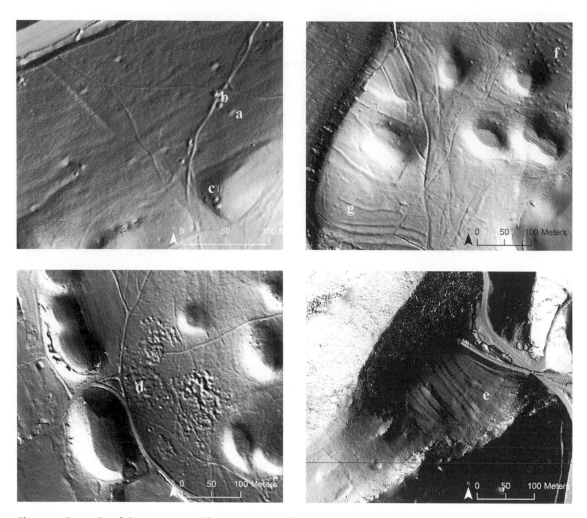

Figure 10. Examples of the appearance of common types of features identified through the LiDAR survey. None of these feature classes are common in survey datasets based on fieldwalking. This poses certain complications in terms of integration and comparison. a: charcoal burning platform, b: limekiln; c: limestone quarry; d: gypsum or slate quarry; e: viniculture terraces; f: field clearance cairns; g: field boundary. *Image: R. Opitz, Lieppec / MSHE C.N. Ledoux.*

The identification of a wide variety of feature types recalls a point raised by Alcock and Rempel (2006): How does (and perhaps more importantly, how should) survey archaeology deal with sites and features that do not fall nicely into a settlement-centric classification – the 'special-purpose sites' and practices other than habitation – the Laconia Survey's (Cavanagh et al. 2002) 'other forms of human activity'? In their case there are not enough of these sites to develop strong spatial or chronological trends, but their presence undeniably adds texture to the landscape and they provide an impetus for thinking beyond spacio-chronological trends. They conclude that work on these sites, the 'wells, threshing floors, burial mounds, kilns, bridges, mills, knapping debris, drainage ditches, pathways, caves, quarries, terraces, shrines and dumps,' (Alcock & Rempel 2006, 42) is both worthwhile and necessary as, 'without the nuance they provide, the questions we can ask of our regional data become unnecessarily limited, reverting largely to the purely economic, the demographic, the functional,' (Alcock & Rempel 2006, 42).

The preliminary results of the Besançon survey show that LiDAR effectively resolves the problem of not having enough data on some types of special-purpose sites. Rapid detection provides an abundance of evidence, allowing us to consider the distribution of these features on the micro-regional scale. The result is a more nuanced understanding of the use of the landscape and of micro-regional development. In future work, there is the potential to include these features in spatial analyses, for example estimating the overall intensity of the use of marginal zones (as illustrated in Georges-Leroy et al. 2008b; Poirier et al. 2008).

In spite of the difficulties of integration, the development of survey-style distributions of sites and features outside the ploughzone will have a significant, enriching impact on our knowledge and understanding of the rural world as a whole. Beyond adding many new sites and features to the record and populating previously empty landscape zones, the LiDAR model can be used to explore and strengthen links between feature, site, distribution and landscape by taking advantage of the spatial continuity of the data to work across multiple scales, an approach we introduce in the next section.

MEASURED AND EXPERIENCED SCALES

A tension exists between theoretical conceptualisations of scale and the treatment of scale within GIS-based spatial analyses (Gaffney & Gaffney 2006; Lock 2009; Kvamme 1999). GIS tend to be used, perhaps because of their database quality, to work on large-scale trends and patterns, such as site distributions and the changing impact of geology or slope as seen over large areas (e.g. Ebert et al. 1996; Allen et al. 1990; Boos et al. 2007). On the other hand, the lived or experienced scale (as conceived in the phenomenological sense, e.g. Tilley 1994; David & Thomas 2008; Ashmore & Knapp 1999) at which people encounter the landscape is of great interest to those invested in individual behaviours, and should not be neglected. Bringing the two together, being multiscale, is easier said than done. To explore how very high resolution topographic models derived from LiDAR might contribute to modelling this relationship we return to the example of lime production.

People working at lime burning sometimes took advantage of the sides of the dolines, treating them as open-faced quarries for easy stone extraction, siting the kiln conveniently in the base of the doline. The location in the landscape for any particular kiln may, on the larger scale, have been chosen to be close to a number of other kilns. Following this decision this kiln becomes part of the process of the creation of a cluster, making the area more attractive for future lime production. It is both the result of and a participant in the process of creating the final pattern. The landscape becomes characterised in part by the sites it attracted.

On another scale, being inside the doline simultaneously takes advantage of and reshapes the natural physical form of the depression. The quarrying, as it provides stone for the kiln, reinforces the steep-sidedness of the doline. The quarrying generally follows the natural contours of the doline, although some squaring off occurs. The form of the doline today is both the attractor for as well as the result of these activities.

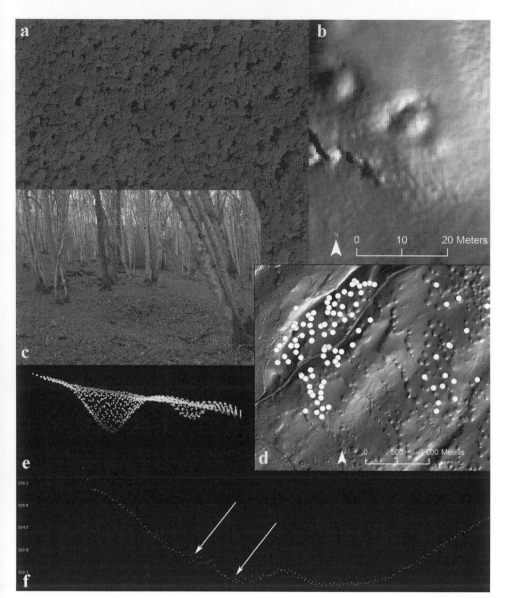

Figure 11. Multiple visualisations of the same data are used to explore relationships between feature, site, landscape and distribution. (a) Air photo of the forested location of the limekiln. (b) Hillshaded DTM of the limekiln. (c) Photo of the remains of the limekiln taken from about 4M away. (d) Yellow dots show the distribution of lime kilns in the local area. (e) The limekiln's appearance in the point cloud, with points coloured by elevation. (f) A profile section through a limekiln situated in the bottom of a doline. *Image: R. Opitz, C. Fruchart, Lieppec / MSHE C.N. Ledoux. See also the full colour section in this book*

This tight-knit connection between feature and context on several scales is partly because when looking at LiDAR, one naturally moves across scales, closely inspecting the physical form of a feature, trying to characterise it, zooming out to look at it from a distance, seeing it together with other features, looking at how it fits with or seems to have altered the terrain. The process is one of simultaneously examining linked displays showing the site and its context in plan, usually as a hillshaded dtm, at a 3D point cloud (fig. 11), and at statistical graphs, adding or removing layers, trying different visualisations to gain an overall understanding of how the place works. In this way, exploiting the connection across scales and visualisations can play an important role in the exploitation and interpretation of the terrain models created with LiDAR data.

CONCLUSIONS: LIDAR AS CONTEXTUAL TOPOGRAPHY

Topographic data, increasingly acquired in great detail and over large areas through LiDAR survey, can and should be more than just another source of new sites and features or an environmental backdrop. Working with these massive topographic datasets gives us an opportunity to rethink our approach to the rural landscape, taking into account more diverse areas and activities. The Besançon surveys have shown LiDAR and fieldwalking to be both interdependent and complementary. The LiDAR survey is also essential as a source of data on the micro-regional scale for forested or otherwise 'marginal' parts of the landscape. The broader implication of having LiDAR data, if it can be well integrated with (micro-) regional surveys, is that it can help bridge the gaps between historical descriptions of the rural world, especially the rural economy, which include a wide range of 'special-purpose sites' and what is commonly found through fieldwalking surveys or through rescue excavation. (For contrasting perspectives see among others Brun 2005; Greene 1991, especially chapter 5; Snodgrass 1991; Witcher 2006; Duncan-Jones 1990; Kehoe 2007.) It also encourages us to work flexibly across scales, thinking continually about sites and their surroundings together.

Thinking across scales, linking an individual feature to a distribution or group of features to long-term changes in the form of the landscape leverages the fluidity of scale and uniformity of the data we have in LiDAR-based terrain models, leading to a more contextualising approach. Hyper-realistic modelling, employing the level of detail and flexibility of visualisation provided through LiDAR models, may be the crux between the statistical and computational, database-centric world of GIS-based analyses, and the experiential, lived-scale, phenomenological approaches to landscape (as suggested by Gaffney & Gaffney 2006, and by the work of Gillings & Goodrick 1996; Gillings 2004; and esp. Cripps, Earl & Wheatley 2006). If so, LiDAR data may play an important role by providing lived-scale data across landscape-scale areas, and the means to move fluidly between these scales.

REFERENCES

Alcock, S. & Rempel, J. 2006.The More Unusual Dots on the Map: 'Special-Purpose' Sites and the Texture of Landscape. In P. Guldager Bilde & V. Stolba (eds.), *Surveying the Greek Chora. The Black Sea Region in a Comparative Perspective, BSS 4*, 27-46. Aarhus, Aarhus University Press.

Allen, K., Green, S. & Zubrow, E. (eds.) 1990. *Interpreting Space: GIS and Archaeology*. London, Taylor and Francis.

Anschuetz, K., Wilshusen, R. & Scheick, C. 2001. An Archaeology of Landscapes: Perspectives and Directions. *Journal of Archaeological Research* 9 (2), 157-211.

Ashmore, W. & Knapp, B. (eds.) 1999. *Archaeologies of Landscape: Contemporary Perspectives*. Maldenn, MA, Blackwell.

Athanassopoulos, E. & Wandsnider, L. 2004. Mediterranean Landscape Archaeology Past and Present. In E. Athanassopoulos & L. Wandsnider (eds.), *Mediterranean archaeological landscapes. Current issues*, 1-14. Philadelphia, University Museum of Pennsylvania.

Banning, E. 2002. *Archaeological Survey*. New York, Kluwer.

Benton, R. 2001. Mediterranean Myopia. *Antiquity* 75 (289), 627–629.

Bertoncello, F. & Nuninger, L. in press. From archeological sherds to qualitative information for settlement pattern studies. In *Beyond the artifact, digital interpretation of the past. Proceedings of the CAA Conference 2004, Prato, 13-17 Feb. 2004, xx-xx.*

Bintliff, J. 1994. The History of the Greek Countryside: As the Wave Breaks. Prospects for Future Research. *Structures Rurales et Societes Antiques, Annales littéraires de l'Université de Besançon* 126, 7-15.

Bintliff, J. & Kuna, M. (eds.) 2000. *The Future of Archaeological Field Survey in Europe*. Sheffield, Sheffield Academic Press.

Bintliff, J., Howard, P. & Snodgrass, A. 2007. *Testing the Hinterland: The Work of the Boeotia Survey (1989-1991) in the Southern Approaches to the City of Thespiai*. Cambridge, McDonald Institute Monographs.

Bommeljé, S. & Doorn, P. (eds.) 1987. *Aetolia and the Actolians*. Utrecht, Parnassus Press.

Boos, S., Hornung, S. Jung, P., & Muller, U. 2007. GIS as a tool for processing hybrid prospection data in landscape archaeology. *International Journal of Humanities and Arts Computing* 1 (2), 137-149.

Boyer, O., Favory, F. & Tourneux F.P. 2003. Morphologie agraire et réseaux d'établissements: de l'analyse relationnelle à la définition de liens. L'exemple de la Vaunage gallo-romaine. In F. Favory & A. Vignot (eds.), *Actualités de la recherche en Histoire et Archéologie agraires, Actes du Colloque AGER V, 19-20 septembre 2000 à Besançon*, 325-337. Besançon, Presses Universitaires Franc-Comtoises.

Brun J.-P. 2005. *Archéologie du vin et de l'huile en Gaule romaine*. Paris, Errance.

Cambi, F. 2000. Quando i campi hanno pochi significati da estrarre: visibilità archeologica, storia istituzionale, multi-stage work. In R. Francovich, H. Patterson & G. Barker (eds.), *Extracting meaning from ploughsoil assemblages*, 72-76. Oxford, Oxbow Books.

Cavanagh, W., Crouwel, J., Catling, R. & Shipley, G. 2002. *Continuity and Change in a Greek Rural Landscape: The Laconia Survey. Vol. 1: Results and Interpretation*. London, British School at Athens.

Cherry, J. 1983. Frogs round the Pond: Perspectives on Current Archaeological Survey Projects in the Mediterranean Region. In D. Keller & D. Rupp (eds.), *Archaeological Survey in the Mediterranean Area* , 375-416. *British Archaeological Reports International Series* 155. Oxford, ArchaeoPress.

Cherry, J. 2003. Archaeology beyond the site: regional survey and its future. In J. Papadopoulos & R. Leventhal (eds.), *Theory and Practice in Mediterranean Archaeology: Old World and New World Perspectives*, 137-160. Los Angeles, Cotsen Institute, University of California at Los Angeles.

Corns, A. & Shaw, R. 2009. High resolution 3-dimensional documentation of archaeological monuments & landscapes using airborne LiDAR. *Journal of Cultural Heritage* 10 (1), 72-77.

Cripps, P., Earl, G. & Wheatley, D. 2006. A dwelling place in bits. *Journal of Iberian Archaeology* 8, 25-39.

Crutchely, S. 2009. Ancient and modern: Combining different remote sensing techniques to interpret historic landscapes. *Journal of Cultural Heritage* 10 (1), 65-71.

David, B. & Thomas, J. (eds.) 2008. *Handbook of Landscape Archaeology. World Archaeological Congress Research Handbooks in Archaeology*. Walnut Creek, Left Coast Press.

Devereux, B.J. et al. 2005. The potential of airborne LiDAR for detection of archaeological features under woodland canopies. *Antiquity* 79 (305), 648-660.

Doyen, B. Decocq, G. & Thuillier, P. 2004. Archéologie des milieux boisés en Picardie. *Revue archéologique de Picardie* 1 (1), 149-164.

Duncan-Jones, R. 1990. *Structure and Scale in the Roman Economy*. Cambridge, Cambridge University Press.

Dupouey, J.-L., Dambrine, E., Dardignac , C. & Georges-Leroy, M. (eds.) 2007. *La mémoire des forêts. Actes du colloque Forêt, Archéologie et Environnement. 14 au 16 décembre 2004*. Paris, ONF – INRA – DRAC de Lorraine.

Ebert, J.I., Camilli, E.L. & Berman, M.J. 1996. GIS in the analysis of distributional archaeological data. In H. Maschner (ed.), *New Methods, Old Problems: Geographic Information Systems in Modern Archaeological Research*, 25-37. *Occasional Paper* 23. Carbondale, Southern Illinois University Press.

Favory, F. 1988. Le site de Lattes et son environnement (France, Hérault), d'après les images aériennes et les documents planimétriques. In *Mélanges d'Histoire et d'Archéologie lattoise*, 15-56. *Lattara*, 1. Lattes, Association pour la recherche archéologique en Languedoc oriental.

Gaffney, C. & Gaffney, V. 2006. No further territorial demands: on the importance of scale and visualisation within archaeological remote sensing. In *From Artefacts to Anomalies: Papers inspired by the contribution of Arnold Aspinall. University of Bradford 1-2 December 2006. Available from http://www.brad.ac.uk/archsci/conferences/ aspinall/ and http://www.brad.ac.uk/archsci/conferences/aspinall/presentations/Gaffney&Gaffney.pdf.*

Gaffney, V., Thomson, K. & Fitch, S. 2007. *Mapping Doggerland: The Mesolithic landscapes of the Southern North Sea*. Oxford, ArchaeoPress.

Georges-Leroy, M., Bock, J., Dambrine, E., Dupouey, J.-L. 2008a. L'apport du laser scanneur aéroporté à l'étude des parcellaires gallo-romains du massif forestier de Haye (Meurthe-et-Moselle). *AGER 18*, 8-11.

Georges-Leroy, M., Tolle, F. & Nouvel, P. 2008b. Analysis of the intensity of agrarian exploitation by spatial analysis of ancient field systems preserved by forest cover. In A. Poluschny, K. Lambers & I. Herzog (eds.), *Layers of perception. Proceedings of the 35th International Conference on Computer Applications and Quantitative Methods in Archaeology (CAA). Berlin, Germany, April 2-6, 2007*, 281 and on CD-Rom. Bonn, Habelt.

Gillings, M. & Goodrick, G. T. 1996. Sensuous and Reflexive GIS: Exploring Visualisation and VRML. *Internet Archaeology 1*. http://intarch.ac.uk/journal/issue1/gillings_index.html.

Gillings, M. 2004. The Real, the Virtually Real and the Hyperreal: The Role of VR in Archaeology. In S. Moser and S. Smiles (eds.), *Envisioning the Past*, 223-239. Oxford, Blackwell.

Goy, C. 2009. Geneuille (Doubs) Un petit établissement rural de l'Antiquité in Archéologie en Franche-Comté. In DRAC (ed.), *Fouilles archéologiques de la LGV Rhin-Rhône: les résultats* . Besançon, DRAC.

Greene, K. 1991. *The Archaeology of the Roman Economy*. Berkeley/Los Angeles, University of California Press.

Kantner, J. 2008. The Archaeology of Regions: From Discrete Analytical Toolkit to Ubiquitous Spatial Perspective. *Journal of Archaeological Research 16*, 37-81.

Kehoe, D. 2007. *Law and the Rural Economy in the Roman Empire*. Ann Arbor, University of Michigan Press.

Kvamme, K. 1999. Recent directions and developments in geographical information systems. *Journal of Archaeological Research 7*, 153-201.

Lock, G. 2009. Archaeological computing then and now: theory and practice, intentions and tensions. *Archeologia e Calcolatori 20*, 75-84.

Nuninger, L. 2003. Exploitation et spatialisation des indices protohistoriques épars en Vaunage (Gard), VIIème-Ier siècles av. J.-C. In F. Favory & A. Vignot (eds.), *Actualité de la recherche en Histoire et Archéologie agraires. Actes du colloque AGER V (Besançon 2003)*, 365-375. Besançon, PUFC.

Nuninger, L., Bertoncello, F., Fovet, E., Gandini, C. & Trément, F. 2008. *The spatio-temporal dynamics of settlement patterns from 800 BC to 800 AD, in Central and Southern Gaul, in 7 millennia of territorial dynamics : settlement pattern, production and trades from Neolithic to Middle Ages: closing colloquium of The ArchaeDyn program ('Spatial dynamics of settlement patterns and natural resources: towards an integrated analysis over a long term, from Prehistory to the Middle Ages')*. Dijon, France.

Pautrat, Y. 2003. Archéologie en forêt: Un patrimoine peu à peu exploré. *Les Dossiers d'archéologie 284*, 18-23.

Poirier, N., Georges-Leroy, M., Tolle, F., & Fovet, E. 2008. "The time-space dynamics of agricultural areas from Antiquity to Modern times." In *7 millenia of territorial dynamics: settlement pattern, production and trades from Neolithic to Middle Ages, ARCHAEDYN ACI "Spaces and territories" 2005-2007 Final conference - Dijon, 23-25 June 2008, pre-proceedings*, edited by C. Gandini, F. Favory and L. Nuninger, 81-94.

Riera, S., Palet, J.M., Ejarque, A., Orengo, H. Miras, Y. Euba, I. & Julià, R. 2010. The Long-Term Shaping of a High Mountain Cultural Landdscape in the Eastern Pyrenees (Madriu-Perafita-Claror Valleys, Andorra and Cadì range, Catalonia): an Integrated Research Program. Oral Presentation: *Landscape Archaeological Conference LAC2010, 26-29 January 2010, Amsterdam*.

Risbøl O., Gjertsen A. & Skare K. 2006. Airborne laser scanning of cultural remains in forests: some preliminary results from a Norwegian project. In S. Campana & M. Forte (eds.), *From Space to Place. 2nd International Conference on Remote Sensing in Archaeology*, edited by , 107-112. BAR International Series 1568. Oxford, Archaeopress.

Sittler, B. 2004. Revealing historical landscapes by using airborne laser scanning. A 3-D model of ridge and furrow in forests near Rastatt (Germany). *International Archives of Photogrammetry. ISPRS 26*, 258-261.

Sittler, B., Weinacker, H., Gütlinger, M. & Koupaliantz, L. 2007. The potential of LiDAR assessing elements of cultural hidden under forests. In Z. Bochenek (ed.), *New Developments in Remote Sensing*, 539-548. Rotterdam, Millpress.

Snodgrass, A. 1991. The Place and Role of the Annales School in the Approach to the Roman Rural Economy. In J. Bintliff (ed.), *The Annales School and Archaeology*, 77-103. Leicester, Leicester University Press.

Tilley, C. 1994. *A Phenomenology of Landscape. Paths, Places, Monuments*. Oxford, Berg.

Walsh, K. & Richer, S. 2006. Attitudes to altitude: changing meanings and perceptions within a 'marginal' Alpine landscape – the integration of palaeoecological and archaeological data in a high altitude landscape in the French Alps. *World Archaeology* 38 (3), 436-454.

Walsh, K., Mocci, F., S. Richer, M. Court-Picon, B. Talon, S. Tzortzis, J. M. Palet-Martinez & C. Bressy 2009. Archéologie et paléoenvironnement dans les Alpes méridionales françaises. Hauts massifs de l'"Argentiérois, du Champsaur et de l'Ubaye (Hautes-Alpes et Alpes de Haute Provence) (Néolithique final – début de l'Antiquité). *Cahiers de Paléoenvironnement* 6, 235-254.

Witcher, R.E. 2006. Agrarian spaces in Roman Italy: society, economy and Mediterranean agriculture. *Arqueología espacial (Paisajes agrarios)* 26, 341-359.

Wilkinson, T.J. 1994. The Structure and Dynamics of Dry-Farming States in Upper Mesopotamia. *Current Anthropology* 35, 483-520.

Wilkinson, T.J. 2004. Off-site archaeology in the area of Mashkan-Shapir. In E. Stone & P. Zimansky (eds.), *The Anatomy of a Mesopotamian City: Surveys and Soundings at Mashkan-shapir*, 402-415. Winona Lake, IN, Eisenbrauns.

5.8 Modelling the agricultural potential of Early Iron Age settlement hinterland areas in southern Germany

Authors

Axel Posluschny[1], Elske Fischer[2], Manfred Rösch[2], Kristine Schatz[3], Elisabeth Stephan[3] and Astrid Stobbe[4]

1. Römisch-Germanische Kommission, Deutsches Archäologisches Institut, Frankfurt am Main Germany
2. Landesamt für Denkmalpflege, Arbeitsstelle Hemmenhofen, Labor für Archäobotanik, Gaienhofen-Hemmenhofen, Germany
3. ArchäoZoologische Dienstleistungen, Bodman-Ludwigshafen, Germany
4. Institut für Archäologische Wissenschaften, Abteilung Vor- und Frühgeschichte, Labor für Archäobotanik, Universität Frankfurt, Frankfurt, Germany
Contact: posluschny@rgk.dainst.de

ABSTRACT

Agriculture was the main basis of daily life in most prehistoric periods in Europe. The possibility of a settlement to produce more than the basic needs in a subsistence economy was in many cases the background of a surplus-based superiority of some settlements over others.

In our paper we will present a GIS and database system with which we model the agricultural potential of settlements within their natural surroundings based on topography and soil quality. Within the framework of the research project 'Early Centralisation and Urbanisation – The Genesis and Development of Early Celtic Princely Sites and their Territorial Surrounding' (http://www.fuerstensitze.de) we have developed a model which is used to calculate the maximum amount of people that can be fed from within the hinterland of both princely sites and 'regular' settlements by cattle and crop. The surrounding itself can be defined by cost-based calculations, creating a hinterland border based on walking time. The model is then used to compare the agricultural potential of different settlement sites as well as of sites from different periods.

It is the aim of our working group to further develop a runtime database file, which can be used to model the agricultural potential from any given archaeological site within a given surrounding and to calculate the amount of people that can be nourished from this hinterland. The database file will be distributed online as free software.

KEYWORDS

GIS analyses, agricultural potential, hinterland, cost distance, calorie expenditure

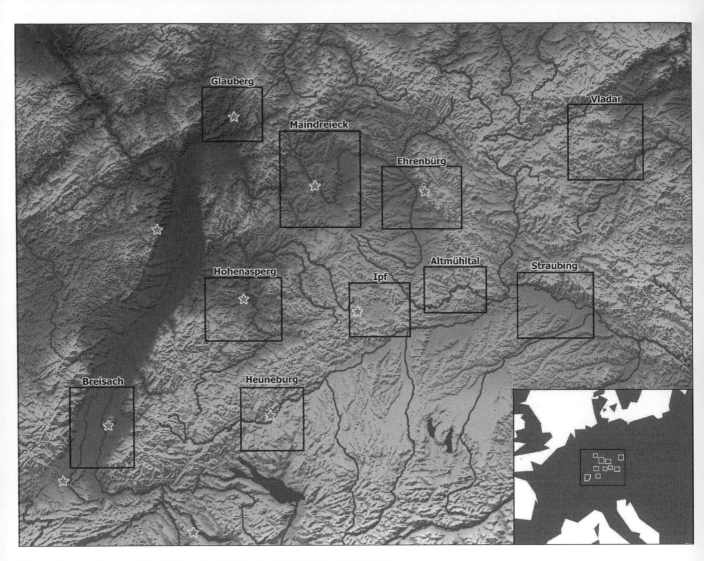

Figure 1. Research areas of the DFG-project *'Fürstensitze'* & *Umland*. – DEM SRTM90.

INTRODUCTION

The project 'Princely Sites' & Environs (*'Fürstensitze"* & *Umland*, funded by the German Research Foundation within the framework of the Priority Programme *'Early Processes of Centralisation and Urbanisation – Studies on the Development of Early Celtic Princely Seats and their Hinterland'*, http://www.fuerstensitze.de) is investigating the so-called 'Princely Sites' or 'Fürstensitze' of the Early Iron Age around 500 BC in south-western Germany, eastern France and comparable places in Bavaria and western Bohemia (http://www.fuerstensitze.de/1121).

Following a definition mainly described by W. Kimmig (1969), 'Fürstensitze' are rich fortified settlements, mainly situated on a hilltop, with large and rich grave mounds in their vicinity and with finds of imported goods, mainly ceramics from the Mediterranean. They seem to be the result of a social and maybe even cultural change or transformation of the protoceltic societies which we still do not understand to its full extent (cf. Schier 1998).

The aim of the project *'Princely Sites' & Environs* (Posluschny 2007) is to investigate the dynamics of settlements and people from the so-called late Bronze Age Urnfield Period to the Early Iron Age Hallstatt and the following Early Latène Period on the basis of the interconnection between man, culture and environment (fig. 1).

THE MEANING OF THE HINTERLAND

One of the main questions in this field of research is the mutual interdependency between the 'Princely Sites' – which we could compare to Christaller's 'Central Places' (Christaller 1933) to a certain extent – and their surrounding area, their hinterland and the settlement places around. The hinterland of prehistoric settlements played an important role for the agricultural livelihood and the economic exploited area in general. The sizes of these hinterland areas may have differed depending on the environment and on the needs of prehistoric people in different periods. Following this thesis we might assume that changes in size and layout reflect a change in settlement behaviour and in the use of natural environment for economic reasons. Both regular settlements and 'Princely Sites' are dependent on their own hinterland as a basis for economic needs. This might be compared with the model of core and periphery or with Thuenen's isolated state and ring-shaped economic model (fig. 2).

The definition of such a hinterland is of greater importance for questions on the economic abilities of the settlements, for questions of subsistence vs. surplus production. The shape and the size of these areas might differ in a regional as well as in a chronological perspective; there may also be differences between 'Princely Sites' and regular settlements. And last but not least, the potential of a settlement to feed its inhabitants and to even produce a surplus is based on the agricultural capacity of each hinterland thus the feeding potential of a settlement could give an idea of how many people might have lived in this settlement at most.

Figure 2. The idea of Thuenen's isolated state (after Rodrigue et al. 2006, Fig. 7.8 [with modifications]). Legend from top to bottom reads as from inside to outside in figure shown.

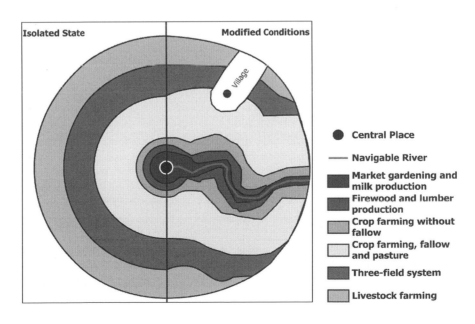

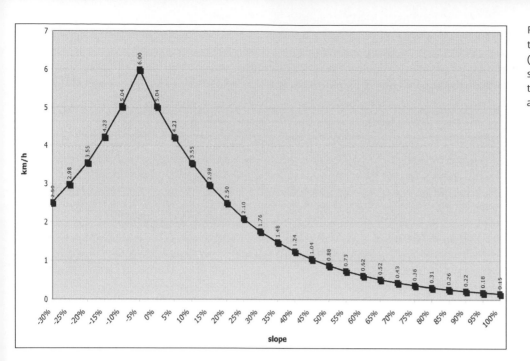

Figure 3. Graph of the walking speed (km/h) based on slope, calculated with the Gorenflo/Gale algorithm.

DEFINING A HINTERLAND

The model to define the potential hinterland of a prehistoric site that we used in our project is based on site-catchment analyses which in general are not new for archaeological research. Site-catchments were used more than 40 years ago (e.g. Vita-Finzi & Higgs 1970), in the earlier days based on Euclidean distances (e.g. Saile 1998; Conolly & Lake 2006, 209-211). With the advanced computer abilities that GIS have to offer and that have been increasing since the 1970s, it soon became possible to not only build site-catchment buffers from Eucledian distances but also to take into account topography and other factors of the natural environment like streams, vegetation or soil properties.

When calculating the agricultural potential the first problem is to define the area of the hinterland itself. Most GIS software offers push-button algorithms to calculate cost distances and least cost paths, generally based on a cost model that defines the costs of passing along in a landscape (cf. Conolly & Lake 2006, 213-226 for a short introduction to cost surface calculations, see also van Leusen 2002, 6-4–6-9 and Herzog & Posluschny in press). The same is true for the calculation of a cost dependent area, based on the maximum vicinity that can be reached within a maximum of time, with a maximum of abstract costs or with a maximum of calorie expenditure. It is not the aim of this paper to discuss the (dis)advantages of the various software algorithms in GIS programmes, nor do we want to go into detail regarding the mathematical background of the cost surface calculations. Various papers have been published on these topics during the last years, especially in the context of the annual CAA conferences (cf. i.a. Herzog & Posluschny in press). This paper concentrates on a case study that makes use of cost surface based calculations and environmental modelling (cf. i.a. Posluschny 2010) and whose greatest advantage is the relative easy way of modelling rather coarse data which is more or less easy to derive.

The background of the calculations that will be used to define a hinterland area for modelling its agricultural feeding potential is the use of a cost surface model that transfers slope into walking speed. The so-called Gorenflo/Gale algorithm is based on empirical data that has been collected from soldiers hiking different types of terrain (Gorenflo & Gale 1990). The result is a model of walking speed, calculated in kilometres per hour (fig. 3). For the use as a friction surface it then has to be recalculated into a cost model with minutes per kilometre as units. The modelling of the potential hinterland that can be reached within a certain time is then calculated within the IDRISI Andes software (module *COSTGROW*, Eastman 2003, 93). To avoid overlapping areas around close sites it was necessary to calculate one buffer for each site at a time so a macro has been built which was able to calculate the cost based buffer areas of more than 300 sites, based on a 25m resolution cost surface grid, using a buffer of 1 hour walking distance within approx. 10 hours.

Another problem when modelling a hinterland or the area of every day extensive use is to define the border of such an area. It is of course not very likely that prehistoric people had strict rules for a limitation of their usable vicinity. But on the other hand, it is very likely to assume that it was not very advisable to use land beyond a certain distance from a settlement simply for economical reasons. Chisholm (1962) and following him Bintliff (1999, 2002) argued – based on cost-benefit ratios – that the land used for agricultural needs, mainly for ploughing, is usually not further away than 1 km which is approximately a 12 or 15 minutes walk. The hinterland that has been used for cattle farming, exploiting forestal resources and so on, but also for a more intensive use in order to be able to produce a surplus, should be no further away than 5 km or 1 hour walking time. The potential models that will be calculated for all settlement sites of the project will therefore be based on a surrounding that can be reached within 15 minutes walking time; some calculations were made for a 1 hour environs.

Figure 4. Cost based hinterland areas within 60 mins. walking time in the area of the Marienberg 'Fürstensitz' (Northern Bavaria). b. Cost based hinterland areas within 15 mins. walking time in the area of the Marienberg 'Fürstensitz' (Northern Bavaria). – DEM D-25 (25 m grid), © German Federal Office for Cartography and Geodesy 2004.

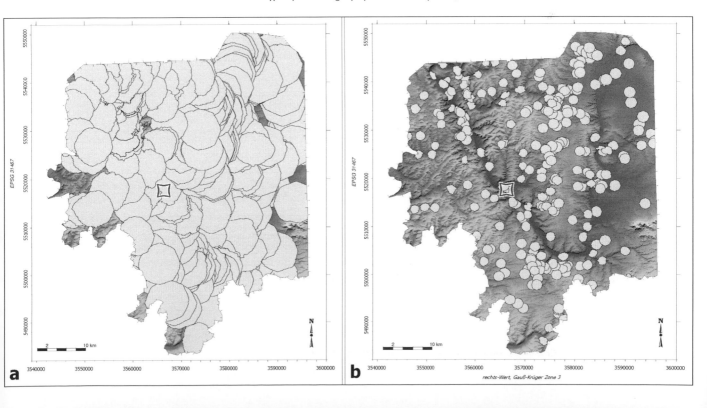

HINTERLAND SIZES AND THEIR MEANING

To test the thesis that the size and layout of the area around a settlement that has been in economical use during prehistoric periods has a meaning for the understanding of the societies under research: first calculations were made to model the area within 1 hour walking time for all regular settlements and all 'Princely Sites' based on the Gorenflo/Gale algorithm within the research area around the 'Fürstensitz' on the Marienberg in Würzburg (fig. 4a).

The overlapping of most of the areas makes clear that the exploited areas might not only belong to one but to several settlements even when we take into account that many of the settlements from one of the periods – which lasted approximately 300 years – were not coexistent. Social interaction as well as some kind of 'political' agreements must have been the basis for contemporaneous settlements, using the same (economic and cultural) hinterland. The picture is significantly different for the areas used for everyday farming activities within a distance of 15 minutes walking (fig. 4b), overlapping is reduced to a minimum and is to be expected for those sites only that are not contemporary existent.

Comparing the sizes of the hinterland areas within 60 minutes walking distance from the 'Princely Sites' with those from the regular settlements (fig. 5) can be used as a means to analyse economic strategies of places with a different social meaning. Figure 5 shows the median values of the 'hinterland' sizes of the regular settlements in each area of research, compared to the size of the hinterland of its 'central place'. This large surrounding area instead of the 15min area, which was used for everyday agricultural activities, was calculated to take into account (or to question) the potential role of the hill forts as surplus producing sites whose wealth could be based on producing and trading agricultural surplus, but which may also have been much larger than the 'normal' sites and therefore might have had to feed a larger amount of inhabitants.

The median values do not differ so much in the interregional perspective as do the sizes of the 'Central Places' surroundings. In general the hinterland areas of the regular settlements are more or less comparable, whilst the 'Fürstensitze' and other important places obviously did differ much more. Within the area of the Nördlinger Ries, with its sites of the 'Fürstensitz' Ipf and the two ditch enclosures of Osterholz, we can see the biggest spread between the regional mean value and the central places values. Only the for-

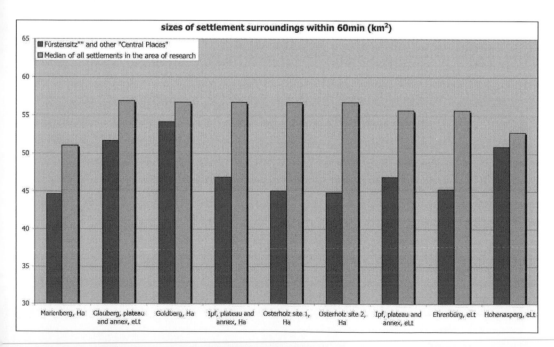

Figure 5. Median values of the hinterland size (km²) of the regular settlements within each research area (grey) compared to the size of the hinterland of the 'central place' itself (black).

tified hillfort Goldberg in the same area can be compared with the regular settlements in that region. The significant difference between the sizes of the surrounding areas of the regular sites compared to those of the 'Princely Sites' cannot be explained by the location of the latter on hilltops. This is because the difference of the site locations has been taken into account by calculating the cost surface surroundings not from the top of the hill with the 'princely sites', but from their outer fortification or the foot of the hill. Also the topographical situation of that specific type of settlement did not play a role because not all of the 'Fürstensitze' are situated on very prominent hilltops in a rather mountainous surrounding. The Heuneburg, for example, lies on a plateau not very much higher than the environs, which is – as are the environs of the other hill forts mentioned in the text – not very much influenced by steep slopes.

QUANTITY AND QUALITY

Not only the size of a settlements hinterland might have played an important role but also the natural resources in terms of soil quality and other factors directly related to agricultural activities.

The Ipf itself has the largest share of soil with low suitability for plant cultivation in its surrounding as well as the smallest share of high-quality soils (fig. 6). In contrast the availability of good or at least medium suitable soils is higher around the ditch enclosures of Osterholz, which is balancing their smaller surrounding areas. The Goldberg site with its large hinterland area had a relatively high percentage of good soils as well.

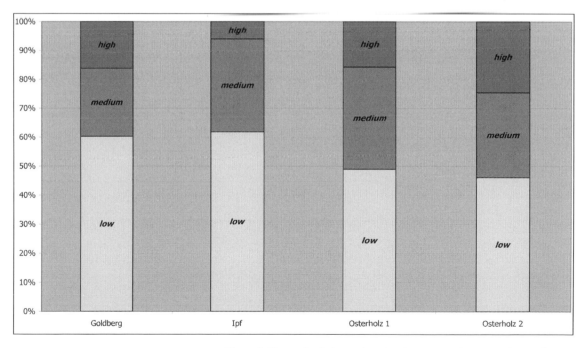

Figure 6. Share of soil classes in a 6omin surrounding around the Ipf 'Fürstensitz' and neighbouring settlement sites.

We know that the people in Celtic times made their living mainly by crop and cattle farming. So large hinterland areas, where the mean values are more or less the same as the value of the hinterland size of the 'special settlement' itself, are the indication for a mainly agricultural based way of living of the people of the 'Central Place'. We can assume that for the Goldberg, while the 'Fürstensitz' on the Ipf itself as well as the ditch enclosures of Osterholz on his foothills seem to have played a different role in the settlement system.

The Ipf is more or less a landmark in both a cultural/ritual way and in an economical way as part of a traffic and trading system, whereas we have some still very weak evidence that at least one of the Osterholz ditch enclosures might have been a place with a ritual meaning.

MODELLING THE AGRICULTURAL POTENTIAL OF HINTERLAND AREAS

The analyses of the hinterland area models have shown that the size of a hinterland has the potential for an assessment of the settlement it belongs to. In order to further investigate the hinterland areas of the Late Bronze and Early Iron Age sites the agricultural potential of the settlements should be modelled. One of the main research questions was to test, whether a 'Princely Site' can be understood as consumer site, being dependent not only on its own hinterland but also on the backing of regular settlements in its vicinity, or as a producer site with a large productivity which gained a surplus that might have been the basis for the wealth of its inhabitants.

Cattle and crops were the main basis of nourishment for prehistoric societies in Middle Europe. Calculating the agricultural potential of a settlement could therefore also be used to estimate the maximum number of inhabitants that could be fed (for a more detailed discussion on the following aspects cf. Fischer et al. 2010; Ebersbach 2004).

Factors of Land Use Classification

Soil is one of the main factors that have an influence on the potential of any hinterland to produce crops or other food. Based on the data from the Geological Survey of Baden-Württemberg we have recalculated six classes of soil, taking also into account climatic factors that had an influence on soil fertility. The data we were using are of course actual data – which might represent lower values than have been available during the Iron Age –, but comprehensive information on the Iron Ages soil and climate is not available. The main outcome of this classification is the allocation of soil values of more than 40 based on the so-called 'Reichsbodenschätzung' (Posluschny 2002, 76) being suitable for crop farming (fig. 7; Fischer et al. 2010). The 'Reichsbodenschätzung' was a project, undertaken in Germany in the 1930s and revised several times afterwards, which aimed to provide a specific value for each field in Germany (based on the type of soil, soil condition, climate...) which should represent the quality of that field for agricultural needs. This value was also used for tax classification based on the expected yield of the field. High values (max 100) represent good soil conditions, low values represent less valuable soils.

Part of the classification that was used for our preliminary modelling are specific soil types and their general fertility as well as their workability. The latter generally had a lower impact on soil fertility in modern times (like the 1930s when the 'Reichsbodenschätzung' was started) than it the Iron Age with its more limited techniques for ploughing so we have chosen rather conservative measures as thresholds.

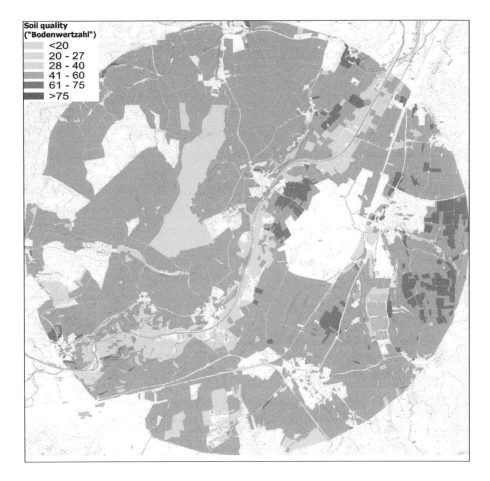

Figure 7. Soil classification in the area around the 'Fürstensitz' Heuneburg (Baden-Württemberg). – Fischer et al. 2010. *See also the full colour section in this book*

Soil quality ("Bodenwertzahl")
- <20
- 20 - 27
- 28 - 40
- 41 - 60
- 61 - 75
- >75

Topography is another parameter, which is part of the model calculation. We have allocated the slope values into different classes and agreed on a rough estimation that land with a slope of more than 10 degrees cannot be ploughed or at least ploughing causes erosion effects that had a significantly negative influence on the land (fig. 8; Fischer et al. 2010).

The combination of soil values and slope classes leads to a classification of the agricultural potential. For crop farming needs we calculated up to five classes of applicability. Areas with a soil fertility value of less than 40 or with a slope of more than 10° have been classified as meadows and woods, suitable for stock farming and a last class represents areas where the information is too sparse to be taken into account (fig. 9).

Food Supply by Crops

The main basis for food supply during Prehistory was mainly crops in different variations. A couple of very intense investigations on the plant remains from several settlement sites from the Bronze and the Iron Age have been undertaken in Baden-Württemberg (fig. 10; Fischer et al. 2010). These investigations give a good overview of the species that where cultivated, their share in the daily food supply and their popularity during different periods and in different areas. On the basis of this knowledge it became clear

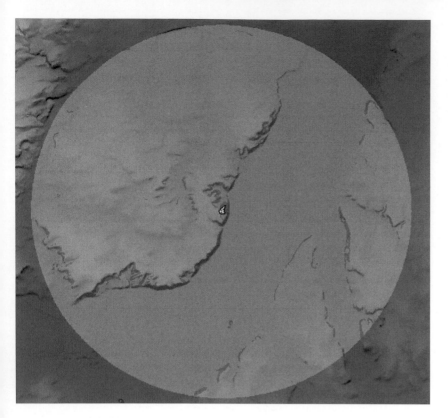

Figure 8. Slope classification in the area around the 'Fürstensitz' Heuneburg (Baden-Württemberg). The dark brown slopes indicate areas with more than 10 degrees of slope which are not suitable for ploughing. – DEM D-25 (25 m grid), © German Federal Office for Cartography and Geodesy 2004. *See also the full colour section in this book*

Figure 9. Combined classification of soil and slope values in the area around the 'Fürstensitz' Heuneburg (Baden-Württemberg). – Fischer et al. 2010. *See also the full colour section in this book*

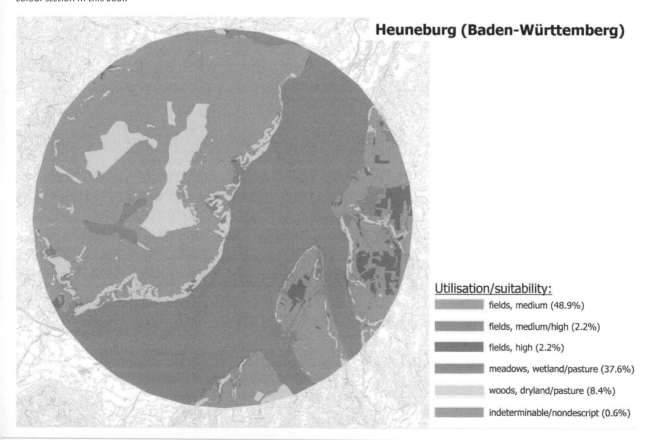

Heuneburg (Baden-Württemberg)

Utilisation/suitability:
- fields, medium (48.9%)
- fields, medium/high (2.2%)
- fields, high (2.2%)
- meadows, wetland/pasture (37.6%)
- woods, dryland/pasture (8.4%)
- indeterminable/nondescript (0.6%)

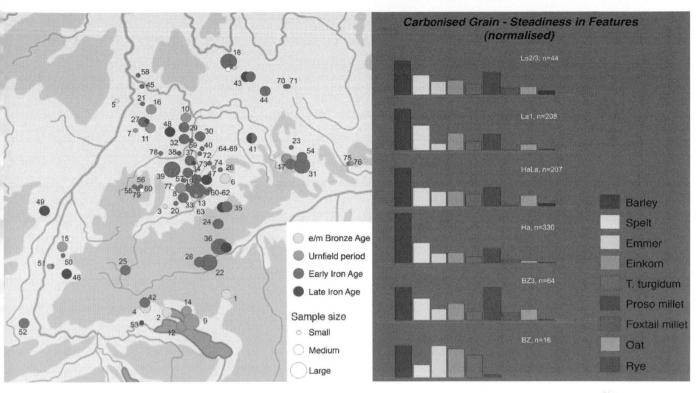

Figure 10. Bronze Age and Iron Age settlement sites with archaeobotanical investigations in Baden-Württemberg. The steadiness of types of carbonised grain is represented in the diagrams for different periods: BZ = Bronze Age, BZ3 = Late Bronze Age/Urnfield Culture, HA = Early Iron Age/Hallstatt Period, HaLa – Early Iron Age/Hallstatt Latène Period, La1 – Early Iron Age/Early Latène Period, La2/3 = Late Iron Age/Middle & Late Latène Period. – Fischer et al. 2010. *See also the full colour section in this book*

that on a very rough scale the nutritional value of the main crop species were quite comparable taking into account the share of the species in the finds from archaeological sites.

The potential of specific soil types to produce crops and other nutritional plants is dependant on various factors; many of them are subject to change due to soil degradation. Nutrients are removed by the growing of plants and have to be replaced depending on the amount of harvested crops per year. This can be done by having an annual change between cultivation and fallow for each field and by putting dung on the fallows. In our model we act on the assumption that cattle is grazing on the fallows and that dung from the stables – if existent at all – is placed manually on the fields as well. The actual model is working on the hypothesis of a more or less sustainable cultivation with a balanced nitrogen budget. Future refinements of the model will try to take into account a small amount of soil degradation as well.

Adding all Factors

The basis for the calculation of the potential of a surrounding of a settlement is the measurement of the area of the potential fields and their classes of soil quality. The cultivation/fallow ratio is a parameter that leads to minimum and maximum values, depending on the chosen ratio. We are well aware that not only grain species have been used as plants for feeding. People used a great variety of leguminous plants or oil seeds as well. Especially in the Hallstatt period, the first phase of the Early Iron Age, *Leguminosae* had a high degree of steadiness within the various settlements. On a general level it is possible to estimate their value in the same way as the grains, taking into account their nourishing values and their share in the plant remains from the archaeological sites. The calorie needs that we are using for our model at the moment can be discussed; in this first attempt it is quite low (1650 kcal/person/day) but of course the input from crop can be amended by other plants like vegetables, mushrooms, nuts and so on and of course by meat as well (fig. 11).

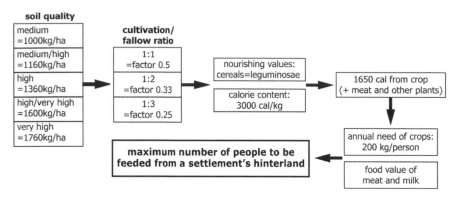

Figure 11. Flow-chart of the model to calculate agricultural potential.

AGRICULTURAL POTENTIAL AND ITS MEANING

Our first modelling attempt is still based on circular surroundings around the investigated sites. The results of cost-based areas will be the basis of our further work. Even in this preliminary stage we can already make some statements on the potential of several sites, which are the starting points for the further interpretation of their meaning in the social and economic network of the Early Iron Age.

From our point of view the gain of crops in the Iron Age is much lower than the gain of Late Neolithic societies and of course also much lower than in modern times, but it is also higher than during mediaeval times. Based on the coarse assumption that the yield of crops on medium suitable soils is 1,000 kilo per hectare per year (based on the results of various field experiments; cf. Fischer et al. 2010; Landesamt 2005), we can conclude that the work that is needed for its production and harvesting equals 110 days of labour. A share of 200 kilo, which is the annual amount of crops food per person, can therefore be produced by that person within 23 days. We can come to the general conclusion that the expenditure of human labour was not the restricting factor for the production of cereals, especially because it is very likely to assume that work with more or less fixed dates like sowing or harvesting might have been supported by other people of the community that have been released from their usual work for those purposes.

Given a ratio for cultivation and fallow of 1:1 first of all we see great differences when comparing the four sites of the Heuneburg, of Walheim, of Hochdorf and of the Ipf (fig. 12). The Heuneburg and the Ipf both are classical 'Fürstensitze' sites whilst Walheim and Hochdorf represent more or less 'normal' settlement sites – even if Hochdorf is a place of a potential higher social or political ranking. The surrounding areas of both 'Princely Sites' could nourish only half the number of people that could be fed from the Walheim and the Hochdorf hinterland. This picture matches with the results of the calculations shown

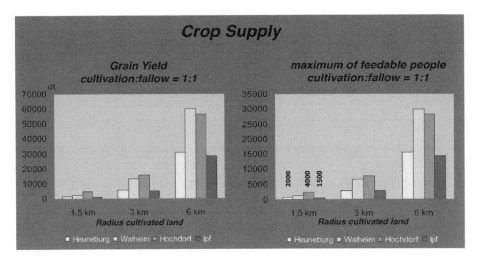

above, indicating that the hinterland of the Ipf was definitely not able to feed as many people as the hinterland of the nearby Goldberg (figs. 5, 6). Speaking in absolute numbers this means that around 1,500 people could live from the Ipfs hinterland, 2,000 from the Heuneburg and 4,000 from the Hochdorf surroundings. It is hard to evaluate the size of the Heuneburg region with its large suburban settlement, but at least for the site of Hochdorf a number of 4,000 inhabitants seems to be relatively high. This means longer periods of fallow, resulting in cultivation: fallow ratios of 1:2 or of 1:3, would have been possible, and allowing the soil to recover from degradation. Alternatively a surplus based on cattle and crops could have been produced as a basis for trade, or to be stored for various purposes like feastings, stockpiling and so on.

The Ipf might have housed 1,500 people but taking into account the sparse knowledge about the settlement structure (von der Osten-Woldenburg 2004, 54; figs. 7-13) it seems to be much more likely that the number of people living there was much lower, possibly less than 800 persons. A surplus production might only have been possible if we calculate a 1:1 cultivation: fallow ratio which then would have led to a faster degradation of soils.

Again a 6 km/1h hinterland was used as a larger area which could produce a surplus and which is also a first estimation that has to be compared with the results of a 15 min hinterland as the main area of everyday agricultural activities in future. The regular settlements seem to be situated in better suited areas for agriculture and it is assumed that their greater agricultural potential was used to produce a surplus which could either be traded, distributed, stored as a reserve, used for social activities connected to ritual feeding and feasting or to support settlements with a not so clear focus on agricultural activities like some of the 'princely sites' – or as a combination of some of these aspects. It is not very likely that they were used from regular settlements to produce larger quantities to feed larger numbers of people because excavations in these settlements have shown that they are usually rather small hamlets.

All these calculations do not yet account for the results of the pollen analyses, which might change the picture to a certain extent by better estimating the amount of land use for cattle, crops and fallows. This is one of our future tasks when refining the model.

ADDING CATTLE TO THE MODEL

Besides the supply of crops animal products were of course used for nutrition as well. Due to the ongoing discussion within our team we can only show first aspects of the potential for producing meat, milk and other animal products.

The general idea of the calculation is pretty much the same as for crops. Based on the size of the potential pastures and the carrying capacity the number of livestock units of 500 kg live weight is calculated. Taking into account that proportions of horse, cattle, sheep/goat and pig were much smaller during the Iron Age than today, the potential numbers of animals were recalculated using the ratios of these species in the investigated settlements. Finally, the ratio of slaughtering, the dressing percentage and the nourishing value of meat and milk are estimated for each species and the food value of milk and meat is determined (fig. 13).

A first result of the model is the conclusion, that animal products could cover only a very low fraction of the daily caloric needs of the people in the settlements. Recent research has shown that game and fish did not play a significant role in the nourishment of Early Iron Age people (Kerth & Wachter 1993). The main use for animal husbandry was the need for proteins in the daily diet as well as for secondary products like wool, dung, leather, bones and so on. In the case of cattle, we also have to take into account that they were used as draught animals.

The amount of animals that are grazing in the vicinity of a settlement and the amount of dung they produced had of course a great influence on the soil, especially on the nitrogen values. Periods of cultivation and of fallow therefore are dependant on these values. Also in very woody areas like around the Heuneburg livestock could balance lower disposability of crops to a certain extent. Refining these influences in our model will be another future task.

Figure 13. Calculation of agricultural potential, based on livestock. – Fischer et al. 2010.

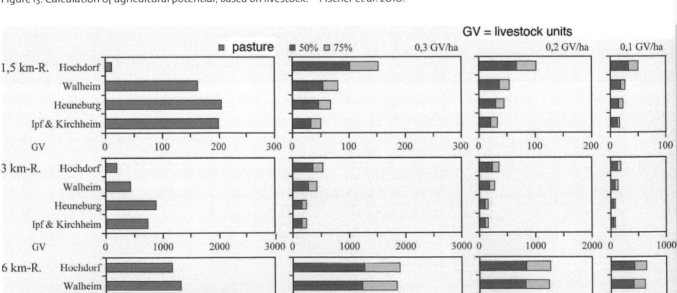

FUTURE WORK

Some aspects still have to be discussed and refined in the model, like the problem of soil exhaustion or the incorporation of the results from the palynological analyses. The model itself is prepared to be rather flexible, so parameters can be adjusted quite easily. Especially the results of various pollen profiles that have been investigated (Fischer et al. 2010) can be used to further refine our knowledge about offsite landscape reconstructions.

It is our aim to build a database software that is able to calculate the agricultural potential of any given hinterland in any given period and area with the possibility to change the parameters of the calculation like the caloric needs, the nutrient removal, the periods of fallows and cultivation and so on. In a first stage we will therefore convert existing Excel spreadsheets into a FileMaker database, which we plan to distribute as a free runtime version. As much of the calculation will be done within a GIS – like the definition of cost based surroundings and the merging of soil and topographic information – we would also like to develop a calculation model that runs in the free GIS software gvSIG.

Within the 'Fürstensitze' Priority Programme we plan to calculate the agricultural potential not only for the 'Princely Sites' but also for all others settlements that are under research in the project 'Princely Sites' & Environs – depending on the availability of sufficient soil data. This would result in the modelling of the agricultural potential of the hinterland of up to 5,000 settlement sites that can be analysed and compared.

CONCLUSIONS

The use of GIS in the research project – though the results are still somewhat preliminary – did not only show its potential for landscape archaeology in general but also to bridge the gap between environment, culture and social behaviour. The way prehistoric people acted in their surrounding environment was of course driven by economic needs to a certain extent. But other factors had an influence on their behaviour as well, thus creating recognisable patterns of different kinds of activities in the landscape.

The 'Fürstensitze' themselves are one of the results of these activities, being a manifestation of both economic and social needs. Even though most of the 'Fürstensitze' of the Early Iron Age seemed to be quite similar in their appearance, a greater variety now becomes obvious, which shows the adoption of man to its environment as well as his use of landscape as a stage for social and cultural interaction.

REFERENCES

Bintliff, J. 1999. Settlement and Territory. In Barker, G. (ed.), *Companion Encyclopedia of Archaeology* Vol. 1, 505-45. Routledge, London/New York.

Bintliff, J. 2002. Going to Market in Antiquity. In Olshausen, E. & Sonnabend, H. (ed.), *Zu Wasser und zu Land. Verkehrswege in der antiken Welt. Stuttgarter Kolloquien zur Historischen Geographie des Altertums 7, 1999* (= Geographica Historica 17), 209-250. Franz Steiner Verlag, Stuttgart.

Chisholm, M. 1962. *Rural Settlement and Land Use*. AldineTransaction, New Brunswick & London.

Christaller, W. 1933. *Die zentralen Orte in Süddeutschland. Eine ökonomisch-geographische Untersuchung über die Gesetz-mässigkeit der Verbreitung und Entwicklung der Siedlungen mit städtischen Funktionen*. Gustav Fischer Verlag, Jena.

Conolly, J. & M. Lake. 2006. *Geographical Information Systems in Archaeology. Cambridge Manuals in Archaeology*. Cambridge University Press, Cambridge.

Eastman, J.R. 2003. *IDRISI Kilimanjaro. Guide to GIS and Image Processing*. Clark University, Worcester, MA.

Ebersbach, R. 2004. Agriculture, Stock Farming and Environment: Adaption and Change during the Neolithic Lakeshore Period (4300–2400 BC cal) in Switzerland. *Anteus* 27, 287-292.

Fischer, E., M. Rösch, M. Sillmann, O. Ehrmann, H. Liese-Kleiber, R. Voigt, A. Stobbe, A.J. Kalis, E. Stephan, K. Schatz & A. Posluschny. 2010. Landnutzung im Umkreis der Zentralorte Asperg, Heuneburg und Ipf. Archäobotanische und archäozoologische Untersuchungen und Modellberechnungen zum Ertragspotential von Ackerbau und Viehhaltung. In Krausse, D. (ed.), *‚Fürstensitze' und Zentralorte der frühen Kelten. Abschlusskolloquium des DFG-Schwerpunktprogramms 1171 in Stuttgart, 12–15 Oktober 2009. Teil II*, 195-265. Konrad Theiss Verlag, Stuttgart.

Gorenflo, L.J. & N. Gale. 1990. Mapping Regional Settlement in Information Space. *Journal of Anthropological Archaeology* 9, 240-274.

Herzog, I. & A.G. Posluschny (in press). Tilt – Slope-dependent Least Cost Path Calculations Revisited. In *On the Road to Reconstructing the Past, in Proceedings 36th Conference on Computer Applications and Quantitative Methods in Archaeology. Budapest, April 2–6 2008*. Budapest, Archaeolingua.

Kerth, K. & N. Wachter. 1993. *Die Tierknochenfunde aus drei nordbayerischen Siedlungen der Hallstatt- und Frühlatènezeit*. Bayerische Vorgeschblätter 58, 61-77.

Kimmig, W. 1969. Zum Problem späthallstättischer Adelssitze. In Otto, K.-H. & J. Herrmann (ed.), *Siedlung, Burg und Stadt. Studien zu ihren Anfängen* (=Festschrift P. Grimm), 95-113. Berlin.

Landesamt für Denkmalpflege (ed.) 2005. Zu den Wurzeln europäischer Kulturlandschaft – experimentelle Forschungen. *Materialhefte zur Archäologie* 73Theiss, Stuttgart.

Posluschny, A. 2002. *Die hallstattzeitliche Besiedlung im Maindreieck. GIS-gestützte Fundstellenanalysen*, BAR International Series 1077. Archaeopress, Oxford (see also with different page numbering: http://archiv.ub.uni-marburg.de/diss/z2002/0092/).

Posluschny, A.G. 2007. From Landscape Archaeology to Social Archaeology. Finding patterns to explain the development of Early Celtic 'Princely Sites' in Middle Europe. In Clark, J.T. & E.M. Hagemeister (ed.), *Digital Discovery. Exploring New Frontiers in Human Heritage. CAA 2006 Computer Applications and Quantitative Methods in Archaeology. Proceedings of the 34th Conference, Fargo, United States, April 2006*, 131-141. Archaeolingua, Budapest.

Posluschny, A.G. 2010. Over the hills and far away? – Cost Surface Based Models of Prehistoric Settlement Hinterlands. In Frischer, B., Webb Crawford, J. & D. Koller (eds.), *Making History Interactive. Computer Applications and Quantitative Methods in Archaeology (CAA). Proceedings of the 37th International Conference, Williamsburg/VA, United States of America, March 22–26, 2009*. BAR International Series 2079, 313-319. Archaeopress, Oxford.

Rodrigue, J.-P., C. Comtois & B. Slack. 2006. *The Geography of Transport Systems*. Routledge, London/New York.

Saile, T. 1998. Untersuchungen zur ur- und frühgeschichtlichen Besiedlung der nördlichen Wetterau. *Materialien zur Vor- und Frühgeschichte in Hessen* 21. Wiesbaden.

Schier, W. 1998. Fürsten, Herren, Händler? Bemerkungen zu Wirtschaft und Gesellschaft der westlichen Hallstattkultur. In Küster, H., Lang, A. & P. Schauer (ed.), *Archäologische Forschungen in urgeschichtlichen Siedlungslandschaften. Festschrift für Georg Kossack. Regensburger Beiträge zur Prähistorischen Archäologie* 5, 493-514. Bonn.

van Leusen, P.M. 2002. *Pattern to Process. Methodological Investigations into the Formation and Interpretation of Spatial Patterns in Archaeological Landscapes*. PhD thesis Groningen University. http://irs.ub.rug.nl/ppn/239009177.

Vita-Finzi, C. & E.S. Higgs. 1970. *Prehistoric Economy in the Mount Carmel Area of Palestine: Site Catchment Analysis. Proceedings of the Prehistoric Society XXXVI*, 1-37.

von der Osten-Woldenburg, H. 2004. Geophysikalische Prospektionen im Umfeld des Ipf. In Krause, R., *Der Ipf. Frühkeltischer Fürstensitz und Zentrum keltischer Besiedlung am Nördlinger Ries*. Archäologische Informationen aus Baden-Württemberg 47, 50-55. Stuttgart.

5.9 Radiography of a townscape. Understanding, visualising and managing a Roman townsite

Authors

Sigrid van Roode[1], Frank Vermeulen[2, 3], Cristina Corsi[2, 4], Michael Klein[5] and Günther Weinlinger[5]

1. Past2Present, Woerden, The Netherlands
2. Universidade de Évora, Évora, Portuga
3. Universiteit Gent, Gent, Belgium
4. Università di Cassino, Cassino, Italy
5. 7 Reasons, Absdorf, Austria

Contact: s.vanroode@past2present.nl

INTRODUCTION

In spring 2009, a European project, short named 'Radio-Past', was launched within the FP7 Marie Curie framework 'Industry-Academia Partnerships and Pathways'. The project, fully titled 'Radiography of the past, integrated non-destructive approaches to understand and valorise complex archaeological sites', aims to join different resources and skills to improve, refine and validate intensive archaeological surveys on complex sites, with a special focus on abandoned ancient urban sites in the Mediterranean. A consortium of seven partners merges academic institutions – University of Évora (P), Ghent University (B), University of Ljubljana (Sl) and the British School at Rome (UK), with private companies: 7Reasons Media Agency (A), Past2Present (NL) and Eastern Atlas (D) – to fulfil the objectives of the programme. Its general European-scale aims can be summarised as follows: 'to open and foster dynamic pathways between public research organisations and commercial enterprises' and 'to stimulate inter-sector mobility and increase knowledge sharing through joint research partnerships in longer term co-operation programmes between organisations from Academia and Industry'.

The Radio-Past project seeks to integrate different methodologies in the widely developed field of non-destructive survey technologies as applied to archaeology, and also to pursue validation of the results through innovative methods of visualisation and the development of strategies for efficient management of the cultural heritage sites studied. It is a main target of this project to allow multiplication of methods and research approaches, and to generate methodological guidelines for archaeological diagnostics. The idea is to develop a standard set of survey approaches, based on a series of already widely used as well as more innovative methods, such as active low-altitude aerial photography, geophysical prospection, LiDAR survey and geomorphological observations, which can in the future be efficiently used in a comparable and integrated way on a wide range of complex sites in Europe. Practically, this work should result in a guide of good practice for many researchers in survey archaeology, which considers with care the suite of survey approaches that are most appropriate for the nature of each site in question.

Furthermore, the project also concurrently targets the development of effective scientific systems for the dissemination of survey results. In particular, the combination of high-resolution fieldwork with computer-based means of mapping and data visualisation, should allow virtual reconstructions of a buried town or large settlement within a relatively short space of time, as opposed to the more traditional excavation-centred approach that could take generations before a broader view of the site becomes available.

With these aims, the project seeks to link up with the EU policies of cultural heritage and landscape management. The core of field research done in the framework of Radio-Past complies fully with art. 3.Ib of the European Convention on the Protection of Archaeological Heritage, better known as the Treaty of La Valletta 1992, where it is stated that 'to preserve the archaeological heritage and guarantee the scientific significance of archaeological research work, each Party undertakes: ... to ensure ... that non-destructive methods of investigation are applied wherever possible'. Cultural heritage management authorities will benefit widely from this approach, as such integrated surveys of complex sites provide them with a very effective tool for gauging the degree of archaeological survival on such sites in their care and choosing appropriate conservation strategies.

Previous initiatives by some of the authors have brought the ancient townsite of *Ammaia* to the centre of the debate about the impact of urbanisation on the Romanisation process and on transformations of ancient landscapes in Southern Europe (Corsi & Vermeulen 2010). We have chosen the abandoned Roman site of *Ammaia* in central Portugal as the 'open laboratory for research and experimentation" within the Radio-Past project, but some research activities by the partner institutions are also carried out in other areas of the Mediterranean (Italy: *Portus* and the Potenza Valley, Greece: Boeotia. See below Web References). In *Ammaia*, the University of Évora (coordinator of the Radio-Past project) is also piloting other projects in partnership with the Portuguese National Research Fund (FCT) and several universities (Cassino, Ghent and Lisbon). The excellent conditions for research, including good site preservation, access and logistics, will help us to develop a research strategy with possible implications for future directions in this field.

A TOOLBOX FOR FIELD SURVEY AND REMOTE SENSING

During the last decade there has been an upsurge in the non-destructive survey of complex, often abandoned urban sites in the Mediterranean area (Vermeulen et al. 2012). More and more archaeologists have started to realise the potential offered by the techniques of wide-scale, intensive survey to map and understand their sites. Large and complex urban sites that had hitherto been studied in a piecemeal approach, which was largely predicated upon the monument-based interests of earlier scholars, are now being subjected to a range of survey techniques to rapidly generate plans of partial, or in some cases, complete townscapes. Although much relevant fieldwork has been generated in recent years, there is need for a more careful strategy in such approaches and for efficient choices of technique, during the process of data gathering in the field as well as in post-field processing and presentation. The Radio-Past team is developing an integrated methodology which firstly involves a wide range of field survey techniques. They are all essentially of a non-invasive nature and particularly adapted to study large and complex sites, such as abandoned towns or villages where most of the present-day use of the terrain is agricultural. These field-oriented techniques include: geomorphological and topographical survey (Vermeulen et al. 2005; Deprez, De Dapper & De Jaeger 2006), intensive and extensive surface artefact collection, vertical and low altitude aerial photography (Corsi & Vermeulen 2008) and a wide range of geophysical prospection techniques.

Topographical, geomatic and geomorphological surveys

The study of large and complex sites and the relationship with the natural components of their setting, is more and more influenced by a true geo-archaeological and geomatic approach, using techniques that combine both methods of the geosciences and of archaeological survey. The integration of a question-driven archaeological approach with the use of a wide array of GIS-based analyses and visualisation tools, the application of intensive geomatic research (GPS-mapping, 3D-scanning and photogrammetry) and geomorphological survey, create a framework which is ideally suited for characterising the location, extent, environmental embedding and erosion history of ancient settlements and towns.

A crucial base for such approaches is the production of a fine Digital Elevation Model (DEM) based on remote sensing, existing (digital) maps and ground geomatics. The latter involve acquiring fine resolution data about the micro topography of the site based on a survey with total station instruments or GPS. Through a set of field observations and activities, such as systematic coring and sampling over the site surface, the geomorphological study of intra-site erosion and palaeo-soil formation can be undertaken.

Innovation can be searched here not only through new integrations of approaches but also by testing the limits and possibilities of the recent developed topographic technique of LiDAR (Light Detection And Ranging) or ALA (Airborne Laser Altimetry). This is a relatively new technique in the toolbox of landscape archaeology. It is a different way of looking at the earth's surface and can produce a high resolution, highly accurate set of surface relief data. The ability to detect height differences of a few centimetres can reveal features on the surface – natural or artificial – which were previously invisible. Its capacity to see below the vegetation enables archaeologists to identify features hidden beneath woodland cover. Moreover, LiDAR can cover large areas of landscape with a resolution and accuracy previously unavailable. A particular challenge is to mount the LiDAR equipment on low altitude, unmanned platforms, which would allow a widespread use of this technique at the highest resolution and in low cost conditions.

Artefact surveys

Of particular importance for the study and evaluation of large and complex sites is the use of an intensive artefact survey approach, characterised by high-resolution fieldwalking, where archaeological material at the surface is collected within grid squares. Intensive artefact surveys are more costly and take more time than extensive surveys, but can provide more comprehensive information on the nature of human activities on the site. In this way activity areas within the site itself can be determined and located. The analysis includes the mapping of the densities of different types of archaeological material. Although material is usually highly fragmented due to ploughing, careful examination of diagnostic pieces can provide chronological information. Statistical analysis of different types of material and different size classes may reveal patterns of past human activities, but also post-depositional disturbance or modern interference with the visibility of the archaeological material on the surface. To understand the pattern of visibility of complex archaeological sites, the analytical field survey is often combined with geomorphological mapping and possibly geophysical prospection (see further). Detailed and intensive artefact surveys can answer questions on the chronology of landscape use, on functional zoning within ancient cityscapes, the distribution of certain classes of material and instruments, palaeo-demography etc. It also provides key information which can guide decisions on further archaeological research on the site, such as excavation.

Aerial photography

Developments in the fields of aerial photography and other types of non-ground-based remote sensing have a serious impact on the potential to study large and complex sites with a non-invasive approach. Detailed investigation of available aerial images combined with intensive monitoring of such sites with the help of low-altitude digital aerial photography can be very productive for the study of ancient urbanisation, site size, site limits, suburban activities, etc. Together with studies where predominant use is made of the more static evidence of existing vertical photography or nowadays also very high-resolution imagery obtained from satellites or airborne radar flights, an ever increasing array of active remote sensing techniques is available for archaeological research or can be further developed. Since the beginning of aerial photography, researchers have used all kinds of devices ranging from pigeons, kites, poles and balloons, to rockets in order to take cameras aloft and remotely gather the aerial data needed for a combination of research goals. To date, many of these unmanned devices are still used, mainly to gather archaeologically relevant information from relatively low altitudes, enabling so-called low-altitude aerial photography (LAAP).

Geophysical prospection

It has been shown on different classical sites in the Mediterranean and beyond that a systematic and integrated wide-scale application of geophysical survey techniques can contribute in a dramatic way to our understanding of ancient urban topography. Geophysical techniques are non-destructive, since all the necessary information is obtained above the ground, which allows research of buried remains without damaging them. Moreover, some of the applications are quite fast and can cover extended plots of land in just a few weeks of fieldwork, sometimes delivering very detailed plans of the archaeological presence in the soil. Common geophysical research designs include the following methods: magnetometer survey, earth resistance measurements, ground penetrating radar and electromagnetic induction survey. As some methods can be slower (e.g. georadar survey vs. magnetic survey), it is more appropriate to apply some instruments to target particular areas of interest or where there is a potential for deeper archaeological deposits. Archaeological features such as brick and stone walls or floors, hearths, kilns and disturbed building material will be represented in the results, as well as more ephemeral changes in soil, allowing the location of foundation trenches, pits and ditches. Results are, however, extremely dependent on the geology of the particular area, and whether the archaeological remains are derived from the same materials, which stresses the need of integration with geological and geomorphological approaches.

The visualisation approach

The reconstruction of ancient landscapes for 3D-visualisation depends on the integration of all available information as well as correct interpretation. Communication between the scientific teams involved in the fieldwork and visualisation technicians is crucial in order to discuss different possible interpretations of the data. This applies to the larger-scale vision of the surrounding landscape, as well as the finer resolution approach towards the reconstruction of a townscape with all its architectural detail.

Available topographic information and environmental studies represent the base for the landscape reconstruction. Digital Terrain Models (DTM) derived from existing topographic or geomorphological maps and/or from ALS (Airborne Laser Scan Data) and DGPS (Differential Global Positioning System) can be used to simulate flow-morphology and erosion, which is then subtracted from the present state DTM in order to assume the ancient landscape topography. Existing information on hydrography and water supply, botanical coverage, land-use, suburban settlement and road systems can then be added to this 'Raw Terrain Model' to arrive at a simulation of the former terrain. The challenging visualisation of such landscapes is done by using special programmes which are capable of displaying millions of parts in the scene with a photo-realistic output using fractal algorithms based on the input of the operator.

The process of the architectural modelling for ancient urban sites includes the generation of building modules in order to create differentiated, large-scale urban structures within an acceptable time-labour frame. Typical local architectural details and styles are integrated in these modules by the use of all existing information from local archaeological research (survey, excavation and material studies), as well as from reference to better preserved sites of the same period and region. These building blocks can be laid out on the digital topographic maps in order to visualise the desired urban structures. As a result, a building typology is generated which will serve as 'filling blocks' for the layout of the whole urban site, maintaining the possibility to make changes and update these models if needed. The reconstruction of prominent, singular buildings, which are better known via stratigraphic excavations or can be considered landmarks of the townscape (e.g. a *forum*, gate or temple), cannot be processed through modulation, although certain parts of their decoration, such as columns, will be reused elsewhere and can be altered if needed.

The visualisation of architecture and landscapes should also be enriched with animated figures to produce a believable image for the viewer. Human motion can be produced by animating virtual characters manually or by recording real actors through a kinematic system called 'Motioncapture'. Poorly animated or ill-conceived characters are immediately recognised by the human eye and can downgrade the quality of the virtual media, even with a surrounding cityscape of outstanding quality. Through instancing, thousands of figures can be implemented into a virtual environment, producing realistic and lively scenery around the reconstructed archaeology. These scenes can then be used to tell a story about the ancient town, allowing the viewer to step into this past world and feed his/her interest in this topic. The attention of the viewer can only be assured with an outstanding quality of form and content. Certain media can be used to catch the attention of the different types of audience. Linear storytelling through books, pictures and films ensures high-quality, interactive worlds, such as 'realtime 3d environments' which often attracts the younger generations.

A CASE STUDY: THE 'OPEN LAB' AMMAIA (PORTUGAL)

Ammaia is a Roman town whose foundation must surely predate the inscription mentioning the *Civitas Ammaiensis* during the reign of Claudius (44/45 AD; *IRPC*, 615: Mantas 2000, 392-393.). It was conferred the status of *municipium* by the time of Lucius Verus, as indicated by another inscription conserved in the

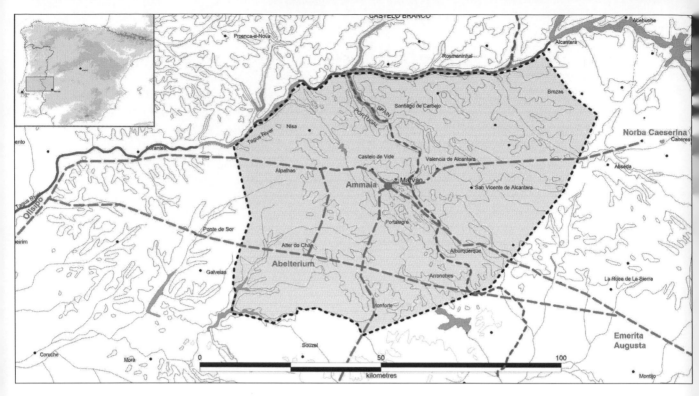

Figure 1. *Ammaia* and its territory at the centre of the road-network in Lusitania (authors).

nearby town of Portalegre (*CIL*, II, 158 = IRCP, 616). The ruins of *Ammaia* are located in the heart of the Natural Park of the Serra de São Mamede, a mountainous area of east-central Portugal extending into Spanish territory (fig. 1). The site is part of the fertile valley of the river Sever (Marvão). At this stage of our research, no settlement traces preceding its Roman foundation have been detected.

Archaeological research started at the site in 1995 with some excavations under the responsibility of the Fundaçao Cidade de *Ammaia*, a private foundation combining public and private institutions. This institution is now the owner of the site, and they are responsible for managing the archaeological park and its infrastructure. Archaeological excavations have been concentrated so far on areas where ruins are still visible above ground. This is clearly the case in the area of the main town gate, Porta Sul (fig. 2, n. 4; fig. 3), where the city street now interpreted as the so-called *cardo maximus* widened into a paved square after passing through an arched double chamber gate fortified by two circular towers. The second main zone of excavation is the *forum* area (fig. 2, n. 1), where the concrete nucleus of the main temple's podium is still visible and where excavation trenches brought to light segments of a *cryptoporticus* and remains of the main bath complex of the town (fig. 2, n. 2). Excavations have also been carried out in zones where some restoration work has been carried out, such as in the area around the building that houses the archaeological museum, the 17th-century farm called Quinta do Deão (fig. 2, n. 3), or where facilities for the archaeological park were planned, such as the visitors car park in front of the museum.

A new programme of excavation was started in 2008, concentrating on the first two campaigns on the bath complex (2008-2009), with the aim of adding to our knowledge of the monument, of the phases preceding the baths and of the transformation and abandonment of the sector. Part of the aim of the new excavation programme is also to do some ground truthing.

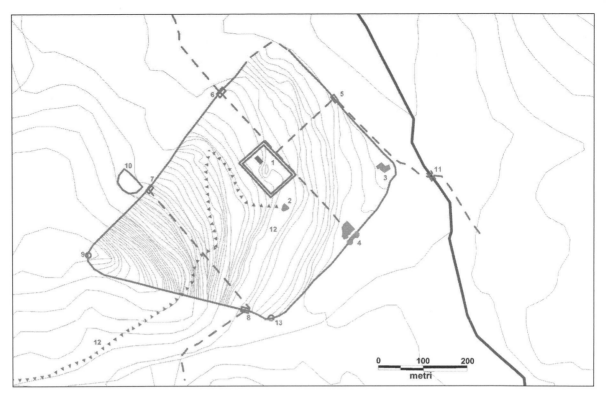

Top

Figure 2. *Ammaia* and the surroundings: archaeological mapping. In grey: archaeological monuments partially preserved and hypothetical monuments; full line: sure topographical elements of the Roman town, dashed line: hypothetical elements. 1. forum 2. baths 3. house complex 4.-8. gates 9. tower 10. theatre 11. bridge 12. Aqueduct 13: tower?

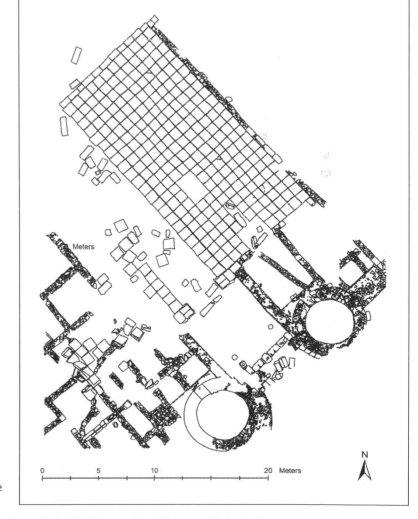

Figure 3. *Ammaia*. Synchronic plan of excavated area of the southern gate (Porta Sul). Fundação Cidade de Ammaia.

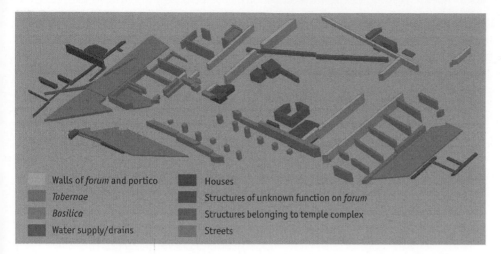

Figure 4. *Ammaia*. 3D reconstruction of the 'buried' structures of the forum detected with the GPR survey (elaboration by L. Verdonck). *See also the full colour section in this book*

Walls of *forum* and portico

Tabernae

Basilica

Water supply/drains

Houses

Structures of unknown function on *forum*

Structures belonging to temple complex

Streets

During the course of the same summer campaigns, some initial geophysical surveys were carried out as preparation for the larger scale, non-destructive survey operations within Radio-Past. In 2008, geophysical prospection was mainly undertaken with GPR by a team from Ghent University, and in 2009, already within the framework of the Radio-Past project, a field collaboration with the University of Southampton (APSS-team) focused on magnetometry. Together, these first campaigns of geophysical survey covered an area of almost 5 hectares.

The results of the 'time slicing' of the processed GPR data (fig. 4) provided the basis for a digital reconstruction of the forum. All visible elements of the survey results, such as the large basilica, the 20 symmetrically positioned shops, the axial temple and a series of monumental structures on the central square can be well reconstructed, by combining the survey data with relevant information from the site and examples from elsewhere. Here, during a small-scale excavation campaign in the summer of 2010, some ground truthing of these results was done, which supplied additional information for the structural and chronological definition of the main architectural phases. New intensive geophysics during the summer campaign of 2010, with magnetometry and electrical resistivity, produced additional high-resolution imagery of this forum area, which at the moment is being processed. This is also the case for the area near the Porta Sul, where in the near future these additional geophysics data could enhance the proposed reconstruction of the excavated gate and its immediate surroundings (fig. 6).

The still ongoing magnetometer survey, intended to fully cover the intramural areas of the city, has already produced a fine map of town structures in some of the central and northern areas of the former town (fig. 5). Clearly visible are the regular grid of city streets, delimiting housing blocks, public spaces (such as the bath complex and a market), workshops and water infrastructures. The results obtained so far give reason to believe that the full intramural town plan can indeed be revealed, limiting the necessity for grand-scale and costly excavation procedures, but at the same time allowing a 3D view of the townscape and opening perspectives on a sustainable touristic exploitation and cultural value of the site.

In parallel with the geophysical investigations and necessary ground truthing operations, which are focused on the reconstruction of the cityscape, many other field operations have been initialised by the Radio-Past team or are being prepared. We can mention here the still embryonic development of robotics to perform high resolution LiDAR coverage of the site, in collaboration with the Instituto Superior Técnico in Lisbon, and terrestrial scanning operations of some of the still standing ruins by an Irish team

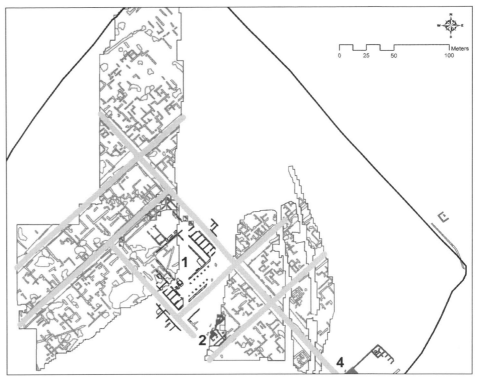

Figure 5. *Ammaia*. Interpretation of geophysics survey and excavated areas with reconstruction of the street network: 1. forum, 2. baths, 4. area of the southern gate. *See also the full colour section in this book*

Figure 6. *Ammaia*. Preliminary 3D reconstruction of the southern gate (Porta Sul) by M. Klein.

from The Discovery Programme. In addition, tests were done with low-altitude aerial photography, using a blimp and a helikite. The helikite is a hybrid between a balloon and a kite. By combining a helium balloon with kite wings, this lighter-than-air device combines the best properties of both platforms without incurring too much of their disadvantages. The helium-filled balloon allows the helikite to take off in windless weather conditions, whereas the kite components become important in case there is wind, as they counteract any unstable behaviour that is characteristic of traditional balloons and blimps flown in windy conditions. Moreover, the construction supports more payload for its size when compared with ordinary aerostats. A Microdrone, which has been recently purchased, could even provide more stability for such low-altitude aerial photography. Initially developed for purely military applications, drones or Unmanned Aerial Vehicles (UAVs) are powered aerial vehicles that do not take a human operator aloft but fly and manoeuvre in the air autonomously or by remote control. Using a drone with four rotors, it is normally possible to lift a digital reflex camera, while a GPS mounted on the instrument provides autonomous waypoint navigation and position hold to take an excellent aerial photograph. Underneath these airborne platforms a construction is mounted that can hold not only normal digital cameras, but also Near Infrared (NIR) or Near-Ultraviolet (NUV) enabled digital reflex cameras, allowing the operators to test a wider range than the visible field and therefore obtaining a maximum of relevant information concerning subsoil features which have an impact on the surface and its vegetation.

Important ongoing or planned work are also a set of geoarchaeological operations. In 2001, a joint team from Universities of Ghent and Cassino started this interdisciplinary fieldwork, mainly meant to understand the general landscape setting of the town, its water supply, the town/country relationship and the wider settlement dynamics. Preliminary results of this work in progress have been published elsewhere (Vermeulen et al. 2005; Taelman et al. 2009; Vermeulen & Taelman 2010). An essential aspect of the research is the reconstruction of the taphonomy of the urban site and its immediate surroundings, including the study of historical erosion and degradation processes and the elaboration of a fine DEM. To obtain this, a campaign of systematic and high density coring in the intra-mural part of the suburban areas is planned for 2011, as well as a continuation of ongoing tests with surface and near-surface artefact collection.

THE ARCHAEOLOGICAL RECONSTRUCTIONS

To apply and refine the visualisation of data sets collected with non-destructive approaches in *Ammaia*, a close collaboration between the scientific researchers who conducted the prospection in the field and the ICT specialists has been set up. A particular challenge for the 3D visualisation work in test-case *Ammaia*, which aims at a 'total' digital reconstruction of the Roman town and its immediate hinterland, is that most of the data sets available will be derived from intensive geophysics surveys. These total-coverage prospection data, essentially geomagnetic measurements, georadar imaging and electrical resistivity mapping, will however be fully integrated with the focused excavation effort. To proceed in relevant phases, the Radio-Past team will first produce detailed digital reconstructions of several specific monuments of the Roman town, e.g. the southern gate (Porta Sul), the *forum* and the baths, while the second phase will reconstruct the total town area.

For the first attempt to produce a 3D reconstruction we chose the well-preserved Porta Sul (fig. 5). Preparation for this work included foreseeing the use of modular structures taken from nearby reference sites, offering more information through standing structures, publications and reconstructions. These were then combined with the archaeological excavation plans and some additional geophysical prospection results. The process of reconstruction started with the gathering of information about the topography of the terrain, observing the standing remains and transforming all existing data into a suitable format. The main application we use to model and animate 3D data is Autodesk 3ds Max, in conjunction with various specialised software applications. Reconstructing the arch was much helped by an old photograph taken before its 19th-century destruction and relocation into a secondary location in the nearby town of Castelo de Vide, where it was brought and integrated into the post-medieval city walls. From the measurements of its proportions we adapted other heights such as those of the city wall and towers. The open space, *intra muros* from the gate, still preserves its layout with massive, well cut (approx. 1 x 1m) granite blocks forming a rectangular square of approximately 23 x12m on each side of the main road through the gate. The main use of this square is still unclear, but it is assumed that a market (*macellum*) was possibly connected to the north of the square. Thanks to clearly visible rounded marks in the granite flooring of the square, a portico could be suggested. Although there is no indication whether these holes were used for posts, pillars or columns, their placements shows that some shelter structure such as tents or roofing could be supposed. Some of the large granite blocks of the square, facing the street, showed marks of a possible basement. This basement could have been used to separate the street and the square place in order to keep dirt and water off its pavement. Finally, less problematic is the reconstruction of the city wall and its towers, as remains of the towers are still standing up to 2,5m and also the city wall is quite well preserved near the gate. As the Romans were very systematic in their elaboration of city walls and gate systems, this model will allow a full digital reconstruction of the town enclosure system.

THE MANAGEMENT PLAN

In order to ensure the sustainability of the project both on a scientific and socio-economic level, a management plan is being developed for the site. The site management plan will be a heritage policy with instruments to ensure the future of *Ammaia*. The heritage policy serves the following objectives:

- outline the roles and responsibilities related to the management of the site;
- create a clear set of guidelines by which to manage the site;
- ensure the physical protection of this heritage site;
- create a sustainable environment to vitalise responsible tourism to *Ammaia*.

Ideally, a heritage policy should search, find and solidify the link between the past and the present. The site itself forms an integral part of contemporary spatial planning as well as the current economic and social landscape. It will therefore not only need to incorporate scientific information, but information on legal aspects as well. As a result, the management plan will be value-based: the various management aspects will be approached from the valorisation of the site.

Several management aspects need to be addressed and researched for the heritage management policy. These will be the building blocks of the heritage policy; together they will form a policy that can be evaluated, adapted and implemented. The various management aspects have been defined in the position paper that has been drawn for this project (van Roode 2008), and include:

- analysis and evaluation of comparable management plans;
- archaeological risk analysis and risk map;
- evaluation system;
- area specific management plan;
- research plan;

- capacity and implementation plan;
- marketing plan;
- integration of website and heritage policy.

CONCLUSIONS

This paper was concerned most of all with presenting the aims and methodology of ongoing concerted research in order to adequately survey, map, interpret, visualise and manage large complex sites, such as abandoned ancient cities, in particular in Mediterranean Europe. Even if more substantial results are still awaited, we were also able to present some preliminary data on the ongoing operations in field lab *Ammaia*. Archaeological data collected there prove that most urban structures of *Ammaia* were developed during the 1st century AD, but that town life lasted until the early Middle Ages. The urban centre of *Ammaia* was delimited by a wall circuit, possibly enclosing some 22 ha, and the town had a regular layout, with main axes connecting the circuit gates (fig. 2). Detailed plans of some of the major monumental areas in the city (the southern gate, forum and bath complex) are now available, and thanks to an integration of data from excavations, multilayered geophysics, micro-topographical measurements, geomorphological fieldwork and some remote sensing, it is now possible to propose the first reconstructions of these crucial urban areas.

From the methodological point of view, the approach of the 'Radio-Past team' aims to be exemplary and wishes to draft guidelines for good practice in the field of landscape archaeology and especially urban survey. Progress in the discipline of landscape archaeology is sought by testing new survey applications and by the holistic integration of a wide set of approaches and multidisciplinary survey techniques. A crucial aspect is the philosophy of merging mostly non-destructive scientific research and heritage-management of buried archaeological sites and their surroundings, while at the same time limiting destructive intervention, such as excavation, to the absolute minimum. The partnership among specialists in different fields and the synergy of resources and competences allows the team to work on a 'total project' based on looking through the surface, as in radiography. But, as in medical diagnostics, this 'scanning' is considered to be only the first step in understanding research on historical landscapes. Scientific research has to face the demands from public and stakeholders and the dissemination of results via effective but still scientifically based multimedia has to be part of the research agenda. The landscape archaeology of the third millennium can in this way play a key role in the protection and valorisation of cultural landscapes, improving the way they are presented to the public and enhancing their sustainable and responsible development.

*The research leading to these results has received funding from the European Community's Seventh Framework Programme (FP7/2007-2013) under grant agreement n° 230679, under the action Marie Curie – People IAPP, with the Project entitled 'Radiography of the past. Integrated non-destructive approaches to understand and valorise complex archaeological sites'.

REFERENCES

Corsi, C. & Vermeulen, F. 2008. Elementi per la ricostruzione del paesaggio urbano e suburbano della città romana di *Ammaia* in *Lusitania*. *Archeologia Aerea* 3, 177-194.

Corsi, C. & Vermeulen, F. (eds.) 2010. *Changing landscapes. The impact of Roman town in the Western Mediterranean. Proceedings of the International Colloquium (Castelo de Vide–São Salvador de Aramenha (Marvão) 15th-17th May 2008)*. Antequem, Bologna.

Deprez, S., De Dapper, M. & De Jaeger, C. 2006. The water supply of the Roman town of *Ammaia* (Northeastern Alentejo, Portugal): a geoarchaeological case study. *Publicações da Associação Portuguesa de Geomorfólogos* 3, 109-133.

Encarnação, J., d' (1984), *Inscrições Romanas do Conventus Pacensis*, Coimbra, Universidade de Coimbra.

Mantas, V. 2000. A sociedade luso-romano do município de *Ammaia*. In Gorges, J.-G. & Nogales Basarrate, T. (eds.), Sociedad y cultura en Lusitania romana. *Mérida, Museo Nacional de Arte Romano*, 391-420.

Roode, S.M. van 2008. Position Paper. Introductory paper to the Heritage Policy and Management Plan of *Ammaia*. *Past2Present Report* 506, Woerden.

Taelman, D., Deprez, S., Vermeulen, F. & De Dapper, M. 2009 Granite and rock crystal quarrying in the Civitas Ammaiensis (north-eastern Alentejo, Portugal): a geoarchaeological case study. *BABesch* 84, 171-182.

Vermeulen, F., De Dapper, M. & Deprez, S. 2005. Geoarchaeological observations on the Roman town of *Ammaia*. *Internet Archaeology* 19 (http://intarch.ac.uk/journal/issue19/corsi_index.html).

Vermeulen, F. 2009. Roman water. A Geoarchaeological Approach to Studying the Water Supply of Moderate Roman Cities. *Zeitschrift für Geomorphologie N.F.* 53, 111-130.

Vermeulen, F. & Taelman, D. 2010. From cityscape to landscape in Roman Lusitania: the municipium of *Ammaia* In Corsi, C. & Vermeulen, F. (eds.): *Changing landscapes. The impact of Roman town in the Western Mediterranean. Proceedings of the International Colloquium (Castelo de Vide – São Salvador de Aramenha (Marvão) 15th-17th May 2008)*. Antequem, Bologna

Vermeulen, F., Burgers, G.-J., Keay, S., Corsi, C. (2012): *Urban Survey in Italy and the Mediterranean*. Oxbow, Oxford.

WEB REFERENCES

http://www.radiopast.eu/
http://www.portusproject.org/
http://www.flwi.ugent.be/potenza/
http://www.nia.gr/Pharos13.htm

5.10 New methods to analyse LIDAR-based elevation models for historical landscape studies with five time slices

Authors
Reinoud van der Zee and Frieda Zuidhoff

ADC ArcheoProjecten, Amersfoort, The Netherlands
Contact: r.van.der.zee@archeologie.nl

ABSTRACT

Light Detection and Ranging (LIDAR) data and derived Digital Elevation Models (DEM) have been available in the Netherlands since 2001. These models have recently become an accepted method within archaeological and historical research. Nevertheless, the use of LiDAR images for landscape characterisation by governmental organisations and institutions that deal with area management is still in its infancy.

The provincial authority of Gelderland commissioned ADC ArcheoProjecten to compile an atlas of LiDAR-based elevation models within its boundaries. Its purpose was to increase the application of the atlas within governmental organisations. To simplify the interpretation of the LiDAR-based elevation models in the atlas, a new method of time-depth was introduced by defining time slices. For each time slice, images from characteristic features were made with different visualisation techniques, depending on the character of the chosen subject. The LiDAR-based elevation images were printed next to historical and topographical maps or photographs to clarify the interpretation. To explain each image, information about the geomorphological, historical or archaeological features was added. The method was tested in four regions, each with different geomorphological and historical characteristics. The results were discussed with the potential users of the LiDAR-based atlas from different governmental organisations in the region. This new method, using the five time slices, proved to be a useful tool for analysing LiDAR-based elevation images. These images provide valuable input to (historic) landscape characterisation, which has the potential to be a substantial resource for heritage-related, archaeological and historical research as well as landscape and urban design.

KEYWORDS

archaeology, geomorphology, historical geomorphology, digital elevation models, LiDAR, landscape characterisation, time depth

INTRODUCTION

LiDAR data and derived digital elevation models (DEM) have been available in the Netherlands since 2001. The detailed measurements of the ground surface at metre and sub-metre resolution generate spectacular images of the relief, revealing the natural landscape, archaeological and historical geographical features as well as features relating to modern human activities (Laan & Van Zijverden 2004; Waldus & Van der Velde 2006; Van Zijverden & Zuidhoff 2009). LiDAR measures with a resolution and accuracy hitherto unavailable, except through labour-intensive field survey or photogrammetry (Bewley et al. 2005). The LiDAR-based images contribute to studies based on geological and historic mapping and archaeological data, and are for example used for landscape biography studies (Van Beek 2009; Bewley et al. 2005; Lewis et al. 2008; Shell & Roughley 2004). Landscape biography or landscape characterisation takes systematic representations of the above-mentioned features and uses processes of assessment and interpretation to characterise the cultural and historical events that established the present-day landscape. It provides a comprehensive overview of the historic landscape in order to provide new, wide-ranging information for conservation, management and development decisions (Lewis et al. 2008). The landscape biography approach is less selective than conventional methods of landscape characterisation, which is the benefit of

Figure 1. A simple DEM of the province of Gelderland. Note that the relief is exaggerated by a factor 5.4 and illuminated from the northwest.

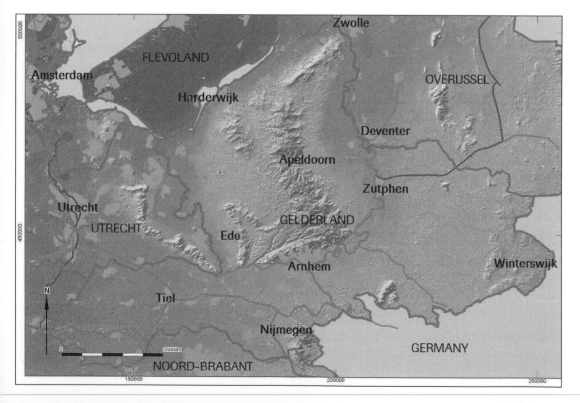

this method (Fairclough 2006a). The process does not identify one area or thing as having greater value than another. Instead it indentifies what distinguishes every place (or a part of the historic environment) from others, what makes it distinctive and what gives it its character.

Despite the fact that the application of LiDAR-based DEMs within the scientific world is widely known, the use of LiDAR images for landscape characterisation by governmental organisations and institutions which deal with area management is still in its infancy. In order to improve the quality of spatial policy and to lay a solid foundation for working with surface relief within heritage management, landscape design and spatial planning, the provincial authority of Gelderland encouraged the application of LiDAR images as mentioned above. To this end, the authority commissioned ADC ArcheoProjecten to compile an atlas of LiDAR-based elevation images within the provincial boundaries (fig. 1). In order to simplify the interpretation of the images, a new method was developed with the introduction of time slicing. This method was applied in four regions. A key, containing the most characteristic natural and man-made features, was also compiled. The project's results were incorporated in a report, as well as being presented at various workshops (Van der Zee et al. 2009).

TECHNICAL BACKGROUND

LiDAR systems transmit laser pulses from an aircraft to the ground and sense the 'echo' pulses when they return. The time that it takes for a pulse to return to the sensor is a measure of the distance between the laserhead and the ground. Early airborne LiDAR systems had problems geo-referencing the laser measurements, but this has improved considerably in recent years (De Boer et al. 2008). It is now even possible to measure surface elevations with an accuracy of ca. 15cm. Between 1996 and 2004, Dutch governmental institutions measured the entire country's surface elevation with LiDAR, at a minimum density of one measurement per 16m^2. Except in areas with large water surfaces and forest cover, the average density was one measurement per 36m^2.

From 2007 onwards, an improved version of the LiDAR system has been developed, which will be available in 2012. For this purpose, every 0.25m^2 of the Dutch surface elevation will be measured with an accuracy of 5cm.

LiDAR-based atlas
In order to construct the LiDAR-based atlas, the provincial authority provided ADC ArcheoProjecten with files containing the raw data. These data consist of elevation points, which are subdivided in sheets. One sheet corresponds to one half map sheet of the arrangement of the Dutch Topographical Service and covers an area of 5,000 x 6,250m (3,125 ha). At the borders the sheets have different dimensions. The entire province contains 212 sheets. In order to diminish the number of sheets and to clarify the atlas, four sheets were consolidated into one new one.

For each sheet, a frame for the raw data was made by using the technique of kriging. Kriging is a geostatistical technique used to interpolate a parameter using geostatistical characteristics of the complete data. In this study the parameter interpolated was the elevation of the landscape as a function of the geographic location. The elevation at an unobserved location was obtained from observations of its value at a nearby location. The interpolation was done with predefined adjustments and a cell dimension of 2.5 x

2.5 m. Subsequently, black-and-white maps and colour-shade relief maps were generated. The same colour pallet was used in all of the maps so that differences between adjacent maps were avoided.

Time Slices and Legend for the LiDAR-based atlas

In order to support the application of the LiDAR atlas for governmental organisations and the interpretation of the LiDAR-based elevation images, the data was simplified in two ways. To begin with, a time-depth method was introduced by defining time slices. A legend for the LiDAR-based atlas was subsequently drawn up.

Nowadays, the landforms and patterns that characterise the landscape of the densely populated area of the Netherlands, including the province of Gelderland, are mostly man-made. Seemingly ancient natural landscapes, for example heathland, are often shaped by human influence (Roymans et al. 2009). Heathland is in fact an unstable phase in the succession of vegetation; one which has been created through human land-use activity. Many different landscapes were presumably created in order to serve the needs of the society (or some part of it) at certain points of time and they have been constantly used and reused (Quigley 2010). To this end the term 'recycled landscape' is used. Within the concept of the biography of the landscape or *landscape characterisation*, the transformation of landscapes from prehistory up to the present are explored (Roymans et al. 2009). In this way the concept of *time depth* is introduced: at each point in time, the landscape is an outcome of the complex interplay between physical constrains, social and economic developments and institutional and political changes. An assumption underlying this approach is that it is not possible to understand the modern landscape without understanding the story of its past development. The introduction of *time depth* allows people to appreciate the trajectory of past change in ways that are useful for guiding future change (Herring 2009).

Within the context of this study a *time slice* can be defined as a predetermined period of time. In each period of time natural and/or human processes continuously modify the relief. They create new features, landforms and patterns, or change existing ones. Based on dominant processes and changes in the landscape, five separate *time slices* were distinguished. It is, of course, possible to distinguish an almost infinite number of *time slices*. However, within the scope of this project the number was restricted, to avoid unnecessary complications. Furthermore, the time slices are not strictly defined periods. Some processes and changes may have taken place in more than one time slice.

The notion of 'landscape' can be defined as an area whose character is the result of the interaction of natural and/or human factors, processes and changes (Fairclough 2006a; Fairclough & Møller 2008). This notion is not exclusively reserved for 'natural' and rural environments, but also for urban and periurban areas (Quigley & Shaw 2010). Evidence for change and for the existence of earlier landscapes exists in the present landscape; the process of landscape characterisation enables us to recognise these former patterns (Herring 2009). Each parcel of land has a current use and one or more previous uses recorded in a LiDAR-based elevation image. The variation in the number of previous uses is, in a sense, a measure of the amount of change the environment has undergone (Quigley 2010). Therefore, a LiDAR-based elevation image is composed of several time slices.

The first time slice consists of features representing the natural landscape. Aeolian river dunes are an example of such a time slice. These dunes are recognised as small, isolated heights in a rather flat, low-lying area. Some of these features are not entirely 'natural', considering the fact that people have had an increasing influence on the processes which change the landscape. Palaeochannels, for example, can be the result of a changing pattern of sediment deposit caused by reclamation of the upstream area. Such features will be considered as representing the natural landscape, however, so as not to unnecessarily complicate this research.

The second time slice consists of features representing the prehistoric landscape. It consists of landforms and patterns, which were used during the initial occupation and reclamation of the area, from prehistoric times up until the Early Medieval period. In this time period people settled on elevated areas with easy access to water. They rarely impacted on the landscape, and man-made features belonging to this period are scarce. Furthermore, the meaning or date of their remains is not always unambiguous. Examples of this are for instance prehistoric cart tracks, which display a pattern of nearly unidirectional lines. These tracks are almost always found on ancient trading routes that were used for long periods of time. Therefore, the tracks that survive in the present-day landscape are much more recent than the original tracks that they have largely obliterated.

The third time slice includes features of the pre-industrial landscape, which evolved from the prehistoric landscape. These features originate from the Late and Post Medieval period. This period is characterised by an increase in the population, new land reclamations and the intensification of agricultural activities. These developments led to the emergence of towns and to the reallocation of rural settlements. An example of this time slice are the open field complexes found on the Late Pleistocene cover sand of the eastern Netherlands. These complexes were developed over a long period of time as a result of fertilisation whereby sods, made up of a mixture of grass, peat and dung, were regularly distributed on the land. These open fields, commonly know as *esdek* fields, appear as egg-shaped hummocks as a result of this practice.

The fourth time slice consists of features representing the industrial landscape. They were influenced by the Industrial Revolution, which took place during the 18th and 19th century. This time slice not only deals with the physical remains of industry itself, but with the concept of 'industrialisation' as a shorthand term for a package of social, economic and technological changes that impacted on the rest of society and on the agricultural landscape (Fairclough & Møller 2008). The Industrial Revolution corresponded with major changes in the agricultural sphere, through improvements in farming techniques and changes in land ownership and land division. The increasing demand for arable land and raw materials influenced the landscape too. Some features are comparable to those of the pre-industrial landscape – the third time slice – but they appear on a larger scale. An example is the reclamation of peat land, which can be recognised in the LiDAR images as elongated high and low-lying strips of land.

The fifth time slice comprises features representing the modern landscape. They appear on a very large scale and have been influenced by mechanisation, industrialisation, urbanisation and mobility, e.g. reallocation by enlargement of fields which can be recognised on LiDAR-based elevation images as a pattern of extended blocks. Later, from about the 1960s onwards, a growing concern arose as to the practice of intensive high agricultural production, usually dressed in 'green' and environmental rhetoric (Fairclough & Møller 2008). The process of improving the environment, and the tendency towards 're-wilding' it creates new landscapes (woodland, heath, green infrastructure) just as much as the intensification period did. In some areas, in more recent years, *post*-industrial agriculture might be recognised, for instance in 'part-time' and 'hobby' farming.

The second way of simplifying the interpretation of the LiDAR-images was to construct a legend for the LiDAR-based atlas. Although the number of distinguishing features seemed to be infinite, the problem was overcome with the use of the 'time-slice' framework. For every time slice five characteristic land-

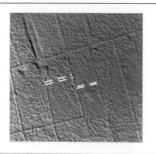
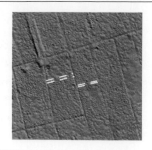

Figure 2. Example of a legend unit.

SAND DITCHES
A pattern of embankments and shallow ditches formed by
the clearing of heatherland and sowing of pine trees. In the
first part of the 19th century, the wood supply was growing,
so large parts of the heatherland were deforested in this
way. Sand was thrown on the seed to protect it against
birds and other animals.

forms were selected. The legend is made up of three columns (fig. 2). The first contains a colour image of the landform, the second is a black-and-white image and the third a definition or a short description.

ANALYSIS OF THE DIFFERENT REGIONS

In order to exemplify the LiDAR-based atlas, four regions with different geomorphological, archaeological and historical-geographical characteristics were chosen (fig. 3). The four different regions were as follows:

– the area surrounding *Aalten – Lichtenvoorde*
– the *Bommelerwaard*
– the lateral moraine of *Nijmegen* and the eastern part of the *Land van Maas en Waal*
– the area surrounding *Vaassen*

All four regions were analysed systematically using the time-slice method. For each time slice three to five images were produced with different visualisation techniques, depending on the characteristics of the chosen subject. Since even the finest grained and most elegantly designed maps are only partial, two-dimensional representations of the landscape (Herring 2009), the LiDAR-based elevation images were printed alongside historical and topographical maps or photographs. These added illustrations helped to clarify the interpretation of the features identified on the images.

To explain each image, information about the geomorphological, historical or archaeological feature was added. The results were discussed with the potential users of the LiDAR-based atlas from different governmental organisations in the region.

CASE STUDY

One of the regions analysed was the landscape surrounding the village of Vaassen. This region is situated in the north of the province Gelderland (figs. 3, 4). The landscape of this region is defined by the sloping border of the Veluwe moraine on the western edge, a plain of melt water deposits on the central part, and

Figure 3. The location of the study areas.

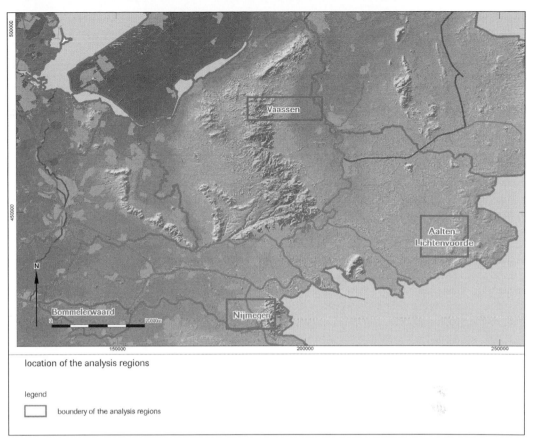

location of the analysis regions

legend

☐ boundery of the analysis regions

Figure 4. A DEM of the region of Vaassen. Note that the relief is exaggerated by a factor 5.4 and illuminated from the north-west.

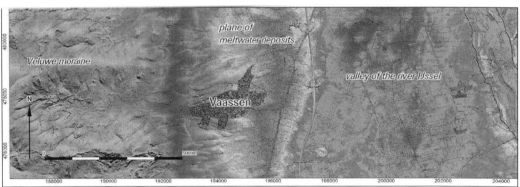

the low-lying valley of the river IJssel on the eastern edge. The region has a varied land usage. The Veluwe moraine is, to a large extent, forested, but there is also some heathland. The plain of meltwater deposits and the valley of the river IJssel are used as farmland. Occupation is concentrated in villages on the plain.

Time slice 1: Natural landscape (figs. 5, 6)

During the cold periods of the Late Glacial, sand dunes were formed from the vast quantities of sand that were blown over the barren landscape of the Netherlands. Nowadays the dunes are vegetated, but they still stand out clearly on the LiDAR-based elevation images.

Figure 5. A DEM of a dunefield in the valley of the river IJssel. Note that the relief is exaggerated by a factor 5.4 and illuminated from the north-west. *See also the full colour section in this book*

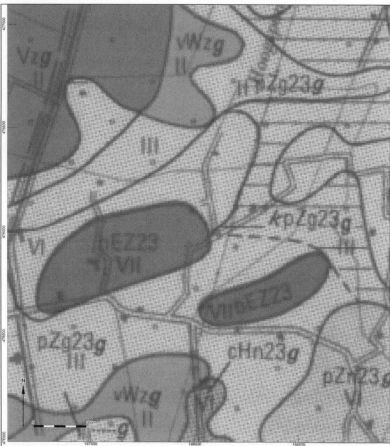

Figure 6. Comparing the soil map of the same area with a DEM, a strong relationship between topographical height and soil type can be recognised. The brown areas represent soils heightened by sods, which are characteristic for cultivated cover sand dunes. *See also the full colour section in this book*

Time slice 2: Pre-historic landscape (figs. 7, 8)

Celtic fields are a characteristic archaeological feature dating to the Iron Age. A Celtic field is an agricultural field system found in north-west Europe, which can be identified by the banks surrounding the parcels of arable land. These banks were formed by systematically dumping crop waste, soil and stones on the edges of each field. In the LiDAR-based elevation images the Celtic fields are clearly visible, even in wooded areas, where they cannot be identified on aerial photographs. The Celtic field system in Vaassen has a surface area of 76 ha (31 acres).

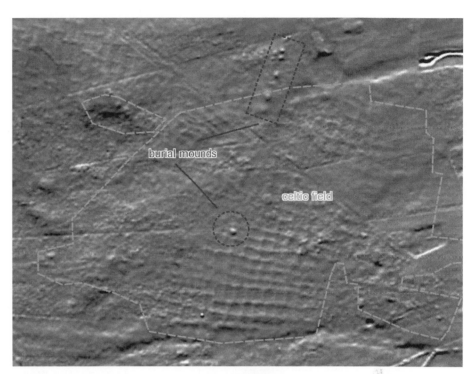

Figure 7. A DEM of a Celtic fields near Vaassen. A black-and-white colour pallet is used to emphasise the characteristic pattern of parcels and banks.

Figure 8. Height difference in the path reveal the presence of Celtic fields, but the vegetation made an overview impossible.

Time slice 3: Pre-industrial landscape (figs. 9, 10)

In the 14th century, count Reinoud van Gelre commissioned the cultivation of an extensive peat bog near the village of Nijbroek. The agreement with the owners was called a *'Cope'*. A system of canals – *'weteringen'* – was designed for drainage of the bog. Numerous ditches, perpendicular to the weteringen, produced long and narrow parcels of ca. 60 x 2,325m. A total of 95 farms were established in the area. The canal system is clearly visible on the LiDAR-based images.

Figure 9. A DEM of the 'cope' cultivation east of Vaassen is represented as a flat and low-lying area. Note that the relief is exaggerated by a factor 5.4 and illuminated from the north-west.

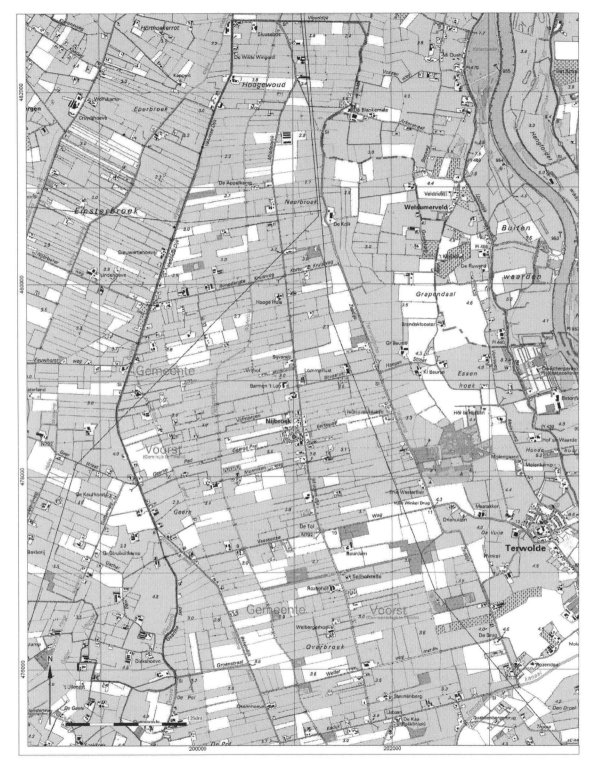

Figure 10. Topographical map of the 'cope' cultivation east of Vaassen.

Time slice 4: Industrial landscape (figs. 11, 12)

To the west of Vaassen lies a network of man-made brooks; *'Sprengenbeken'*, which is a good example of the stream systems of the Veluwe. These were dug between the 14th and the 18th century to obtain groundwater for watermills. In Vaassen there are four of these man-made brooks with three cross-connections, some of which were constructed as aqueducts. Along the brooks as many as seventeen watermills were built.

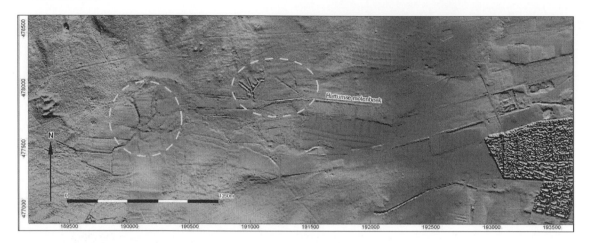

Top

Figure 11. A DEM of the stream systems near Vaassen. Note that the relief is exaggerated by a factor 5.4 and illuminated from the northwest.

Bottom

Figure 12. A topographical map dating from the 18th century gives an overview of the stream systems near Vaassen.

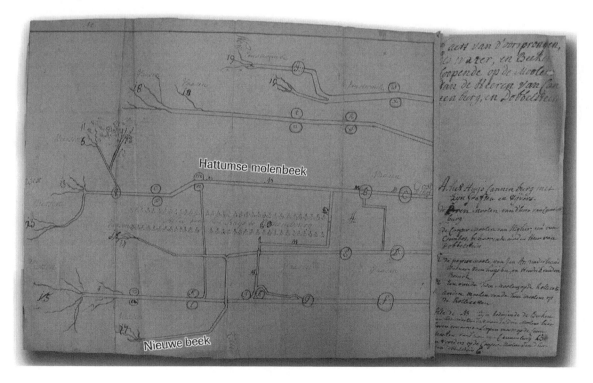

Time slice 5: Modern landscape (figs. 13, 14)

In 1949, fear of an invasion by the Soviet Union led to the development of a line of defensive works along the river Rhine from Switzerland to the North Sea. In the Netherlands, defence works that could inundate an area of 120 kilometres by 10 kilometres were built along the river IJssel: the 'IJssellinie'. The illustration shows the location of a floating dam. Today only the entrances of the dam remain.

Figure 13. A DEM of the IJssellinie. Note that the relief is exaggerated by a factor 5.4 and illuminated from the north-west. *See also the full colour section in this book*

Figure 14. The IJssellinie on a topographical map from 1976. *See also the full colour section in this book*

Assessment of the LiDAR-based images in heritage practices

The new method of time slicing has proved to be a useful tool for time-depth analysis of LiDAR-based elevation images. During various workshops we found that these methods inspire potential users of the atlas, even those who are not very familiar with LiDAR. They discovered new landforms and patterns, which were subsequently discussed. The images can also give valuable input to (historic) landscape characterisation, which has the potential to be an immensely important resource for archaeological, historical and other heritage related research, as well as landscape and urban design. The images enhance existing input deduced from geological and historic mapping and archaeological data. An example of the use of the LiDAR-based images by urban design is the concept of a public park in a newly developed urban expansion ('VINEX wijk') to the west of the city of Utrecht: Leidsche Rijn. The park is situated at a fossil stream ridge of the river Rhine. With the help of amongst others the LiDAR-based images, the course of the channel during the Viking age, spanning the late 8th to 11th centuries, was reconstructed. The architect of the park used this reconstruction for the design of the water in the Park (www.utrecht.nl/ wonenenwijken/leidscherijn/vikingrijn). By the reconstruction of the water course, a distinctive feature of the pre-industrial landscape survives in the present-day urban landscape.

Other examples whereby cultural and historic features are used in spatial development are the projects undertaken within the framework of the Belvedere Memorandum (1999). The Belvedere strategy aims to involve cultural historians early in planning processes and to provide architects, urban and rural planners and administrators with effective, usable (and understandable) information. The LiDAR-based images can be a powerful source of inspiration for spatial design within these projects.

Not only in the Netherlands, but also in other European countries, the use of the LiDAR-based images has proved a useful tool for exploring the (pre)historic landscape and landscape design. The Stonehenge LiDAR survey is a good example of this. The survey had its origin in the requirement for English Heritage to develop new approaches for investigating the historic environment (Bewley et al. 2005). As well as contributing to the archaeological record itself, there was also a need to provide an archaeological context to the proposed roadworks around the Stonehenge monument, and to the design and location of a new visitor centre. Within the framework of The Loughcrew Project, LiDAR data was used as a tool for exploring the prehistoric landscape, creating images by draping aerial photography over the LiDAR digital surface model (Shell & Roughley 2004). The ability to generate detailed visualisations and to move through the digital landscape allows us to gain a more comprehensive insight into the prehistoric landscape.

ACKNOWLEDGEMENTS

We would like to thank the reviewers Menne Kosian and Philip Verhagen for their critical comments and Wesley van Breda for improving the English. Bert Groenewoudt and Henk van der Velde contributed by providing literature.

REFERENCES

Belverdere Memorandum. 1999. *A Policy document examining the relationship between cultural history and spatial planning*. The Hague, VNG Uitgeverij.

Beek, R. van. 2009. *Reliëf in Tijd en Ruimte. Interdisciplinair onderzoek naar bewoning en landschap van Oost Nederland tussen vroege prehistorie en middeleeuwen*. Wageningen, Wageningen University.

Bewley, R.H., S.P. Crutchley & C.A. Shell. 2005. New light on an ancient landscape: LiDAR survey in the Stonehenge World Heritage Site. *Antiquity* 79, 636-647.

Boer, A.G. de, W.N.H. Laan, W. Waldus & W.K. van Zijverden. 2008. LiDAR-based surface height measurements: applications in archaeology. *British Archaeological Reports* S1805.

Deprez, S. & D. Taelman (eds.) 2009. Ol'man river. Geo-archeological aspects of rivers and river plains. *Archaeological Reports Ghent University (ARGU)* 5.

Fairclough, G. 2006a. From assessment to characterisation: current approaches to understanding the historic environment. In J. Hunter & I. Ralston (eds.), *Archaeological Resource Management in the UK: An Introduction*, 2nd edition, 253-75. Sutton Publishing, Stroud.

Fairclough, G., & P.G. Møller (eds.) 2008. *Landscape as heritage. The Management and Protection of Landscape in Europe, a summary*. Action COST A27 'LANDMARKS'.

Herring, P.C. 2009. Framing Preceptions of the Historic Landscape: Historic Landscape Characterisation (HLC) and Historic Land-Use Asssesment (HLA). *Scottish Geographical Journal* 125 (1), 61-77.

Laan, W.N.H. & W.K. van Zijverden. 2004. Landscape reconstructions and prospective modeling in archaeological research-using a laser altimetry based DEM and digital boring database. In W. Börcher (ed.) *Abstracts der 9. Internationalen Workshop Archäologie und Computer, 2004 de Stadtarchäologie Wien, 3-5 November*.

Lewis, H., C. Gallager, W.A. van Breda, G. Mulrooney, S. Davis, R. Meehan, J. Turner, A. Brown, L. Guinan & C. Brady. 2008. An integrated comprehensive GIS model of landscape and land-use history in the Boyne River valley and its catchments. *INSTAR grant 16666. Final report to the Heritage Council*.

Quigley, P. & M. Shaw. 2010: Characterization in an Urban Setting: The Experience of the Black Country. *The Historic Environment* 1 (1), 4-28.

Quigley, P. 2010. Recycled landscape. The Legacy of 250 Years in the Black Country. An Analysis of the Black Country Historic Landscape Characterisation. *English Heritage Project Number 3638 Main Second Report 2010*.

Shell, C. & C. Roughley. 2005. Exploring the Loughcrew landscape: a new airborne approach. *Archaeology Ireland* Summer, 22-25.

Waldus, W. & H. van der Velde (eds.). 2006 . *Archeologie in vogelvlucht: een onderzoek naar de toepassingsmogelijkheden van het AHN voor de archeologie*, Geoarchaeological and Bioarchaeological Studies 6. Amsterdam: Vrije Universiteit Amsterdam.

Zee, R.M. van der, F.S. Zuidhoff & A.G. de Boer. 2009. *Gelderland in Reliëf. Legenda en verhalen over het landschap. Handleiding bij de Atlas van het Actueel Hoogtebestand Nederland*. ADC ArcheoProjecten rapport 2059, Amersfoort.

Zijverden, W.K. van, P.F.B. Jongste & F.S. Zuidhoff. 2009. Landscape and occupation, long term developments in the Dutch river area during the Bronze Age. In Dapper, M. de, F. Vermeulen, S. Deprez & D. Taelman, *Ol'man river. Geo-archaeological aspects of rivers and river plains. Archaeological reports Ghent University* 5. Ghent, Ghent University.

WEBREFERENCE

http://www.ahn.nl
http://www.gelderland.nl/bodem

How will landscape archaeology develop in the future?

6.1 The future of landscape archaeology

Author

Andrew Fleming

School of Archaeology, History and Anthropology, The University of Wales Trinity Saint David, Carmarthen, United Kingdom
Contact: andrewfleming43@btinternet.com

ABSTRACT

This overview of landscape archaeology/landscape history (LAH) is presented in the form of a SWOT analysis (Strengths, Weaknesses, Opportunities and Threats). LAH is no longer just an array of field skills and methods, however holistically deployed. The discipline has had to respond to postmodernism and to embrace cognitive approaches, and it now needs to become more richly theorised, building on its reconstructive potential to develop an archaeology of landscape. Strengths discussed here include the flexibility of landscape archaeology, and its capacity to operate at different scales; it is suggested that LAH's ultimate importance may lie in its potential contribution to human historical ecology. Under 'weaknesses', the dangers of localism are considered. It is argued that the success of the 'Annales' school has shown how the microcosm may encapsulate the macrocosm. As the author's own work on medieval roads demonstrates, a focus on apparently local matters may engender trains of thought which open up much wider issues (the dividend of empiricism). In considering 'threats', this paper argues that the postmodern challenge has introduced an unhelpful and unnecessarily polarised debate. It is important to acknowledge the potential of the 'digital revolution', although computer-driven methodologies cannot be expected to supplant more traditional insights and ways of working.

KEYWORDS

landscape archaeology, postmodernism, SWOT analysis, human ecology

INTRODUCTION

It would be a brave, not to say foolhardy person who would venture to predict how landscape archaeology will have developed in ten or twenty years' time; in the words of the old joke, I had to tell my Amsterdam audience that owing to unforeseen circumstances, the clairvoyant would not be making an appearance. In any case, these days, people in England who wish to be taken seriously never say 'in future'; instead, Margaret Thatcher's children say 'going forward', to indicate how focused and businesslike they are. And of course there are no 'problems' anymore, just 'challenges'. Approaches and buzzwords fashionable in the world of business studies have spread into our universities; for instance, in recent years some British academics have found themselves under instructions to carry out a SWOT analysis. SWOT is an acronym, meaning Strengths, Weaknesses, Opportunities and Threats. One way of trying to foresee the future, in some sense, is to carry out a SWOT analysis. Before speaking at LAC2010, I naively thought that in attempting such a procedure in relation to an academic discipline, I might be something of a pioneer. However, I had barely sat down after giving my paper at LAC2010 when Graham Fairclough informed me that a SWOT analysis for Landscape already exists, in the pages of an official report. Be that as it may; the SWOTs selected here will be very much a personal choice, and I will make no attempt at rigid definitions or comprehensive treatment. In other words, I will not be 'ticking all the boxes' (to use another cliché of management-speak).

Theoretical perspectives

In Britain, a much sharper distinction is drawn between 'landscape archaeology' and 'environmental archaeology' than was apparent in Amsterdam. Furthermore, we islanders also have to deal with the introduction of postmodern perspectives into the archaeology of landscape (see below), a phenomenon which has spread beyond our own shores to some extent. Postmodernism was practically invisible at LAC2010. Much the same could be said for the ethnographic perspectives which sometimes inform it. Although from a British viewpoint this has been a little disconcerting, I regard the setting up of LAC2010 as an important initiative. If an international forum for multidisciplinary and interdisciplinary landscape archaeology can become firmly established, there will be plenty of scope in future for a more diverse 'coverage' of the subject or an examination of particular themes and perspectives.

When I first started my fieldwork, in the mid-1970s, I thought of landscape archaeology (the term had only recently been invented) mostly as a set of specialist skills which could provide a service to the mainstream discipline of archaeology. These skills could help an excavator to put an archaeological site into its local context, like the work done by environmental archaeologists. And landscape archaeology could provide information for the landscape historian, supplementing the documentary record or disputing it. By the mid-1990s, when I published a discursive piece about the role and character of landscape archaeology (Fleming 1996), I had moved away from these rather modest claims. In that article, I noted that research into landscape history ideally requires the application of a range of methodological skills and specialist knowledge normally beyond the capacity of a single scholar. In this sense, landscape archaeology could be portrayed as a 'holistic' discipline, feeding into a landscape history which is not only rooted in a uniquely complex and interdisciplinary methodological mix but also constitutes a distinctive, insightful and refreshing way of studying and presenting long-term human history.

However, at about the time when I wrote these words, a few British archaeological theorists were beginning to develop what appears to be a devastating critique of conventional landscape archaeology, claiming that it has roots which mean that it is politically compromised, and that its over-empirical approach makes it banal and unimaginative. These postmodernists have sought to replace traditional approaches by new methodologies, such as phenomenology, and new 'ways of telling' (for references to this critique, see Fleming 2006). By the turn of the 21st century, landscape archaeology had become the area in which theorists wanted to engage with archaeological data. As readers of Anglophone literature may be aware, I find the rhetoric and the field methods of these theorists highly problematic (Fleming 1999, 2005a, 2006, 2008). Nevertheless, their intervention has been valuable in several ways. No one should be shouted down for seeking more imaginative approaches in archaeology. And the cognitive approach has proved very stimulating. For most readers, this will be rightly associated with the writing of Ian Hodder (e. g. 1982a, 1982b, 1989), although here it is also worth acknowledging two progenitors of this stream of thought within modern archaeological theory – Edmund Leach (1973) and Frances Lynch (1975). In my view, the years of theoretical debate have clearly demonstrated this: if *landscape archaeology* may be described as a set of investigative skills, which feed into an essentially reconstructive landscape history, *the archaeology of landscape* is a much more open subject, rather an exciting one. As an arena for theoretical debate, it is not difficult to see landscape's attractions. As Howard Morphy put it: 'it is useful to have a concept that is free from fixed positions, whose meaning is elusive, yet whose potential range is all-encompassing' (1993, 205).

Certainly the archaeology of landscape is under-theorised by practitioners of landscape archaeology. Most of us are too absorbed in our current projects, in simply *doing* landscape archaeology, to take much interest in wider theoretical perspectives; at the time when landscape archaeology was introduced, explaining its basic field methodologies seemed a more pressing concern. Earlier writers who discussed 'field archaeology' – for example, O.G.S. Crawford – had not been very explicit on this topic. From a philosophical point of view, empiricism may seem banal. But in landscape archaeology, discovery is immensely exciting. So is the Sherlock Holmes style 'detective work' (also, incidentally, the object of contemporary theoretical disdain) which mixes induction and deduction in ways which are often highly productive. Nevertheless, I believe that we can no longer treat 'landscape archaeology' simply as a set of methodologies, a sub-discipline of archaeology and a supplier of information to a reconstructive landscape history, without considering the much wider issues raised by the concept of 'the archaeology of landscape'. That would be disingenuous, and also very limiting. So in presenting a brief SWOT analysis, I have to consider 'traditional' or conventional landscape archaeology, but I will also have in mind a more complex and difficult question: how might landscape archaeology contribute to a more richly theorised 'archaeology of landscape'?

STRENGTHS OF LANDSCAPE ARCHAEOLOGY

What are the *strengths* of landscape archaeology? One's first instinct, I think, is to try to explain how we may complement the work of the excavator. But there is much more to the archaeology of landscape than 'archaeology beyond the edge of the trench'. And the archaeology of landscape is not quite the same thing as 'off-site archaeology', in Rob Foley's phrase (1981). It was once fashionable to say that the entire landscape is just one big archaeological site. And it certainly *is* a strength of our discipline that landscape archaeologists may work without boundaries. To be more accurate, we have the capacity to regard boundaries of any kind – whether they are geographical constraints, detectable archaeological features, or concepts created by archaeologists – as flexible, variable, permeable, and of greater or lesser importance according to context. I refer here, of course, to questions of *scale*.

Paradoxically perhaps, my own understanding of the variability of scale, and the nuances which the concept may assume in different contexts, deepened when I was working in Britain's most distant archipelago – St Kilda, which lies some sixty kilometres off the coast of the Western Isles of Scotland. Depopulated in 1930, St Kilda's only inhabited island was small, remote, and often unreachable outside the summer months. As the title of my book (Fleming 2005b) acknowledges, despite these apparently severe geographical constraints – or more probably *because* of them – it is impossible to discuss the history of St Kilda and its little community without implicating the wider world. Scale, of course, does not much concern questions of *measurement*, like size or distance; it is about perceptions – in the case of St Kilda, those of islanders and outsiders, elites and commoners, writers and dreamers, as well as those with hard-edged political agendas. Visitors to St Kilda were fascinated by the little community which they encountered there; their preconceptions and reactions in turn influenced the attitudes of the islanders and the way they presented themselves to the world. Today, St Kilda and its history still stimulate the imagination, as recent visitors to Scotland will be aware. My own discussions of scales of perception were of course heavily indebted to the rich St Kilda literature (albeit produced by outsiders). But although *St Kilda and the wider world* is text-aided, I hope it demonstrates one way of achieving what Matthew Johnson has advocated in his book, *Ideas of Landscape* – moving from doing landscape archaeology in a reconstructive sense towards engaging with the archaeology of landscape. And if islands, paradoxically, do not have boundaries, conversely there is something to be said for thinking about the 'islandness' of land-locked, 'mainland' communities – such as the one which has occupied the upland valley of Swaledale (North Yorkshire) in the Pennines of northern England (Fleming 2010).

The opportunities provided by written texts are not, of course, available to prehistorians, and for those dealing with protohistory, they are limited. In these areas of our work, I think we must try to make more of landscape archaeology's other strengths – above all perhaps, its perpetual engagement with the immense diversity of the surface of the earth, the physical properties of the land. Archaeologists have an almost instinctive tendency to seek and recognise patterns, correlations, variations in density and distribution, gaps – and then to interpret and explain them. The most dynamic and potentially eloquent feature of this matrix of potential correlations is the landscape itself, always available to be treated as a constant or a variable, a constraint or an opportunity, a taskscape or a canvas for the imagination. Landscapes change, over various timescales. And talking of canvasses, despite the well-documented origins of the use of the word 'landscape' to describe the scenery depicted in paintings, we should never think of the landscape as essentially the 'background' to human affairs. We have developed an understandable aversion for geographical determinism; yet geography has always played a significant role in human affairs.

Landscape is very definitely a player in the drama of history. Here I am not referring merely to climate, relief, drainage and so on. If the landscape often seems disconcertingly alive to the alert field archaeologist, that is because it *is* alive, and always has been. This is not simply because it teems with flora and fauna, or contains cultivated 'crops', domestic livestock, prey for hunters and fishermen, or even living artefacts, such as pollarded trees. Once a human has entered the scene, as historical participant or scientific observer, the landscape is full of potential. We humans, with our interesting brains, are – and have been – opportunists in ways which are literally unthinkable for other species. The landscape is the canvas for our imaginations, past and present. This is not, as it might seem, a plea for the release of the unfettered imagination of the landscape archaeologist; we have heard plenty of those in recent years. But it *is* a call for landscape to be allotted a role in historical studies which matches its potential. Personally, I hope that in future we will be more concerned to work with the *ecology* of long-term human history, in a nuanced fashion, without worrying about ill-considered charges of geographical determinism. If landscape can contribute to a mainstream narrative of history, it must surely work with human ecology, over *la longue durée*, in Fernand Braudel's phrase, and concern itself with identifying and understanding the critical variables in human history. I am well aware that 'human ecology' is still something of a Cinderella subject, diffusely addressed and represented in the literature, regarded as problematic in some respects. It may seem contentious, not to say yawn-inducing, in terms of how it is defined and distinguished from other areas of enquiry, and it is certainly open to what one might call cybernetic abuse, particularly the inappropriate or banal application of totalising theories or models. Yet when I read a paper such as David Siddle's, on 'Goats, marginality and the dangerous other' (2009), I feel that I am in historical territory which an archaeologist of landscape should have no difficulty in recognising. I am talking here about high historical relevance; for if we will not address the historical dimensions of issues of human ecology now, in the face of the problems which humanity faces as a species – the dominant species on this planet – when *will* we address them? In Amsterdam it was interesting to hear Mats Widgren's praise for the continued willingness of American colleagues to address global issues scientifically.

WEAKNESSES

What about weaknesses? One troubling question concerns the intellectual status of landscape history. If 'history' is primarily about the thoughts and actions of human beings, what kind of beast is 'landscape history?' What serious claim might it have upon our attention? How could such an apparently dehumanised version of history be considered mainstream, or indeed worth pursuing? Julian Thomas has indeed argued that conventional landscape archaeology ignores or sidelines people; he has claimed that Mick Aston's landscape archaeology results in 'a huge Heath-Robinson apparatus, within which human beings have the metaphysical status of the ghosts in the machine' (1993, 26). Although there are reasons for this apparent state of affairs (Fleming 2006) this is certainly an area of concern. Here, however, I want to address a slightly different question. When landscape archaeologists and landscape historians speak of their 'research area', we are often talking about an area in the literal, geographical sense; the focus of interest is a parish, a small region, a valley, or perhaps a block of hills. It may be apparent that both the subject and the object of study are defined geographically; arguably, everything is of interest, and the exclusion of any significant body of material from the narrative might detract from the 'holistic' character of landscape history. But does this state of affairs not make us essentially 'local historians'? Are we not at risk of being considered 'amateurs' in the original sense of the word? Many places are fascinating in their own way; enthusiasm is undoubtedly infectious; it *is* possible for good to emerge from obsession, or close engagement. But how far is this particular intense focus intellectually challenging, or relevant to wider human concerns?

Writing in my festschrift, Rob Young & Jane Webster (2008) have urged me not to worry about this.

They enthuse about local participation and community archaeology. They explain the impact of a landscape archaeology project on the imaginations of local people, how they are excited and energised by a project which starts them thinking seriously about the history of their locality for the first time. They learn to take and share responsibility for local historical narratives, and how to make their own contributions. I applaud this, of course; but I do not find their thesis a satisfactory response to my concerns about the intellectual standing and wider relevance of what we are doing.

It is possible that I should be more relaxed about all this. By definition, local projects will not produce an overview. However, we are all conscious, I hope, of the historical perspective represented by the Annales School; we may well have read, for example, Le Roy Ladurie's *Montaillou* (1978). In England, Eamonn Duffy (2001) has used events in a single Devon parish as an exemplar of the Roman Catholic-Protestant transition in England; a comparable approach could be applied within landscape history. But we can't expect the microcosm to encapsulate the macrocosm just like that. If it is to do so, a certain kind of academic alchemy is necessary; research areas may have to be carefully selected, and 'local' projects will need to be informed by wider knowledge. That said, the most empirical of approaches may produce insights which lead to enquiries of much more general relevance. Let me provide a brief example. I have recently been working on a medieval road in the mountains of mid-Wales, a road connecting two 12th-century Cistercian monasteries (Fleming 2009). Much of the road takes the form of a narrow, constructed terrace, dug into the hillside. At first I was concerned with the physical appearance and 'behaviour' of this road, regarding it as a 'target' for the landscape archaeologist. But soon I had to think about the 'interpretation' of the Monks' Trod, as it is called, and to explore its political and social context. I had to ponder the strategic significance of this road, and comparable, potentially medieval roads. I have come to regard them as instruments of elite control, facilitating relatively swift, purposeful movements of men on horseback. So I have had to consider the nature of the 'horse culture' of medieval elites, and its potentially dynamic role in the formation and maintenance of socio-political relationships and the integration of early polities. And it seems that the more 'important' of these long-distance, horseworthy roads would have had considerable psycho-social significance. They represented and symbolised the relationships which – in the ideal world – integrated regional social hierarchies, and the unwritten social contract between those at the top of the hierarchy and those lower down the social scale. Important long-distance roads were not simply the means of getting from place to place. They were ever present in the landscape and in its mythology. They had dynamic historical roles. The landscape historian who writes the biography of a road will encounter much more than a string of localities. Empiricism, then, produces dividends.

That said, I must acknowledge that not all landscape historians start with localities. They also develop overviews of particular topics from a much wider geographical perspective. To apply the 'compare and contrast' approach over a 'large' region often brings rich and more widely applicable insights, as shown for example by Tom Williamson's study of the origins of medieval open fields in eastern England (2003). Sustained work of this kind, such as that of both Barnes & Williamson and Rackham in eastern England (Barnes & Williamson 2006; Rackham 1976, 1980), not only provides methodological leadership. It may also bring into existence a 'reference' or 'lead' region, which may well affect the way the histories of less thoroughly investigated regions are perceived; on balance, this is probably a good thing.

OPPORTUNITIES

This brings me to the third item of my SWOT analysis – opportunities. I am not going to say very much here, because it seems to me self-evident that there are many exciting things to do in the field of landscape archaeology. We are still making discoveries which are not simply additions to the corpus of 'sites and monuments'; they challenge us to think about past perceptions of the world. The so-called 'Seahenge' on the east coast of England is a good example; the finding of a large inverted tree-trunk at the centre of a ring of large upright posts must set the imagination racing (Pryor 2001). And other challenges to our perceptions of landscape will come from excavated sites – for example, the recent discovery that some of the cattle associated with the Neolithic horizon of the Stonehenge area came from much further west, from south-west England or Wales (M. Parker Pearson, personal communication). But we need to do more than respond to the puzzles set by random encounters with intriguing new data sets. One of the strengths of landscape history is its holistic character, its potential to integrate and develop many strands of enquiry and argument. Landscape history has numerous dimensions, and I hope that in the future we will be able to assemble more *teams* of specialists who are able to bring a multidisciplinary approach, and ideally an *inter*disciplinary approach, to our enquiries. Disciplinary fragmentation is the enemy; we need fusion rather than fission. And if we are to convince those who control the distribution of research money, we may have to consider our contemporary relevance as well as our traditional preoccupations (see above). Archaeology has the potential to organise cognate disciplines in order to gain a greater understanding of the furtherance of long-term human history, and it is surely the archaeology of *landscape* which brings us the holistic approach we require for this endeavour. Here I am referring not only to multidisciplinary and interdisciplinary work and ways of thinking. Landscape archaeology combines theoretical insights and different ways of exploring the archaeological record with a close focus on past cultural contexts and horizons, in rather a special way; its matrix of enquiry is richer than that achieved by treating landscape studies as optional or peripheral.

THREATS

But as with all branches of historical enquiry, we need to achieve the right mix of theory and practice. This brings me onto threats. In Britain, in the 1990s, a group of influential postmodern theorists had identified landscape archaeology as an area ripe for colonisation. Using whatever weapons came to hand, they tried to expel or marginalise the natives, who turned out to be curiously passive, but also quite stubborn. Quite a few of them carried on as usual. But one old character had the temerity to point out that some of the farms established by the colonists were not very well adapted to the terrain; he questioned the purpose of the new fences which have been erected across the land (Fleming 2006). We do not yet know the outcome of this confrontation; recently, things have gone a bit quiet (but see Barrett & Ko 2009).

The chosen weapon of these theorists is a polarising, adversarial rhetoric; they seek to create and foment opposition between theory and practice, between old and new approaches, between investigative and performative approaches, to name only three of numerous alleged oppositional dividing-lines. I do not know how many readers have perused *Stone Worlds*, an account of a campaign of such theoretically-informed fieldwork on Bodmin Moor, an upland area in south-west England (Bender, Hamilton & Tilley 2007). The book recounts an episode in which Tilley, one of the theorists, arrives at an excavation trench just before it is to be backfilled. An excavator has formed a poor opinion of Bourdieu's *Outline of a Theory of Practice*. Having used it contemptuously as a coffee-pot stand for most of the duration of the excavation, he has now thrown it into the trench, and urinated on it. When Tilley comes onto the scene, he promptly 'neutralises' the gesture by indulging in a symbolic performance: he takes his own trowel (which has been little used) and throws it into the trench, to be buried along with the book. A little cairn is then built over these relics, on the bottom of the trench (Bender et al. 2007, 273-275).

I do not think theory and practice should be in this kind of oppositional, not to say antagonistic, relationship, and in that sense I am identifying a threat to the future health of the discipline. I would use words like nuanced, exploratory, responsive and empirical to describe the relationship between theory and practice in the archaeology of landscape; we should not have to get involved in battles between different 'isms'. If there is any value in a 'performative' archaeology of landscape, it should not be promoted at the expense of the discipline's investigative potential. There is certainly a place for rhetoric, but it should not displace argument, or critical standards. If we go into the field having already decided on the story we are going to tell, we are in trouble.

Another potential threat, perhaps, is that we may be over-influenced by the all-powerful computer. As Verhagen so ably pointed out in Amsterdam, there has been a digital revolution in landscape archaeology (Verhagen, this volume). These days, even an old-fashioned field archaeologist like me knows that a raster is not an Afro-Caribbean gentleman with an impressive hair style. We have discovered the immense potential of Geographical Information Systems, not only for information storage and display, but also for seeking patterns and examining potential correlations in complex data spread over wide areas. Using the computer allows us to get the best out of geophysics. When we use LiDAR, we may if we wish place the sun where denizens of the northern hemisphere have never seen it – low in the northern sky. We can fly over the surface of land at any angle to the horizon; unlike a fighter pilot, we can stop the flight whenever we want to and examine the ground surface. We can ask a computer to reconstruct past patterns of vegetation according to different parameters, to 'reconstruct' a monument from a pattern of subterranean holes, to show us what the North Sea may have looked like in the days when it was Doggerland (Gaffney et al. 2007, 2009; Coles 1998). We may still worry about whether we can see the wood for the trees – but these days, at least, we are able to see *through* the crowns of trees even when they are in leaf. A few clicks of a mouse may give us a compelling sense of *control*, similar to the feeling we get when we deploy a grand, totalising theory with apparent success.

At the end of the day, computers are only obeying orders. They will never be good at thinking laterally; they cannot arrange for those chance encounters which allow the imagination to play the maverick. You can equip a computer with a webcam, but it is still as blind as a bat. For as far into the future as I can see, we will depend on the interaction between the human eye and the human brain. More than one archaeologist whom I met in Amsterdam at LAC2010 insisted that everything picked up on a LiDAR scan has to be checked in the field (or 'ground-truthed' – to use an awful phrase sometimes heard in Britain). The computer cannot do this, and it certainly cannot replicate the unique observational journey undertaken by the field archaeologist in his or her lifetime. I still think that some of the best insights arise not from computer-generated simulations but from the landscape as experienced on the ground. By all means let us provide our students with up-to-date equipment and engage them in the latest theoretical debates; but if we immerse them in the field experience of landscape archaeology, we will be providing them with the best equipment of all.

One of our senior archaeologists, Professor Charles Thomas, once said: 'You don't need all this theory nonsense; all you need is a countryman's eye and a good pair of boots'. I do not agree with him; but I know exactly what he meant.

ACKNOWLEDGMENTS

This paper benefitted from reviews by Mats Widgren and Jos Bazelmans.

REFERENCES

Barnes, G. & T. Williamson. 2006. *Hedgerow history: ecology, history and landscape character*. Oxbow, Oxford.

Barrett, J. & I. Ko. 2009. A crisis in British landscape archaeology? *Journal of Social Archaeology* 9 (3), 275-294.

Bender, B., C. Tilley & S. Hamilton. 2007. *Stone worlds: narrative and reflexivity in landscape archaeology*. Left Coast Press, Walnut Creek, CA.

Coles, B. 1998. Doggerland: a speculative survey, *Proceedings of the Prehistoric Society* 64, 45-81.

Duffy, E. 2001. *Voices of Morebath: reformation and rebellion in an English village*. Yale University Press, New Haven.

Fleming, A. 1996. Total landscape archaeology: dream or necessity? In Aalen, F. (ed.) *Landscape study and management*, 81-92. Department of Geography, Trinity College Dublin/Office of Public Works, Dublin.

Fleming, A. 1999. Phenomenology and the megaliths of Wales: a dreaming too far? *Oxford Journal of Archaeology* 18 (2), 119-125.

Fleming, A. 2005a. Megaliths and post-modernism: the case of Wales. *Antiquity* 79, 921-932.

Fleming, A. 2005b. *St Kilda and the wider world: tales of an iconic island*. Windgather Press, Bollington.

Fleming, A. 2006. Post-processual landscape archaeology: a critique. *Cambridge Archaeological Journal* 16 (3), 267-280.

Fleming, A. 2008. Don't bin your boots! *Landscapes* 8 (1), 85-99.

Fleming, A. 2009. The making of a medieval road: the Monk's Trod routeway, Mid Wales. *Landscapes* 10 (1), 77-100.

Fleming, A. 2010 [1998]. *Swaledale: valley of the wild river*, Oxbow Books, Oxford,.

Foley, R. 1981. Off-site archaeology: an alternative approach for the short-sited. In Hodder, I., G. Isaac & N. Hammond (eds.), *Pattern of the past: studies in honour of David Clarke*, 157-83. Cambridge University Press, Cambridge.

Gaffney, V., K. Thomson & S. Fitch. 2007. *Mapping Doggerland: the Mesolithic landscapes of the southern North Sea*. Archaeopress, Oxford.

Gaffney, V., K. Thomson & S. Fitch (eds.) 2009. *The rediscovery of Doggerland*. York, Council for British Archaeology.

Hodder, I. 1982a. *Symbols in action*. Cambridge University Press, Cambridge.

Hodder, I. (ed.) 1982b. *Symbolic and Structural Archaeology*. Cambridge University Press, Cambridge.

Hodder, I. (ed.) 1989. *The meanings of things: material culture and symbolic expression*. Unwin Hyman, London.

Johnson, M. 2006. *Ideas of landscape*. Blackwell, Oxford.

Leach, E. 1973. Concluding address. In Renfrew, C. (ed.), *The explanation of culture change: models in prehistory*, 761-771. Duckworth, London.

Le Roy Ladurie, E. 1978. *Montaillou: cathars and catholics in a French village, 1294-1324*. Scolar Press, London.

Lynch, F. 1975. The impact of landscape on prehistoric man. In Evans, J.G., S. Limbrey & H. Cleere (eds.), *The effect of man on the landscape: the Highland Zone*, 124-126. Council for British Archaeology, London.

Morphy, H. 1993. Colonialism, history and the construction of place: the politics of landscape in Northern Australia. In B. Bender (ed.), *Landscape: politics and perspectives*, 205-43. Berg, Oxford.

Pryor, F. 2001. *Seahenge: a quest for life and death in Bronze Age Britain*. Harper Collins, London.

Rackham, O. 1976. *Trees and woodland in the British landscape*. Dent, London.

Rackham, O. 1980. *Ancient woodland*. Edward Arnold, London.

Siddle, D. 2009. Goats, marginality and the 'dangerous other'. *Environment and History* 15, 521-536.

Thomas, J. 1993. The politics of vision and the archaeologies of landscape In B. Bender (ed.), *Landscape: politics and perspectives*, 19-48. Berg, Oxford.

Tilley, C. 1994. *A phenomenology of landscape*. Berg, Oxford.

Williamson, T. 2003. *Shaping medieval landscapes: settlement, society, environment*. Windgather Press, Bollington.

Young, R. & J. Webster. 2008. Love letters, love stories and landscape archaeology. In P. Rainbird (ed.), *Monuments in the landscape*, 228-38. Tempus, Stroud.

6.2 Look the other way – from a branch of archaeology to a root of landscape studies

Author

Graham Fairclough

formerly English Heritage, London, United Kingdom
Contact: graham.fairclough@newcastle.ac.uk

ABSTRACT

This paper explores Landscape Archaeology's location within the broader interdisciplinary field of landscape research beyond archaeology. A variety of factors are already creating a widening field of landscape research, including the increasingly integrative role of the concept of landscape as promoted, for example, by the European Landscape Convention, a growing questioning (within more general trends in society towards holistic thinking) of the traditional divides between disciplines, and the scale of social and environmental problems perceived to be confronting the world which require comprehensive views such as landscape offers. There are similar trends in heritage management concerned with the social role and value of heritage (the Faro Convention), the distinction (or lack of) between the present and the past, and the social relevance of archaeological work.

It will be argued that, by presenting the continuum of (pre)history (as represented by past material culture) as a part of present-day landscape not merely as a pointer to understanding past environments or landscapes, Landscape Archaeology could become an important part of broader landscape research in addition to being a sub-discipline of archaeology. Landscape archaeology can bring special and unique expertise to landscape studies, and can in its turn benefit from exposure to the different horizons, theories and aims of other landscape disciplines. Working more closely with other landscape disciplines and practices studies would also help landscape archaeology to develop greater social relevance through the unifying framework of landscape.

KEYWORDS

interdisciplinary studies, perception, landscape research, social relevance, European Landscape Convention

INTRODUCTION

The organisation of a first international conference devoted exclusively to 'landscape archaeology' implied a coming of age for a discipline which (although its many sub-types vary in age, with some countries having longer traditions than others) is nevertheless mainly relatively young. The conference could not represent every part of Landscape Archaeology's broad scope, and there have been substantial sessions devoted to Landscape Archaeology at other international conferences, notably EAA and WAC. In its special focus, aspiration and consciousness, however, LAC2010 represented an important milestone and it is to be hoped that it will become as established in future years as, for instance, PECSRL, Ruralia, EAA or CAA already are. A continuing forum for Europe- (or world-) wide meditation on the character, theories, methods and aims of Landscape Archaeology is much-needed, not least because a well-defined understanding of one's own disciplinary position is the essential prelude to interdisciplinary fusion.

The closing session of LAC2010 was entitled 'How will landscape archaeology develop in future?', thus consciously encapsulating an increased sense of identity, mission and purpose. It was to this part of the conference that the present paper was originally addressed. Its aim was neither to applaud nor to challenge the growing sense that Landscape Archaeology is cohering as a single multidisciplinary sphere of research and practical activity, but to ask how a maturing Landscape Archaeology should work in a wider context than archaeology, and particularly how it should locate itself in relation to the very much larger realm of interdisciplinary landscape studies that extends far beyond archaeology, and even beyond historical studies more generally. The underlying question is whether Landscape Archaeology exists to use the idea of landscape only to study the past (which will mainly interest historical disciplines) or also (or mainly) to use archaeology to study the landscape of the present-day, in both its materiality and mentality, and thus to connect to all landscape disciplines and to a wider public.

There seems little doubt that Landscape Archaeology will grow further as an important sub-section of archaeology, and the case can be made that all types of archaeology should make a greater use of the idea of landscape. This paper will look in the other direction, however, outwards from archaeology, to suggest that if the discipline is to gain further momentum, social relevance and practical influence over the shaping of future landscapes, then it must play a more central part in the larger interdisciplinary field of landscape research as a whole. As will be discussed later, and as any archaeologist who has spent time in multidisciplinary gatherings of landscape researchers will have seen, many landscape disciplines do not always seem to notice or value the methods, insights and results of 'archaeologically-based' landscape research as much as we might hope, nor do they see a relevant use for them. Perhaps some of them do not consider that landscape archaeologists really work with the landscape idea, or we think they study the 'wrong sort' of landscape, but for whatever reason the constant restatement to other disciplines and policymakers of the relevance of archaeological views to landscape is often without success. Landscape Archaeology needs additional ways of seeing if it is to look in both directions, both back towards its parent discipline, and forwards towards the larger academic community beyond that is starting to form around the landscape concept. At the risk of pushing the metaphor too far, 'coming of age' may mean leaving home.

LARGER COMMUNITIES AND THE UNIVERSALITY OF LANDSCAPE

Landscape research does not stop at the borders of archaeology. When we enter the landscape field we open our minds to other concepts. We find new ways of seeing and of interpretation and in a sense (and to an extent) we leave behind our interest in finding out about the past and look to the present as well. We also inevitably leave behind some of our scientific detachment and begin to engage more with perception and subjectivity. We are obliged to enter into different types of discussions about the future and about politics, governance and spatial planning or land management; conventional ideas that the aim of archaeological resource management is protection give way in the world of landscape to more nuanced ideas about managing change, preservation by development and future-oriented approaches (Bloemers 2002; Fairclough 2006a; Bloemers et al. 2010). In short, we find ourselves adapting to being part of a very large and differently-focused community of science and practice. We also find ourselves being drawn to engage in different ways with the public at large.

Landscape, although often casually characterised as being about and championing the local, is in fact global or universal. It is universal culturally, in that most nations have concepts of landscape, whether landscape/landschaft, paysage/paessagio, krajina, maisema or tájatas. This appears at the very birth of nations, evidenced for example in the way that 'landscape' is very visible in the great national epics (from, for example, the Mabinogion and the Táin Bó Cúailnge, through the Icelandic sagas and the Arthurian cycle, to the Canterbury Tales, the Lusiads and Don Quixote, and including both tales of Ulysses). It is universal socially, chronologically and spatially. It is universal too as a subject of research, indeed it is almost the archetypical interdisciplinary theme, a subject for the physical sciences as much as the humanities, for social sciences as much as ecology, and it spans domains and disciplines. The same cannot be said of many of Archaeology's other sub-disciples, because not all disciplines find it worthwhile to work in the past, or to study human action and agency, for example, or indeed to study material culture – the common ground is missing. Landscape however is quintessentially a common ground. All disciplines that deal with the physical world or with people and land will find landscape relevant because it integrates environment and society, history and culture. They will find in landscape a useful frame for their work and a valuable set of perspectives and tools.

The field of interdisciplinary landscape studies is in practice an emerging 'super-discipline' (although its creation may involve taking a post-disciplinary stance). Its growth is encouraged by interest in the European Landscape Convention at academic and policy level, not least in the forum that the Council of Europe's ELC conferences, workshops and publications (Council of Europe 2002-2010) has provided for interdisciplinary meeting and collaborations. It is also nurtured by the growing numbers of European-funded or coordinated networks and trans-frontier projects that have taken landscape as both subject and integrative framework (for example, Clark et al. 2003; Hernik 2008; Orejas et al. 2009; Pungetti et al. 2010). Many landscape disciplines are on the verge of a further coalescence under the aegis of strategies and policies currently forming within the European Science Foundation and the COST Programme.

The Science Policy Briefing published on the 10th anniversary of the European Landscape Convention – '*Landscape in a Changing World: Bridging Divides, Integrating Disciplines, Serving Society*' (ESF-COST 2010) – places landscape research as a fully integrative and interdisciplinary field of research into the European Research Area. It bases its case that landscape research can claim a central place in the European Research Area on the high relevance of interdisciplinary landscape research to major social, economic and cultural challenges facing the pan-European region. The key word there is 'interdisciplinary'. As LAC 2010 showed, Landscape Archaeology already crosses the boundaries of many disciplines (primarily in the historical sphere and its related geo-science fields), but landscape is a valid and existing field of research for almost disciplines in the arts and the humanities, the physical and natural sciences, and in social sciences. In such a large arena, Landscape Archaeology might appear to be a small player, but as will be seen below, it can make significant contributions, most notably by the study of long-term transformation, human agency through time and the human-nature interface that are at the heart of landscape.

It was suggested above that a number of different principles will need to be absorbed by landscape archaeologists if they are part of this wider more socially-embedded community. We cannot continue simply to learn more and more about 'the past at landscape scale'. The most important change of mentality is to take to heart the aspects of landscape summed up in the Council of Europe's definition (COE 2000). Landscape is an area 'as perceived by people'. Landscape is not just a natural, physical space, but it as much urban as rural, built as much as 'natural' – in short, always a cultural construct in which human perception is a prerequisite, and without which 'landscape' is merely the environment or the natural world. 'Land', 'environment', 'territory' or 'region' are not synonyms for landscape; they can contribute to the construction of landscape perceptions, but landscape is the bigger concept, because it encompasses perception as well as materiality, and because whilst spatial it is paradoxically free of spatial constraints. Not accepting some of these ideas alongside Landscape Archaeology's currently main focus on the material dimensions of landscape is an obstacle to interaction and synergy with other landscape disciplines.

Secondly, the words 'as perceived by' inevitably and inherently mean 'in the present day'. Landscape Archaeology however is traditionally occupied with understanding the past (including its focus on environmental history and time depth as simple chronological sequence rather than matrix). Such an approach can fail to engage with other landscape disciplines which firmly locate their landscape interest in the present or even (such as spatial planning and landscape architecture) the future. More importantly there can be a 'disconnect' with the needs and interests of policymakers and practitioners, those who change the present-day landscape and create tomorrow's landscape. For such people, past-centred interpretations of landscape may inform heritage management or provide raw material for tourism but they are unlikely to be as helpful for positive spatial planning or environmental policy.

CURRENT POSITIONS

If a 'coming of age' is a good time to reflect on what comes next, we should begin with mapping our current situation, placing Landscape Archaeology in its wider world. An over-simplified diagram might place the main landscape disciplines on a spectrum between 'environment' at one end (those disciplines which study the landscape primarily in physical terms) and 'perception' or 'representation' at the other (those which focus on cultural and human issues).

Everything on that long spectrum must be considered to be part of landscape research. Towards the first end of the spectrum stand many of the natural sciences and the physical sciences such as geology and geomorphology, the earth sciences, some branches of ecology and, within landscape archaeology, palaeo-environmental and related subjects. Towards the other end might stand disciplines such as the social sciences, psychology, cultural and human geography, disciplines concerned with landscape as symbolism, meaning and representation, or the designing and creating disciplines, for example much landscape architecture and artistic disciplines such as performance, because landscape is first and foremost a lived, experienced thing, hence the value of using phenomenological approaches in landscape archaeology.

In the middle ground we might place those disciplines that offer to bridge the divide most effectively, some types of landscape ecology for instance with its recognition that the current natural world is the product and past and continuing human management (or as a landscape archaeologist might say, recognition that biodiversity is a culture construct), and the core aspects of landscape archaeology, able to stand in both natural and human camps, to deal with time as well as space, to be primarily concerned with the change and creation that lies deep at the heart of the human-natural relationship that is landscape.

Landscape Archaeology occupies other middle grounds too: between empirical practice and reflexive theory, between humanities and sciences, and between research and design, knowledge and action. Landscape is a 'bridge' between many disciplines, but landscape archaeology is naturally interdisciplinary, already well-versed in the difficulties and challenges of multidisciplinary work because it draws theory and methods from many different parts of archaeology together, from archaeological science to its more humanistic branches. Through landscape, it also possesses strong connections to other historical fields, notably historical geography but also urban morphology and anthropology. The range of interdisciplinary connections that Landscape Archaeology possesses as a branch of archaeology, and as part of the 'tree' of historical studies, becomes much narrower when we look the other way to see Landscape Archaeology not as a large part of archaeology but as a small part of landscape studies as a whole. Yet whilst we might think that Landscape Archaeology is a very small branch of that larger tree, there is reason to think that it might come to be seen as an important root of a future integrated super-discipline.

OPENING A DOOR - OPPORTUNITIES FOR EXPANSION, GROWTH AND CHANGE

Now is a good time to contemplate a future for Landscape Archaeology within this wider field. Landscape Archaeology has its own increasing maturity and identity, as these proceedings of LAC 2010 show. It is developing serious theoretical discourses, it now uses a plethora of techniques and methods, with new IT-led horizons notably in remote sensing and spatial computing, and it can show a critical mass of past results and models, which among other things offers a platform for constructing scenarios. This is a strong basis for interdisciplinary collaboration and for connecting with policy.

There are also landscape-related trends beyond archaeology that pull in this direction, notably the integrative influence of the European Landscape Convention's (ELC's) national and pan-European implementation. The ELC has created a new level of international and interdisciplinary discourse on landscape. Integrative pressure comes also from European-funded research programmes and networks, as part of the 2000 Lisbon Strategy, the Ljubjlana Process, and within the Horizon2020 agenda for the European Research Area and the knowledge economy. Landscape studies need to make their case more strongly in this latter arena, and the ESF/COST Science Policy Briefing on landscape research referred to above will only carry weight and win success if enacted on a truly interdisciplinary basis in which the humanities has an equally weight to the natural sciences. It offers a platform not a panacea. The Joint Programming Initiative on cultural heritage, also adopted at the end of 2010, will be highly relevant, offering landscape as a wider forum linked to social values.

Strengthening those trends is a deeper questioning of the traditional divides between disciplines; it is going too far to suggest that centuries of fragmentation of science and 'philosophy' into many separate, bounded and indeed defended sciences and disciplines will be rapidly reversed, but it is being questioned more frequently and its drawbacks in terms of helping to address major social or environmental challenges are becoming increasingly clear. The maxim 'Governments have problems; Universities have departments' seems to be heard more and more often. For an inherently integrative field like landscape, the benefits of specialisation are relatively quickly outweighed by the disbenefits of fragmentation, dissonance and conflict. These divides, especially between humanities and science, are becoming serious obstacles to progress in both landscape research and landscape policy. Climate change is obviously one issue that cannot be addressed through disciplinary reductionism, but neither can many social and demographic issues.

Another relevant consideration is the issue of sustainability. Sustainable development in relation to landscape might be perceived differently, as a primarily cultural aspiration and need, and only secondarily as a natural, environmental or 'green' one. The common emphasis of 'green' rhetoric on people's carbon footprints, on personal consumer choices, on individual recycling, car sharing and so on serves only to underline how sustainable development issues revolve first and foremost around human decisions, individual and collective choice and above all lifestyle. The strong connection between lifestyle and landscape perception means that ecological and environmental anxieties conceal questions of consumption, action and behaviour, and therefore lead to a need for an integrated understanding of landscape as a human perception of the human-nature interaction.

Sustainability is often described as a 'tripod' with environmental, economic and social legs, but traditionally it is the environmental leg, with its focus on the fundamentals of water, air and soil and its ethical and sentimental issues of biodiversity, that is given priority in policy and public perception. Landscape allows us to think in a slightly different way. All the practical concerns of sustainability collide in the nexus of human agency, production and consumption, demographics and identity, individual and collective aspiration that creates landscape and is summarised by the words 'as perceived by people'. Landscape offers an arena to debate all these issues, a common ground between science and humanities, between place and nature, and between place and globalism, short-term and long-term change, past and future, private and public. Asking landscape to help to address all these may be unrealistic, but it surely has a part to play, and Landscape Archaeology should be able to become more socially and politically relevant.

But why choose 'landscape' as a unifying frame for addressing these big issues when there are other concepts with equal or greater power? Part of the answer is that landscape has appeal to almost everyone, academic and lay, rich and poor, town dweller or farmer, resident or visitor, 'native' or 'migrant'. It also offers other advantages for integration: landscape is an interdisciplinary meeting place because so many disciplines use the idea of landscape, it is an integrative tool because there are many theories, problems, goals that cross disciplinary borders and allow landscape to be a common analytical starting point, a discourse, a frame or filter, a common language. As a way of seeing and therefore a way of acting and performing, it combines materiality and mentality, and thereby connects place with people, science with humanities, past with present, real with virtual. It can provide a loaded rhetorical device; or it can be a blank screen onto which people can project many other social, ideological and environmental needs and ambitions. Finally, it is a shared territory, connecting readily to concepts of commons and collectivities, public realm in cities and access in the country; there is thus a politics of landscape.

Many disciplines use or study landscape and they inevitably find common ground in the concept. Take landscape architecture: the LE:NOTRE network (http://www.le-notre.org/) of landscape architecture schools examined over 20 'neighbouring disciplines' whose interests and expertise intersect with the practice of landscape architecture (Bell et al., 2011). These include fields such as agronomy (and agricultural history), cultural geography, environmental psychology, fine arts (and garden history), historical geography, landscape archaeology, landscape ecology, regional and spatial planning, urban design, urban (and other forms of) sociology, health and well-being. Landscape archaeology will possess an equally extensive set of neighbours.

VIEWS FROM OUTSIDE

But how do our disciplinary neighbours (and our audience or 'customers', those we wish to influence, for example politicians, research funders, planners, developers) regard Landscape Archaeology and its results? How relevant does the wider landscape community think Landscape Archaeology is to their work? It is not easy to know, because landscape archaeology, whether theory, methods or results, is mainly invisible in the published discourse of other disciplines (but then, it is noticeable that the bibliographies of any discipline's work on landscape tend not to look beyond its own literature). Recent overviews of the development of landscape studies rarely give much space to landscape archaeology. Wylie, on the evolution of the landscape branch of Geography (2008), for example, treats in detail only early historical geographical or topographical approaches or more recent phenomenological standpoints. Only 2 or 3 chapters out of 20 in *Landscape Interfaces* (Palang & Fry 2003) are from a historical or archaeological standpoint. Do other disciplines (and politicians and practitioners) think Landscape Archaeology is really about 'landscape'? Do they come to our conferences? Read our papers and books? Do our results find the right audiences? If not, why not? – does Landscape Archaeology use the wrong manner of presentation or do its results have too limited relevance to present and future landscape problems or too poor a resonance in popular mentality? Do we produce landscape stories, narratives, results to which ordinary people can connect? Is Landscape Archaeology too scientific, insufficiently attuned to stories and narratives? That landscape archaeology has much to contribute to interdisciplinary studies cannot be in doubt, as the Dutch BBO programme amply demonstrated (Bloemers et al. 2010), but the message seems not to be heard often enough.

A gentle critique of Landscape Archaeology that might be put into the mouths of other landscape researchers could include the complaint that Landscape Archaeology is not really about 'landscape' at all but is just a form of environmental history, concerned only with the past (what happened in the Iron Age) rather than with the past in the present (how the 'Iron Age' still happens in today's landscape; for an example of archaeologists doing this see Nord 2009). Or, that it is too little concerned with people and their experience of landscape, whether in the past or the present. Or, that it does not sufficiently balance perception with materiality, that archaeologists treat landscape as a thing, not an idea. Some landscape archaeologists even speak as if landscape has its own history, independent of society, let alone of individual people. There can also be criticism that Landscape Archaeology seeks to tell the story of land not of landscape, where the environment is the star of the film and humans are just passive extras.

Furthermore, there are landscape disciplines whose practitioners would accuse Landscape Archaeology of misusing the idea of landscape by treating it purely as a matter of scale or size: some landscape archaeology appears to be little more than conventional archaeology carried out over a large area. Archaeologists sometimes reach for the landscape-word ('Roman industrial landscape') whenever more than three or four sites are excavated in close proximity. Even the concept of 'landscape scale' often appealed to as a method by landscape archaeologists (and by some ecologists) can be seen as a basic misunderstanding of the idea of landscape. Scale is an important issue in landscape studies, but landscape is not itself a scale issue; landscape ways of seeing can be used at many scales. The scale problem lies behind the tendency of landscape archaeology (and landscape history) to work at regional level, to write regional history instead of landscape biography, to study cultural zones or settlement and label them as 'landscapes', as if landscape was only territory. Territories (whether townships, manors or political territories, transhumance zones, or modern territories of landscape character areas or pays, national or regional 'natural parks', or even tourism-branded regions) are important building blocks and tools of landscape understanding, but they are not the whole of landscape.

There is a lot of merit in the argument that understanding the past is a valid, fruitful and necessary goal in its own right and that curiosity-driven research leads to practical use and social benefit or relevance, but sometimes those arguments need to be tied more closely to questions of audience, strategy and applicability. There is perhaps another way forward.

RECONFIGURING LANDSCAPE ARCHAEOLOGY

The potential contribution of Landscape Archaeology to wider landscape studies is not to be underestimated. There are many aspects of interdisciplinary landscape research and policy to which Landscape Archaeology is very well-placed to contribute, which can be crudely (over)simplified as time, people and change. Foremost is the place of time in landscape (time depth, long-term processes, biography, narrative), a strong understanding and evidence of landscape's dynamism (change, continuities and discontinuities, adaptations), and the ability to offer a long-term and complex, multifactor explanation, merging human agency and environmental factors. In addition, its distinctive methods enable Landscape Archaeology to bring different ways of seeing to the interdisciplinary field: a vertical gaze on landscape, for example, from its use of remote sensing, sense of stratigraphy and map-based GIS work, producing a layered landscape that challenges the visual ideology of much landscape description.

Landscape archaeology's strong tradition of slow 'hands-on' fieldwork and visual analysis can provide simultaneously (and rarely) both internal (from a more subjective standpoint) embodied and external (objectified) analytical views of landscape. Finally, but not least, archaeology is a humanistic discipline as much as it is a scientific discipline: it can study the land, and environmental change through time, but ultimately its subject is human beings, their agency and their ways of thinking. Whilst perhaps the most difficult understanding to reach, it is cognition, which of course lies at the core of all landscape, that archaeology ultimately strives to capture.

At present, however, these contributions are mainly only potentialities. As suggested in the previous section, the ways in which Landscape Archaeology is currently framed or how it interprets and presents its findings, seems to restrict their relevance to other parts of landscape studies. Some of the characteristics of Landscape Archaeology stand in the way of interdisciplinary progress through a reluctance to appreciate how other disciplines theorise landscape. It is however possible to imagine that many characteristics that might seem to be 'flaws' to an observer from another landscape discipline – that is, that 'landscape' archaeology is too fixated on the past, is too environmentally-focused, takes too little account of the immaterial and of the place and role of people, their actions and perceptions, that it confuses 'land' with landscape and thus fails to see the meaning of 'scape', and that it has a naive view of scale – can be converted to strengths when reframed.

The traditional archaeologist's concern with the past not the present, for instance, can be redirected within a landscape frame to provide the understanding and appreciation of time depth and long-term explanation that many landscape disciplines seem to lack and that perhaps Landscape Archaeology can best provide. To do this however requires our past-oriented narratives to be accompanied by new present-focused and future-oriented narratives so that our understanding of the past landscape and of the historic dimension of the present landscape is linked to the contemporary landscape, not stranded in anachronistic 'periods' (Fairclough 2007). The very concept of 'periods', anachronistic in itself, is in terms of landscape an obstruction to the appreciation of the continuous flow of time, change and continuity. Landscape research should begin and not end with the present-day landscape. It can focus on character and biography, providing more synthesis and interpretation and less description. It can replace period-based studies with descriptions and analyses of sequence, defining time depth not as a sequence of layers but as a continuous flow; landscape is a way of seeing all surviving remains of the past, all periods of time, folded together into the present day. This is probably closer to how the lay public see the past, as being present, not passed.

The weakness of treating landscape simply as a scale issue is easily remedied by adopting more sophisticated theories of multiple scales; Landscape Archaeology's expertise in working at different scales will translate easily to interdisciplinary work (e.g. Fairclough 2006b). The regional scale for example can become a useful frame for landscape studies that seeks to identify long-term and large-scale transformation and continuity; it can provide laboratories for trans-frontier contrast and comparisons; it can also provide a way to capture public perceptions of place and landscape and add them to scientific insights.

The tendency of Landscape Archaeology to descend into being merely an environmental history (a story of land not landscape) can be counteracted by collaboration with those disciplines that have expertise in taking human perception into account, including other parts of archaeology and history that focus on human actions. Landscape Archaeology can thus provide the chronology and the stage for one half of the ELC's human/nature interaction. The criticism that landscape archaeologists do not sufficiently balance perception with materiality and that landscape is studied as a thing not an idea might not be turned into a positive contribution so easily, but it represents the reciprocal benefits of interdisciplinary research. This is one of the areas where archaeology can learn from other disciplines.

Changing Landscape Archaeology so that it fits its existing and well-developed practice and results into conceptual frames more accessible to other disciplines, will help to develop further the multidimensional Landscape Archaeology that already largely exists. Three broad types of landscape archaeology can be defined; they could all be used simultaneously in a single project but rarely are, it seems

- A first type of archaeological practice (the oldest) that uses landscape as a tool (notably merely a scale) for understanding the past, through (e.g.) long-term narratives and the ability to work at supra-site scales;
- A second type that studies the mentality of landscape in the past (lived, embodied, perceived landscape; mind not matter – how past people perceived landscape);
- A third type which seeks to identify, explain and describe the present-day historic character of landscape, how the past makes today's perceived landscapes.

In the first type it is the results, when properly framed and presented, that will provide the link to other disciplines; Landscape Archaeology can offer long-term explanatory models, and cross-time and cross-cultural comparisons.

The second type however offers more – not simply description and explanation but a conceptual common ground. It provides the chance to compare how and why people perceived landscape at different times, and the opportunity to explore in many cultural contexts the mechanisms of non-material landscape construction. From it we might work with social scientists or psychologists to form a better understanding of landscape social construction that would have significant contemporary relevance, not least in the sphere of landscape and social resilience to change.

The third of these types, however, has most to offer to an interdisciplinary landscape research because it fits relatively straightforwardly into the ELC-type view of landscape. By characterising historically the landscape that exists today it has the strongest potential for social relevance, and the greatest potential for integration with other disciplines' present-day appreciation and understanding of landscape. It gains interdisciplinary appeal simply through topical and contemporary relevance; this is archaeology studying where people live now (see, for example, in their different ways, Penrose et al. 2007; Blur & Santori-Frizell 2009). It also offers processes that can be shared with other disciplines to provide the main integrative force towards other landscape disciplines. What is required to inspire interdisciplinary work are common agendas, which are readily provided by practical issues such as climate change response or changes to lifestyle and society caused by new levels and types of mobility and lifestyle, or the issue of collective multiple identities, so often linked to land, territory and landscape.

CLOSING WORDS

This paper has tried to suggest that whilst applauding the recent growth and maturity of Landscape Archaeology we need also to be aware that it currently stands only at the edge, or too often outside, the larger field of interdisciplinary landscape research. Landscape Archaeology is qualified to contribute to and benefit from that larger community of practice. To do so, however, it needs to look away from, as well towards, archaeology and the past, and to develop a keener awareness of landscape debates and issues in other disciplines. New ways of thinking about and seeing landscape will be required, as will a willingness to redirect some of our aims and methods to different goals and achievements. Landscape Archaeology's theorisation of landscape remains relatively basic, and too single-mindedly focused on materiality, environmental change and the distant past. Ironically for a discipline which claims to study people and their work, when it comes to landscape, archaeologists often seriously understate the role of human beings in making and perceiving landscape. We fail even to understand time except as a sequence of layers (and perhaps 'landscape' is telling us that archaeology as a whole is in the process of outgrowing its origin in geology and stratigraphy). We do not make enough connections between past and present and too often fail to help others understand the present-day world in a historically-informed landscape frame.

The growing trend for interdisciplinary collaboration, in landscape as other topics, is evidenced by initiatives such as the EC's joint programming initiatives designed to co-ordinate both disciplines and national funding streams; one of them concerns the cultural heritage and provides another connection. All of the ESF principal domain committees have emphasised in recent Position Papers the need to cross disciplinary boundaries and to reach out to other domains. This was the context for ESF/COST to commission a transnational and cross-discipline group of landscape specialists to prepare the Landscape Science Policy Briefing (ESF/COST 2010). The project was one of the first examples of such an attempt (although see now as RESCUE 2012, which delivers the same messages) 'Landscape in a Changing World', was published to coincide with the 10th anniversary of the Council of Europe's Landscape Convention – the ELC (the Florence Convention, COE 2000) as a strong statement of the need for more concerted efforts to be made to promote landscape research in concert with policy and action designed to address current social and environmental challenges.

Coinciding publication of the SPB with that anniversary is significant because the European Landscape Convention has been instrumental in forging the context for increased interdisciplinarity (and trans-disciplinarity) in landscape studies. It promotes landscape as a public good and a tool to be used in all sectors of policy and practice. It can be twinned with the Council's Faro Convention (2005) on the Value of Cultural Heritage for Society, which broadens definitions of cultural heritage to include intangible as well as the tangible, the perceptual as well as the physical, and action and performance, custom, behaviour, identity, and thus can provide another framework for this recalibration of landscape archaeology. Both conventions argue that landscape and heritage (which are inextricably entwined, each necessarily includes the other) are central not peripheral to 'real life', and through identity, community, place and belonging are cornerstones of society's very construction (Council of Europe 2009). Both Conventions therefore provide a framework and support for leading Landscape Archaeology into integrated landscape research and developing greater social relevance for it.

Some of the contributions that Landscape Archaeology should be able to make to integrated landscape research have been suggested above. The other side of the coin is the broader and more plural view of landscape and objectives and purpose that Landscape Archaeology would gain. A greater practical relevance for Landscape Archaeology would be developed in terms of contributions to the decision being taken about the shape of future landscapes, particularly in the face of global drivers for change and environmental change. Finally, a focus on integrated landscape studies, notably the connection between past and present and people and place, would give Landscape Archaeology a new level of social relevance.

All that is needed is for landscape archaeologists to recognise that landscape is a widely shared concept, an idea more than it is a thing, a tool more than a subject of study, and to cross a few boundaries toward other disciplines. Landscape Archaeology may be a branch of archaeology but it should be a root of landscape studies.

ACKNOWLEDGMENTS

This paper benefitted from reviews by Henk Baas and Matthew Johnson.

REFERENCES

Bloemers, J.H.F. 2002. Past- and Future-oriented archaeology: protecting and developing the archaeological-historical landscape of the Netherlands. In Fairclough, G.J. & Rippon, S.J. (eds.), *Europe's Cultural Landscape: archaeologists and the management of change*, EAC Occasional Paper no 2, 89-96. Europae Archaeologiae Consilium and English Heritage, Brussels/London.

Bloemers, J.H.F, Kars, H. & van der Valk, A. (eds.) 2010. *The cultural landscape and heritage paradox: Protection and development of the Dutch archaeological-historical landscape and its European dimension*. Amsterdam University Press, Amsterdam.

Blur, H. & Santori-Frizell, B. (eds.) 2009. *Via Tiburtina: Space, Movements and Artefacts*. Swedish Institute in Rome, Roma.

Clark, J., Darlington, J. & Fairclough, G.J. (eds.) 2003. *Pathways to Europe's Landscape*. EPCL/EU: Heide.

Council of Europe. 2000. *European Landscape Convention*, European Treaty Series 176: http://conventions.coe.int/Treaty/en/Treaties/Html/176.htm accessed 5 November 2010.

Council of Europe. 2002 – 2010. *Proceedings of the First to the Ninth ELC Workshops*, http://www.coe.int/t/dg4/culture-heritage/heritage/Landscape/Publications_en.asp (accessed 5 November 2010).

Council of Europe. 2009. *Heritage and Beyond*, Strasbourg: Council of Europe Publishing http://www.coe.int/t/dg4/cultureheritage/heritage/identities/beyond_en.asp) (accessed 19 December 2009)

ESF/COST. 2010. *Landscape in a Changing World - Bridging Divides, Integrating Disciplines, Serving Society*, Science Policy Briefing 41, ESF: Strasbourg / Brussels. http://www.esf.org/publications/science-policy-briefings.html; http://www.cost.esf.org/library/newsroom/spb (accessed 3 November 2010)

Fairclough, G.J. 2006a. A new landscape for Cultural Heritage Management: characterisation as a management tool. In Lozny, L. (ed.), *Landscapes Under Pressure: Theory and Practice of Cultural Heritage Research and Preservation*, 55-74. Springer, New York.

Fairclough, G.J. 2006b. Large Scale, long duration and broad perceptions: Scale issues in Historic Landscape Characterisation. In Lock, G. & Molyneaux, B. (eds.), *Confronting Scale in Archaeology: Issues of Theory and Practice*, 203-215. Springer, London.

Fairclough, G.J. 2007. The contemporary and future landscape: change & creation in the later 20th century. In McAtackney, L., Palus, M. & Piccini, A. (eds.), *Contemporary and Historical Archaeology in Theory, Paper for the 2003 and 2004 CHAT conferences* (Studies in Contemporary and Historical Archaeology 4, BAR International Series 1677), 83-88. Archaeopress, Oxford.

Hernik, J. (ed.) 2009. *Cultural Landscape – Across Disciplines*, Interreg IIIB/Cadses. Bydgoszcz, Krakow.

Bell, S., Sarlov-Herlin, I and Stiles, R. (eds) 2011: *Exploring the Boundaries of Landscape Architecture*, Routledge, Abingdon, 83 -114

Nord, J.M. 2009. Changing Landscapes and Persistent *Places*. Acta Archaeological Lundensia, Series in Prima Quarto 29. Lund University, Lund.

Orejas, A., Mattingley, D. & Clavel-Lévêque, M. 2009. *COST A27 - From Present to Past through Landscape*. CSIC, Madrid

Palang, H. & Fry, G. (eds.) 2003. *Landscape Interfaces: Cultural Heritage in Changing Landscapes*, Landscape Series 1. Kluwer Academic Publishers, Dordrecht.

Pungetti, G., Kruse, A. & Rackham, O. (eds.) 2010. *European Culture Expressed in Agricultural Landscapes, The Eucaland Project*. Palombi Editori, Roma.

Penrose, S. 2007. *Images of Change – an archaeology of England's contemporary landscape*. English Heritage, London.

RESCUE 2012: *Responses to Environmental and Societal Challenges for our Unstable Earth (RESCUE), ESF Forward Look* – ESF-COST 'Frontier of Science' joint initiative. European Science Foundation and European Cooperation in Science and Technology, Strasbourg and Brussels. http://www.esf.org/publications.html (accessed 25 February 2012).

Wylie, J. 2008. *Landscape* (Key Ideas in Geography). Routledge, Abingdon.

6.3 The past informs the future; landscape archaeology and historic landscape characterisation in the UK

Author

Peter Herring

English Heritage, London, United Kingdom
Contact: peter.herring@english-heritage.org.uk

ABSTRACT

In helping society both understand its past and design a sustainable future, the landscape archaeologist's role is widely inclusive. If all society matters, then all landscape matters and all stories have relevance. All disciplines and all actors are drawn in.

Focusing on how landscape archaeology is practised in the UK, this paper explores how we currently tease out and present those myriad stories and how we might adjust our methods to extend our range. Inquisitive, theoretical, empirical and phenomenological approaches all contribute much while their methods are rigorous and transparent and their outputs are clear.

In addition, characterisation generalises from and so extends the benefit of more particular landscape archaeology while creating a scheme of provisional interpretation that serves as a framework for further research. Historic landscape characterisation developed alongside the drafting of the European Landscape Convention and shares many of its principles and aims. It encourages inquisitiveness, debate and argument over stories and their interpretations, and over strategies and plans.

KEYWORDS

characterisation, future, inquisitiveness, inclusivity, framework, action

Ceaselessly the river flows, and yet the water is never the same, while in the still pools the shifting foam gathers and is gone, never staying for a moment... (Chōmei 1212, 1)

THE RESPONSIBILITIES OF THE LANDSCAPE ARCHAEOLOGIST

If archaeology involves study of people (present and past) through their material and if landscape is perception of an area that has been affected by both natural and human actions (Council of Europe 2000), then landscape archaeology should necessarily be widely inclusive in terms of subject, method and discipline. Archaeologists help society, communities, and individuals appreciate that landscape is more than environment: that it is doubly cultural – a cultural product culturally perceived. They also continually confirm that landscape is fundamentally a product of change, and so help prepare society and its members to cope with and take control of further change.

Inheritance, continuity and legibility of earlier forms and features, as in the 13th century poet Chōmei's ceaselessly flowing but constant river, are widely valued landscape qualities. The newly emergent Mole's delight in *The Wind in the Willows* at discovering the 'sleek, sinuous, full-bodied animal' that was his local river will be shared by those enchanted by the implications that while landscape constantly changes it yet retains as timedepth the increasingly fragmentary chronicles of previous change. On the riverside Mole 'trotted as one trots, when very small, by the side of a man who holds one spellbound by exciting stories: and when tired at last, he sat on the bank, while the river still chattered on to him, a babbling procession of the best stories in the world, sent from the heart of the earth...' (Grahame 1908, 2). How then can undertaking landscape archaeology not be one of the most enjoyable ways of spending time? We may listen to those stories, but we also revel in writing them, unpicking time's tangles by using a satisfying mix of disciplined method and imaginative theorising, dwelling in the landscape and with past communities as we do so (Ingold 1993).

But who does landscape archaeology, for what purposes, and for whom? And if consideration establishes a narrow range in any of those three, what does that mean for the futures of landscape archaeology, landscape and society, especially if society is at least partly defined by its relationship with place and the past? Responsibilities come with the pleasures and privileges of the landscape archaeologist.

All but a small handful of the Amsterdam papers either used or championed approaches that were at least multidisciplinary (several disciplines working constructively together, but each employing its established approach) and many were interdisciplinary (novel approaches in which established disciplines fertilise each other by crossing their boundaries). If landscape archaeology were trans-disciplinary (recognising how landscape study draws on a wide range of disciplines and becomes a means to interconnect each), we might want to greatly extend the range of disciplines we draw in to inform us, and to reach out to and inform (see Fairclough, this volume). Most of the papers were also reporting on fairly straightforwardly positivist and past-oriented research that contributes greatly to deepening understanding of the development of human society and its dynamic relationship with place (Rippon 2004, 4). But only a few used their landscape archaeology as a means of challenging the established assumptions, ideologies, structures and ranges of academic archaeology and history.

A few papers pictured the landscape archaeologist positioning themselves Janus-like on the threshold that is our present-day inherited world, looking both backward and forward. Landscape archaeology can be overtly future-oriented, feeding understanding gained from the past into plans, policies and strategies for future change (Herring 1998; Fairclough et al. 1999; Bloemers 2002) and into strengthening the identities of those individuals and communities and their commitments to themselves, to place and to society. Such an approach encourages landscape archaeology to be multi-targeted and self-critical, with the needs or demands of future management and social health helping to drive the extension and deepening of understanding of the past. Landscape biography, as developed in the Netherlands, provides a good example (Roymans et al. 2009).

As elsewhere, the historic environment sector in the UK has been adapting for some time its approaches, developing proactive heritage resource management that complements designation and protection (e.g. English Heritage 1997). As well as encouraging and informing more positive land management we now also set out to guide change, preferably as early in its design as possible. We aim to retain the fabric and legibility of many more aspects of the past, recognising the range and fluidity of systems of valuing and so including the locally and personally significant, better enabling our successors to build identity from the past and from place.

In themselves, of course, the past has no narrative and place has no biography; it is people, landscape archaeologists among them, with their own motivations, assumptions and prejudices, who record, analyse and interpret in order to compose, narrate and edit stories of the particular. They may also generalise and characterise from them, stimulating historical argument. Evidence, interpretation and understanding all being partial, one of the landscape archaeologist's key aims, exemplified in Amsterdam, is to improve the art of guessing when working with incomplete, equivocal material that might be misread, misinterpreted or misunderstood, or, more interestingly, differently understood (Arnold 2000, 12). We benefit from varying approaches, drawing ideas, models, material and stimulus from those who lay emphasis on working from the field – surveying, recording, abstracting, interpreting, modelling, testing – and those who privilege theory, but who also survey, record, abstract, interpret and test. In doing so we reconcile the threads of landscape archaeology recently represented by boots and brains (Johnson 2007a; Fleming 2007a; Johnson 2007b).

The development of Historic Landscape Characterisation, initially in England, is one of a number of responses to the recognition that landscape is everywhere, that all is historic, and that all is valued. The European Landscape Convention definition (summarised earlier) captures its ubiquity and its interest well and insists that we draw in everything, whether natural, rural, urban and peri-urban, whether dry land or water, and whether everyday or outstanding (Council of Europe 2000; Fairclough 2003). So historic characterisation, expanded on below, includes the sea (Hooley 2007), the present-day (Penrose 2007), the built and the urban (Lake & Edwards 2006; Thomas 2006), and, since biodiversity is fundamentally cultural in location and form, also the 'natural' (Herring 1998).

The European Landscape Convention also insists that all perceptions matter, those of all communities and individuals, with all their varying approaches to and relationships with place. And so all landscape matters. Change may be assessed by some people in some circumstances as positive and enhancing or as disturbing and damaging, but others can see it differently; and landscape itself is fundamentally transitory or dynamic, always evolving, developing, decaying and metamorphosing. Change is a characteristic of landscape as much as an impact upon it. Landscape has no original, pristine, authentic or traditional form, or conception. Knowledge and reflection change personal and community perception of it. Recognition that landscape never has been and never will be finished or complete should empower and inspire, being a reason to value, care and act, and so use the past to better inform future change (Fairclough 2006, 194-196).

If all this means we are buffeted by conflicting perceptions of landscape itself, then our experience of landscape archaeology may be similar. It too constantly changes and like landscape also differs from place to place, as LAC2010 in Amsterdam again demonstrated, interestingly reflecting varied established academic and heritage management approaches to place and past, some peculiarly national or regional.

BRITISH LANDSCAPE ARCHAEOLOGY

Landscape archaeology, as practiced by UK archaeologists, both at home and abroad and throughout the 20th century, but especially in its last quarter, has developed widely accepted systematic methods. As mentioned, interpretation is usually the result of a dialogue between theoretical model building, responding to a range of local, regional or national narratives, and empirical observation, fieldwork, investigation and recording, usually involving application of a range of other disciplines. Testing hypotheses involves critical use of the wide range of sources and methods displayed in Amsterdam.

UK landscape archaeology has consequently developed sets of increasingly sophisticated procedures, devices and languages (visual and verbal). Since John Aubrey's Avebury survey, analytical or critical fieldwork has held a special place in British archaeology. Field surveyors happily spend months or years producing plans, maps and elevations that are simplified and interpreted representations, usually in two dimensions, of complex three-dimensional worlds. In the field or on the computer screen they can employ a range of symbols, codes and lines, including hachures, to show earthworks, the humps and bumps that we have learnt to record, analyse and interpret. What they choose to record, and which symbols they choose to employ and at what scale, resolution and tolerance depends on a range of decision-making systems and processes.

The inquisitive method of modern analytical fieldwork has been set out and summarised by three of its subtlest performers, Mark Bowden (1999), Paul Everson and Graham Brown. It involves, 'questioning relationships, conformity and non-congruity by careful observation of detail on the site, and recording accordingly; perhaps most distinctively, anticipating change rather than stability; looking to explain what the earthworks indicate – often through dynamism in the landscape up to the present day – rather than comfortable pigeon-holing into simple archaeological categories such as "deserted medieval village"' (Everson & Brown 2010, 62).

Andrew Fleming's work on the Dartmoor reaves (extensive Middle Bronze Age boundary patterns subdividing thousands of hectares of the south-west English granite upland) provides an example of what thoughtful landscape archaeology can achieve. A dialogue developed between inquisitive observation of local detail, including horizontal stratigraphy, morphological variation and responses to topography, and the realisation from that of broader patterns whose interpretation was informed by still wider Bronze Age studies. Local and regional narratives were constructed that fed into a developing theory, itself tested by further observation and excavation. The complex explanatory model created on Dartmoor – of community-level organisation of households and neighbourhood groups – has been applied as a framework for study of other areas and has influenced understanding of MBA Britain. Its explanation of the development and decline of a mid second-millennium BC hierarchical rural society (Fleming 1988) has properly been subjected to detailed scrutiny, both in the field (e.g. Bruck et al. 2003) and in the way it has fed into other models (e.g. Johnston 2001; Herring 2008). It has consequently been revisited and revised (Fleming 2007b).

W.G. Hoskins, a great pioneer of British landscape history, but rather in thrall to and constrained by what documents offered, may have expected archaeologists to have maintained a supporting role in relation to the historian, and to have confined themselves to their traditional method of excavation (Palmer 2007, 2; Everson & Brown 2010). Instead they took him at his word and agreed that the landscape is, 'the richest historical record that we possess' (Hoskins 1955, 14). Hoskins might have been disappointed that increasing numbers of those who study the materiality of landscape place their consequent narratives outside his rather romantic overarching scheme. Not all accept his view of Britain as a deeply rural landscape whose society and economy reached its pinnacle in the doughty yeomen who succeeded the medieval peasant communities, a world remorselessly diminished by subsequent modernism and industrialism (Bender 1992; Johnson 2007a). Many consider the landscape and landscape archaeology of industry and modernity to be as relevant and as exciting as that of prehistoric or medieval life (Palmer 2007; Penrose 2007; Hicks 2008).

The enriched understanding developed from interdisciplinary landscape archaeology also enables schemes of historical change scenarios, agencies and motivations to be developed that may not be immediately apparent from the narrower sources and methods of traditional studies of history and prehistory, and that may challenge established ways of thinking. For example, detailed survey and analysis allows reconstruction of ways of exploiting, moving around and perceiving parts of medieval Cornwall, whether a lord's castle in a deer park cut out of farmland formerly shared by peasants, or a hamlet and its subdivided fields set within extensive common grazing (Herring 2003; 2006). We already routinely reconstruct, however cautiously, past ways of being and thinking when we interpret certain landscape components as representational of meanings about economy and society. We might attach to landscape features and patterns oppositional concepts such as 'communal and individual', 'owned and held', 'mine and hers', 'power and impotence' to help order our thoughts. If interested in personal motivation in stimulating and delivering change, an agency that the formal and bureaucratic sources of much medieval history tends to underemphasise, then we can consider where individuals (peasant, lord, servant, alien, priest, child, etc.) may have placed themselves on a scale between such oppositions.

It is not such a great step to then consider how such features and patterns would have reflected and constrained how such individuals and communities experienced place and built up senses of belonging and identification. Oppositions now might be more universal, such as 'included and excluded', 'understood and unknown', 'inherited and new-made', 'safe and frightening'. Behaviour patterns influenced by such concerns would make their mark on the landscape both physically (through lines of movement and areas of avoidance, etc.) and perceptually (through development of meanings, stories and namings, etc.) (Altenberg 2003; Franklin 2006).

Recalling that landscape is a matter of perception, the next step to adopting a more personal and subjective approach to landscape archaeology is smaller still. Post-processualist perspectives like phenomenology encourage archaeologists to use their understanding of a place to imagine more or less tentatively what it might have felt like to dwell there, to work there and to move and look around. How past people might have been influenced by a 'sense' of place can provide useful insights to set alongside other approaches to the narrative, history or biography of those people and that place, provided that practitioners are systematically critical and transparent in their treatment of such evidence and clear and careful in their presentation of assumptions and proofs (Tilley 1994; Fleming 2005).

ISSUES SHARED BY ALL LANDSCAPE ARCHAEOLOGISTS

A landscape archaeologist might therefore be characterised as restlessly querying, aware that partial and equivocal evidence, limitations of knowledge and constraints on resources means that most pronouncements are open to constructive debate. Most set out the principles they use in observing (or overlooking), including (or excluding), emphasising (and downplaying), but the user of their record is still obliged to take much on trust, however aware all parties are that the primary data-collecting stage of landscape archaeology, like all archaeology, is as dependent on systems of prioritisation and valuing as it is on the survival of remains and the ability to recognise and interpret their meaning.

Further scope for uneven treatment and thus contention or contestation lies in the process of analysis of the landscape archaeologist's records, whether by the person or team who prepared them in the field (and thus enjoy the privilege and memory of on-site experience, observation, recording and interpretation) or those secondary users able to read and understand the meaning conveyed by their codes, symbols and language. As noted, the language of recording and reporting can be technical, abstract and highly particular, and of course may alienate as well as connect. It represents to some 'a rhetoric of authority' and an uncertainty that method and interpretations are as neutral and value-free as they might seem. Of course fieldwork also, 'has its own social, political and emotional context... a network of powers both facilitating and inhibiting the manner in which the past is understood' (Bender et al. 2007, 27-8). These factors often make it difficult to communicate the landscape archaeology discipline's results beyond its borders.

Landscape archaeology, being resource hungry, is also necessarily selective. Relatively small blocks of land are covered, often selected with particular themes or periods in mind, despite many practitioners' aims of applying the 'total landscape archaeology' that treats the marks of all periods as equally important (Bowden 1999, 18), and usually in those parts of the landscape that have well-preserved remains visible at surface, however unrepresentative they may be.

Landscape archaeology's impact on local, regional or national narratives has been disappointingly insubstantial compared with landscape history, and compared with what society may expect from it. The reason may lie in a tendency for much (but not all) landscape archaeology to emphasise and celebrate the particular. Practitioners appreciate how each place's development depends on a uniquely peculiar mix drawn from the wide range of variables that influence change (or continuity) and thus form: geology, topography, society, economy, memory, temperament, chance, etc. The narrative or biography, teased out by critical scrutiny, can be as peculiar as the place, and discipline and imagination are required to identify points of commonality with time and space beyond it, to make the study of wider relevance.

This problem was recognised nearly a quarter of a century ago by David Austin, who wished the best of his fellow landscape archaeologists when noting that, 'Whether consciously or unconsciously, we have opted to study communities and blocks of landscape with the ultimate intention of comparing them so that we may draw conclusions about general trends... Our overall objective, therefore, seems to be a series of analytical but narrative accounts of a range of different landscapes within different cultural contexts...' (Austin 1989, 245).

Austin's call for greater clarity in landscape archaeology's wider purpose and for more overt use of generalising modelling to make the particular more widely relevant was buttressed by recognition that, 'The more we have probed into the methodology and objectives of landscape archaeology.... the greater the range of potential information there is and the more unwieldy it has become' (ibid., 244).

THE DEVELOPMENT OF HISTORIC LANDSCAPE CHARACTERISATION

To help cope with such comprehensive and intricate views of the historic landscape we can apply to it a classification of material culture into typologies on the basis of distinguishing attributes, an approach with a long archaeological tradition. In historic landscape characterisation attributes are selected to ensure that the classification meets the needs of a wide range of applications, both past and future oriented, which would include furthering landscape archaeology research (see Aldred & Fairclough 2003 for consolidated method).

Indeed, historic landscape characterisation was developed directly from the UK tradition of landscape archaeology; the earliest large-scale historic landscape characterisations – Bodmin Moor (1993) and Cornwall (1994) – were devised by a team of landscape archaeologists employed by the local authority's archaeological field unit who had long experience of undertaking detailed field survey and archaeological interpretation and assessment for the purposes of informing land management and planning (Herring & Johnson 1997; Herring 1998, 2007, 2009). It had been appreciated that many of the historic landscape patterns surveyed were repeated across Cornwall.

- Small-scale late prehistoric brick-like fields of Zennor and Morvah were also found in other West Cornish parishes (Johnson 1985).
- Enclosed formerly open medieval strip fields recorded as archaeological remains on Bodmin Moor were also visible, fossilised in later field patterns, in much of lowland Cornwall (Herring 2006).

Figure 1. A Cornish corner whose beauty and interest depends on the varying legibility of thousands of intersecting stories. Evening geese return to slate and corrugated–iron-roofed farm buildings perched above a valley busy a century ago with active china-clay works, their pits, heaps and buildings now overwhelmed by secondary woodland. The hilltop village Trewoon was originally a pre-Norman farming hamlet (Cornish *tre*) established on an open Downland (*goon* = 'woon'). Beyond are the action- and meaning-packed coast and sea. Recording and interpreting this complex world is the landscape archaeologist's challenge; the sustainable management of change within it is their goal.

- Rectilinear early modern fields with ruler-straight boundaries recorded on Bodmin Moor and in Zennor were present on all of Cornwall's downlands (Herring et al. 2008).

Separately, English Heritage and the Countryside Commission were concerned to make available a more comprehensive representation of the historic landscape to place alongside and be capable of being integrated with other depictions and interpretations of landscape and place (Fairclough et al. 1999).

These two strands, the national and the local, the heritage-management-focused and the partnership-oriented, came together in the Cornish historic landscape characterisation. This, like all its successors, uses only systematic sources, largely modern and historic maps and aerial photographs, supplemented by other sources like place names and habitat types, as it characterises the attributes of the present landscape through a historical interpretation of morphology using the understanding of social and economic meanings developed by landscape archaeology (Rippon 2004, 4). Attributes might include the following.

- Patterns, such as those of different field, boundary or drainage systems, or of streets in a town.
- Morphologies, such as the straightness, sinuosity or curvilinearity of field boundaries, often diagnostic of period and particular economic and social conditions (communal, individual, intake, etc.), or the external boundaries of woodlands, rough ground, etc., or the plots within a settlement.
- Forms, to draw in distinctive structures, earthworks, buildings etc. To distinguish industrial, commercial, military, settlement, recreation, ornamental types, etc.
- Scale, such as typical size of enclosures in a field system.
- Ground cover, such as forms of rough ground, woodland, horticulture, etc.
- Whether active or abandoned (of industry, military, recreation etc). Condition and survival may also be recorded.
- Broad and narrower interpretations of use and period, based on analysis of the other attributes.
- Confidence (certain, probable, possible) that the characteriser has in those interpretations.

Those interpretations enable the characteriser to place each unit of the historic landscape characterisation, usually a GIS polygon, into a broad type and then into a subdivision of this (a type or sub-type). There is no fixed method of historic landscape characterisation in England though its core elements have been consolidated (Aldred & Fairclough 2003) and Table 1 lists the broad types that most historic landscape characterisations work with – the subdivisions of these are usually more locally defined, reflecting the historic particularity of regions and sub-regions.

By defining and interpreting the time depth detectable within it, historic landscape characterisation makes available to all users the link between landscape's present form and its historic development. It does not, however, privilege periods, but accepts that a late 20th-century factory imposed on a medieval derived field system makes that landscape predominantly late 20th-century industrial in character. By using GIS to support the historic landscape characterisation and by layering in 'previous character' to the attributes database, a user might be able to strip away the late 20th century to reach earlier landscape types like fields, woods, rough ground, etc., and thus appreciate how much of the latter is still detectable, or legible, in the former.

Table 1. Broad Historic Landscape Character Types employed in England.

Enclosed land
Unenclosed or unimproved land
Woodland
Orchards and horticulture
Valley-floor uses
Industrial
Military
Ornamental
Recreational
Settlement
Civic and commercial
Communications
Waterbodies

While greatly extending the breadth or area of reach of our understanding, historic landscape characterisation simplifies or flattens landscape complexity (see Austin 2007, 103), although those who create and use historic landscape characterisation also recognise that it contains and makes more widely accessible that landscape complexity. By being undertaken by just one or two characterisers historic landscape characterisation necessarily imposes particular views of the meaning of that landscape complexity (ibid., 103-104). However, historic landscape characterisation's method anticipates this concern by being systematic and transparent about the sources it uses (through metadata protocols) and, unlike much other landscape archaeology, sets out the confidence the characteriser has in the interpretations they make. It is therefore set up as a provisional framework of historical understanding of the whole landscape, not just selected parts of it.

That framework may initially be one person's perception of place, but it is also capable of stimulating and hosting multiple alternative perceptions and ways of valuing. Heritage values being both manifold and fluid (as nicely set out in English Heritage 2008), historic landscape characterisation takes care not to impose, as attributes, fixed or official grades of value.

The setting out of sources and interpretative assumptions means that the anticipated contestations of types and their meanings, and any extensions through deepening, are themselves also framed. It is partly through contestation that historic landscape characterisation stimulates further landscape archaeology research. Its curators should expect and indeed encourage querying, testing and revision, and should be prepared to undertake these themselves. They might also expect alternative historic landscape characterisations to be created in which other attributes are recorded or other interpretations are set out and justified. As historic landscape characterisation is never complete and its 'data' are as partial as those of any other form of landscape archaeology or history, it serves to confirm that nothing is absolutely fixed. Seen then as a scheme of interpretation, historic landscape characterisation is necessarily a spatially expressed research agenda. Many applications also require it to be supplemented, extended or deepened to become adequate for the task in hand.

EXTENDING HISTORIC LANDSCAPE CHARACTERISATION

Historic landscape characterisation has therefore already spawned secondary work. The 1994 Cornwall historic landscape characterisation showed that much of the lowland part of the county has irregular field patterns whose forms suggest development from those small-scale open fields associated with hamlets (see above). Divergence from these patterns contributed to the design of traditional field-survey-based landscape archaeology projects in St Keverne, where fields derive from later prehistoric coaxial systems and at Godolphin, where a 17th-century enclosure of a deer park produced the large and regularly shaped so-called 'barton' fields of the estate's home farm (see Herring 2007 for these and other similar secondary projects). Correlations with other historic environment information, such as individual archaeological sites, create dialogues of interpretation that deepen understanding of both the historic landscape characterisation and the site types. In north Cornwall Bronze Age barrows and later prehistoric enclosed settlements both have distributions that follow the patterning of certain historic landscape characterisation types (Herring 2009, fig. 4). Understanding of early Christian territories and remains have also been fruitfully informed by comparison with historic landscape characterisation (Turner 2006).

Elsewhere the patterns found in historic landscape characterisations have themselves been further analysed. In Devon equivalents of the French *pays*, areas with common ways of being and carrying on, were identified (Turner 2007, 113-127), while in Cornwall again, a model of the dispositions of cultivated,

Figure 2. One of several late medieval carvings on the wooden rood screen at Sancreed church, west Cornwall, that depict individuals facing, like the Roman god Janus, and like increasing numbers of landscape archaeologists, both forward and back, looking into the past and the future. In this case the figure, a triciput, is also in the present looking out and thus, like us, responsible for bridging the two. *See also the full colour section in this book*

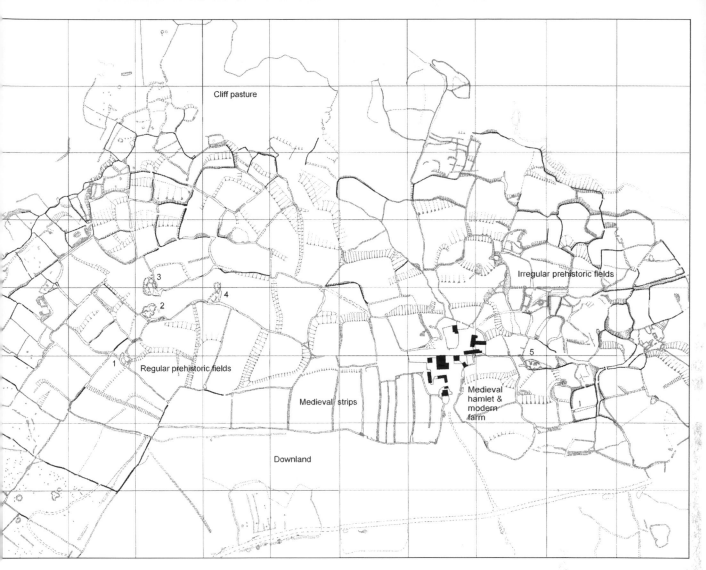

Figure 3. Extract from 1983 measured survey of Bosigran tenement in Zennor, west Cornwall, prepared, analysed and interpreted for and by the National Trust to inform management and presentation of the complex archaeological remains within a working farm. The report, a form of conservation management plan whose format was devised by the Trust, included a comprehensive landscape narrative or biography that included the several prehistoric, medieval, post-medieval and modern settlements and associated fields. Roman period courtyard houses, four in a hamlet with shared regular fields (1-4) and a single one (5) in the midst of its own irregular fields, are shown. Statements of significance as well as practical recommendations relating to repair, consolidation and future land use were also prepared (Herring 1987). North to top; 100m grid.

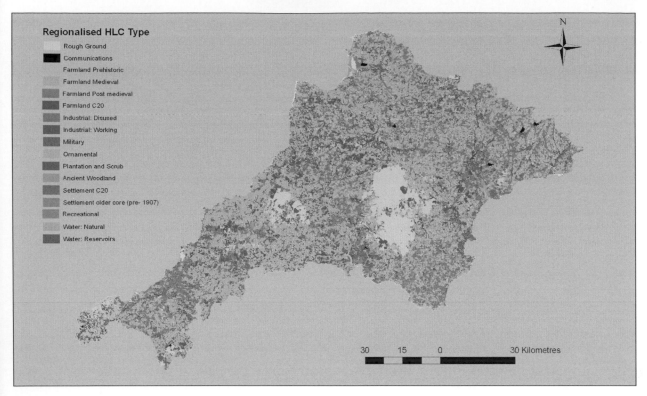

Regionalised HLC Type

- Rough Ground
- Communications
- Farmland Prehistoric
- Farmland Medieval
- Farmland Post medieval
- Farmland C20
- Industrial: Disused
- Industrial: Working
- Military
- Ornamental
- Plantation and Scrub
- Ancient Woodland
- Settlement C20
- Settlement older core (pre- 1907)
- Recreational
- Water: Natural
- Water: Reservoirs

30   15   0   30 Kilometres

Figure 4. The Cornwall and Devon HLCs combined and simplified to create a regionalised characterisation. Patterns in and relationships between the several phases of enclosed land, rough ground and settlement suggest numerous regional and more local landscape archaeology research issues. Closer examination of the detail of each parent HLC would identify many more. (Derived from material that is the copyright of Cornwall Council and Devon County Council.) *See also the full colour section in this book*

wooded and rough grazed land in medieval Cornwall was reconstructed through interpretation of historic landscape characterisation types (Herring 2007, fig. 2). Extending this to include neighbouring Devon (fig. 5) throws up a range of issues that can be pursued through further deepening of historic landscape characterisation or through other forms of landscape archaeology. These might include the following.

- Establishing whether farming systems differed where large areas of seasonal rough grazing were either adjacent or distant.
- Considering whether the several linear areas of rough ground acted as cultural barriers to communities on either side.
- Exploring the strategies adopted by communities in areas with little or no woodland.

Figure 4 combines, simplifies and so regionalises the Cornwall and Devon historic landscape characterisations, interpreting elements of the latter (completed in 2006) to make it compatible with the earlier and simpler Cornwall characterisation. It effectively illustrates how present landscape is wholly human.

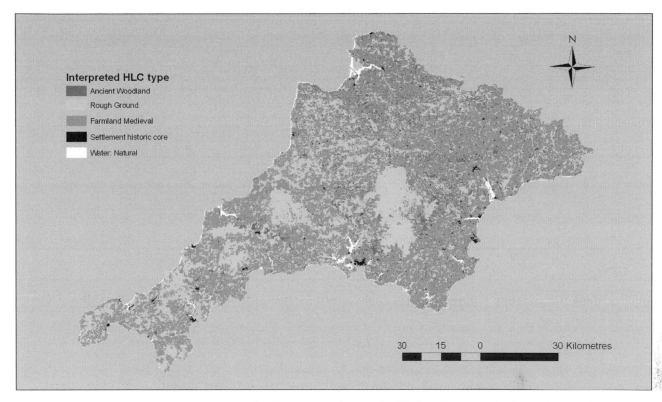

Figure 5. The Cornwall and Devon HLCs further simplified and interpreted using understanding developed by landscape history and archaeology to create a model of the likely disposition of main land use zones in the later medieval period. Farmland (both enclosed and in open strip fields) dominated most of the peninsula, but would have been broken up by more extensive rough ground (used for summer grazing and as a source of household fuel) and woodland. Most (but not all) of the historic cores of modern towns would have also been the main later medieval towns of the region, so these have been retained in this image to help the reader see how agriculture worked in relation to centres of population and commerce. Like fig. 4, the patterns and relationships within this characterisation suggest numerous research issues that can be further pursued by landscape archaeology. (Derived from material that is the copyright of Cornwall Council and Devon County Council.)

The figure also confirms that historic landscape characterisation is as noted above a spatially organised research framework. Examination sends up swarms of queries that can be pursued by landscape archaeology, landscape history, anthropology, ecology, sociology, economy, and so on.

Numerous examples demonstrate that the core methods and principles of historic landscape characterisation can be successfully applied to any type of place, including cities and towns (Thomas 2007), the sea – its surface, its water column, its floor and beneath its bed – as in English Heritage's Historic Seascapes Characterisation (Hooley 2007), and militarised land that many would regard as wilderness (Dingwall & Gaffney 2007).

In England the county-level historic landscape characterisation has largely been supported, guided and funded by English Heritage, but undertaken by local authorities, ensuring regional and local design and application. In doing so it, and other forms of historic characterisation, helps enable strategic and development control planners meet the requirements of national planning policy that is increasingly directed towards the maintenance and enhancement of local distinctiveness and character (see a review of such policy in Homes and Communities Agency & English Heritage 2009).

In Scotland a closely similar characterisation method termed Historic Land-use Assessment is being undertaken for the whole country by the Royal Commission on Historic Monuments (Scotland) in close collaboration with Historic Scotland (Dyson-Bruce et al. 1999; Dixon 2007; Herring 2009). In Wales and Ireland work closely related to historic landscape characterisation has been undertaken for designated areas and individual counties (Alfrey 2007; Oxford Archaeology & George Lambrick Heritage Consultancy 2010). Historic landscape characterisation has also been undertaken in Texas and Amazonian Brazil (Dingwall & Gaffney 2007; Megan Val Baker, personal communication) as well as parts of mainland Europe (Fairclough & Rippon 2002; Nord 2009). Other novel interdisciplinary approaches to landscape may not use historic landscape characterisation per se, but have independently developed very similar principles notably around the relationships between the past and the present and future (e.g. Blur & Santori-Frizell 2009).

USING HISTORIC LANDSCAPE CHARACTERISATION

In terms of future-oriented application, historic landscape characterisation can be used strategically as a spatial framework in which to assess the historic environment's sensitivity to types of change or its capacity to accommodate it. This allows the historic environment sector to provide reasonable advice when contributing to the design of change, for example at a master-planning stage (e.g. Homes and Communities Agency & English Heritage 2009). Approaches to sensitivity, capacity and opportunity modelling typically involve assessing the varying vulnerability of historic landscape types to the effects of different change scenarios (Herring 2009).

Historic landscape characterisation mapping and text are designed to be readily accessible and understandable, in format and language that encourage dialogue with a range of partners. Joint working with communities and other agencies that have interests in planning for the future is likely to develop further. For example, strategic management is more realistic when historic landscape characterisation and landscape archaeology help ecologists appreciate that the environment is not purely natural, but the outcome of long-term dialogue between nature and human communities. Historic landscape characterisation and landscape archaeology bring understanding of long-term change trajectories and the effects of distant and recent decision-making to models of ecosystem services, helping ensure that decisions and policy based on them are appropriate and sustainable.

Through informing spatial strategy and sustainability by combining place and time, the cultural and the semi-natural, historic landscape characterisation helps society use the past when looking forward. The way it does this may be reactive, informing responses to others' plans. But historic landscape characterisation can also be used more proactively when the needs of the historic environment become the principal driver for change.

Historic Environment Action Plans (HEAPs) require spatial representation of historical understanding, such as that provided by historic landscape characterisation. Prepared for selected historic types or for areas, HEAPs draw on landscape archaeology's multidisciplinarian character by being developed in partnership with all interested agencies, communities and individuals. A typical HEAP considers the vulnerability of archaeological, historical and semi-natural components to change scenarios etc., assesses their value to a range of communities, agrees objectives, sets targets and designs SMART (specific, measurable, achievable, realistic and time-based) actions, with delivery roles of partners set out. By not concentrating on the historic environment's designated elements, but instead on what a range of partners consider as the historic environment's needs, HEAPs may be seen as more inclusive than the Biodiversity Action Plans on which they were first modelled (Herring 2007).

CONCLUSION: RETURNING TO CHANGE

So, while concluding that landscape archaeology will become increasingly concerned with the future (as well as the past), we close as we began, with inclusivity. A healthy and sustainable landscape archaeology may be expected to draw in diverse communities (within academia and without) and learn to accommodate very different sets of ideas, values, approaches and applications, while maintaining standards and rigour. Its increasingly diverse outputs, of which historic landscape characterisation is one, may also be expected to help reduce the sense of regret that accompanies the routine perception of change as disturbance and loss. Both society and landscape may no longer be seen as on an endless slide from a state of rural Arcadian perfection, but may instead be appreciated for what they are now, early in the 21st century (see for example English Heritage 2004). Landscape archaeology's key role may be to not only demonstrate how their rich diversity, the outcome of long, medium and short-term change, came into being, but also how that understanding can most effectively inform a sustainable future.

ACKNOWLEDGEMENTS

It is a pleasure to thank Graham Fairclough for constructive comments on the text, Bryn Tapper for help with the maps, Sam Turner for comments on them, Cathy Parkes for challenging discussions and Megan Val Baker for seeking the Janus image. This paper benefitted also from reviews by Mats Widgren and Jan Kolen.

REFERENCES

Aldred, O. & G. Fairclough. 2003. *Historic Landscape Characterisation: Taking Stock of the Method*. English Heritage and Somerset County Council, London.

Alfrey, J. 2007. Contexts for Historic Landscape Characterisation in Wales. *Landscapes* 8 (2), 84-91.

Altenberg, K. 2003. *Experiencing Landscapes; a study of space and identity in three marginal areas of medieval Britain and Scandinavia*. Almqvist & Wiksell International, Stockholm.

Arnold, J.H. 2000. *History, a very short introduction*. Oxford University Press, Oxford.

Austin, D. 1989. The Excavation of Dispersed Settlement in Medieval Britain. In Aston, M., D. Austin & C. Dyer (eds.) *The Rural Settlements of Medieval England*, 231-246. Blackwell, Oxford.

Austin, D. 2007. Character or Caricature? Concluding Discussion. *Landscapes* 8 (2), 92-105.

Bender, B. 1992. Theorising landscapes, and the prehistoric landscapes of Stonehenge. *Man* (N.S.) 27, 737-755.

Bender, B., S. Hamilton & C. Tilley. 2007. *Stone Worlds: Narrative and Reflexivity in Landscape Archaeology*. Left Coast Press, Walnut Creek.

Bloemers, J.H.F. 2002. Past- and future-oriented archaeology: protecting and developing the archaeological-historical landscape in the Netherlands. In Fairclough, G. & S. Rippon (eds.) *Europe's Cultural Landscape: archaeologists and the management of change*, 89-96. Europae Archaeologiae Consilium and English Heritage, Brussels.

Blur, H. & B. Santor-Frizell (eds.) 2009. *Via Tiburtina*. Swedish Institute at Rome, Rome.

Bowden, M. (ed.) 1999. *Unravelling the landscape, an inquisitive approach to archaeology*. Tempus, Stroud.

Bruck, J., R. Johnston & H. Wickstead. 2003. Excavations of Bronze Age field systems on Shovel Down, Dartmoor. *Past, the Newsletter of the Prehistoric Society* 45.

Chōmei, K. 1212. *The ten foot square hut and Tales of the Heike*, translated in 1928 by A.L. Sadler. Angus & Robertson, Sydney.

Council of Europe. 2000. *European Landscape Convention, European Treaty Series* 176. Council of Europe, Florence.

Dingwall, L. & V. Gaffney (eds.) 2007. *Heritage Management at Fort Hood, Texas. Experiments in historic landscape characterisation*. Archaeopress, Oxford.

Dixon, P. 2007. Conservation not Reconstruction: Historic Land-Use Assessment (HLA), or Characterising the Historic Landscape in Scotland. *Landscapes* 8 (2), 72-83.

Dyson Bruce, L., P. Dixon, R. Hingley & J. Stevenson. 1999. *Historic Land-Use Assessment (HLA): Development and potential of a Technique for assessing Historic Land-Use Patterns*. Historic Scotland, Edinburgh.

English Heritage. 1997. *Sustaining the historic environment*. English Heritage, London.

English Heritage. 2004. *Change and Creation, Historic Landscape Character 1950-2000*. English Heritage, Swindon.

English Heritage. 2008. *Conservation Principles, policies and guidance for the sustainable management of the historic environment*. English Heritage, Swindon.

Everson, P. & G. Brown. 2010. Dr Hoskins I presume! Field visits in the footsteps of a pioneer. In Dyer, C. & R. Jones (eds.) *Deserted Villages Revisited*, 46-63. University of Hertfordshire Press, Hatfield, UK.

Fairclough, G. 2003. 'The long chain': archaeology, historic landscape characterisation and time depth in the landscape. In Palang, H. & G. Fry (eds.), *Landscape Interfaces: Cultural Heritage in Changing Landscapes*, landscape series 1, 295-318. Kluwer Academic Publishers, Dordrecht.

Fairclough, G. 2006. Our Place in the Landscape? An Archaeologist's Ideology of Landscape Perception and Management. In Meier, T. (ed.), *Landscape Ideologies*, 22, 177-197. Archaeolingua (Series Minor), Budapest.

Fairclough, G., G. Lambrick & A. McNab. 1999. *Yesterday's World, Tomorrow's Landscape*. English Heritage, London.

Fairclough, G. & S. Rippon (eds.) 2002. *Europe's Cultural landscape: archaeologists and the management of change, EAC Occasional Paper 2*, Europae Archaeologiae Consilium & English Heritage, Brussels/London.

Fleming, A. 1988. *The Dartmoor Reaves: investigating prehistoric land divisions*. Batsford, London.

Fleming, A. 2005. Megaliths and post-modernism: the case of Wales. *Antiquity* 79, 921-932.

Fleming, A. 2007a. Don't Bin Your Boots! *Landscapes* 8 (1), 85-99.

Fleming, A. 2007b. *The Dartmoor Reaves: investigating prehistoric land divisions*. Windgather Press, Macclesfield.

Franklin, L. 2006. Imagined Landscapes: Archaeology, Perception and Folklore in the study of Medieval Devon. In Turner, S. (ed.), *Medieval Devon and Cornwall*, 144-161. Windgather Press, Macclesfield.

Grahame, K. 1908. *The Wind in the Willows*. Methuen, London.

Herring, P. 1987. *Bosigran, Zennor. Archaeological Assessment*. Cornwall County Council, Truro.

Herring, P. 1998. *Cornwall's Historic Landscape, presenting a method of historic landscape character assessment*. Cornwall County Council, Truro.

Herring, P. 2003. Cornish Medieval Deer parks. In Wilson-North, R. (ed.) *The Lie of the Land*, 34-50. Mint Press, Exeter.

Herring, P. 2006. Cornish strip fields. In Turner, S. (ed.), *Medieval Devon and Cornwall*, 44-77. Windgather Press, Macclesfield.

Herring, P. 2007. Historic landscape Characterisation in an Ever-Changing Cornwall. *Landscapes* 8 (2), 15-27.

Herring, P. 2008. Commons, fields and communities in prehistoric Cornwall. In Chadwick, A. (ed.), *Recent Approaches to the Archaeology of Land Allotment*, 70-95. British Archaeological Reports (International Series), 1875, Oxford.

Herring, P. 2009. Framing Perceptions of the Historic Landscape: Historic Landscape Characterisation (HLC) and Historic Land-Use Assessment (HLA), *Scottish Geographical Journal* 1251 (), 61-77.

Herring, P. & N. Johnson. 1997. Historic landscape character mapping in Cornwall. In Barker, K. & T. Darvill (eds.) *Making English Landscapes*, 46-54. Oxbow, Oxford.

Herring P., A. Sharpe, J.R. Smith & C. Giles. 2008. *Bodmin Moor: an archaeological survey, Vol 2, The post-medieval and industrial landscapes*. English Heritage, Swindon.

Hicks, D. 2008. Review of T. Rowley, The English Landscape in the 20th Century. *Landscapes* 9 (1), 86-90.

Homes and Communities Agency & English Heritage. 2009. *Capitalising on the inherited landscape, an introduction to historic characterisation for masterplanning*. Homes and Communities Agency, London.

Hooley, D. 2007. England's Historic Seascapes – archaeologists look beneath the surface to meet the challenges of the ELC. *Landscape Character Network News* 26, 8-11.

Hoskins, W.G. 1955. *The Making of the English Landscape*. Hodder & Stoughton, London.

Ingold, T. 1993. The Temporality of the Landscape. *World Archaeology* 25, 152-174.

Johnson, M. 2007a. *Ideas of Landscape*. Blackwell, Oxford.

Johnson, M. 2007b. Don't Bin Your Brain! *Landscapes* 8 (2), 126-128.

Johnson, N. 1985. Archaeological Field Survey, a Cornish Perspective. In Macready, S. & F.H. Thompson (eds.) *Archaeological Field Survey in Britain and Abroad*, 51-66. The Society of Antiquaries, Occasional Paper (New Series) VI, London.

Johnston, R. 2001. *Land and Society: the Bronze Age cairnfields and field systems of Britain*. PhD thesis, University of Newcastle.

Lake, J. & R. Edwards. 2006. Farmsteads and the landscape: towards an integrated view. *Landscapes*, 7 (1), 1-36.

Nord, J.M. 2009. *Changing Landscape and Persistent Places*, Acta Archaeological Lundensia (Series in Prima Quarto 29), Lund.

Palmer, M. 2007. Introduction: Post-Medieval Landscapes since Hoskins – Theory and Practice. In Barnwell, P.S. & M. Palmer (eds.) *Post-medieval Landscapes*, 1-7. Windgather Press, Macclesfield.

Penrose, S. 2007. *Images of change, an archaeology of England's contemporary landscape*. English Heritage, Swindon.

Rippon, S. 2004. *Historic Landscape Analysis, deciphering the countryside, Practical handbooks in Archaeology* 16. Council for British Archaeology, York.

Roymans, N., F. Gerritsen, C. van der Heijden, K. Bosma & J. Kolen. 2009. Landscape Biography as Research Strategy: The Case of the South Netherlands Project. *Landscape Research* 34 (3), 337-359.

Thomas, R.M. 2006. Mapping the Towns: English Heritage's Urban Survey and Characterisation programme. *Landscapes* 7 (1), 68-92.

Tilley, C. 1994. *A Phenomenology of Landscape:Places, Paths and Monuments*. Berg, Oxford.

Turner, S. 2006. *Making a Christian landscape, the countryside in early medieval Cornwall, Devon and Wessex*. University of Exeter Press, Exeter.

Turner, S. 2007. *Ancient Country: The Historic Character of Rural Devon*. Devon Archaeological Society Occasional Paper 20, Exeter.

6.4 'Landscape', 'environment' and a vision of interdisciplinarity

Author

Thomas Meier

Institut für Ur- und Frühgeschichte und Vorderasiatische Archäologie, Ruprecht-Karls-Universität, Heidelberg, Germany
Contact: thomas.meier@zaw.uni-heidelberg.de

KEYWORDS

environment, landscape, interdisciplinarity, history of archaeology, humanities, science

ABSTRACT

In this paper, presented at the end of the Amsterdam-LAC2010 conference, I felt obliged to react to the contributions delivered so far. Thus, I am starting with a few words on the term 'environment', which was first used at the beginning of the 20th century. Its use encapsulates an epistemological division between an individual being and its a-/biotic surroundings and therefore should be used for any research based on that fundamental division. Therefore, 'environmental archaeology' is the proper term for all those archaeological approaches based on scientific methods, as science in itself is rooted in the analytical division between humans and the world.

On the other hand, the word 'landscape' has a very old epistemology and history of meaning. In the Middle Ages its emphasis was on a politically defined body of people and, on a secondary level, on the land inhabited by them, i.e. it was the people who made the land. During early modern times the word acquired an additional aesthetic notion incorporating social imaginations of beauty and nature. Therefore – despite its quite shapeless use in actual academia – 'landscape archaeology' is a reasonable term for all research in the social construction of space.

This separation of 'environmental archaeology'and 'landscape archaeology' is not meant to perpetuate the grand divide between science and humanities. But this is an attempt to establish a clear-cut terminological clarification in order to enable an understanding of different disciplinary epistemologies as a necessary component of interdisciplinary cooperation. While actual multidisciplinary work aims at an exchange of disciplinary results, I am presenting a model of interdisciplinarity, which takes into account the presuppositions of participating disciplines as well. This approach asks for greater consciousness of different epistemologies and it is with this aim that I am proposing a clear-cut terminology for 'environmental archaeology'and 'landscape archaeology'.

'LANDSCAPE' – A SEXY WORD?[1]

Nowadays the word 'landscape' is in. It obviously sounds sexy to archaeologists in 2010. Starting some years ago, there were a growing number of archaeological publications proudly bearing 'landscape' in their titles. Simultaneously the word 'environment' is losing its prominent position on the front page of archaeological books and papers. Does this reflect a new type of research, a new topic in archaeology – or is it just one of the fashionable sound bites of the new millenium? In my eyes there are some indications of this last suggestion; the word 'landscape' today at least partly acting as an envelope for anything. For example, looking at the papers and posters of the Amsterdam conference on 'landscape archaeology' there are presentations on quarternary geology, taphonomy, the microhistory of nature, deterministic and possibilistic approaches to the culture-nature dichotomy, the ecological impact of ancient economies, survey techniques, settlement structures, communication routes, the social dimension of space and phenomenology, and, finally, on the heritage aspect of how to deal with ancient landforms – to name only a few. Altogether this mixture looks quite disparate I wonder whether it really makes sense to summarise such different topics and methods by using the single term 'landscape archaeology'? Actually we seem to be back in 1996, when Robert Johnston observed: 'by allowing landscape to mean relatively anything and have all possible contextual value, it loosens all definition and effectively has no interpretative value.' (Johnston 1998, 317).

Of course, I would welcome 'landscape' as an umbrella term, a kind of unifying concept for many different strands of research to engage in a closer, interdisciplinary way of cooperation (Gramsch 2003; for a more precise unifying concept in landscape research cf. van der Valk &Bloemers 2006). I expect most of us will agree on the need for such interdisciplinarity when dealing with such a complex matter like the world in its historic dimension. However, looking at practice, most of the work in so-called 'landscape archaeology' is still predominantly mono-disciplinary, sometimes in more or less successful cooperation with another discipline, which more often is multi- than interdisciplinary (cf. Tress et al. 2003, 2006; Potthast 2011). But if we are going – and we are just at the beginning of going – to unify our research under one umbrella, and if we wish to avoid an incidental mixture of everything I think it will be an essential prerequisite of coherent cooperation to know the parts which we are going to unify. For the sake of such conscious clarification I do not see any benefits in the actual hyper-fluent use of the term 'landscape', but I am arguing for a more concise terminology as it was broadly in use half a decade ago: With the terms of 'landscape archaeology'and 'environmental archaeology' there are two distinct expressions, which by traditions of etymology, daily use and scholarly meaning are and should be much more than two sexy words.

ENVIRONMENT

Just over a century ago, in 1906, Alfred Schliz pointed out for the first time the highly significant cor-relation of settlements of early Neolithic Linearbandkeramik-culture and Loess-soils (Schliz 1906; cf. Friederich 2003). After World War I, Ernst Wahle, among others, quite deterministically stated that the structures of prehistoric settlement and culture were the results of natural conditions (Wahle 1915, 1920). He tried to show that prehistoric settlement down to the Bronze Age was strictly determined by the natu-ral existence of the so-called *Steppenheide*, the most Western offspring of eastern European steppe-vege-tation (Gradmann 1901; 1906). Somewhat later, in the early 1930s, Cyril Fox published his 'Personality of Britain' arguing for natural factors dividing Britain in two parts of lowland and highland Britain with two quite different cultural trajectories (Fox 1932). While a correlation of Linearbandkeramik and Loess-soils is still under discussion, Cyril Fox looks much too deterministic today and the *Steppenheide* theory has long been out of fashion (cf. Ellenberg 1963; Clark 1974, 43). These examples, however, show that an inter-relation of nature and culture has been under discussion for a century or more, that it has been answered in different ways and that it is an interdisciplinary question from its very beginning.

From the 1950s onwards, the general question of any relation between nature and culture stayed alive. In German archaeology Herbert Jankuhn and his concept of '*Siedlungsarchäologie*' defined some kind of systematic approach arguing within the current deterministic and possibilistic modes of expla-nation of its time – though never stating that he was doing so. At the time this approach was generally focused on economic needs, technical skills and population dynamics influencing society's ability to in-teract with natural factors (Jankuhn 1952/1955; 1977).

In Anglo-American archaeology Lewis Binford and his processual approach began to understand prehistoric people and nature in new ways. In terms of theory this was a much more holistic approach than Jankuhn's '*Siedlungsarchäologie*', as Binford attempted to set out a totally new theoretical framework for archaeology as a whole. Now, archaeology was meant not only to describe, but to explain – especially it was designed to explain cultural change in an explicit and testable way (Binford 1962). Binford's interest in an interrelation of culture and nature was not very new at that time; especially his definition of culture as an extra-somatic adaptation to environment was well known (Moran 1990; Pantzer 1995, 6f., 20-24). However, Binford's focus on cultural change along with his definition of culture as a means of adaptation to chang-ing natural factors focused archaeological research on environmental conditions and thrust them into the very centre of processual archaeology. Thus, some subsystems or factors at the intersection of culture and nature seemed to be of greater importance than others: technology, economy and population dynamics especially being the favourite subjects of New Archaeology. At the very least they owed their favourite role to New Archaeology's explicitly systemic framework of argumentation. These same factors were of spe-cial relevance to Jankuhn's *Siedlungsarchäologie* as well, but within a processual approach they gained a much greater importance, becoming the central screws of the system. Moreover, the methodological focus on empirical research and testable hypotheses furthered those components of the cultural system, which were countable and measurable: again, these being technology, material culture and some aspects of econ-omy and population dynamics (Clarke 1978; Bernbeck 1997, 35-129; Johnson 1999, 12-84).

All of these approaches, regardless of *Siedlungsarchäologie* or processual archaeology, are aimed at cate-gorically separating culture and humans from the world around. Thus, humans are analytically standing outside, they are external observers of that globe called the world (Ingold 1993) (fig. 1). By regarding themselves as external observers, humans restrict the world to an objectively existing prediscursive container (physical space) with life happening within it. The world around becomes a stage on which the grand play of humankind as biological, economical, political and cultural beings develops. For decades and decades it was exactly this vision of human-world-interrelation which dominated the geographical perspective on the world, most clearly expressed by the model of geographical (and descriptive) layers developed by Alfred Hettner in 1927. Geographers being regarded as specialists of space, this model highly influenced neighbouring disciplines like archaeology as well.

This approach is not to be criticised in itself, as the separation of man and the world around is an analytical tool according to specific research interests. We should be aware, however, that all knowledge produced by this tool is valid only with respect to this categorial separation. This is especially true for all ecological thinking logically based on discriminating between beings and the world around them. In ecology the world around is called the 'environment', meaning the totality of all ecological factors influencing a species or a single being, while an ecosystem is a system which comprises all creatures and their environment as well as all interactions (cf. Odum 1975). In ecological thinking 'environment' is an objective, inter-individual term and, in its daily use, has more or less adopted the ecological meaning of the word (cf. Harvey 1993; Winiwarter 1994). This ecological meaning of environment, its categorial separation of beings and the world as well as its concentration on the natural parameters of the animate and inanimate world, more or less corresponds to the world around of *Siedlungsarchäologie* and processual archaeology. Thus in my view it makes good sense to call this line of research 'environmental archaeology'– as was usual until fairly recently: 'environmental archaeology' meaning any approach arguing within an ecological framework, focusing on environmentally relevant sub-systems and mainly based on scientific methods.

LANDSCAPE

I do not want to go into any of the well known details regarding the criticism of processual archaeology. Basically this critique rejects the aim of universal cultural laws and is designed to strengthen the hermeneutic model of the humanities against a supposedly reductionist deductive model of scientific reasoning (Hodder 1984; 1985). While the first point only addresses the question of why we do archaeological research at all, the second criticism is the really fundamental one: Criticising the deductive approach to be reductionism easily leads to the conclusion that empirical sciences are irrelevant for the study of human culture, while the hermeneutical approach may be regarded as unscholarly by scientists.

By first waging a fierce battle, then by building an Iron Curtain between processualists and post-processualists, followed by a deaf-mute disinterest in the other party, this criticism has finally culminated in the development of a bundle of post-processual archaeologies (Bernbeck 1997, 271ff.; Johnson 1999, 98-115). While quite different in many respects these archaeologies nevertheless have some points in common:

– All of them are based in an epistemological model of hermeneutics.
– All of them take a constructivist point of view, i.e. reality is not an empirical, objective entity, but socially constructed.

This focus on hermeneutics and constructivism has quite far-reaching consequences on the vision of humans and on the aims and methods of archaeology: Post-processual archaeologists concentrate on the individual and its perceptions of the world in its specific cultural context. Thus, the fields of ideology and religion, aesthetics and social structures especially – mainly neglected by processual archaeology – are now becoming the centre of interest.

While Binford claimed to research the whole of the cultural system with all its subsystems, and while post-processual archaeology renewed this holistic claim, both strands of archaeological research almost complementarily compare with each other.

Table 1

processual archaeology	post-processual archaeology
humans are passive	humans are active
culture adapts systemically to external stimuli (environment)	social rules are negotiated between actors
humans are subordinated to rules and aims of society	social structure is constructed by individuals
aim: cross-cultural generalisation	aim: cultural context
methods: science, systems theory	methods: humanities, hermeneutics, constructivism

Among other consequences this post-processual mode of thinking heavily touches upon the concept of space: space is no longer thinkable as a prediscursive container or a stage for the theatre of mankind, but now it is socially constructed and meaningful as well. Sociologist Martina Löw (2001, esp. 152ff.) has developed a convincing theory of constructed space, which, moreover, is practicable in archaeology. Löw starts by defining space as the relational arrangement of objects and humans in a place. Thus, space is formed by the practice of arranging objects and humans – called spacing – and by positioning symbolic markers to make this ensemble visible. To constitute space, however, it is necessary not only to position these constituents, but to synthesise them in a mental process as well, i.e. to perceive an arrangement as space. Such a synthesis follows cultural patterns of imagination (*Vorstellung*) and experience. Therefore, according to Löw, space is permanently constituted as well as changed by social practice. On the one hand space is structuring action, on the other hand space is simultaneously constituted by action and perception. Thus – to turn it theoretically – space is no longer a matter of ontology (a prediscursive container) but a matter of epistemology (a social construct).

Such a space socially constituted by humans is totally different from the concept of 'environment' with humans standing outside the world looking upon it as a globe, with a strict separation between culture and nature, humans and their environments. When humans create space themselves they become parts of their surroundings, they are not separated from or outside the world any longer, and the world changes from a globe to a lifeworld (Ingold 1993). In practice, research on the social construction of space is mostly focused on the actor and is therefore small-scale, while the 'global approach' normally covers a wider area. However, in this respect, the discrimination between these two spatial approaches has no epistemological implication on scale.

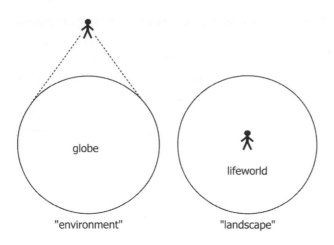

Figure 1. Humans and the world (after Ingold 1993 with modifications).

The word 'space' would be an option for this kind of life world, but it may provoke misunderstanding, as it is an appropriate term for physical space as well. Accordingly, I would suggest calling such kinds of socially constituted spaces 'landscapes'. However, as the contributions to the LAC 2010 in the beginning of this paper exemplify, 'landscape' in the academic community actually means everything and nothing and the word has followed very different histories of meaning in different disciplines (cf. Jones 2003; Cosgrove 2004; Meier 2006, 24f.; for the archaeological history of the term Darvill 2001; Gojda 2003, 40f.; for history and geography cf. Meinig 1979; Groth & Bressi 1997; for an example of a largely differing disciplinary use of the term with a specific natural meaning cf. the earth sciences in specific geomorphology: e.g. Migoń 2010). Thus, it makes little sense to refer to any specific disciplinary meaning of 'landscape' but to bear in mind that the word had a very strong etymological tradition before academics got their hands on it and, moreover, that it still enjoys a vivid afterlife in the common sense of the word. In this extra-academic etymological tradition of 'landscape', social, political and aesthetic aspects are especially emphasised throughout its history. To very briefly summarise the history of its meaning (cf. Müller 1977; Schenk 2001) the suffix 'scape' is derived from Germanic *skapjan* (> *skapi-, *skapja-, *skafti*), meaning 'to shape'; this is illustrated by Old-English *gisceap* meaning 'shape, form, composition' and Old Nordic *skap* meaning 'composition, condition, manner'. Stringently *landskapr* means the manner or fashion of a land, i.e. the practice of the people inhabiting an area. In Old German the oldest evidence of *lantscaf/ lantskepi* dates from around 830 meaning – like in Old English – a greater area or a region, which primarily is defined politically. Thus, in the early Middle Ages landscape does not essentially denote a physical space, but either the custom of a region or the area inhabited by a politically-defined body of people. This focus on the inhabitants of a land and especially on those of political influence dominates the meaning of *lantschaft* in Middle High German as well and it is still active today in some parts of north-western Germany (e.g. 'Ostfriesische Landschaft') or Switzerland (e.g. 'Basel Landschaft'). It took until early modern times that landscape's meaning of a political body increasingly shifted back on the area inhabited by that people.

Based on this newly gained relation to space the term *landschaft* started a new career during the 16th century (for a detailed analysis of the formative early modern use of the term cf. Drexler 2009). With the formation of landscape painting its meaning shifted to a picture showing a bucolic scene, meaning an envisioned detail of nature. Painter Hans Sachs (1494-1576) for example used the word *landschaft* to describe a panorama structured in fore- and background (Müller 1977, 9; cf. Groh & Groh 1991; Schramm 2008). In this specifically artistic sense of 'painting of a detail of nature' the word *landskip* was introduced into Dutch and English with its first evidence in 1598 (Schama 1995, esp. 18-21; Olwig 2002; Cosgrove 2004, 61). Thus, in the beginning of early modern times 'landscape' gains a pronounced artistic connotation, which it keeps for the future; from now on the term obviously makes an aesthetic statement and thus carries social imaginations of space. Altogether the etymology and history of meaning of the word 'landscape' in its earlier sense emphasises the idea that it is the people, who make up the land. In its later sense the word is more expressive of an (aesthetic) quality. At any rate, 'landscape' obviously encapsulates the social constructedness of space. It is this meaning, which is still active in the popular use of the word – regardless of all academic discussions and re-formations of its meaning. And it is this meaning to which the European Landscape Convention is affiliated, stating in its first article:

> '"Landscape" means an area, as perceived by people, whose character is the result of the action and interaction of natural and/or human factors' (http://conventions.coe.int/Treaty/en/Treaties/Html/176.htm [accessed 12 March 2009]; cf. Déjeant-Pons 2002; Fairclough 2006).

The first part of this definition means an area first of all becomes a landscape through the perception of its people. There is no such thing as a natural landscape of and within itself – unless we take 'nature' as a cultural construct as well (cf. Eisel 1986; Radkau 1994; Cronon 1995; Meier 2006, 18-20). We are immediately reminded of Martina Löw's synthesising effort, which is necessary for the creation of space. With regard to the history of the term 'landscape', in my eyes 'landscape archaeology' is the proper term to refer to any academic approach which concentrates on the social construction of space.

A VISION OF INTERDISCIPLINARITY[2]

Processual and post-processual archaeologies, due to their theoretical framework and their tools, put their hands on quite different sub-systems of historic societies. But at first glance there seems to be no logical justification for dividing past societies along the frontiers created by modern epistemological tools. However, research is organised along historically-rooted disciplinary trails and these trails so far are our only ways of approaching the historic world. Combined research involving more than one discipline is useful when our interest in the past invokes a complex, close-to-holistic image of it, rather than one which requires unidisciplinary details. So far it is mainly multidisciplinary research with a number of disciplines working on what they believe to be the same topic each of them more or less acknowledging the results of the others (for terminology cf. Tress et al. 2003; 2006; Potthast 2011). This kind of vague research community undoubtedly has its benefits as it brings together different disciplines into a dialogue. However, it may be much too assertive to call these connections baulks or even bridges of interdisciplinarity, as quite often disciplinary results are combined by means of a book cover only. Interdisciplinarity in the proper sense of the word requires a jointly negotiated question and jointly designed close cooperation, which takes effect during the progress of work and which may include newly developed approaches and methods. Under these circumstances, closer inspection shows that many of these multidisciplinary baulks, bridges and spines have deep cracks, especially those bridging the grand divide between the two cultures of science and humanities, while others on each side of this divide are more stable and established.

Reasons for these cracks are not caused by the object of our research, but by home-made problems, which are twofold, i.e. theoretical and social.

Academic differentiation in the last two centuries has produced a bulk of disciplinary rationalities – each discipline following its own presuppositions and obeying its own rules. Though, as Jürgen Mittelstraß emphasises, all of these disciplinary rationales are based in a single common academic rationality (Mittelstraß 1991; 1999), disciplinary practice is not oriented towards a common ground, but aims at developing a highly specialised toolkit to resolve highly differentiated disciplinary questions. Finally, we are now ending up with a set of different and partly exclusive theoretical frameworks and with a set of highly specialised methods, each of them engineered to deal with a specific aspect of past societies but which prove partially impossible to combine with other methods and theories. Additionally, none of them possess the potential to deal with past realities as a whole. This toolkit not only comprises a number of explicit methods and theories, but is more closely based on a great number of silent presuppositions which have proved useful or simply have been handed down from teachers to students over generations and thus have become canonical. At the usual level of multidisciplinarity results obtained by such toolkits are exchanged between disciplines, at best including some clarification about their methods and theories but remaining silent about such deeply rooted presuppositions. This kind of cooperation requires a considerable level of superficiality by its participants, as closer examination usually reveals that the methods and theories of different disciplines are based on very different and fairly inconsistent grounds. It also shows that the exchange of hypotheses and tests runs into self-fulfilling circular structures instead of hermeneutic spirals, and that the rationales of the participants are inconsistent, e.g. on basic points, what may be counted as an argument or what is needed to falsify a hypothesis.

Interdisciplinarity, therefore, necessarily has to communicate not only its results, but also the presuppositions and rationalities of the participating disciplines. It is for this reason that in the previous chapters I have made an attempt at the terminological clarification of 'environmental archaeology' and 'landscape archaeology' as the understanding of different epistemologies is furthered by a well defined terminology.

At the end of an interdisciplinary adventure all the sections of the disciplinary processes involved have to be consistent with each other. Interpretations of one discipline may provide feedback on the presuppositions of another discipline. It may even be that the data of one discipline may contradict the presuppositions of another and thus influence its interpretation. These interactions are alternating, resulting in a multicircular, intertwining structure best described as interdisciplinary hermeneutics. The final interdisciplinary results as well as the intermediate disciplinary results are approached by permanent communication and negotiation between the disciplines, including inductive as well as deductive elements. Given the permanent feedback loops and the intensive demand for communication and negotiation that is required by trying to conform to the desirable perspectives of true interdisciplinarity, I fear that a successful outcome for an interdisciplinary approach may be highly improbable.

Moreover, disciplines are social spaces of power as well. Therefore, interdisciplinarity not only needs a theoretical framework, but is subject to social conditions. Students – assuming they want to be successful in terms of employment – and researchers – assuming they want to be successful in terms of citations and

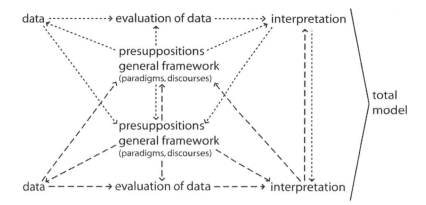

Figure 2. A model of interdisciplinarity.

funding – have to internalise disciplinary orders. Under these circumstances they need to correctly learn and practice disciplinary methods, especially as academic teachers and peers are in a position to enforce the truth and acceptance of disciplinary practices by excluding the disobedient from further participation in the academic field. Obedience to this disciplinary toolkit along with its presuppositions is deeply related to social position within the discipline (Bourdieu 1984; 2000). Even the mechanisms of exclusion themselves formally refer to the rightful obedience to theories and practices (and sometimes presuppositions), making the disciplinary toolkit the most powerful instrument of disciplinary discourse as it is the yardstick to judge true from false (Foucault 1966). This social practice is fundamentally thwarted by the pan-academic ideal of deep questioning. As in a postmodern world the act of questioning enables the questioner to deconstruct any presuppositions, methods and theories, it is vital to disciplinary discourse that basic and central aspects of disciplinary practice and its toolkit are excluded from such questioning. Actually it is quite easy to achieve such socialisation, as academic rewards are organised in an overwhelmingly disciplinary way (Weimann 2005).

Interdisciplinarity as sketched above, however, is challenging these disciplinary requisites. Interdisciplinarity demands that we question and transgress disciplinary borders by thoughts, words and deeds. It requires us to test the arguments of other disciplines against one's own material or to work on 'foreign' material with one's own methods. From the point of view of a disciplinary order of discourse interdisciplinarity asks for conscious blunder (cf. Winiwarter 2002, 210f.). Within a discipline such blunder is ignored at its best, but if things turn worse, it is honoured by trouncing. Therefore, successful interdisciplinarity primarily requires courage if you want to work in this desert of disciplinary vacancy. Secondly it requires openness in order to challenge disciplinary certainties and, thirdly, it requires high skills of negotiation. In short: interdisciplinarity is a specific form of academic social behaviour.

ACKNOWLEDGEMENTS

I am cordially thankful to Sjoerd Kluiving and to Guus Borger for valuable comments on this paper. Jude Jones and Matthew Johnson have been so kind to help me very much with the English version.

REFERENCES

Bernbeck, R. 1997. *Theorien in der Archäologie*. Francke: Tübingen/Basel.

Binford, L. 1962. Archaeology as anthropology. *American Antiquity* 28, 217-225.

Bourdieu, P. 1984. *Homo academicus*. Les Ed. de Minuit: Paris.

Bourdieu, P. 2004. Teilnehmende Objektivierung. In Elke Ohnacker & Franz Schultheis (eds.), *Schwierige Interdsizi- plinarität. Zum Verhältnis von Soziologie und Geschichtswissenschaft*. Westfälisches Dampfboot: Münster [first engl.: Participant objectivation. Breaching the boundary between anthropology and sociology: How? Lecture in London 2000].

Clark, G. 1974. Prehistoric Europe: The economic basis. In Gordon R. Willey (ed.), *Archaeological researches in retro- spect*, 31-57. Winthrop: Cambridge.

Clarke, D.L. 1978. *Analytical archaeology*². Methuen: London.

Cosgrove, D. 2004. Landscape and Landschaft. *German Historical Institute Bulletin* 35, 57-71.

Cronon, W. 1995. The trouble with wilderness; or, Getting back to the wrong nature. In William Cronon (ed.), *Uncommon ground. Rethinking the human place in nature*, 69-90. W.W. Norton & Company: New York/London .

Darvill, T. 2001. Traditions of landscape archaeology in Britain: issues of time and scale. In Timothy Darvill/Martin Gojda (eds.), *One land, many landscapes*, 33-45. BAR International Series 987. Hadrian Books: Oxford.

Drexler, D. 2009. *Landschaft und Landschaftswahrnehmung. Untersuchung des kulturhistorischen Bedeutungswandels von Landschaft anhand eines Vergleichs von England, Frankreich, Deutschland und Ungarn*. Dissertation Technische Universität München, Südwestdeutscher Verlag: München. mediatum2.ub.tum.de/doc/738822/738822.pdf (accessed 20 September 2010).

Déjeant-Pons, M. 2002. The European landscape convention, Florence. In G. Fairclough, S. Rippon, D. Bull (eds.), *Europe's cultural landscape: archaeologists and the management of change*, 13-24. Europae Archaeologiae Consilium (EAC: Brussels).

Eisel, U. 1986. Das 'Unbehagen in der Kultur' ist das Unbehagen in der Natur. Über des Abenteuerurlaubers Behaglichkeit. *Konkursbuch* 18, 23-38.

Ellenberg, H. 1963. *Vegetation Mitteleuropas mit den Alpen in kausaler, dynamischer und historischer Sicht*. Ulmer: Stuttgart.

Fairclough, G. 2006. Our place in the landscape? An archaeologist's ideology of landscape perception and manage- ment. In T. Meier (ed.), *Landscape ideologies*, 177-197. Archaeolingua: Budapest .

Foucault, M. 1966. *Le mots et les choses. Une archéologie des sciences humaines*. Gallimard: Paris.

Fox, C. 1932. *The personality of Britain. Its influence on inhabitant and invader in prehistoric and early historic times*. AMS Press, Inc.: Cardiff.

Friederich, S. 2003. Die erste Verbreitungskarte zur bandkeramischen Kultur. In J. Eckert, U. Eisenhauer, A. Zimmer- mann (eds.), *Archäologische Perspektiven. Analysen und Interpretationen im Wandel. Festschrift Jens Lüning*. Inter- nationale Archäologie. Studia honoraria 20, 27-31. Leidorf: Rahden/Westfalen.

Gojda, M. 2003. Archaeology and landscape studies in Europe. Approaches and concepts. In J. Laszlovszky & P. Szabó (eds.), *People and nature in historical perspective* 35-51. Central European University Department of Medieval Studies/Archaeolingua: Budapest.

Gradmann, R. 1901. Das mitteleuropäische Landschaftsbild nach seiner geschichtlichen Entwicklung. *Geographische Zeitschrift* 7, 361-377, 435-447.

Gradmann, R. 1906. Beziehungen zwischen Pflanzengeographie und Siedlungsgeschichte. *Geographische Zeitschrift* 12, 305-325.

Gramsch, A. 2003. Landschaftsarchäologie – ein fachgeschichtlicher Überblick und ein theoretisches Konzept. In J. Kunow & J. Müller (eds.), Landschaftsarchäologie und geographische Informationssysteme. Prognosekarten, Besiedlungsdynamik und prähistorische Raumordnungen. Forschungen zur Archäologie im Land Branden- burg 8, Archäoprognose Brandenburg 1 35-54. Brandenburgisches Landesamt für Denkmalpflege und Archäo- logisches Landesmuseum: Wünsdorf.

Groh, R. & Groh, D. 1991. Von den schrecklichen zu den erhabenen Bergen. Zur Entstehung ästhetischer Natur-erfahrung. In R. Groh & D. Groh, *Weltbild und Naturaneignung. Zur Kulturgeschichte der Natur*, 92-149. Suhrkamp: Frankfurt am Main .

Groth, P. & Bressi, T.W. (eds.) 1997. *Understanding ordinary landscapes*. Yale University Press: New Haven/London.

Harvey, D. 1993. The nature of environment: The dialectics of social and environmental change. *The Socialist Register* 29, 1-51.

Hettner, A. 1927. *Die Geographie. Ihre Geschichte, ihr Wesen, ihre Methoden*. Hirt Verlag: Breslau.

Hodder, I. 1984. Archaeology in 1984. *Antiquity* 58, 25-32.

Hodder, I. 1985. Postprocessual archaeology. *Advances in Archaeological Method and Theory* 8, 1-26.

Ingold, T. 1993. Globes and spheres. The topology of environmentalism. In K. Milton (ed.), *Environmentalism: the view from anthropology*, 31-42. Routledge: London. Reprinted in Ingold, T. 2000. Globes and spheres. The topology of environmentalism. In Ingold, T. *The perception of the environment. Essays on livelihood, dwelling and skill*, 209-218. Routledge: London/New York.

Jankuhn, H. 1952/1955. Methoden und Probleme siedlungsarchäologischer Forschung. *Archaeologia Geographica* 2, 73-84.

Jankuhn, H. 1977. *Einführung in die Siedlungsarchäologie*. De Gruyter: Berlin/New York.

Johnson, M. 1999. *Archaeological theory. An introduction*. Blackwell: Malden/Oxford/Melbourne/Berlin.

Johnston, R. 1998. The paradox of landscape. *European Journal of Archaeology* 1, 313-325.

Jones, M. 2003. The concept of cultural landscape: Discourse and narrative. In H. Palang & G. Fry (eds.), *Landscape interfaces. Cultural heritage in changing landscapes*, 21-51. Landscape Series 1. Springer: Dordrecht.

Löw, M. 2001. *Raumsoziologie*. Suhrkamp: Frankfurt.

Meier, T. 2006. On landscape ideologies: an introduction. In T. Meier (ed.), *Landscape ideologies*, 11-50. Archaeolingua: Budapest.

Meier, T. 2009. Umweltarchäologie Landschaftsarchäologie. In S. Brather, D. Geuenich & C. Huth (eds.), *Historia archaeologica. Festschrift Heiko Steuer*, 695-732. Reallexikon der Germanischen Altertumskunde Ergänzungs-band 70. de Gruyter: Berlin/New York.

Meier, T. & Tillessen, P. 2011.Von Schlachten, Hoffnungen und Ängsten: Einführende Gedanken zur Interdiszipli-narität in der Historischen Umweltforschung. In T. Meier & P. Tillessen (eds.), *Über die Grenzen und zwischen den Fächern. Fächerübergreifende Zusammenarbeit im Forschungsfeld historischer Mensch-Umwelt-Beziehungen*. Workshop Frauenchiemsee 2006, 19-44. Archaeolingua: Budapest.

Meinig, D.W. (ed.) 1979. *The interpretation of ordinary landscapes. Geographical essays*. Oxford University Press: New York/Oxford.

Migoń, P. (ed.).2010. *Geomorphological landscapes of the world*. Springer: Dordrecht/Heidelberg.

Mittelstraß, J. 1991. Geist, Natur und die Liebe zum Dualismus. Wider den Mythos von zwei Kulturen. In H. Bachmaier & E.P. Fischer (eds.), *Glanz und Elend der zwei Kulturen. Über die Verträglichkeit der Natur- und Geisteswissenschaften*, 9-28. Konstanzer Bibliothek 16. Universitäts-Verlag Konstanz: Konstanz.

Mittelstraß, J. 1999. Krise und Zukunft der Geisteswissenschaften. In H. Reinalter (ed.), *Natur- und Geisteswis-senschaften – zwei Kulturen?*, 55-79. Arbeitskreis Wissenschaft und Verantwortlichkeit 4. Studien-Verlag: Innsbruck/Wien/München.

Moran, E.F. (ed.) 1990. *The ecosystem approach in anthropology. From concept to practice*. University of Michigan Press: Ann Arbor.

Müller, G. 1975. Zur Geschichte des Wortes Landschaft. In A.H. von Wallthor & H. Quirin (eds.), *'Landschaft' als inter-disziplinäres Forschungsproblem*. Kolloquium Münster 1975, 4-12. Veröffentlichungen des Provinzialinstituts für Westfälische Landes- und Volksforschung des Landschaftsverbandes Westfalen-Lippe. Aschendorffsche Verlagsbuchhandlung: Münster.

Odum, E.P. 1975. *Ecology. The link between the natural and the social sciences*. Holt, Rinehart and Winston: New York.

Olwig, K. 2002. *Landscape, nature and the body politic. From Britain's Renaissance to America's new world*. University of Wisconsin Press: Madison.

Pantzer, E.H.M. 1995. *Settlement archaeology und Siedlungsarchäologie. Zum Vergleich amerikanischer und europäischer Forschungsstrategien.* Dissertation Hamburg.

Potthast, T. 2011. Terminologie der fächerübergreifenden Zusammenarbeit: Kurzer Problemaufriss und ein Vorschlag zur Verständigung über n> 1 – Disziplinaritäten. In T. Meier & P. Tillessen (eds.), *Über die Grenzen und zwischen den Fächern. Fächerübergreifende Zusammenarbeit im Forschungsfeld historischer Mensch-Umwelt-Beziehungen.* Workshop Frauenchiemsee 2006, 11-15. Archaeolingua: Budapest.

Schama, S. 1995. *Landscape and memory.* Harper Collins: London.

Schenk, W. 2001. *Reallexikon der Germanischen Altertumskunde 17*, 2nd edition, 617-630 s.v. Landschaft. de Gruyter: Berlin/New York.

Schliz, A. 1906. Der schnurkeramische Kulturkreis und seine Stellung zu den anderen neolithischen Kulturformen in Südwestdeutschland. *Zeitschrift für Ethnologie*, 312-351.

Schramm, M. 2008. Die Entstehung der modernen Landschaftswahrnehmung (1580-1730). *Historische Zeitschrift* 287, 37-59.

Tress, B., Tress, G. & Fry, G. 2003. Potential and limitations of interdisciplinary and transdisciplinary landscape studies. In B. Tress, G. Tress, A. van der Valk & G. Fry (eds.), *Interdisciplinary and transdisciplinary landscape studies: potential and limitations*, 182-191. Alterra Green World Research, Landscape Centre: Wageningen.

Tress, B., Tress, G. & Fry, G. 2006. Defining concepts and the process of knowledge production in integrative research. In B. Tress, G. Tress, G. Fry & P. Opdam (eds.), *From landscape research to landscape planning. Aspects of integration, education and application*, 13-26. Springer: Dordrecht .

Valk, A. van der & Bloemers, T. 2006. Multiple and sustainable landscapes. Linking heritage management and spatial planning in the Netherlands. In W. van der Knaap & A. van der Valk (eds.), *Multiple landscape: Merging past and present in landscape planning.* Workshop Wageningen 2004, 21-33. Ponsen & Looijen: Wageningen.

Wahle, E. 1915. Urwald und offenes Land in ihrer Bedeutung für die Kulturentwicklung. *Archiv für Anthropologie* 41 (13), 404-413.

Wahle, E. 1920. Die Besiedelung Südwestdeutschlands in vorrömischer Zeit nach ihren natürlichen Grundlagen. *Bericht der Römisch-Germanischen Kommission 12*, 1-75.

Weimann, J. 2005. Integration zwecklos: Interdisziplinäre Umweltforschung als Verbundprojekt selbständiger Disziplinen. In S. Baumgärtner & Ch. Becker (eds.), *Wissenschaftsphilosophie interdisziplinärer Umweltforschung.* 53-71 Ökologie und Wirtschaftsforschung 59. Metropolis-Verlag: Marburg.

Winiwarter, V. 1994. Umwelt-en. Begrifflichkeit und Problembewußtsein. In G. Jaritz & V. Winiwarter (eds.), *Umweltbewältigung. Die historische Perspektive*, 130-159. Verlag für Regionalgeschichte: Bielefeld.

Winiwarter, V. 2002. Disziplinäre (Um-)weltbilder. Zur Verständigung zwischen Biologie und Geschichtswissenschaft. In V. Winiwarter & H. Wilfing (eds.), *Historische Humanökologie. Interdisziplinäre Zugänge zu Menschen und ihrer Umwelt*, 197-221. Facultas: Wien.

NOTES

1   This chapter is a very short essence of my paper Meier 2009, which provides arguments and references in full detail.

2   This chapter shortly summarises the paper Meier & Tillessen 2011, which provides a more detailed argument and full references.

6.5 Landscape studies: The future of the field

Author

Matthew Johnson

Dept. of Anthropology, Northwestern University, Evanston, Illinois, USA
Contact: matthew-johnson@northwestern.edu

ABSTRACT

In this paper, I will go back to basics in landscape archaeology. How do we use the evidence of the land-scape to find out about the past? I reflect on the distinctive nature of landscape studies as distinctive not as 'art' or as 'science', but as a field-based discipline, in which the [claim of an] apprehension and encoun-ter with a reality 'out there' in the field is placed at centre stage.

Historically, field-based disciplines have a stress on what is claimed as 'direct experience', particu-larly in terms of their vernacular imagination – in other words, the colloquial language and everyday values deployed in field practice. Field-based disciplines rest, in part, on the urge to go out and see for oneself, rather than rely on others' reports.

'Direct experience', however, is a problematic concept. Most crucially, the claims of any discipline to be 'scientific' must rest not on the amount of direct experience gathered, but rather in the way that expe-rience is brought to bear on concepts and theories about it, and vice versa. If landscape archaeology is to be rigorous and scientific, it must abandon rhetorical appeals to an untheorised category of direct experi-ence and reflect more seriously on the relation of evidence to inference.

INTRODUCTION

This paper has as its aim the setting out of issues of knowledge construction and knowledge evaluation in landscape archaeology. It forsakes some of the more detailed and advanced dicussions of the field (for which see Johnson 2007; David & Thomas 2008; Tilley 1994, 2004; Bender 1998; Thomas 1999; Ingold 2000; Bender et al. 2008; Smith 2003). Instead, it goes back to basics: asks how landscape archaeologists find out about the past, and in particular, how they make a judgment as to whether their interpretations are 'good' or 'bad', are or are not supported by evidence. My desire is to strengthen these methods of evaluation and to make landscape archaeology a more rigorous empirical science.

I want to be as clear and precise as possible in what follows. Landscape archaeology, for reasons we shall explore, is an area of research that is full of woolly thinking. The field has been beset by a lack of conceptual clarity. In particular, thinking on landscape archaeology, particularly in Europe, is riddled with false, rhetorically loaded choices – between supposedly 'theoretical' and 'practical' approaches, between supposedly subjective and objective positions, between supposedly 'scientific' and 'humanistic' positions and interests, between those who supposedly stick to the evidence and those who supposedly go beyond it, between those who are supposedly positivist/empiricist and those who are supposedly postmodernist. These false choices most frequently surface not in analytical discussions but in chance, along-the-way comments by practising landscape scholars (for example, Liddiard & Williamson 2008) and are rarely explicitly set out (credit must be given to Fleming 2006 for putting into print much of this implicit discourse). The alternatives offered up can be given different moral evaluations (for Fleming: good/bad; for Tilley 2004, bad/good), but they are false choices nevertheless.

These false choices are seductive because, having been set up, an appeal is then made to have a 'debate' between them. Such a 'debate' has the surface appearance of openness, equality and respect between different positions. But a debate thus set out is bound to fail, because it rests on simplistic binary oppositions that are themselves symptomatic of a failure to understand the underlying issues. Alternatively, and more seductively, these two supposed traditions are framed in such a way as to present them as complementary or independent. Such a framing is just as damaging, because it implies that scholars in one tradition can get on with their work without having to worry about any critical evaluation that might be made by the other. Such a framing, then, is implicitly and crudely relativist, in that it ducks the issue of relative evaluation of two (falsely framed) traditions.

It is time to set these false choices, rhetorical appeals and implicit relativism aside and instead go back to basics, to think clearly and dispassionately about the nature of landscape archaeology and the way it constructs knowledge. This is what this paper tries to do.

SCIENCE

Archaeology is the study of past human societies from the material remains that are left in the present. Landscape archaeology, then, is the study of past human societies through the traces left in present landscapes.

Archaeology attempts to undertake this study of the past in a scientific and rigorous manner. Archaeology rightly sees itself as an empirical science which routinely and habitually seeks evidence to confirm or to disconfirm the statements about the past that it makes. However, it is equally true that the terms 'science', 'rigour' and 'evidence' are not simple or common-sense terms. In reality, science, rigour and evidence are very slippery, difficult and complex ideas. If they were easy, obvious and simple ideas, then the discipline of the philosophy of science would have nothing to talk about and the project of developing archaeological theory would have been successfully concluded a long time ago. Instead, the definition of science and the question of what counts as evidence continue to be hotly contested topics, however much we would like them to be simple (Wylie 2002).

I take a pragmatist view of the definition and practice of science. Pragmatism is a school of philosophy developed in 19th-century North America, by (among others) Charles Sanders Peirce, William James, Dewey, and latterly Richard Rorty (Stuhr 2005 is an introduction; Rorty 1999; Baert 2005 is a good discussion of the application of pragmatist philosophy to the human sciences). Pragmatism asks: what are the practical consequences of a particular belief? In what ways does a belief lead us to think and act differently? Pragmatism distrusts the niceties of metaphysical assertions and avoids appeals to abstract Platonic Truth. Pragmatists instead prefer to think about practical action to change the world. Pragmatism is, I believe, particularly useful for landscape archaeology as it directs attention not simply to the abstract intellectual properties of an idea, but to critical scrutiny of the specific ways theories lead us to do archaeology differently (Mrozowski & Preucel 2010).

Science is, among other things, a way of learning about the world. Good science should enable and require us to learn more, and unexpected, things about the world. Conversely, a symptom of bad science is the absence of learning and the absence of unexpected results. If landscape archaeology has been a field of woolly thinking, I think it is also true that in many areas it has become a field where learning about the world is less and less likely to happen.

In the sense of science I am proposing here, the enemy of science is empiricism. Empiricism is a word that is apt to confuse, since it has several meanings: here I mean by empiricism the belief that the data just speak for themselves, or that the way to understand what data 'mean' or how they should be interpreted is self-evident. Empiricism in this sense is overtly disavowed by most serious scholars (though see Johnson 2011 for a discussion of the way it nevertheless lurks within much archaeological interpretation). But the whole thrust of the practice of landscape archaeology is to privilege empiricism, through its stress on the direct gaze, on bodily movement ('muddy boots'), and on trust within a narrow community (see below). The problem is compounded by a persistent misreading: when the epistemological difficulties with such a empiricist position are pointed out, as I did in *Ideas of Landscape* (Johnson 2007), this is treated as a rejection of the process of empirical method as a whole. I must insist: A truly critical and fruitful empirical method starts with a questioning of its means of inference, not an assumption that those means of inference lie beyond 'theory'.

Recent pragmatist philosophers have stressed that good science, in the sense of learning unexpected things about the world, is most likely to happen when a diversity of stakeholders are involved. Conversely, where the practice of science is socially restricted, learning is less likely to happen. A socially restricted practice of science is therefore impoverished and lessened in value. This point has been made in relation to the development of a proliferation of scientific standpoints, for example feminist, postcolonial and Indigenous thought (Baert 2005, 161; see also the discussion in *American Antiquity* 75 (2), particularly Silliman 2010 and Wilcox 2010).

Let us then ask the pragmatist question: How, in practice, do scientists find out about the world? It is worth turning to the early history of the 'scientific revolution' to think about this question. 17th-century thinkers rested claims of the distinctive nature of scientific enquiry on its direct observation of the natural world. There was a cultural and political element to this claim; British empiricists in particular couched the claim rhetorically in a rejection of medieval Catholic theology. Early philosophy of science was in this sense heavily influenced by Protestantism. The medieval Catholic Church saw the world as a complex system of signs and allegories, to be explained as part of the divine mystery. For natural scientists, many of whom were also committed Protestants, the world was to be seen as it really was, through direct experience, in a fundamentalist manner (Shapin 1994, 1996).

However, philosophers of science were immediately aware of a problem of trust. Science depended not simply on direct experience, but on the reliance of the experience of others, the acceptance of their work as a secure base upon or alongside which to work. In the 17th century, 'trust' was defined socially and also legally. Thus, the 17th century notion of 'trust' excluded women and lower social orders; indeed, anyone who needed to work for a living and therefore could not be counted as being disinterested was disqualified. At the same time, many of these scholars also being trained as lawyers, issues of trust and authority were settled using metaphors and practices from the law courts to settle questions of 'truth' and what counted as 'fact' (Shapiro 1999).

In time, early modern scientists and natural philosophers resolved this question of trust by developing the idea of the repeatable experiment. Arguably, the repeatable experiment worked well and continues to work for the laboratory-based natural sciences. However, for emergent disciplines such as geography, geology and archaeology, the main or primary location of observation or direct experience was, in these early phases, not in the laboratory but in the field.

THE FIELD

In landscape archaeology, a central arena of everyday practice is 'the field'. The encounter with primary data in the field remains central in the hearts and minds of archaeologists. 'Direct field experience' is routinely cited as a primary determinant of evidence. A routine device in the praise of archaeologists is to praise the length and arduous nature of their time in the field. Conversely, abuse is directed at 'clean-booted historians' (Crawford 1953, 198).

Fieldwork is, without doubt, much more than the gathering of data. It is a bodily, aesthetic and sensual experience as well as a dry gathering-up of information (Ingold 2008). Archaeologists love being 'in the field' and they do it as often as possible. A routine complaint about university administration is that it keeps archaeologists away from the field. Student assessments of courses routinely call for 'more field trips'.

The 'field experience' is different from what happens in the laboratory in several crucial respects, respects which speak to the cultural practice and sensibility of field disciplines as much as they do to formal or abstract properties of knowledge. First, the field produces a different set of bodily practices. Rather than being cooped up like Dr Frankenstein in an isolated, stuffy laboratory in a dark basement, the field scientist practices his or her craft in the open, through strenuous action, like Indiana Jones. Second, the field produces a different sensibility, and set of terms of praise or denigration: muddy boots, toughness in terms of immunity to climatic vagaries of heat or cold. Fieldwork has been argued by feminists to be gendered masculine and valued accordingly (Gero 1985; Joyce 1994). Third, the field introduces an aesthetic sensibility: the object of study is often not merely complex and fascinating, but beautiful as well. And with this beauty comes a string of other cultural assumptions and practices relating to the conception of Nature, again, an exceptionally complex concept, 'perhaps the most complex word in the [English] language' (Williams 1976, 184), from Romanticism to modern ecological sensibilities (Bate 1991).

Above all, the field experience elevates the primacy of the gaze. The direct gaze is a central part of field practice – 'I saw it myself' becomes the final arbiter. It is easy, too easy, to slip from an entirely proper insistence on the importance of field practice to the notion that the gaze, or direct field experience, is somehow therefore beyond theory, or is somehow theory-free, or can be appealed to as an alternative to 'theory'.

It is a symptom of its very deep discursive nature that the notion of direct experience as prior to theory cuts across the usual theoretical divisions:

> There can be no substitute for the human experience of place – of being there – and it is *only after this that the various technologies of representation come into play*... our attempts at thick descriptions of place contrast with the standard mode of thin technicist archaeological description which effectively dehumanises the past and makes it remote and sterile because such technical descriptions are based on abstracted Cartesian conceptions of space and time (...) (Tilley 2004, 221; my italics)

Such a critical slippage, between field experience as important and field experience as beyond or an alternative to theory, is assisted by the weight of the cultural freight that comes with appeals to the primacy of the gaze. Such appeals are not simply or only associated with the Enlightenment (Foucault 1976, 89); they go much deeper still into the very roots of European culture. This primacy is deeply embedded in the history of British empiricism and its legal context and background: in the 17th century, it is precisely this direct gaze that had a legal origin, in the testimony of witnesses in the courts, as noted above. Again, the all-encompassing gaze was a central element in practices of 18th-century European Romanticism (Urry 2002).

There is a large and growing literature that critically analyses archaeological practice, particularly excavation methods (Lucas 2001; Hodder 1999; Edgeworth 2006). It is interesting that relatively little of this literature engages with the field practice of landscape archaeology, as opposed to excavation. The debate between Gillings, Pollard and Bowden (Gillings & Pollard 1998; Bowden 2000) over the nature of representations of the Avebury landscape is interesting in Bowden's insistence, in reply to Pollard and Gillings' critique, that maps and plans are in fact critical interpretive practices – in other words, Bowden's position is that a critical approach is implicit in traditional techniques. He thus arguably mounts a successful critical defence of traditional field practice, but invites the obvious question: why is it then implicit and rarely spoken about rather than explicit and critically discussed?

Field experience is essential to the craft of landscape archaeology. This does not mean, however, that field experience is therefore immune from critical enquiry. Indeed, the status of landscape archaeology as an empirical science makes such critical enquiry all the more essential. To repeat, scientific rigour demands that the links made between evidence and inference are examined closely and critically, not assumed or placed beyond the domain of theory. To place field experience in some category beyond theoretical reflection, as if it were somehow transcendent or ineffable, is to insulate it from critical scrutiny. Rhetorical appeals to the primacy of raw field experience are, therefore, unscientific, and approaches which divide views of landscape into 'practical' and 'theoretical' are misleading.

I want to dwell on this point, because such appeals have a high visceral or emotive charge because of the cultural freight associated with field practice discussed above. Most damagingly, a false choice is invited between those who supposedly place fieldwork at the centre of their method and those who supposedly do not. It is precisely this emotive emphasis on the field that leads to the central element of woolly thinking I want to discuss, namely the empiricist fallacy that data gathered will tell us about the past without the benefit of intervening theory.

In and of itself, 'the field' tells an archaeologist nothing about the past.

If you stand in the middle of a muddy field or on top of a hill surrounded by ancient earthworks and listen very carefully, you will hear nothing about the past. If you walk across many more muddy fields, you will still hear nothing. Further, if you take lots of photographs, you will still hear nothing. If you draw what is around you, however accurately and meticulously you do the drawing, and however carefully you measure in your observations, you will still hear nothing. When Barbara Bender writes that '"the evidence" does not of itself deliver an understanding' (Bender 1998, 7), this is not a dangerously postmodern or relativist sentiment to be deplored (as others have alleged, cf. Fleming 2006); it is a statement of the blindingly obvious.

Much modern landscape archaeology is preoccupied with ever more complex techniques for mapping field observations of different kinds, many of which will be found explicated in the pages of this volume. These techniques are exciting, and offer tremendous potential. However, they do not, in themselves, tell us more about the past. If you rectify air photographs in new ways, you still hear nothing about the past. If you use LiDAR techniques to produce ever more complex mapping of traces in the present, you will still hear nothing.

Science is not about accumulating lots of empirical observations. I can leave my office and count the blades of grass in the lawn; I can do this very carefully, and my back and eyes will ache at the end of the exercise. However, without a clearly defined and formulated set of research questions linking propositions to data, I will not be doing science, either in a pragmatist or in any other sense.

The accumulators of evidence par excellence are the ley-line hunters. Ley-line hunters get out in the field. They spend their time walking across the landscape, and directly observing it with their own eyes. They make systematic recordings of their observations which are careful and accurate. However, they are, of course, completely misguided – in part, because of their uncritical reliance on raw experience. Anyone can 'see' the evidence for a ley line: walk along one marked on a map, and you will find its existence confirmed over and over again as the route yields further evidence of its ancient nature as the walker discovers banks, ditches and other ancient features (Willamson & Bellamy 1983).

What makes landscape archaeology into a scientific and rigorous discipline is not then, primarily, the raw experience of fieldwork. Rather, it is the processes of inference and deduction that are used to translate between data gathered in the field to general statements about the past, and back again. If these processes of back-and-forth translation are faulty, or have not been clearly thought through, then no amount of field observation or muddiness of boots will turn archaeology into a science, any more, with apologies to David Clarke, than a wooden leg turns a man into a tree. This very basic point was absolutely central to early New Archaeology. It remains fundamental to any kind of critical and scientific archaeological practice, and deserves to be repeated over and over again, in part due to its counter-intuitive nature (Clarke 1976; Binford 1983).

The problem with much of landscape archaeology is that 'raw field observation' is placed as primary, and placed beyond theoretical reflection. Such a placement resonates with the great love and passion for fieldwork that landscape archaeologists have, for the reasons discussed above. However, by this placement, the question 'how do we use the results gathered to tell us about the past?' is bypassed.

Instead of addressing this problem of inference in a clear and direct way, archaeologists – of all theoretical stripes and affiliations, empiricist and postmodernist, Marxist and culture-historical, 'traditional' or avant-garde' – simply end up telling stories about the past. Monographs on landscape archaeology frequently consist of lovingly-gathered data, beautiful air photographs, complex maps showing clusters of settlements. Then, in the final chapter, we get a narrative about the past. The reader is led to believe that this narrative is empirically informed, because the preceding bulk of the book has been taken up with so much data. But if one stands back and asks the question 'how do these data inform our judgements – how does it allow us to judge this statement about these convincing past processes and that judgement wanting?' things are often not so clear.

HUMAN BEINGS

If landscape archaeology needs to address issues of inference, what kind of past is it attempting to make inferences about? By definition, such a past has to be a human one. Landscape archaeology is exactly that: it is not geology, nor is it physical geography. Archaeology is the study of past human life. Landscape archaeologists, then, make statements about human subjectivity (their view of their world), whether they like it or not.

In human sciences, we have to think about human beings and their view of their world. Consider for a moment the alternatives. First, the proposition that we do not have to bother: mental attitudes are irrelevant to understanding long-term human impact on landscapes. Second, past peoples were just like us: human subjectivity in the past was just like it is in modern, bourgeois society in the present. The first proposition is just about tenable, and is held by a few hard-line cultural ecologists and behaviourists; it remains nevertheless very much a minority pursuit. The second proposition is not seriously tenable for more than a few seconds.

If this proposition is accepted, then a necessary part of rigorous and responsible science is a questioning and thinking-through of human subjectivity in the past, and in particular a rejection of the assumption that human subjectivity can be dealt with as a matter of common sense. This is precisely the task that phenomenology sets itself. If there is a bogeyman that has been uncritically deployed to denigrate others and rhetorically reinforce empiricist approaches, it is phenomenology: thus Liddiard & Williamson (2008, 525), citing Fleming, write: 'phenomenologists (...) believe (...) that the rigorous testing of the ideas that are derived from it is either impossible or unnecessary (...) having to a significant extent freed themselves from traditional concerns about the verification of ideas, many archaeologists of this persuasion have effectively given themselves permission to say more or less whatever they liked about the past (Fleming 2006, 269)'. Ironically, phenomenology as a philosophical project sets out quite explicitly to *problematise* the uncritical imposition of modern ideas of human subjectivity on to the past. It is true that in practice, some work under the phenomenological banner has tended to reproduce rather than question modern subjectivities. However, this is an issue with particular scholarly applications of the theory, not the theory itself. Fleming's criticism, and Liddiard & Williamson's uncritical repetition of that criticism, betrays an inattentive reading of the theoretical literature, and betrays an assumption that 'rigorous testing' is a clear and unproblematic procedure that needs no further definition or explication beyond being noted as a 'traditional concern' – in short, that it is somehow beyond theory.

I have been critical of some aspects of phenomenology in the past (Johnson 2007), but here I want to acknowledge that its central aim is unavoidable. Or to put it another way, *we are all phenomenologists*, insofar as we acknowledge that the exploration of different ways of living in and experiencing the world is a necessary and legitimate part of what we do as archaeologists. The only question at stake is: how, by what methods, do we do so?

The dismissive comments made about phenomenology by some critics are a classic example of how a questionable rhetoric of appeal to direct experience has taken over from reasoned and informed argument. I explored above how much of landscape archaeology, particularly of historic periods, ends up telling stories about the past, however lovingly large bodies of empirical material are brought together and presented, because the critical links between evidence and inference are not examined. It follows, then, that no *a priori* judgement can be made as to the extent to which such traditional studies are more or less evidentially based than any other form of landscape archaeology, for example explicitly phenomenological studies. The epistemological criteria by which traditional, culture-historical arguments are designated 'cautious' whereas theoretically informed arguments are 'wild' or 'speculative' are, then, entirely unclear; it is difficult to believe that they rest on more than an untheorised gut feeling, hardly the basis for rigorous scientific evaluation.

The response to critics of explicitly phenomenological approaches should be: if you are unhappy with the philosophical basis of phenomenology, tell us your alternative. Do you either a) reject any possibility of exploring past human experience or b) simply impose your own assumptions onto the past? If you do neither, will you set out the philosophical basis of your approach and open it up to critique? It is striking that the most recent and authoritative discussion of the issues surrounding phenomenology (Barrett & Ko 2009) does precisely this. Barrett & Ko forsake easy platitudes and engage in a much deeper – and more fruitful – exploration of the underlying theoretical issues.

THE WAY FORWARD: A PRAGMATIST VIEW

I have the following suggestions, then, for the future of landscape archaeology.

First, abandon the highly rhetorical use of theoretical bogeymen and the invitation of false choices. Every time the reader encounters such rhetoric, the question to ask is: what is your alternative? Are you seriously asserting that fieldwork practice is somehow beyond theory? If you are not making such a naïve move, then what is your alternative, and how do you justify that alternative?

Second, maintain and extend the critical examination of field practices. In recent years, there have been a series of developments that render the direct gaze problematic and expose its underlying theoretical geology. The Visualisation in Archaeology project (www.viarch.org.uk) has gathered together and presented much of this work, which includes reflexive approaches to computer 'reconstruction' (Earl & Wheatley 2002), critical reflection on air photography (Hauser 2007; Barber 2010); and critical histories of landscape archaeology (Stout 2008; Johnson 2007).

Third, open the practice of landscape archaeology up to different groups and stakeholders. We have seen above how good science is aided by the opening up of method to a diversity of groups. Landscape archaeology, then, is intimately connected to issues of access to the countryside, of multiculturalism, and of diversity and inclusion. Such issues have been raised, for example, by Barbara Bender's work at Stonehenge (Bender 1998), and developed in the last decade by heritage agencies like English Heritage (Fairclough, this volume).

Such developments are not in opposition to science: rather, they expand and deepen scientific enquiry. They render it more critical and more rigorous by exploding the narrow circle of trust. Questions over science, and over the social and political context of science, are not an either/or option. The work of a generation of feminist standpoint theorists (for example Gillian Rose in geography: Rose 1992) has shown how good science is also inclusive and diverse. Science needs a diversity of hypotheses to drive its learning forward and a diversity of stakeholders to inform its thinking.

Approaches which emphasise a diverse and inclusive view of science, and question a naïve empiricism, are judged by their ciritics against an implicit yardstick of a tried and tested method. This tried and tested method cannot be set out or defined, because it is non-existent. A neutral, value-free method of doing field science is an impossibility, as the VU University of Amsterdam acknowledges as a matter of official policy: 'we take the view that academic work cannot be divorced from the concerns of society, in terms of standards, values, philosophy and religion (http://www.vu.nl/en/about-vu-amsterdam/mission-and-profile/index.asp, accessed 1 January 2010)'.

CONCLUSION: IN PLACE OF A 'DEBATE'

If a value-free science is an impossibility, it follows that the way forward is not in 'debate': there is no debate to be had. Rather than a 'debate', what we find around us is a deepening crisis over the place of academic knowledge in a Europe in which landscape and nationalism have been and are linked in disturbing ways. This linkage between the study of landscape and its wider social, political and cultural context provides the reason, I suggest, why heritage organisations are at the forefront of new approaches, whereas Universities and the Academy have become increasingly inward-looking in their theoretical discourse; see for example recent work by English Heritage (Fairclough 2001; Fairclough et al. 2008).

I wrote above that there is no debate to be had. This may be true in a strictly intellectual sense. Of course, landscape archaeologists, as a group, do have a choice ahead. If we want, we can cling to a myth of academic detachment that has long since been comprehensively discredited, and we can watch our work not simply become more and more narcissistic and irrelevant, but also increasingly reactionary, uncritical and incurious, too easily harnessed by others to unsavoury nationalist agendas, but most crucially, become less scientific, to generate fewer and fewer new insights about the past even as we work harder and harder to accumulate more and more data. And we can defend that choice with a shrill rhetoric of untheorised objectivity and increasingly embattled and desperate denunciation of postmodernists, feminists and other bogeymen outside the narrow circle of 'people like us'. But that is not really a choice.

Landscape archaeology makes a claim to be a responsible and rigorous field science. If it is to substantiate this claim, it needs to think carefully and self-critically about how it moves from evidence to inference. The amassing of evidence in itself, however carefully and arduously done, is not the single defining or primary feature of science; theoretical reflection on the means of inference is at least equally important. It is this point which has been lost in some recent commentaries on the field, and which needs to be placed at the centre of everything that we do.

REFERENCES

Barber, M. 2010. *A History of Aerial Photography and Archaeology: Mata Hari's Glass Eye and Other Stories*. London, English Heritage.

Baert, P. 2005. *Philosophy of the Social Sciences: Towards Pragmatism*. Oxford, Polity.

Bate, J. 1991. *Romantic Ecology: Wordsworth and the Environmental Tradition*. London, Routledge.

Barrett, J. & Ko, I. 2009. A phenomenology of landscape: a crisis in British landscape archaeology? *Journal of Social Archaeology* 9 (3), 275-294.Bender, B. 1998. *Stonhenge: Making Space*. Oxford, Berg.

Bender, B., Hamilton, S. & Tilley, C. 2007. *Stone Worlds: Narrative and Reflexivity in Landscape Archaeology*. Walnut Creek, Left Coast Press.

Binford, L.R. 1983. *In Pursuit of the Past*. London,Thames & Hudson.

Bowden, M. 2000. Virtual Avebury revisited: in defence of maps and plans. *Archaeological Dialogues* 7, 84-88.

Clarke, D.L. 1968. *Analytical Archaeology*. London, Methuen.

Crawford, O.G.S. 1953. *Archaeology in the Field*. London, Phoenix House.

David, J. & Thomas, J. 2008. *Handbook of Landscape Archaeology*. Walnut Creek, Left Coast press.Earl, G. & Wheatley, D. 2002. Virtual reconstruction and the interpretative process: a case-study from Avebury. In Wheatley, D., Earl, G. & Poppy, S. (eds.) *Contemporary Themes in Archaeological Computing*, 5-15. Oxford, Oxbow Books.

Edgeworth, M. (ed.) 2006. *Ethnographies of Archaeological Practice: Cultural Encounters, Material Transformations*. Lanham, Altamira.

Fairclough, G. 2001. *Europe's Cultural Landscape: Archaeologists and the Management of Change*. EAC Occasional Paper 2. Brussels, Europae Archaeologiae Consilium.

Fairclough, G., Harrison, R., Jameson, J.H., & Schofield, J. (eds) 2008. *The Heritage Reader*. London, Routledge.

Fleming, A. 2006. Post-processual landscape archaeology: a critique. *Cambridge Archaeological Journal* 16 (3), 267-280.

Foucault, M. 1976. *The Birth of the Clinic*. London, Tavistock.

Gero, J. 1985. Socio-politics of archaeology and the woman-at-home ideology. *American Antiquity* 50, 342-350.

Hauser, K., 2007. *Shadow Sites: Photography, Archaeology, and the British Landscape 1927-1951*. Oxford, Oxford University Press.

Hodder, I. 1999. *The Archaeological Process: An Introduction*. Oxford, Blackwell.

Ingold, T. 2000. *The Perception of the Environment: Essays on Livelihood, Dwelling and Skill*. London, Routledge.

Johnson, M.H. 2007. *Ideas of Landscape*. Oxford, Blackwell.

Johnson, M.H. 2011. On the nature of empiricism in archaeology. *Journal of the Royal Anthropological Association* forthcoming.

Joyce, R. 1994. Dorothy Hughes Popenhoe: Eve in an archaeological garden. In Claassen, C. (ed.) *Women in Archaeology*, 51-66. Philadelphia, University of Pennsylvania Press.

Liddiard, R. & Williamson, T.M. 2008. There by design? Some reflections on medieval elite landscapes. *Archaeological Journal* 165 (1), 520-535.

Lucas, G. 2001. *Critical Approaches to Fieldwork: Contemporary and Historical Archaeological practice*. London, Routledge.

Rorty, R. 1999. *Philosophy and Social Hope*. Harmondsworth, Penguin.

Silliman, S. 2010. Courage and thoughtful scholarship: a response to McGhee. *American Antiquity* 75 (2), 217-220.

Stuhr, J.J. 2005. *American Pragmatism: An Introduction*. Oxford, Blackwell.

Rose, G. 1992. *Feminism and Geography: The Limits of Geographical Knowledge*. Cambridge, Polity.

Pollard, J. & Gillings, M. 1998. Romancing the stones: towards a virtual and elemental Avebury. *Archaeological Dialogues* 5 (2), 143-164.

Preucel, R.W. & Mrozowski, S. (eds.) 2010. *Contemporary Archaeology in Theory: the New Pragmatism*. Oxford, Blackwell.

Scott, J. 1996. The evidence of experience. In McDonald, T.J. (ed.) *The Historic Turn in the Human Sciences*. Ann Arbour, University of Michigan Press, 379-406.

Shapin, S. 1996. *The Scientific Revolution*. Chicago, Chicago University Press.

Shapin, S. 1994. *A Social History of Truth: Civility and Science in 17th-Century England*. Chicago, University of Chicago Press.

Shapiro, B.J. 1999. *A Culture of Fact: England, 1550-1720*. Ithaca, Cornell University Press.

Smith, A.T. 2003. *The Political Landscape: Constellations of Authority in Early Complex Societies*. Berkeley, University of California Press.

Thomas, J. 1999. *Time, Culture and Identity: An Interpretative Archaeology*. London, Routledge.

Tilley, C. 1994. *A Phenomenology of Landscape*. London, Routledge.Tilley, C. 2004. *The Materiality of Stone*. Oxford, Berg.

Urry, J. 2002. *The Tourist Gaze*. London, Sage.

Wilcox, M. 2010. Saving indigenous people from ourselves: separate but equal archaeology is not scientific archaeology. *American Antiquity* 75 (2), 221-227.

Williamson, T. & Bellamy, E. 1983. *Ley Lines in Question*. London, Heinemann.

Wylie, A. 2002. *Thinking From Things: Essays in the Philosophy of Archaeology*. Berkeley, University of California Press.

LANDSCAPE AND HERITAGE SERIES

Landscape & Heritage Series (LHS) is a new English-language series about the history, heritage and transformation of the natural and cultural landscapes, and built environment. The series aims at the promotion of new directions as well as the rediscovery and exploration of lost tracks in landscape and heritage research. These two theoretically oriented approaches play an important part in the realization of this objective.

LHS welcomes monographs and edited volumes, theoretically oriented approaches and detailed case studies, dealing with one or a combination of the above mentioned topics. For more information about the series, please visit www.aup.nl or www.clue.nu

PUBLISHED IN THIS SERIES

Proceedings
The Cultural Landscape & Herritage Paradox by Tom Bloemers et al. (eds.)
ISBN 978 90 8964 155 7